World Soils Book Series

Series editor

Prof. Alfred E. Hartemink
Department of Soil Science, FD Hole Soils Laboratory
University of Wisconsin–Madison
Madison
USA

Aims and Scope

The *World Soils Book Series* brings together soil information and soil knowledge of a particular country in a concise and reader-friendly way. The books include sections on soil research history, geomorphology, major soil types, soil maps, soil properties, soil classification, soil fertility, land use and vegetation, soil management, and soils and humans.

International Union of Soil Sciences

More information about this series at http://www.springer.com/series/8915

James G. Bockheim · Alfred E. Hartemink

The Soils of Wisconsin

Springer

James G. Bockheim
Department of Soil Science
University of Wisconsin–Madison
Madison, WI
USA

Alfred E. Hartemink
Department of Soil Science
University of Wisconsin–Madison
Madison, WI
USA

ISSN 2211-1255 ISSN 2211-1263 (electronic)
World Soils Book Series
ISBN 978-3-319-52143-5 ISBN 978-3-319-52144-2 (eBook)
DOI 10.1007/978-3-319-52144-2

Library of Congress Control Number: 2016963779

Printed on acid-free paper

This Springer imprint is published by Springer Nature
The registered company is Springer International Publishing AG
The registered company address is: Gewerbestrasse 11, 6330 Cham, Switzerland

This book is dedicated to Dr. Francis D. Hole (1913–2002), eminent pedologist at the University of Wisconsin–Madison. He received his doctorate in soil science and geography at the UW Madison in 1943 and joined the Soils Department in 1946. Thirty years of soil mapping and soil research led to the publication of Soils of Wisconsin in 1976. It has been a standard reference book on soils of Wisconsin and is now a collector's item. Jim Bockheim had the pleasure of sharing field trips, including biking to the UW Arboretum, with Dr. Hole for a 27-year period. Dr. Hole was a highly creative pedologist, a relentless teacher of soil science, as well as a musician and poet:

Darkle, darkle, little grain,
I wonder how you entertain
A thousand creatures microscopic.
Grains like you from pole to tropic
Support land life upon this planet
I marvel at you, crumb of granite!

Foreword

Wisconsin is a diverse state where forests, prairie, farmland, and urban areas form a pleasant mix across the landscape. Each of these ecosystems is in part determined by the soils that nourish and shape what we see aboveground. To understand the history and economic development of Wisconsin, one must understand its soils. This book on the soils of Wisconsin provides information and insights based on decades of research and soil investigations conducted by scientists across the state.

I know something of the importance of this topic because I am the child of a soil scientist. My father graduated in 1943 from the University of Missouri and spent much of his career on agricultural and economic development in the upper Midwest. Driving trips always included lectures on the countryside outside the car windows, particularly the effects of glaciation and soil chemistry on the crops or woods that we were seeing. I admit that I did not always listen as closely as my father might have wished. Fortunately, this book provides a detailed soils history and analysis here in Wisconsin, more than filling in all that I missed during those family drives.

We need to think about the soil as a natural resource, about which we need up-to-date and relevant data for managing and planning purposes. Farms and related agricultural business in Wisconsin generate over $80 billion per year and employ more than 400,000 people. The long-term economic and ecological sustainability of these farm operations is dependent on deep knowledge of the soil resources. We know that soil information is also essential for managing our wilder regions such as forests, marshes, and grasslands. And the development of habitat around towns and cities requires soil knowledge.

I am proud that the University of Wisconsin harbors the oldest and one of the most renowned soil science departments in the world, and the contributions of its faculty and students have been key to the long-term research on Wisconsin soils. Thanks to Jim Bockheim and Alfred Hartemink for their long-term commitment to this research and for pulling together the detailed information contained here.

Soils are an essential and too often underappreciated asset for this state. It is my sincere hope that this book will help document the importance of the soil for the well-being and prosperity of all of us in Wisconsin.

Rebecca M. Blank
Chancellor
University of Wisconsin–Madison

Preface

The first book solely focusing on the soils of Wisconsin was published in 1927 by A.R. Whitson—professor of soil science at the University of Wisconsin–Madison. He described the function of soils; climate, future agriculture, and forestry; and origin and classification of Wisconsin soils and gave detailed descriptions of 23 soil series from nine soil regions. In 1947, Dr. Charles Kellogg, who later became Director of the Soil Survey Division of the USDA, Soil Conservation, published a general overview of the soils of Wisconsin as a *Preliminary Study of the Profiles of the Principal Soil Types of Wisconsin*. In 1976, the University of Wisconsin Press published *Soils of Wisconsin* that was prepared by Dr. F.D. Hole. This book contained chapters on factors of soil formation, processes of soil formation, properties of Wisconsin soils, classification of soils, and soil associations within ten soil regions of the state and an appendix with soil descriptions and analytical data. This book included a soil map at a scale of 1:710,000.

This present book in a sense constitutes the fourth edition of the *Soils of Wisconsin*. The authors have benefitted from several decades of new research and from the digital age in which official soil descriptions, laboratory characterizations, soil classification data, and the Web Soil Survey are available which enabled us to provide an up-to-date analysis of the soils of the state of Wisconsin. Despite the demand for soils information, only 3 (Michigan, Minnesota, and Wisconsin) of the 50 states in the USA have a general soil map published after 1999; and only 5 states (Arkansas, Florida, Louisiana, North Carolina, and Washington) have a book published after 1999 compiling soils data for the state.

We hope this book will serve as a written and digital account on the soils from Wisconsin, as a reference book for soil science and related courses and as a compendium for land users, policymakers, and anyone interested in our soils. On Wisconsin!

Madison, WI, USA
November 2016

James G. Bockheim
Alfred E. Hartemink

Acknowledgements

This work is based on decades of soil research and the data collected by soil scientists with the Wisconsin Geological and Natural History Survey (WGNHS) and the many surveyors and scientists from USDA Natural Resources Conservation Service (NRCS) at Madison, Wisconsin, and Lincoln, Nebraska. Without their dedicated surveying, mapping and research this book could not have been written. WGNHS allowed use color images depicting the natural resources of Wisconsin. We are grateful to David Evans, Michael Notaro, and Yakun Zhang for producing the soil, land use, and climatic maps. Unless indicated, the soil profile and landscape pictures were taken by Alfred Hartemink. Jim Bockheim acknowledges his wife, Julie, as a steady source of encouragement. We are grateful to Dr. Robert Doe, Suresh Rettagunta, and Corina van der Giessen of Springer for help in producing this book and to the chancellor of the University of Wisconsin–Madison for her support and the foreword to this book.

Contents

About the Authors

Dr. Jim Bockheim was professor of Soil Science and of Forest and Wildlife Ecology and at the Nelson Institute for Environmental Studies at the University of Wisconsin from 1975 until his retirement in 2015. He has conducted research in forest soils and pedology throughout the state and in the Upper Peninsula of Michigan. His previous books include *Pedodiversity* (2013; with J.J. Ibáñez); *Soil Geography of the USA: a Diagnostic-Horizon Approach* (2014); *Cryopedology* (2015); and *The Soils of Antarctica* (2015).

Dr. Alfred Hartemink has been a professor of Soil Science at the University of Wisconsin since 2011 where his research focusses on digital soil morphometrics, soil mapping, and soil C. Prior to his current position, he was senior researcher at ISRIC–World Soil Information (The Netherlands) and coordinator of the GlobalSoilMap project. He has worked as pedologist and soil fertility specialist in Tanzania, DR Congo, Indonesia, Kenya, Australia, and Papua New Guinea. He was (Deputy) Secretary General of the IUSS between 2002 and 2014 and has edited and written several books.

Summary

This is the fourth book on of the soils of Wisconsin published over the past 90 years.[1] This version benefits from an additional 40 years of insights, soil mapping and research in the state, the development of geographic information systems, and the access to digital data by the Natural Resources Conservation Service.

In Chap. 1, it is noted that soils of Wisconsin have been investigated for 120 years. The number of soil series identified in the state has increased exponentially from a dozen in the early 1900s, to more than 700 in 2016. Ten broad soil regions have been identified in Wisconsin based on geographic location, topography, parent materials, soil texture, vegetation, and major soil groups. The soils can also be divided into 11 Major Land Resource Areas (MLRAs) that are based on a variety of natural resources and land use.

Since the early 1960s, the soils have been classified using *Soil Taxonomy* (ST). This hierarchical system delineates soils into orders, suborders, great groups, subgroups, families, and series. Diagnostic surface and subsurface horizons (natural soil layers containing defined properties) are of key importance in classifying the soils, of which there are five surface horizons and seven subsurface horizons in soils of Wisconsin. Seven[2] of the 12 orders are present in Wisconsin, including from most to least (on an area basis): Alfisols (mildly acidic, clay-enriched soils formed under forest), Spodosols (strongly acidic, iron-enriched soils formed under forest), Entisols (young soils), Mollisols (soils with a thick organic-enriched topsoils formed under prairie vegetation), Histosols (organic soils formed in wetlands), and Inceptisols (weakly developed soils).

Wisconsin has a long history of soil studies, and this is reviewed in Chap. 2; some of the major achievements of Wisconsin soil scientists are discussed. The geologist T.C. Chamberlin prepared the first map of soils in Wisconsin (1882); F.H. King wrote one of the first soil science textbooks in 1895 and was the first chair of the Department of Soil Science at the University of Wisconsin–Madison (1889); H.H. Bennett conducted research in the Driftless Area and was the first director of the Soil Erosion Service (1933), later to the Soil Conservation Service (1935) and then the Natural Resources Conservation Service (1994); S.A. Wilde wrote the earliest forest soils textbook in the USA (1946); and F.D. Hole published the first color soils map of Wisconsin and the third version of *Soils of Wisconsin* (1968). He also was instrumental in getting the Antigo Silt Loam recognized as Wisconsin's state soil.

Chapter 3 discusses the factors that influence the nature, properties, and distribution of soils in Wisconsin, including climate, organisms, topography, initial material, time, and human activities. Climate and vegetation have had pronounced effects on the distribution of soils in the state. Spodosols exist under mixed coniferous and deciduous forests in the northern third of the state; Hapludalfs (Alfisols) are present under broad-leaved forests dominated by oaks in the southern third of the state; and Glossudalfs (Alfisols) are present in a broad zone extending northwest to southeast across the state, to the north of the "tension zone." The line separating soils in the frigid and mesic soil-temperature classes (8 °C mean annual soil temperature at the

[1]First version by A.R. Whitson, *Soils of Wisconsin* (1927); second version by R.J. Muckenhirn and N.P. Dahlstrand, Soils of Wisconsin (1947); and third version by F.D. Hole, *Soils of Wisconsin* (1976).

[2]A relict Ultisol has been identified in Wisconsin. The Ultisols are strongly leached soils with relatively low native fertility that are often under forest and commonly occur in southeastern USA.

50 cm depth) extend across the tension zone. Prairies, possibly encouraged by periodic fires, exist in southern and western Wisconsin and are underlain by Mollisols. Marshes are found throughout the state and feature organic soils (Histosols). The poorly developed Inceptisols exist in areas where bedrock is close to the surface and where drainage is restricted. Poorly developed Entisols occur mainly on sandy materials in former lake basins and outwash plains. The topographic factor is the main factor distinguishing soils at the landscape scale.

Nearly 80% of the state is covered with glacial deposits that differ in texture, composition, thickness, and age. The upper 1 m of soil is primarily loess (34%), followed by till (14%), alluvium (13%), outwash (12%), organic sediments (10%), and other materials (17%). The "Driftless Area" has not been glaciated but is covered with loess (windblown silt-enriched material) and outwash from the Wisconsin River. There is a strong relationship between the soils and parent materials. Glossudalfs occur on clayey glaciolacustrine deposits and silty till; Spodosols occur on pitted outwash; Entisols exist on sandy lake sediments and pitted and unpitted outwash; and Hapludalfs occur on loess and calcareous drift in southern Wisconsin.

Most of the soils of Wisconsin are derived from drift of Late Wisconsinan age (9500–30,000 year BP). However, soils have been developed on pre-Late-Wisconsinan to Illinoian drift (>30,000–300,000 year BP) and on pre-Illinoian drift between 780,000 and 2,400,000 year BP. The red clay pediment material in the Driftless Area may be even older. Finally, human activities influenced the soils through cultivation, irrigation, fires, clearcutting, urbanization, and other land-management practices. The history of human impacts on soils in Wisconsin extends back approximately 13,500 years ago, but became intensified during the Late Woodland Tradition (1600–500 year BP) when fires were used to clear land and further intensified in the mid-1800s when European settlers arrived and land clearing and large-scale crop production begin.

Soil processes are discussed in Chap. 4. Ten soil-forming processes are common in Wisconsin soils: argilluviation (movement of clay); biological enrichment of base cations (retention of Ca, Mg, and K in the system by vegetation); gleization (the effects of poor drainage on Fe and Mn compounds); cambisolization (early stages of soil formation whereby the structure and color of the parent material are altered); paludification and ripening (the accumulation and decomposition of organic materials in organic soils); melanization (the accumulation of well-humified organic matter near the soil surface); podzolization (the accumulation of Fe and Al compounds complexed with humic substances); base cation leaching (translocation and removal of Ca, Mg, and K from leaching); ferrallitization (residual accumulation of Fe compounds in a clayey matrix from previous weathering episodes); and pedoturbation (mixing of soils by plants, animals, frost, and other mechanisms).

A discussion of the general soil regions of Wisconsin is given in Chap. 5. This includes modifications in soil regions and the recognition of more than twice as many soil series as when F.D. Hole prepared his version of *Soils of Wisconsin*.

Chapter 6 discusses the diagnostic horizons and soil taxa. There are 8 epipedons (surface horizon) in *Soil Taxonomy* of which 3 are common in Wisconsin: ochric, mollic, and histic. An ochric epipedon is the most common diagnostic surface horizon (76% of state's land area) and averages 24 cm in thickness. A mollic epipedon is found in 10% of the land area; it averages 39 cm in thickness. Histic epipedons cover 9% of the state and average 95 cm in thickness. Only six soil series in the state have an umbric epipedon, which averages 36 cm in thickness. The argillic horizon is the most common diagnostic subsurface horizon in the soils of Wisconsin. It covers 60% of the land area and averages 62 cm in thickness. The spodic horizon covers 18% of the state area and averages 30 cm in thickness. The third most extensive diagnostic subsurface horizon is the cambic horizon, which covers 9% of the state area and averages 48 cm in thickness. The glossic horizon occurs in conjunction with an argillic horizon or with argillic and spodic horizons, covering 25% of the state's land area and averaging 33 cm in thickness. The albic horizon covers 35% of the state soil area and averages 13 cm in thickness. Only ten soil series in Wisconsin contain a fragipan, which averages 62 cm in thickness and covers 1.2% of state's land area. Fragipans are restricted to the northern tier of counties in the state.

Wisconsin soils are included in seven orders, 15 suborders, 32 great groups, 142 subgroups, 425 families, and 741 soil series. More than half of the soil series were originally reported in Wisconsin; 40% occur only in Wisconsin; and 18% are unique and are considered endemic soils. More than half (52%) of the soils have a frigid soil-temperature regime, and 42% have a mesic soil-temperature regime. Eighty-seven percent of the soil series have a mixed mineralogy, meaning that the soils contain a variety of minerals in the fine-earth (<2 mm) fraction. Nearly half (48%) of the soils are in loamy particle-size classes, 22% are sandy, 18% are silty, 10% are clayey, and 2% are comprised of organic materials. Half (50%) of the soils have a superactive cation-exchange capacity, meaning that they have clay-size minerals with a high nutrient-holding capacity. About 83% of the soil area of Wisconsin contains soils that are very deep (>150 cm) to a restricting layer or bedrock, and 40% of the soil area is well drained.

In Chap. 7, we recognize 16 taxonomic soil regions (primarily great groups) in Wisconsin, including (from most to least in area) Hapludalfs (25%), Glossudalfs (16%), Haplorthods (15%), Haplosaprists–Haplohemists (9.6%), Udipsamments–Udifluvents (8.3%), Argiudolls (5.5%), Aquolls (3.0%), Aqualfs (2.6%), Aquods (2.4%), Hapludolls (2.0%), and six other regions that collectively comprise 8.6% of the state area (Paleudalfs, Udepts, Aquepts, Aquents, Quartzipsamments, and Fragiorthods).

Chapters 8 through 13 review the distribution, properties and processes, and use and management of each of the seven soil orders present in Wisconsin, along with order, suborder, and great group maps and images and data of representative soil series.

The paleosols ("soils of the past") of Wisconsin are described in Chap. 14. Paleosols may be buried, relict (remain at the surface as ground soils), or exhumed (previously buried material eroded from the surface). These soils can be divided into five age groups: (i) Holocene burials; (ii) relict soils on pre-Late Wisconsinan to Illinoian till; (iii) relict soils on pre-Illinoian till; (iv) loess over clayey pedisediments containing a buried soil over sandstone bedrock; and (v) loess over clayey pedisediments containing a buried soil over dolomitic or limestone bedrock. The oldest paleosols in Wisconsin include the red clays (*terra rossa*) of mid-Miocene age on saprolite from dolomitic limestone in the Driftless Area that has been covered by loess.

Endemism refers to plant or animal species or soils that occur naturally and are confined to a particular geographic area (Chap. 15). About 132 soil series in Wisconsin qualify as being endemic. Fifteen of these soils (11% of total endemic soils) occur in Bayfield County, which contains a diversity of parent materials, elevations, and proximity to Lake Superior. Two-thirds (67%) of these soils are considered rare; i.e., each has an area of less than 10,000 ha; 58% may be considered endangered and only occur in Wisconsin.

Chapter 16 focuses on the distribution of soils in the state in relation to past and present climates. By 2050, Wisconsin will experience a temperature increase of 3.3–3.8 °C and an approximate 25% increase in winter precipitation. A key climatic factor in Wisconsin pertains to the amount of snowfall. Soils will cool as the climate warms in the Great Lakes region, because of a thinner snowpack which will reduce the insulation from soil freezing. The impact of warming on fire frequency will influence soil-forming rates in Wisconsin. It will also result in land use changes that affect soil properties.

In Chap. 17, we examine the yield potential of Wisconsin soils. We observe highly significant correlations between economic parameters, such as agricultural land value sales and adjusted gross income to soil great groups on a county-wide basis.

In the final chapter, we identify issues that will need resolution in the near future that pertain to the mapping, classification, and organization of soil databases in Wisconsin.

1 Introduction

1.1 Definition of Soil

There are many definitions for soil ranging from the utilitarian to a description that focuses on material. Soil has been recognized as (i) a natural body, (ii) a medium for plant growth, (iii) an ecosystem component (iv) a vegetated water-transmitting mantle, and (v) an archive of past climate and processes. In this book, we follow the definition given in the Keys to Soil Taxonomy (Soil Survey Staff 2014, p. 1) that the soil "is a natural body comprised of solids (minerals and organic matter), liquid, and gases that occurs on the land surface, occupies space, and is characterized by one or both of the following: horizons, or layers, that are distinguishable from the initial material as a result of additions, losses, transfers, and transformations of energy and matter or the ability to support rooted plants in a natural environment."

1.2 The Soils of Wisconsin

With an area of 169,639 km^2, Wisconsin is the 23rd largest state in the USA. Soils in the state have been investigated for over 120 years (Chap. 2). The most recent comprehensive examination of the soils of the state, *Soils of Wisconsin*, was prepared in 1976 by F.D. Hole. The first part of his book contained chapters describing soil-forming factors, soil-forming processes, properties, and soil classification. The second part described 121 soil associations in 11 soil regions of the state. The regions were largely based on physiography, vegetation, and texture of the soil parent materials. The last part of the book examined properties and occurrence of major soil series in Wisconsin, along with major soil topographic sequences and soil associations.

At the time of Hole's book, there were about 300 soil series described in Wisconsin; today, 740 soil series have been recognized. In the mid-1970s, soil surveys existed for about 50% of the state; in 2006, the completion of mapping in Iron County gave 100% soil-map coverage for the state.

1.3 Major Soil Regions of Wisconsin

Beginning with *Introduction to the Soils of Wisconsin: a general scheme of classification of the principal soils of Wisconsin* (Hole and Lee 1955) and continuing through 1993 with the general soil map by Madison and Gundlach (1993), the state has been divided into general regions based on physiography and texture of soil parent materials (e.g., Hole and Lee 1955; Hole 1968) or geographic location, broad vegetation type, and soil texture (Madison and Gundlach 1993) (Table 1.1). The Natural Resources Conservation Service (2006) has divided the USA into Major Land Resource Areas (MLRAs) based on by topography, other landscape features, hydrologic units, resource concerns, resource uses, and human considerations affecting use and soil and water conservation treatment needs. The MLRAs generally coincide with soil regions as delineated by Hole (1976) (Table 1.2).

1.4 Classification of Wisconsin Soils

From 1905 until 1955, the classification of soils in Wisconsin was limited to soil series and parent material texture, although a national soil classification scheme had been available since 1928 (Marbut 1927; Baldwin et al. 1938). In 1956, F.D. Hole used great soil groups (Baldwin et al. 1938) in his soil survey of Waukesha County. The great soil group system also was used by the Wisconsin Geological and Natural History Survey. Beginning in 1970, the Soil Conservation Service (now Natural Resources Conservation Service used the *Seventh Approximation* (Soil Survey Staff 1960) for classifying soils of Wisconsin; this system was the precursor to *Soil Taxonomy*.

All soils in Wisconsin are now classified using *Soil Taxonomy* (ST) (Soil Survey Staff 1999, 2014) and are widely used throughout this book. *Soil Taxonomy* is a hierarchical classification system that classified soils based

J.G. Bockheim and A.E. Hartemink, *The Soils of Wisconsin*,
World Soils Book Series, DOI 10.1007/978-3-319-52144-2_1

Table 1.1 Three approaches for subdividing Wisconsin soils into regions

Letter	Soil region (Hole 1976)	MLRA	NRCS (2006)	Letter	Soil region (Madison and Gundlach 1993)
A	Soils of the southwestern ridges and valleys	105	Northern Mississippi Valley Loess Hills		*Soils of southwestern and western Wisconsin*
				A	Forested, silty soils
B	Soils of the southeastern upland	95B	Southern WI & Northern IL Drift Plain	Am	Prairie, silty soils
				Dr	Forested soils over sandstone
C	Soils of the central sandy uplands and plains	89	Wisconsin Central Sands		
					Soils of southeastern Wisconsin
D	Soils of the western sandstone uplands, valley slopes, and plains			B	Forested, silt soils
				Bm	Prairie, silty soils
E	Soils of the northern and eastern sandy and loamy reddish drift uplands and plains	95A	Northern Wisconsin drift plain		
					Soils of central Wisconsin
F	Soils of the northern silty uplands and plains	90A	WI and MN Thin Loess & Till, Northern Part	C	Forested, sandy soils
		90B	WI and MN Thin Loess & Till, Southern Part	Cm	Prairie, sandy soils
G	Soils of the northern loamy uplands and plains	93B	Superior Stony & Rocky Loamy Plain & Hills, Eastern Pt.	Fr	Forested, silty soils over igneous/metamorphic rock
		94D	Northern Highlands Sandy Drift		
H	Soils of the northern sandy uplands and plains	91B	WI and MN Sandy Outwash		*Soils of northern and eastern Wisconsin*
		94B	Michigan Eastern Upper Peninsula Sandy Drift	E	Forested, red, sandy, and loamy soils
I	Soils of the northern and eastern clayey and loamy reddish drift uplands and plains	92	Superior lake plain	Er	Forested, red, sandy, and loamy soils over dolomite
				F	Forested, silty soils
J	Soils of the stream bottoms and major wetlands			G	Forested, loamy soils
				H	Forested, sandy soils
				I	Forested, red, clayey or loamy soils
					Statewide
				J	Stream bottom and major wetland soils

on the properties as contained of diagnostic surface (epipedons) and subsurface horizons. For classification purposes, the upper limit of the soil is defined as the boundary between the soil (including litter layer) and the air above it. The lower limit is 200 cm. The definition of the classes (taxa) is quantitative and uses well-described methods of analysis for the diagnostic properties. The formation of the soil is not explicitly used in the system, and the soil is classified "as it is" using morphometric observations in the field coupled with laboratory analysis and other data. The nomenclature in *Soil Taxonomy* is mostly derived from Greek and Latin sources and is not that different from the classification of plants or animals.

Soil Taxonomy classifies soils, from broadest to narrowest levels, into orders, suborders, great groups, subgroups, and families. Families occur in one or more soil series, which along with soil associations (combinations of soil series) constitute the primary soil-mapping units.

There are eight diagnostic surface horizons (epipedons) in *Soil Taxonomy* and five of them occur in Wisconsin: folistic,

Table 1.2 Abbreviated descriptions of diagnostic horizons in *Soil Taxonomy* as applied to soils of Wisconsin

Horizon	Brief description
Epipedons	
Folistic	Thick organic soil materials (unsaturated)
Histic	Thick organic soil materials (saturated)
Mollic	Thick, dark colored, highly base saturated, enriched in organic C
Ochric	Pale, low organic matter, or thin
Umbric	Same as mollic but less base saturation
Subsurface horizons	
Albic	Subsurface horizon ≥ 1 cm thick; light colored; eluvial; leached of clay, Fe oxides
Argillic	Accumulation of illuvial, high-activity silicate clays; the presence of clay bridges orcoatings
Cambic	Minimal but visible subsurface development; does not meet requirements of other horizons
Fragipan	Subsurface horizon that is brittle; reversibly cemented; slakes in water; breaks with characteristic snap; restricts rooting
Glossic	Subsurface horizon reflecting the degradation of an argillic horizon with zones of E and Bt materials
Ortstein	Subsurface layer >25 mm thick with cemented spodic materials
Spodic	Illuvial accumulation of organic matter & Al (with or without Fe); dark colored; low base saturation (acidic)

histic, mollic, umbric, and ochric (Table 1.2). The histic and folistic horizons contain primarily organic materials; the histic epipedon is saturated for prolonged periods during the year, the folistic epipedon is saturated for less than 30 cumulative 30 days in most years. The folistic epipedon occurs in thin soils on sloping bedrock, and the histic epipedon occurs in soils of wetlands. The mollic and umbric epipedons occur in mineral soils and are thick, dark colored, and enriched in organic matter (Fig. 1.1). The mollic epipedon is enriched in base cations, such as calcium, magnesium, and potassium, whereas the umbric contains low amounts of these cations. The ochric epipedon is thin, pale, and often low in organic matter.

Seven of the 20 diagnostic subsurface horizons identified in *Soil Taxonomy* are present in the soils of Wisconsin (Table 1.2). The albic horizon has a bleached, ashy appearance. The argillic horizon is enriched in clay that has moved down the profile from percolating water. The cambic horizon shows minimal development other than soil structure and color. The spodic horizon is acid, dark reddish brown in color, and has an accumulation of humus complexed with aluminum and iron. A glossic horizon reflects the degradation of an argillic horizon and has zones of E and Bt materials, e.g., E/B and B/E. There are two kinds of hardpans in Wisconsin: the fragipan is brittle, reversibly cemented, and breaks with a characteristic snap when dried; ortstein is dark reddish brown and is cemented by iron. Photographs of the subsurface horizons are given in chapters describing soils in each of the seven orders represented in Wisconsin. The formative elements used in constructing soil names are given in Appendix A.

Soil orders are delineated primarily on the basis of the diagnostic horizons. Seven of the 12 orders in *Soil Taxonomy* occur in Wisconsin: Alfisols, Entisols, Histosols, Inceptisols, Mollisols, Spodosols, and one Ultisol (Fig. 1.2). Alfisols are base-enriched forest soils with an argillic horizon; Entisols are very poorly developed recent soils with only a diagnostic epipedon; Histosols are organic soils; Inceptisols are juvenile soils that contain a diagnostic epipedon and/or a cambic horizon; Mollisols are dark-colored, base-enriched grassland soils; Spodosols are acid, dark reddish brown soils with subsoil accumulation of aluminum and iron; and Ultisols are base-depleted soils with subsoil accumulation of low-activity clay (Table 1.3).

Suborders are distinguished on the basis of soil climate for five of the seven orders occurring in Wisconsin, the Alfisols, Inceptisols, Mollisols, Spodosols, and Ultisols. Soil parent materials are used to differentiate among the Entisols; and the degree of organic matter decomposition is used for the Histosols. There are 15 suborders of soils in Wisconsin. Great groups are distinguished from a variety of soil characteristics; there are 32 great groups of soils in Wisconsin.

Of the 740 soil series identified in Wisconsin, 43% are Alfisols, followed by Mollisols (20%), Spodosols (15%), Entisols (11%), Inceptisols (7%), Histosols (4%), and Ultisols (0.1%) (Fig. 1.3). A list of all soil series recognized in Wisconsin, along with their soil-forming factors, is given in Appendix B.

Fig. 1.1 Examples of five diagnostic surface horizons (epipedons) in Wisconsin, including the mollic (*upper left*), umbric (*upper center*), ochric (*upper right*), histic (*lower left*), and folistic (*lower right*). Histic epipedon photograph by J.W. Williams and folistic epipedon photograph by M.B. Miquez

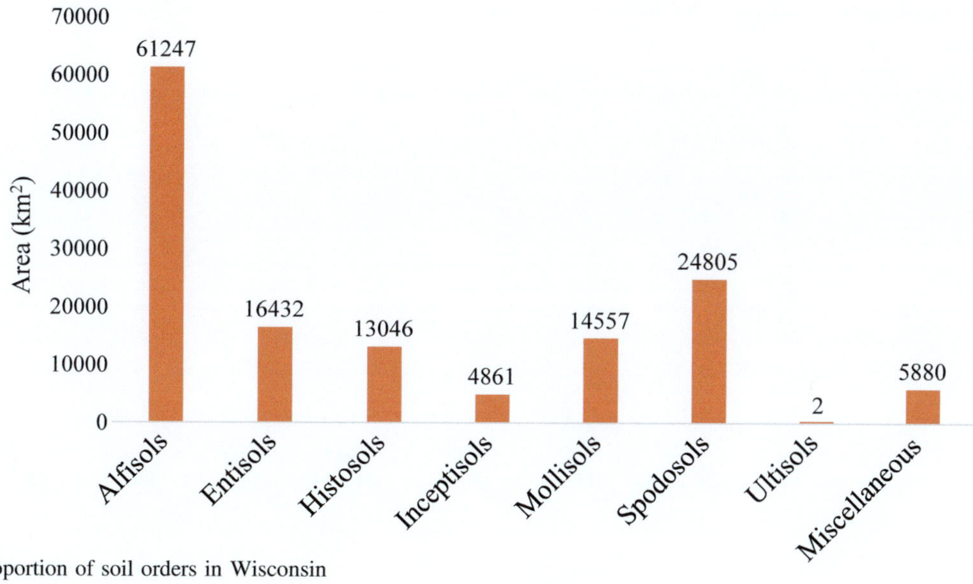

Fig. 1.2 Area proportion of soil orders in Wisconsin

Table 1.3 Brief descriptions of soil orders occurring in Wisconsin

Order	Brief description
Alfisols	Moderately leached soils that have relatively high native fertility. These soils have mainly formed under forest and have a subsurface horizon in which clays have accumulated. They contain $\geq 35\%$ base cations
Entisols	Soils of recent origin. The central concept is soils developed in unconsolidated parent material with usually no genetic horizons except an A horizon. All soils that do not fit into one of the other 11 orders are Entisols. Thus, they are characterized by great diversity, both in environmental setting and land use
Histosols	(From Greek *histos*, "tissue") soils that are composed mainly of organic materials. They contain at least 20–30% organic matter by weight and are more than 40 cm thick
Inceptisols	(From Latin *inceptum*, "beginning") are soils that exhibit minimal horizon development. They are more developed than Entisols, but still lack the features that are characteristic of other soil orders
Mollisols	(From Latin mollis, "soft") are the soils of grassland ecosystems. They are characterized by a thick, dark surface horizon. This fertile surface horizon, known as a mollic epipedon, results from the long-term addition of organic materials derived from plant roots
Spodosols	(From Greek *spodos*, "wood ash") are acid soils characterized by a subsurface accumulation of humus that is complexed with Al and Fe. These soils typically form in coarse-textured parent materials and have a light-colored E horizon overlying a reddish brown spodic horizon
Ultisols	(From Latin *ultimus*, "last") are strongly leached, acid forest soils with relatively low native fertility. These soil have a clay-enriched B horizon and a base saturation that is $\leq 35\%$

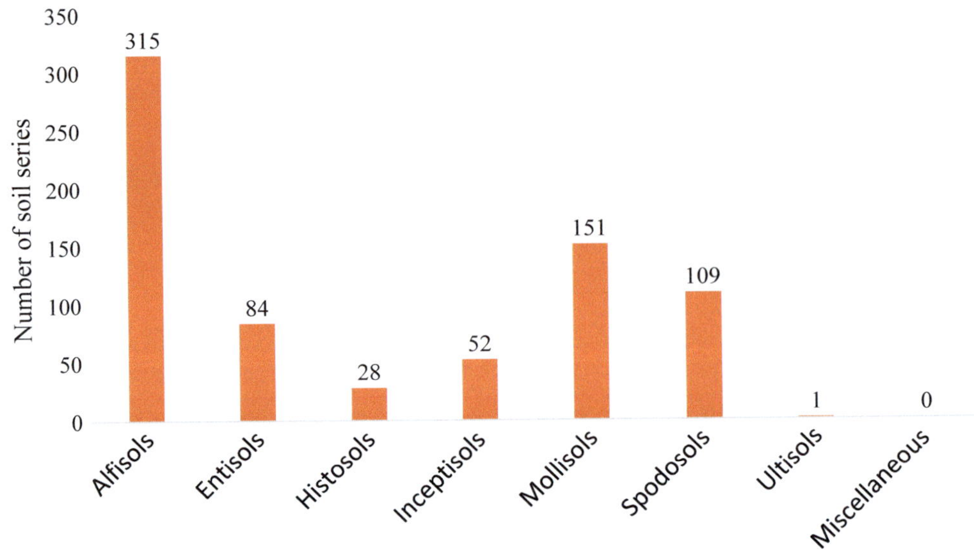

Fig. 1.3 Proportion of soil series by soil order in Wisconsin

1.5 Summary

The soil is a natural body that occurs on the land surface and is characterized by horizons that are distinguishable from the initial material as a result of additions, losses, transfers, and transformations of energy and matter. From 1905 until 1955, the classification of soils in Wisconsin was limited to soil series and parent material texture, despite that a national soil classification scheme, the great soil group system, had been available since 1928. The great soil group system was used by the Wisconsin Geological and Natural History Survey until Hole's book in 1976, when the *Seventh Approximation* was used. Prior to Hole's book, the state was divided into general regions based on physiography and texture of soil parent materials. Soils in Wisconsin are now classified using *Soil Taxonomy*, which is a hierarchical system that classifies soils based on natural properties as contained in what are known as diagnostic surface (epipedons) and subsurface horizons. Seven of the 12 orders in *Soil Taxonomy* are recognized in Wisconsin, including from greatest to least in area, Alfisols, Spodosols, Entisols, Mollisols, Histosols, Inceptisol, and Ultisols. There are 15 suborders and 32 great groups of soils in Wisconsin. Five of the nine diagnostic surface horizons are used in Wisconsin, including the folistic, histic, mollic, umbric, and ochric. Seven of the 20 diagnostic subsurface horizons are present in Wisconsin soils, including the albic, argillic, cambic, glossic, spodic, fragipan, and ortstein.

2 History of Soil Studies

2.1 Introduction

The word "Wisconsin" originates from the name given to the Wisconsin River by one of the Algonquin-speaking American Indian groups living in the region at the time of European contact (Wisconsin Historical Society 2014). The Algonquian word for Wisconsin and its original meaning have grown obscure. Interpretations vary, but most implicate the river and the red sandstone that lines its banks. One leading theory is that the name originated from the Miami word *Meskonsing*, meaning "it lies red," a reference to the setting of the Wisconsin River as it flows through the reddish sandstone of the Wisconsin Dells (Fig. 2.1). French explorer Jacques Marquette was the first European to reach the Wisconsin River, arriving in 1673, and calling the River *Meskousing* in his journal. This spelling was later changed to *Ouisconsin* by other French explorers, and over time this became the French name for both the Wisconsin River and the surrounding lands. English speakers changed the spelling to its current form when they began to arrive during the early nineteenth century. The current spelling was made official by the legislature of Wisconsin Territory in 1845.

2.2 Early Soil Investigations

All aspects of soil have been researched in Wisconsin, and there is a vast body of literature that goes back to the late 1800s. To list a few examples that shaped Wisconsin soil science: the agricultural physics work of F.H. King, the seminal work of M.L. Jackson on silicate crystal chemistry and radioactive isotopes, the soil fertility "revolving fund" concept of A.R Whitson (Whitson and Walster 1918), the plant mineral nutrition work of E. Troug, and the ordination concept of Hole and Hironaka (1960) that still needs to be assimilated in the soil science community (Hartemink 2012). All these are renowned nationally and internationally, and there are several remarkable findings and characters in the soil science of Wisconsin—some of them have been described by Beatty (1991), and Table 2.1 summarizes some of the main historical events for Wisconsin.

2.2.1 T.C. Chamberlin

Soils have been studied in Wisconsin for over a century, beginning with the publication of T.C. Chamberlin's *General Map of the Soils of Wisconsin* (Table 2.1; Fig. 2.1) (Hartemink et al. 2012). Chamberlin was the chief geologist for Wisconsin, and between 1873 and 1877 he published with several co-authors four voluminous books (3035 pages) titled "Geology of Wisconsin." Chamberlin introduced the glacial stages of North America and produced an Atlas that includes the first soil map of Wisconsin (Fig. 2.2).

Chamberlin's map, which was the first soil map developed in the USA (Brevik and Hartemink 2010), shows eight soil textural groups: sandy soils, sandy loams, calcareous sandy loams, prairie loams, clayey loams (three types), and humus soils (Chamberlin 1882). Chamberlin had a strongly geologic view on soils, and he considered that the character of the soil will depend upon the nature of the rock, the degree of weathering, and amount lost by leaching and gained by vegetation or capillary action from beneath (Chamberlin 1877). He evidently recognized the difficulties in mapping soils as he wrote:

> Map of soils. There are few natural formations more difficult to map than soils. There is an almost infinite gradation of varieties between which there are no hard-and-fast lines, and it is nearly or quite impossible to represent these gradations on a map.

It is not clear how this first soil map of Wisconsin was made, but it was most likely based on extensive travels on horseback through the state and Chamberlin's knowledge of the surficial geology of the region. The impact of Chamberlin's soil map has not been assessed, but it could be that his map was an argument in the establishment of the national

J.G. Bockheim and A.E. Hartemink, *The Soils of Wisconsin*,
World Soils Book Series, DOI 10.1007/978-3-319-52144-2_2

Fig. 2.1 Sandstone bluffs at the Wisconsin River near the Dells. Pine trees are common and many soils are shallow with a folistic epipedon

Table 2.1 Historical events in the study of Wisconsin soils

Year	Event
1882	First soil map in USA of WI by T.C. Chamberlin
1889	Department of Agricultural Physics established, UW-Madison
1895	First soil science book published in USA by F.H. King, *the soil—its structure, relations, and fundamental principles of management*
1902	First published soil survey, Janesville Area
1906	First published county soil survey, Racine Co.
1909	Renamed Agricultural Physics into Department of Soil Science, UW-Madison
1919	Wisconsin Geological & Natural History Survey established, began mapping the first of 50 counties in the state
1926	Second general soil map of WI, A.R. Whitson
1927	Publication of first book of WI soils, *Soils of Wisconsin*, A.R. Whitson
1930	*Preliminary study of the profiles of the principal soil types of Wisconsin*, (included soil map) C.E. Kellogg
1947	Publication of *Soils of Wisconsin*, R.J. Muckenhirn & N.P. Dahlstrand
1965	Renamed Department of Soils into Department of Soil Science
1968	Color version of map, Major Soil Regions of WI, F.D. Hole
1973	Publication of F.D. Hole's *Soils of Wisconsin*
1980	*Soil Guide for Wisconsin Land Lookers*, F.D. Hole
1983	Antigo silt loam established as Wisconsin state soil
1993	Revision of soil regions map of WI, F.W. Madison & H.F. Gundlach
2000	Soil survey digitized and available as SSURGO
2006	Completion of mapping in WI with Iron Co.

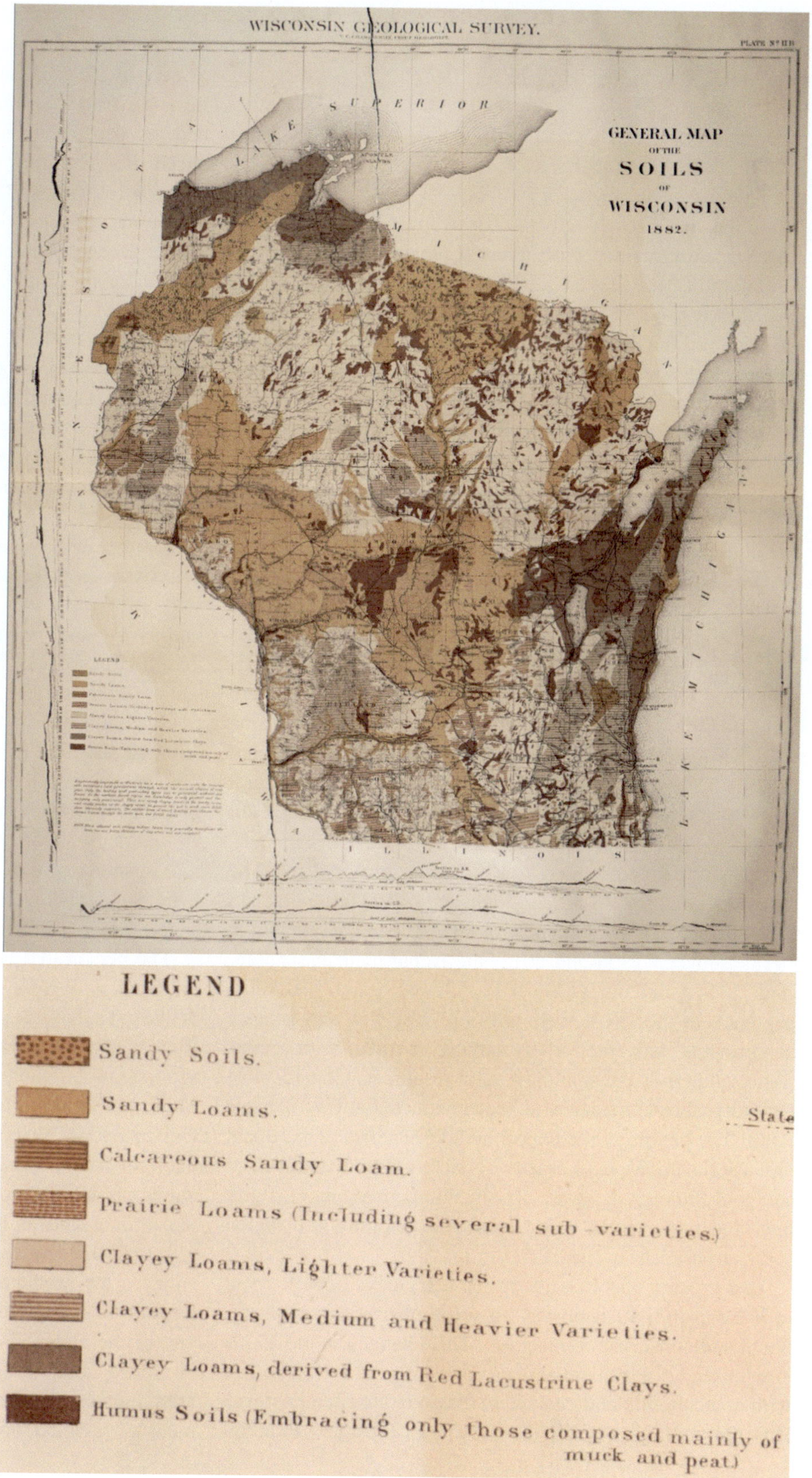

Fig. 2.2 First soil map and its legend in the USA prepared by T.C. Chamberlin (1882) for Wisconsin

soil survey in 1899. After Chamberlin's map, it would take 40 years before the next reconnaissance soil map of the entire state was produced (Whitson 1927).

2.2.2 Soil Research

When the first French settlers arrived in 1634, most of the state was still covered with forest (Campbell 1906). At the beginning of the twentieth century, most of the natural forest had been logged (Whitson 1927). Soil science as a subject for study did not receive much attention in the early years of settlement. Systematic soil research was more or less started by F.H. King of the University of Wisconsin, who wrote one of the first soil science text books for the USA, *The Soils—Its Nature, Relations, and Fundamental Principles of Management* (King 1895). A section of a terminal moraine near Whitewater and an early topographic soilscape is reproduced in Fig. 2.3 from King's book from 1895.

The Department of Agricultural Physics was established in 1889, with F.H. King as its first chair, at a time when scientific studies relating to agriculture were beginning to be recognized as valuable. It was one of the founding departments for University of Wisconsin College of Agriculture and Life Science (CALS), which was established in the same year. King Hall was built in 1894. In 1909, the Department of Soils was established from the Department of Agricultural Physics, and in 1915 the Soils Annex was erected. King Hall is now part of the UW Department of Soil Science (Fig. 2.4). The Department celebrated its 125-year history in 2014. The first student with a BS degree graduated in 1905, the first MS in 1906, and the first student with a PhD degree graduated in 1918. Between 1906 and 2015, the Department has graduated 2364 students (1296 BS, 577 MS and 491 PhD degrees). In the USA, there are only very few soil science departments left; the others having merged with crops, water, and environmental or natural resources. Considerable pedological research has also been conducted at UW Stevens Point. The pedology section of the UW-Madison department has a long tradition of research, teaching, and outreach in soil morphology, genesis, classification, mapping, geography, and more recently, in digital soil mapping and digital soil morphometrics (Fig. 2.5).

Soil erosion was massive in the early days of settlement in Wisconsin, but there was no sense of urgency in developing soil conservation practices. That changed in the 1930s in which a US President and a soil scientist were confronted with an economic and ecological disaster. That scientist was H.H. Bennett, whom with the backing of President Franklin Roosevelt, established in 1933 a New Deal agency: the *Soil Erosion Service*. It was succeeded by the *Soil Conservation Service* in 1935. Bennett ("Big Hugh") grew up on a cotton farm in North Carolina, and his soil survey work in Virginia and the paper by T.C. Chamberlin on *Soil Wastage* influenced his research direction. As he wrote later about Chamberlin's paper: "....it fixed my determination to pursue that subject to some possible point of counteraction" (Helms 2010). Bennett developed the idea of a soil conservation district for the implementation of soil and water conservation. One of the first soil conservation districts was the Coon Creek watershed in the non-glaciated part (the Driftless Area) of Wisconsin. As the land is sloping and much of the forest was logged, there was massive soil erosion in the Driftless area, and over 60% of the cropped land had lost 10–15 cm of its topsoil (Clark 1940). The Coon Creek watershed in Wisconsin has become a landmark for soil erosion research (Fig. 2.6).

Another area of soil research that has been developed in Wisconsin is research on soils under forest. Research on soils in forests was well developed in Germany in the mid-1900s, and German scientists brought the study of soils into the forestry discipline. Studies were conducted on tree–soil interactions and the role of soils in increasing wood production. Also in the USSR, soils under forest received research attention (Tiurin 1930). In the USA, some forestry schools' students took classes in soils, but these were usually given in the agricultural departments. This lasted until the 1940, after which several national institutions took a leading role in soils research in the USA. These were Yale, Cornell, Penn State, Duke, and the University of Wisconsin (Gessel and Harrison 1999).

The person that brought soil research under forest to the forefront in Wisconsin was S.A. Wilde. Sergei Alexander Wilde (1898–1981) was born in 1898 near Moscow, Russia, where he was introduced early to the boreal forest ecosystems he came to love. He arrived at Ellis Island in May 1929 and joined the Soils Department of the University of Wisconsin in 1934 (Fig. 2.7). He aimed to interpret forest soils as carriers of definite floristic associations, as media for the growth of nursery stock or forest plantations, and as dynamic systems that react to different forms of silvicultural cuttings. His primary aim was to enhance the production of wood without depleting the soil fertility or contaminating the environment. Wilde authored one of the classical and most widely used reference books on forest soils (Wilde 1946). His work on the site–soil requirements for successful establishment, and development of planted species is acknowledged as a classical

Fig. 2.3 Section of a terminal moraine near Whitewater, and relation between soils and their landscape position; both illustrations from one of the first textbooks on soils in the USA (King 1895)

research effort (Gessel and Harrison 1999). Wilde's investigations had an enormous impact on the forest economy of the state. Practicing foresters in Wisconsin, managing private and public lands, state and federal forest services, private nurseries, the pulp and paper industry, and owners of Wisconsin's private woodlots benefited from the insight and approach to sound forest management (Wilde et al. 1949). His work in the field of forest soils research included important studies on soils, woody-plant nutrition, tree–mycorrhiza relationships and reforestation. Wilde was not only able to bring some soil science into forestry, but also brought forestry and vegetation knowledge back into

Fig. 2.4 King Hall in the early 1900s, now Department of Soil Science, University of Wisconsin-Madison. In addition to studying soil physics, F.H. King studied windmills, round bars, and a whole range of practical and theoretical subjects. He can be viewed as the Benjamin Franklin of US soil science

soil science. He was a vivid reader, a musician, and also wrote poems:

'Ode to Forest Soils'
With a face most hair and glassy stare
Like a dog whose day is done
I juggle some pink stuff in a flask,
as the drops fall one by one.
Ours not to question why—
Ours but to weigh and dry,
What does it matter that someone has a buffer
We are the ones who must titrate and suffer.
Aye, bury me deep in the mouldy earth
And sprinkle me well with lime,
Where I can rest in deep sweet peace
Until the endpoint of time.
No more shall I cringe to the crash of glass
Nor gag at the brimstone's smell
I've a cleaner, sweeter haven
In the very depths of the gley horizon.

2.2.3 Soil Survey and Mapping

The Soil Conservation Service, now the Natural Resources Conservation Service (NRCS), published the first published soil survey for Wisconsin of the Janesville area in 1902. The Wisconsin Geological and Natural History Survey was established in 1919 and began mapping its first of 50 of Wisconsin's 72 counties.

The first book of Wisconsin soils, *Soils of Wisconsin*, was written by A.R. Whitson in 1927. Whitson included a general soil map of the state in 1926 (Fig. 2.8). In 1930, Charles Kellogg, who was instrumental in developing *Soil Taxonomy*, published his *Preliminary Study of the Profiles of the Principal Soil Types of Wisconsin*. The bulletin provided detailed accounts of 25 soil series and a soil zone map for Wisconsin (Fig. 2.9). This map shows the Gray-Brown

Fig. 2.5 Pedology classes at UW Madison: Kemp station (above) and field investigations at O'Briens farm in Brooklyn and the Wallendal farm in Westfield. Fieldwork includes the use of paper maps, digital information on iPads, and proximal soil sensors such as vis-NIR and pXRF

Fig. 2.6 Agricultural fields and forest in the Driftless area of Wisconsin

(Podzolic) Forest Zone with intermixed prairie soils, a Transitional Zone, and a Podzol Zone. We recognize Alfisols (Hapludalfs) in the Gray-Brown Forest Zone, Alfisols (Glossudalfs) in the Transition Zone, and Spodosols in the Podzol Zone.

In 1968, F.D. Hole published the first soils map of Wisconsin. This was followed by a third version that was included in the books *Soils of Wisconsin*, by F.D. Hole in 1976 (Fig. 2.10). A popularized version, named *Soil Guide for Wisconsin Land Lookers*, was published in 1980. A "land looker" was defined as a speculator who travels around looking for parcels of land that can be developed.

In 1993, F.W. Madison and H.F. Gundlach revised the soil regions map of Wisconsin (Fig. 2.11). Table 2.2

Fig. 2.7 Sergei ("Doc") Wilde (1898–1981) in 1935 (with axe) and 1970. Professor of Soil Science and Forestry at the University of Wisconsin-Madison, 1934–1969. He was born in the western part of Moscow to parents of Tartar and Dutch ancestry, and his childhood home was about a mile from the Kremlin. Doc Wilde is one of the founders of the forest soils discipline

contains the legends of the maps by Chamberlin (1882), Whitson (1927), Hole (1968), and Madison and Gundlach (1993) (Hartemink et al. 2012).

In 2000, the NRCS began scanning and digitizing soil surveys. These surveys were made available at the state level as STATSGO and SSURGO. The State Soil Geographic (STATSGO) dataset is a broad-based inventory of soils and non-soil areas that occur in a repeatable pattern on the landscape and that can be cartographically shown at a scale of 1:250,000. The SSURGO database contains information that can be displayed in tables or as maps and is available for most areas in the USA. The information is collected at scales ranging from 1:15,840 or 1:20,000 and is intended for natural resource planning and management. In 2006, soil-map coverage was completed for Wisconsin with the completion of the soil map of Iron County.

2.2.4 The State Soil

Since the official start of the soil survey program in the USA, many thousands of soil series have been mapped and named. The soil series is the most homogenous category, and as a class it is a group of soils or polypedons that have horizons similar in arrangement and in differentiating characteristics (Soil Survey Division Staff 1993). The series name is typically taken from a town, village, or stream near the area where the soil is first defined. In the 1970s, some soil scientists started to advocate the concept of State Soils. They drew a parallel that many states already have natural state symbols like, for example, a bird, flower, tree, or rock.

It is not known who introduced the State Soil concept, but according to R.W. Arnold (pers. comm., 2011) it was probably Professor F.D. Hole from the University of Wisconsin-Madison. Nebraska was the first to have a State Soil (Holdredge silt loam) in 1979, followed by Wisconsin in 1983 and Vermont in 1985. By 1991, 10 states had a State Soil (Quandt and Watts 1995), and presently each state has selected a State Soil, but only twenty have been legislatively established. In Wisconsin, the State Soil shares the same level of distinction as official state flowers (viola), rock (red granite), tree (sugar maple), and bird (robin).

The establishment of the State Soil of Wisconsin (Antigo silt loam—Typic Glossudalfs) was a long and tedious process and it took Hole, with the help of several key legislators, seven years to accomplish his aim (Devitt 1988). The Antigo soil is not the world's richest soils, but it is not the world's poorest either. The Antigo silt loam occurs nearly entirely in Wisconsin (a small portion occurs in Minnesota) and is important for crops (in particular potatoes), pastures, and forestry (Fig. 2.12).

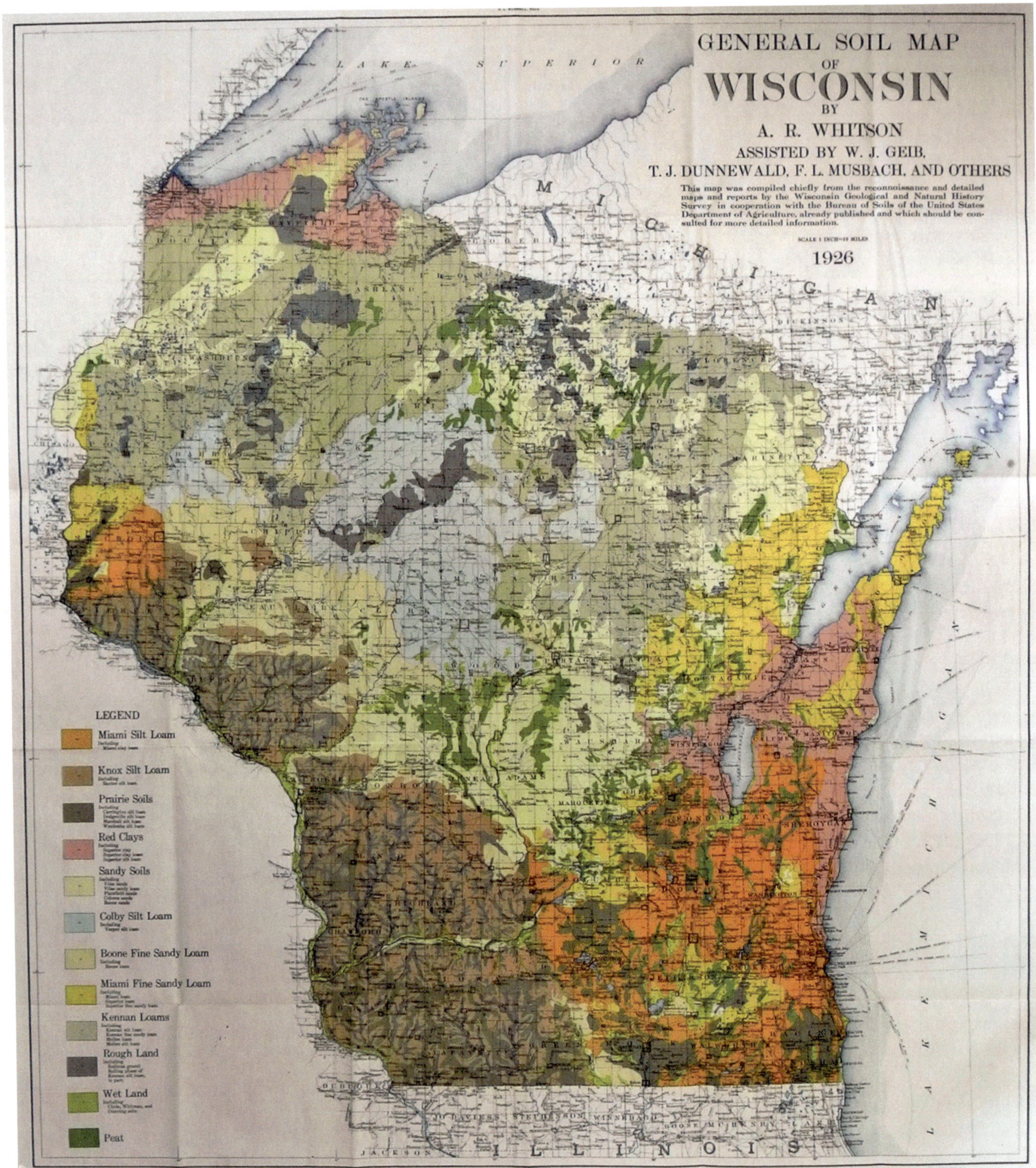

Fig. 2.8 Second soil map of Wisconsin, prepared by A.R. Whitson in the 1920s

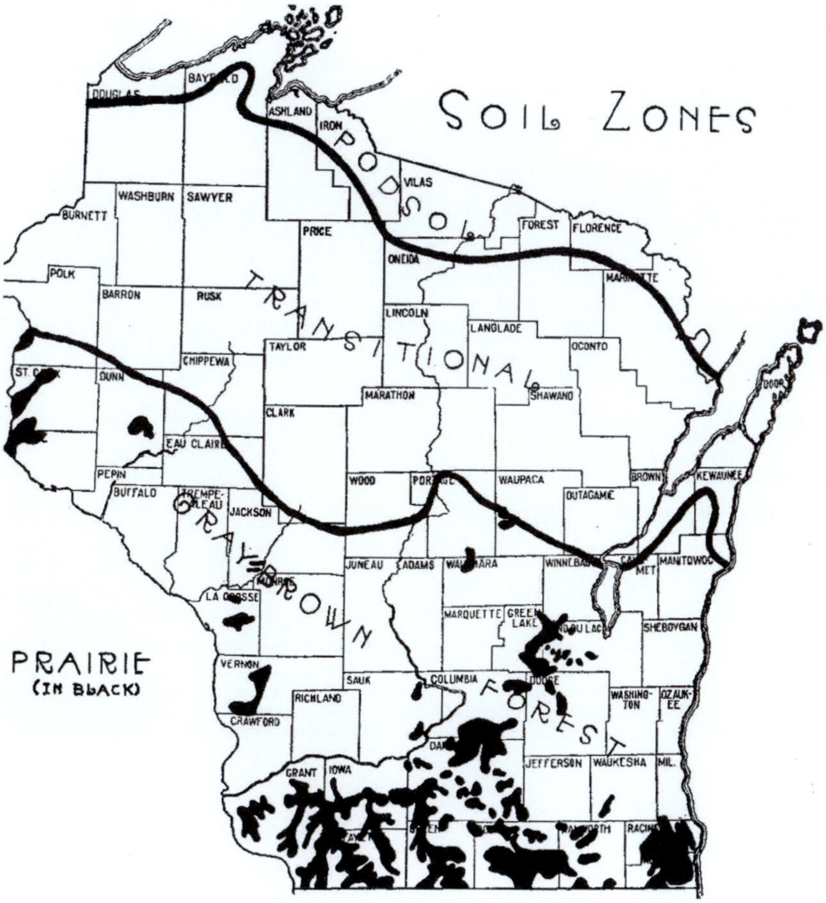

Fig. 2.9 Map prepared by C.E. Kellogg in 1930 showing the main soil zones in Wisconsin

Fig. 2.10 Francis Hole (1913–2002) and his book (1976)

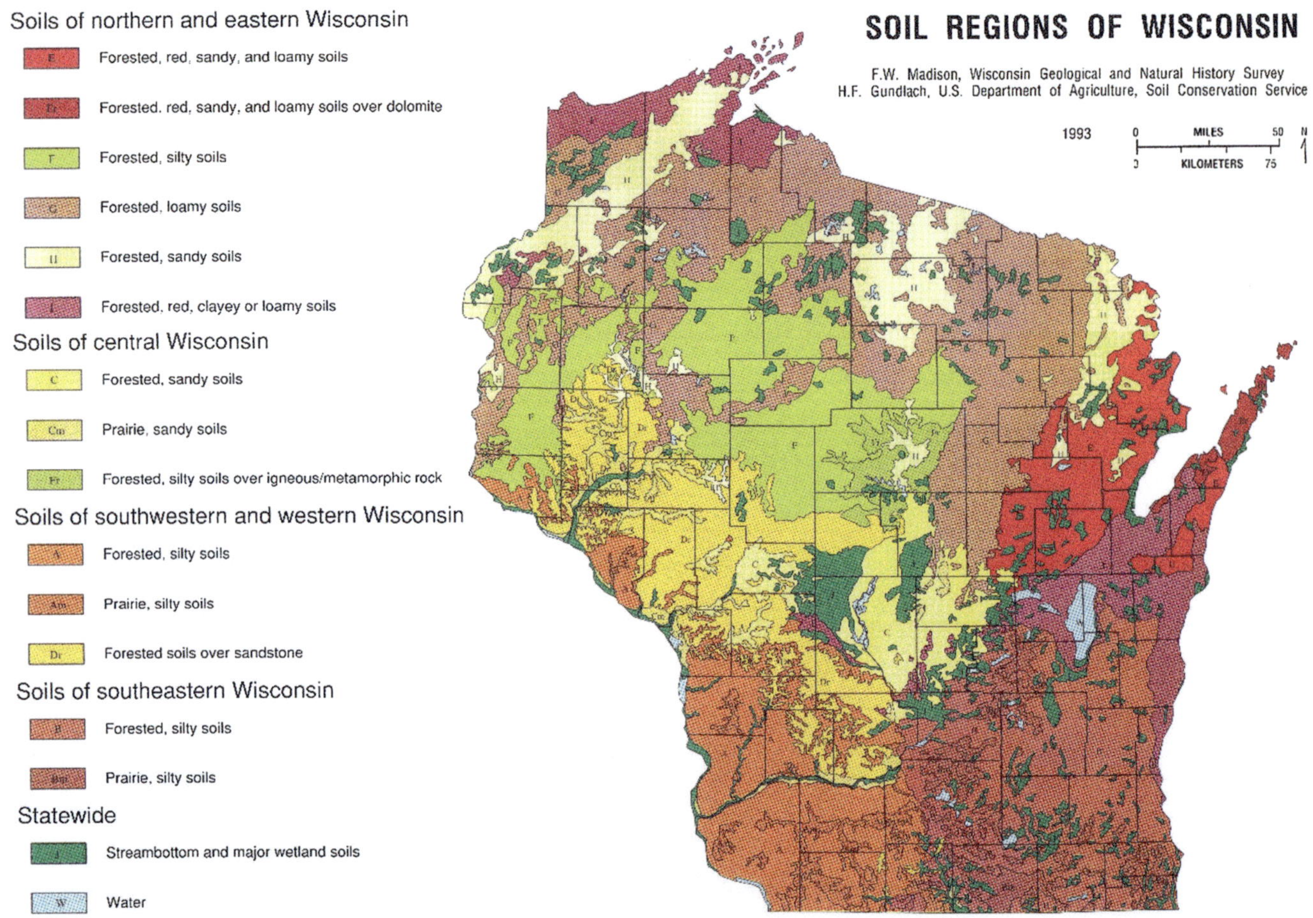

Fig. 2.11 Fourth general soil map of Wisconsin, prepared by F.W. Madison and H.F. Gundlach

Table 2.2 Legends of the Wisconsin soil maps from Chamberlin (1882), Whitson (1927), Hole (1965), Madison and Gundlach (1993) from Hartemink et al. (2012)

1882 T.C. Chamberlin	1926 A. R. Whitson	1976 F.D. Hole	1993 F.W. Madison and H.F. Gundlach
Sandy soils Sandy loams Calcareous sandy loam Prairie loams (including several sub-varieties) Clayey loams, lighter varieties Clayey loams, medium and heavier varieties Clayey loams, derived from red lacustrine clays Humus soils (embracing only those composed mainly of muck and peat)	Miami silt loam Knox silt loam Prairie soils Red clays Sandy soils Colby silt loam Boone Fine Sandy loam Miami fine Sandy loam Kennan loams Rough land Wet land Peat	**Soils of the southwestern ridges and valleys** Forest and prairie soils; Alfisols, Mollisols, Entisols; Gray-Brown Podzolics, Brunizems, Lithosols, and Humic Gley soils **Soils of the southeastern upland** Forest, prairie, and wetland soils; Alfisols, Mollisols, Entisols, Inceptisols, Spodosols, Histosols; Gray-Brown Podzolics, Brunizems, Lithosols, Regosols, Humic Gleys, Podzols, and Bog soils **Soils of the central sandy uplands and plains** Forest, prairie and wetland soils; Alfisols, Entisols, Mollisols, Spodosols, Inceptisols, Histosols; Gray-Brown Podzolics, Regosols, Brunizems, Humic Gleys, and Bog soils **Soils of the western sandstone uplands, valley slopes, and plains** Forest and wetland soils; Alfisols, Entisols, Inceptisols, Mollisols, Spodosols, Histosols; Gray-Brown Podzolics, Regosols, Lithosols, Humic Gleys, Podzols, and Bog soils **Soils of the northern and eastern sandy and loamy reddish drift uplands and plains** Forest and wetland soils; Alfisols, Entisols, Inceptisols, Mollisols, Spodosols, Histosols; Gray-Brown Podzolics, Regosols, Lithosols, Brunizems, Humic Gleys, Podzols, and Bog soils **Soils of the northern silty uplands and plains** Forest, prairie, and wetland soils; Spodosols, Alfisols, Mollisols, Inceptisols, Histosols; Podzols, Gray-Brown Podzolics, Brunizems, Podzols, Humic Gleys, and Bog soils **Soils of the northern loamy uplands and plains** Forest and wetland soils; Spodosols, Alfisols, Entisols, Inceptisols, Histosols; Podzols, Gray-Brown Podzolics, Regosols, Lithosols, Acid Brown Forest soils, Humic Gleys, and Bog soils **Soils of the northern sandy uplands and plains** Forest and wetland soils; Spodosols, Entisols, Alfisols, Histosols; Podzols, Regosols, Gray-Brown Podzolics, Brown Podzolics, and Bog soils **Soils of the northern and eastern clayey and loamy reddish drift uplands and plains** Forest and wetland soils; Alfisols, Mollisols, Spodosols, Inceptisols, Histosols; Gray-Brown Podzolics, Gray Wooded soils, Podzols, Humic Gleys, and Bog soils **Soils of stream bottoms and major wetlands** Stream bottom, marsh and bog soils; Entisols, Histosols, Mollisols, Spodosols, Inceptisols, Alfisols; Alluvial soils, Bog soils, Regosols, Humic Gleys, Podzols, Brunizems, and Gray-Brown Podzolics	**Soils of northern and eastern Wisconsin** Forested, red, sandy, and loamy soils Forested, red, sandy, and loamy soils over dolomite Forested, silty soils Forested, loamy soils Forested, sandy soils **Soils of central Wisconsin** Forested, sandy soils Prairie, sandy soils Forested, silty soils over igneous/metamorphic rock **Soils of southwestern and western Wisconsin** Forested, silty soils Prairie, silty soils Forested soils over sandstone **Soils of southeastern Wisconsin** Forested, silty soils Prairie, silty soils **Statewide** Stream bottom and major wetland soils Water

Fig. 2.12 Wisconsin's state soil, the Antigo silt loam (Haplic Glossudalfs) and landscape (irrigated potato field near Antigo, Wisconsin). The Antigo series consists of very deep, well-drained soils formed in 50–100 cm of loess or silty alluvium and in loamy alluvium and in the underlying stratified sandy outwash

2.3 Summary

Soils have been studied in Wisconsin since the 1880s, beginning with the publication of T.C. Chamberlin's *General Map of the Soils of Wisconsin.* It was the first soil map developed in the USA and showed eight soil groups based on texture of the parent material and plant cover type. Systematic soil research was more or less started by F.H. King of the University of Wisconsin, who wrote of the first soil science text books for the USA. The first book of Wisconsin soils, *Soils of Wisconsin,* was written by A.R. Whitson in 1927. In 1930, Charles Kellogg, who was to become a key in developing *Soil Taxonomy*, published his Preliminary *Study of the Profiles of the Principal Soil Types of Wisconsin.* In 1968, F.D. Hole published the first color soils map of Wisconsin, a map that is still used. This was followed by a third version of *Soils of Wisconsin*, by F.D. Hole in 1976. In 1983, the Antigo silt loam (Haplic Glossudalfs) was legislatively established as Wisconsin's state soil. In 2006, traditional soil mapping was completed for Wisconsin.

3 Soil-Forming Factors

3.1 Introduction

The expression of a soil results from five factors operating collectively: climate, organisms, relief, parent material, and time (Fig. 3.1). The factors are interacting over time and cause a range of soil processes (e.g., illuviation) that result in a diversity of soil properties (e.g., high clay content in the subsoil). Human activities that result in soil changes are often considered a sixth factor. Following the "Russian school of soil science," Kellogg (1930) illustrated the importance of geology, climate, and native vegetation on the distribution of soils in Wisconsin. Nygard et al. (1952) related the general distribution of soils in the northern Great Lakes region to climate, vegetation, and parent materials. The following is a review of the role of soil-forming factors in the development of Wisconsin soils.

3.2 Climate

3.2.1 Previous Work

Climate influences the soil development through differences in mean annual, seasonal, and extremes in temperature and moisture. Hole (1976) emphasized the importance of climate in the distribution of soils in Wisconsin: Spodosols are present in the north (Fig. 3.2); a transition zone with bisequal soils occurs in the central region, and Alfisols and Mollisols are predominant in the south (Fig. 3.3). The line separating soils in the frigid and mesic soil-temperature classes (8 °C mean annual soil temperature at the 50-cm depth) extends across what is known as the "Wisconsin Tension Zone" (Curtis 1959) (Fig. 3.4). The tension zone occupies an area of 18,500 km^2 and has a high biodiversity, with both northern and southern plants and animals occurring within it. The boundary between the mesic and frigid soil-temperature regimes roughly approximates the mean July temperature isotherm of 21 °C.

3.2.2 Current Climate

The Wisconsin State Climatology Office has a collection of data that extend back into the nineteenth century. There are presently nearly 200 stations that provide maximum and minimum temperature readings along with precipitation data on a daily basis.

Wisconsin lies in the temperate continental climatic zone. According to the Köppen–Geiger climate classification system, Wisconsin has a Dfa climate in the southern part of the state and a Dfb in the north. Southern Wisconsin has hot, wet summers and cool-to-cold, wet winters, whereas

J.G. Bockheim and A.E. Hartemink, *The Soils of Wisconsin*,
World Soils Book Series, DOI 10.1007/978-3-319-52144-2_3

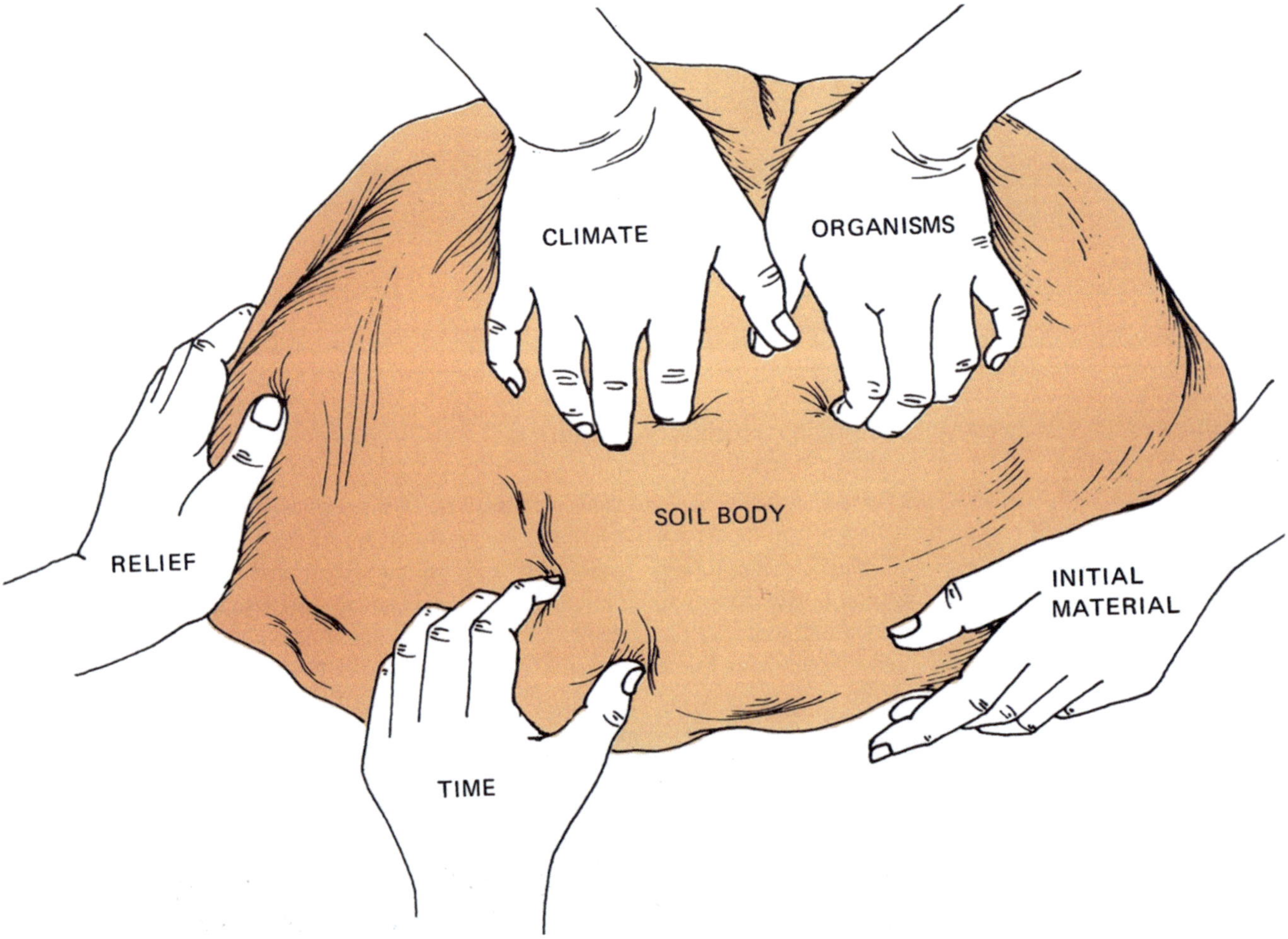

Fig. 3.1 The Five soil-forming factors (From Hole 1980). Climate and organisms are often considered active factors, whereas relief, time, and initial or parent material are considered more passive soil forming factors

northern Wisconsin has warm, wet summers and cold, wet winters. The mean annual temperature ranges from 4.0 °C in Ashland County in the far north to 9.4 °C in Grant County along the Mississippi River in southwestern Wisconsin (Fig. 3.5). Mean annual precipitation ranges from 740 mm in Florence County in the far northeast to 960 mm in Green County in southcentral Wisconsin (Fig. 3.6). Mean annual snowfall ranges from 81 cm in southern Rock County near the Illinois line to 425 cm in northern Iron County near the Upper Peninsula of Michigan (Fig. 3.7).

3.2.3 Past Climates

Wisconsin's climate has changed dramatically over the past 2.1 million years (Quaternary Period). A dozen or so glaciations covered most of the state, except for the Driftless Area in southwestern Wisconsin (see Sect. 3.5.3). The mean annual temperature was about 5.6 °C colder during the last glaciation, the Wisconsinan, approximately 14,000–20,000 years ago (Weaver et al. 1998). During the Holocene Climatic Optimum, approximately 9000–5000 years ago,

Fig. 3.2 Spodosol under Hemlock at Kemp station in Oneida County in North Wisconsin

Fig. 3.3 Alfisol developed in outwash (*top picture*) and Mollisol developed in loess (*bottom picture*)

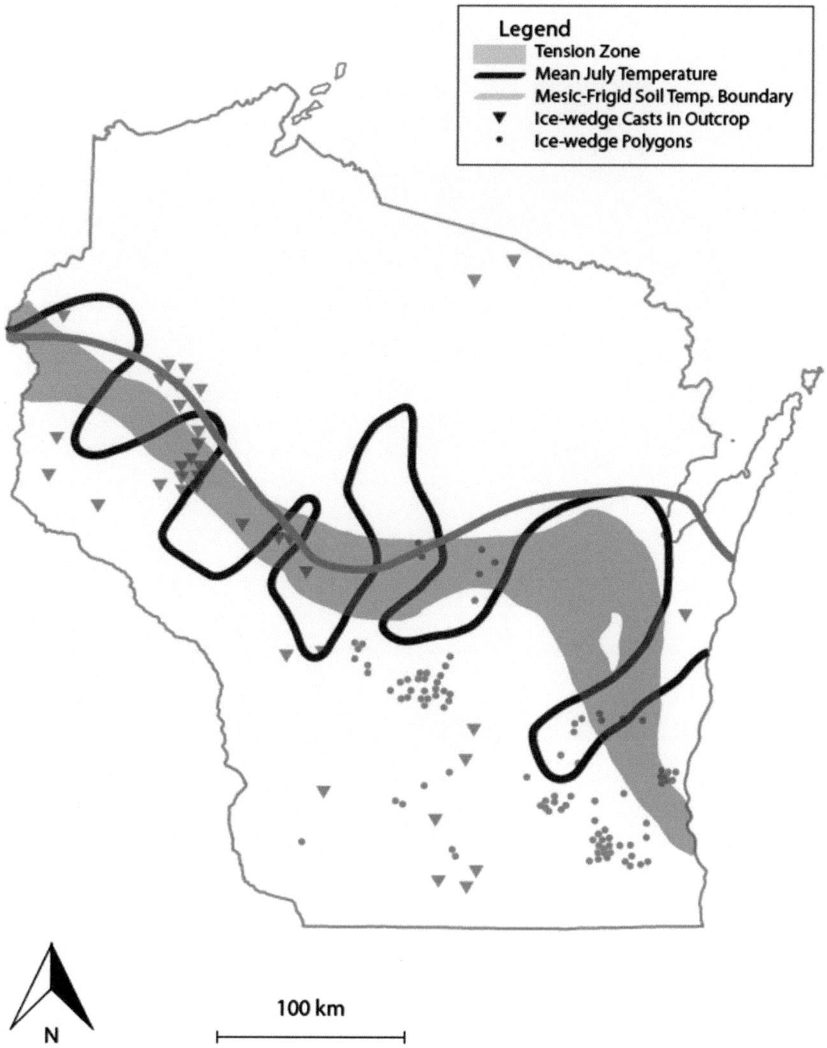

Fig. 3.4 Wisconsin's tension zone (from Bockheim and Schliemann 2013). The solid line is the 21 °C isotherm for July

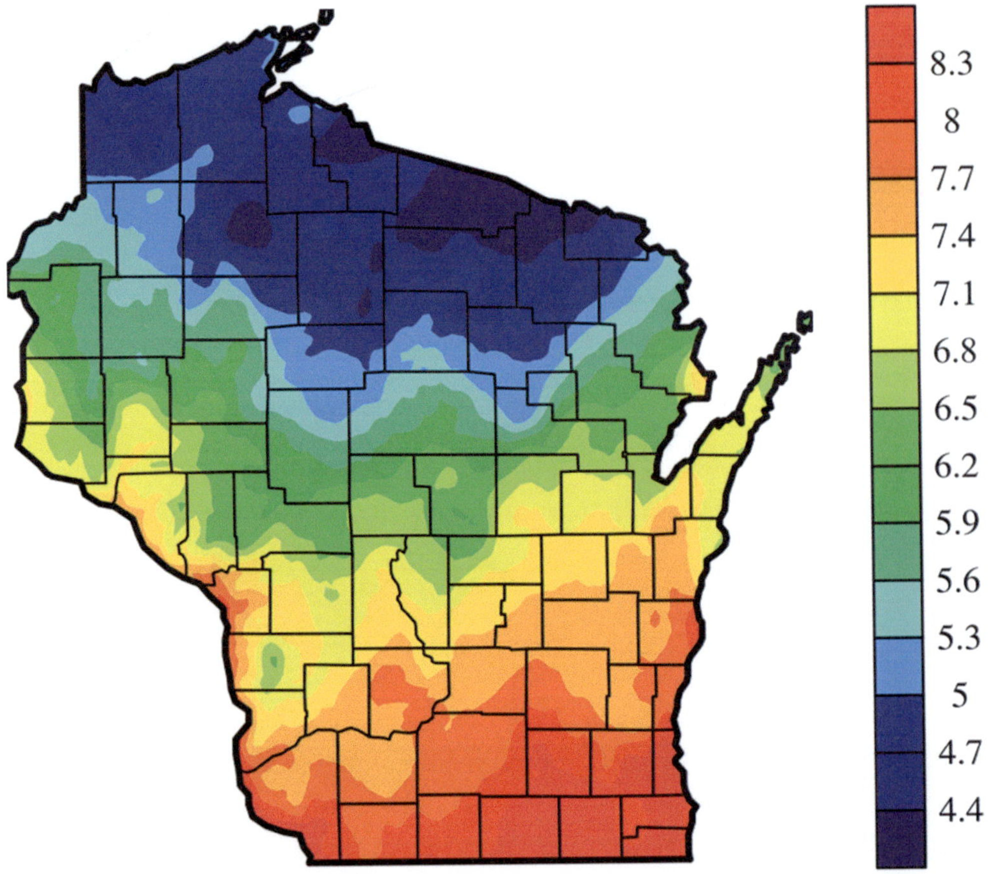

Fig. 3.5 Mean annual air temperature patterns (°C) in Wisconsin over the period 1950–2010. Data from the PRISM Climate Group (http://prism.oregonstate.edu)

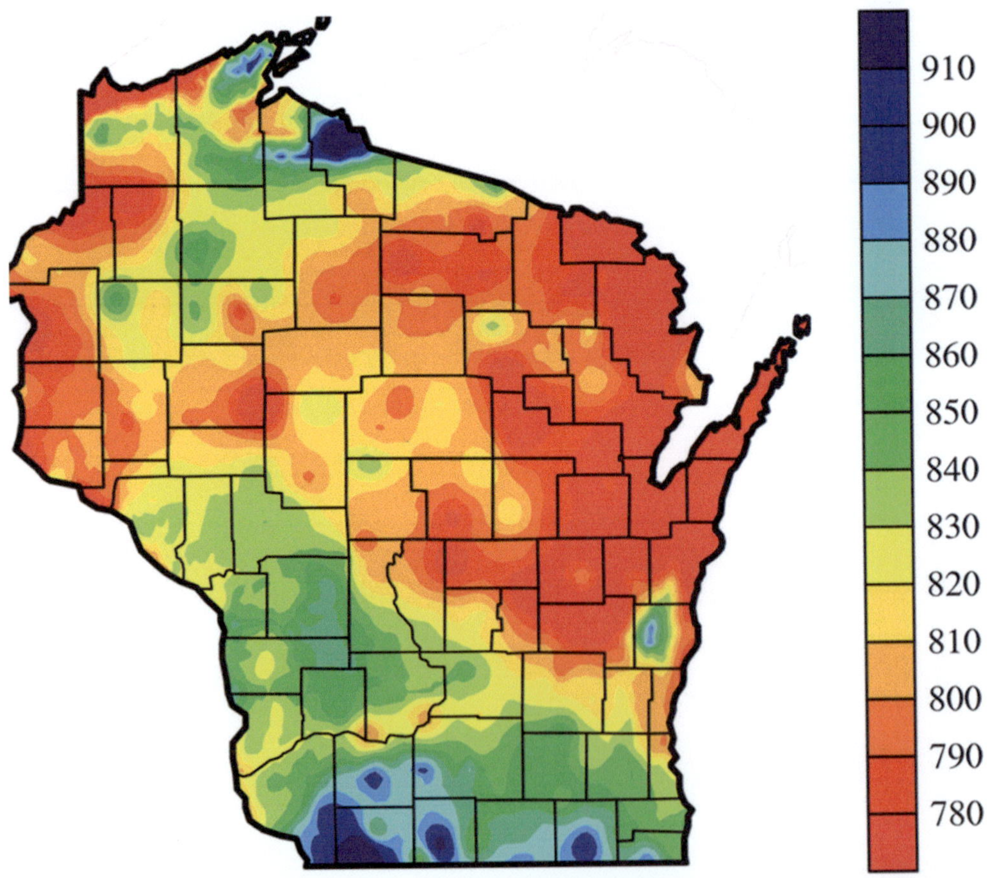

Fig. 3.6 Mean annual precipitation patterns (mm) in Wisconsin over the period 1950–2010. Data from the PRISM Climate Group (http://prism.oregonstate.edu)

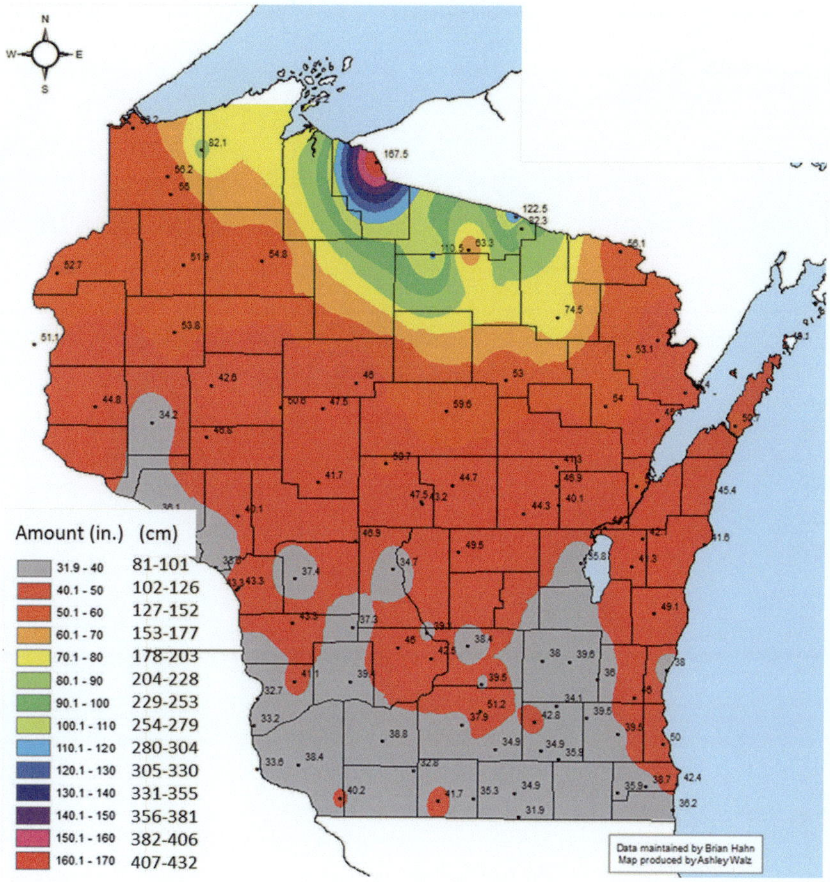

Fig. 3.7 Annual snowfall patterns (inch and cm) in Wisconsin over the interval 1981–2010. Values at specific points are reported in inches

the mean annual air temperature may have been 4 °C warmer than at present (Baker et al. 1992). These climate changes are evidenced in the fossil pollen record (Graumlich and Davis 1993), buried soils (paleosols) (see Chap. 14), and relict periglacial features such as patterned ground (Black 1964). The impact of the past climates is still present in the soils and includes, for example, sand-wedge casts in B and C horizons of some soils (Fig. 3.8), and the widespread presence of fragipans in northern Wisconsin that may have been formed partly under permafrost conditions (Hole 1976).

3.2.4 The Changing Climate

The Wisconsin Initiative on Climate Change Impacts (WICCI) Task Force prepared a comprehensive report in 2011, describing changes in Wisconsin's climate over the

Fig. 3.8 Sand-filled frost wedge in a Typic Argiaquolls, O'Briens farm, Dane County

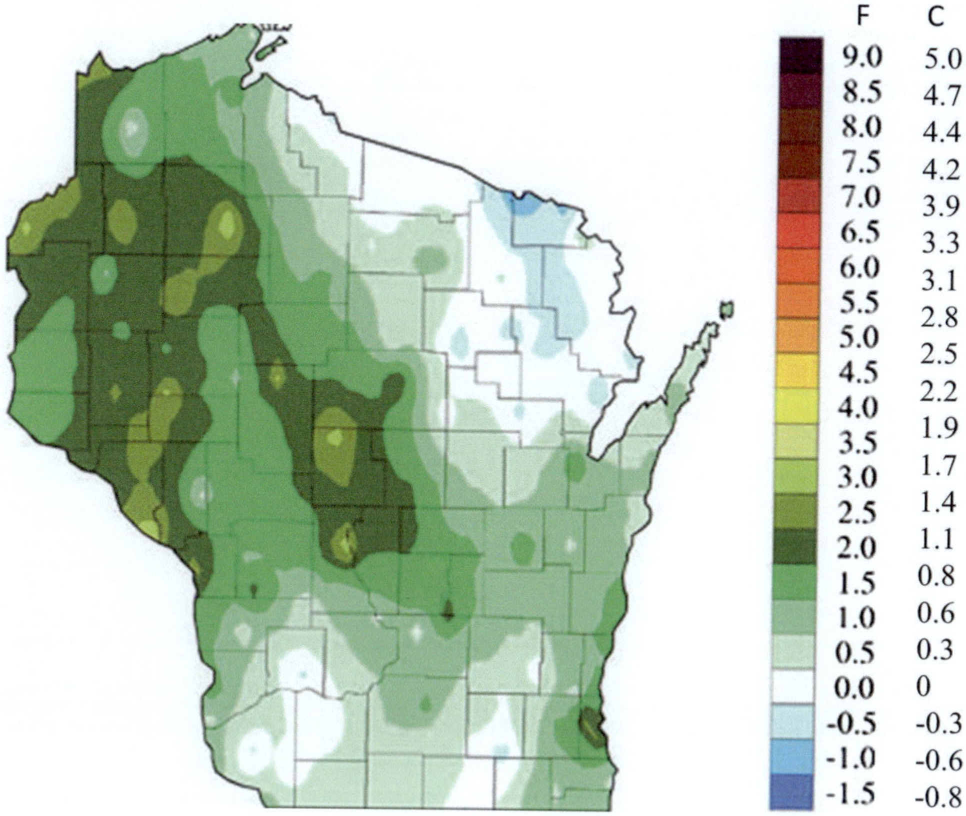

Fig. 3.9 Changes in mean annual air temperature (°F and °C) in Wisconsin over the period 1950–2006 (WICCI 2011)

period 1950–2006. During this period, the mean annual air temperature increased up to 1 °C in primarily the western part of the state (Fig. 3.9). Wisconsin winters have warmed more than any other season in recent decades, particularly in the northwestern part of the state. The length of the growing season has increased by as much as four weeks in parts of the state. The mean annual precipitation has increased by up to 178 mm over the past 56 years, especially in western and southcentral Wisconsin (Fig. 3.10).

Using climate simulation models, the task force projected up to a 4 °C increase in average annual temperature throughout the state by 2055 (Fig. 3.11). In view of the strong influence of climate on soils, we can expect changes in soil properties and distribution. It will also result in changing land use with increased areas under corn and soybeans in the northern part of the state. Such change in land use will also alter the soils. This is further discussed in Chap. 16.

3.3 Organisms

3.3.1 Past Work

Plants and animals influence soil development. Hole (1961) provided one of the first classifications of pedoturbations (soil mixing) from plants, animal, and humans. Nielsen and Hole (1963) compared the influence of prairie and forest vegetation on the distribution and cycling of organic matter on long-term plots in the University of Wisconsin Arboretum. Although the total amount of organic matter in the two ecosystems was similar, 54% was present in the

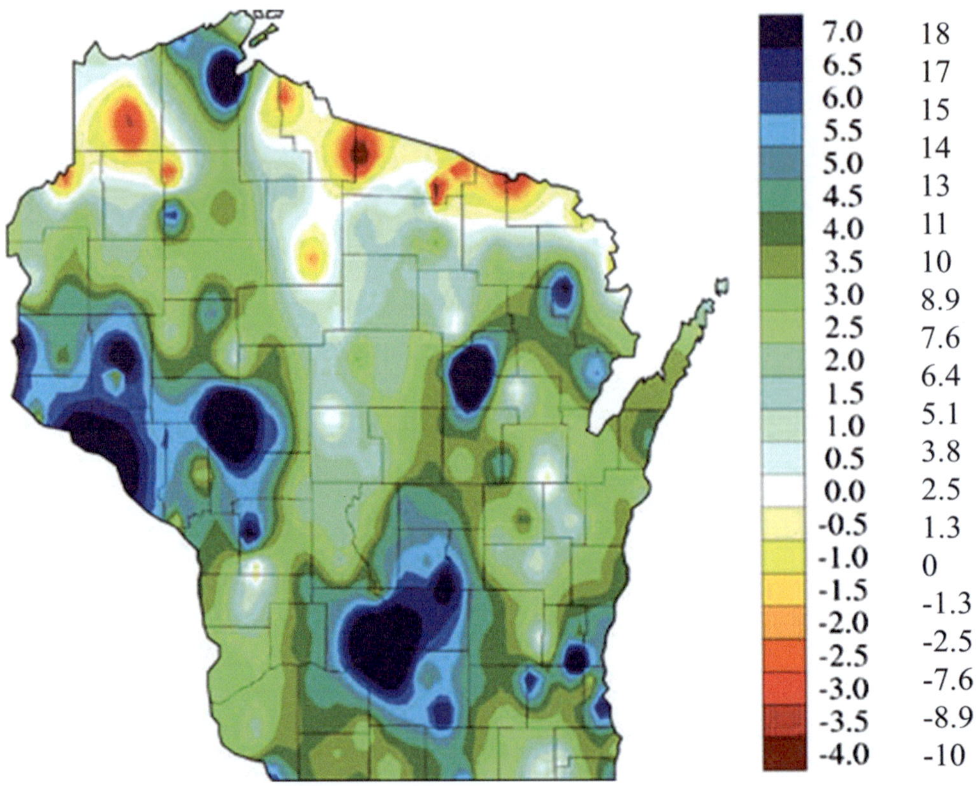

Fig. 3.10 Changes in mean annual precipitation (cm) in Wisconsin over the period 1950–2006 (WICCI 2011)

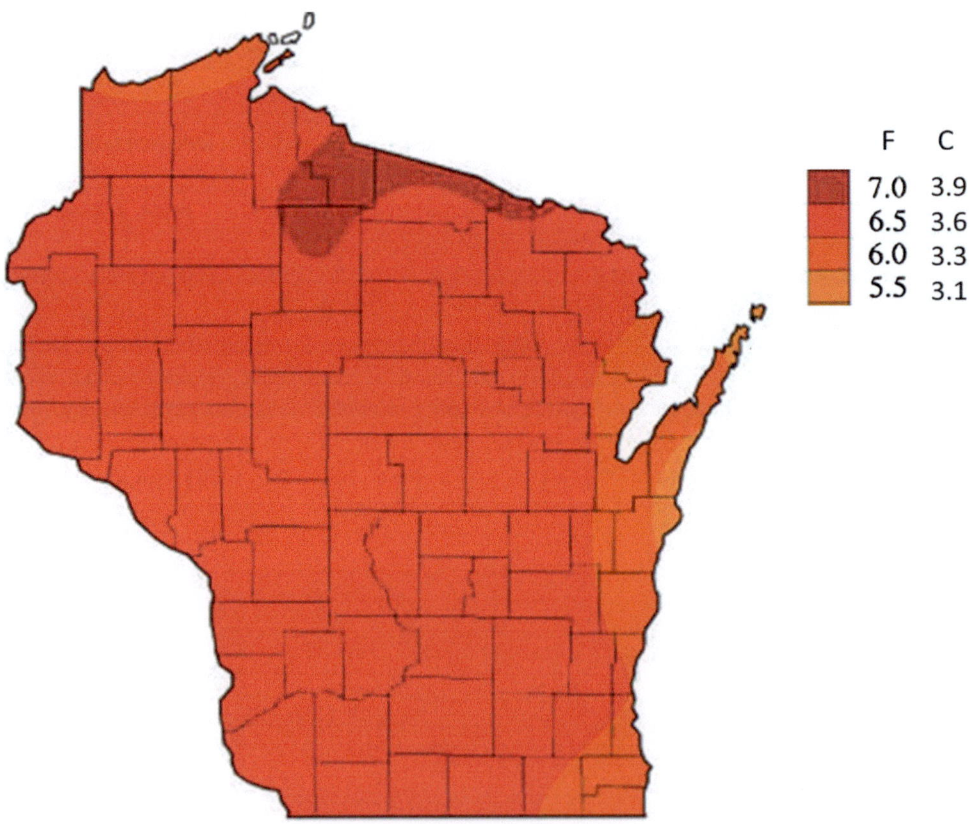

Fig. 3.11 Projected increase in mean annual air temperature in Wisconsin (°C) over the period 1980–2055 (WICCI 2011)

Table 3.1 Comparison of vegetation classification systems used in Wisconsin

Curtis (1959)[1]	% Area	Cottam and Loucks (1965)[2]	USFS (no date)	This study	% Area
Southern Mesic Forest	9.8	Southern Mesic Forest	Beech, sugar maple, basswood, red oak, white oak, black oak	Mixed hardwoods	7.1
			Sugar maple, basswood, red oak, white oak, black oak		
Southern Xeric Forest	4.0	Southern Oak Forest	Oak—white oak, black oak, bur oak	Mixed oaks	1.4
				Oak-hickory	11.4
Southern Lowland Forest	1.2	Lowland Hardwood	Lowland hardwoods—willow, soft maple, box elder, ash, elm, cottonwood, river birch	Wet hardwoods	3.2
Northern Mesic Forest	33.6	Northern Mesic Forest	Hemlock, sugar maple, yellow birch, white pine, red pine	Mixed forest	16.8
			Sugar maple, yellow birch, white pine, red pine	NHW (Northern Hardwoods)	18.4
			Aspen, white birch, pine	NHW-H (Northern Hardwoods-Hemlock)	9.8
Northern Xeric Forest	6.5	Pine Forest	White pine, red pine	Mixed pines	0.6
Northern Lowland Forest	6.4	Conifer Swamps	Swamp conifers—white cedar, black spruce, tamarack, hemlock	Wet conifers	2.3
				Wet mixed forest	6.1
Boreal Forest	1.9	Boreal Forest	White spruce, balsam fir, tamarack, white cedar, white birch, aspen	Boreal forest	1.6
Prairie	6.0	Prairie	Prairie	Prairie	5.8
Oak Savanna	20.7	Oak Savanna	Oak openings—bur oak, white oak, black	Oak savanna	3.5
Pine Barrens	6.7	Pine Barrens	Jack pine, scrub (hill's) oak forests and barrens	Oak-pine	6.8
Sedge Meadow	3.2	Sedge Meadows	Marsh and sedge meadow, wet prairie, lowland shrubs	Marsh	5.1

[1]Curtis (1959). The Vegetation of Wisconsin: an Ordination of Plant Communities. Univ. of Wisconsin Press, Madison, WI
[2]Cottam and Loucks (1965). Early Vegetation of Wisconsin. Univ. of Wisconsin-Extension, Geological and Natural History Survey
[3]US Forest Service. Wisconsin Original Vegetation. Great Lakes Ecological Assessment (http://www.ncrs.fs.fed.us/gla/histveg/wi-orveg.htm)

aboveground biomass of the mixed oak forest and 90% was present in the soil of the tallgrass prairie (Table 3.1).

Hole (1975) showed the effects of different forest vegetation on the development of B horizons in soils of the Menominee Tribal Lands, Wisconsin. Broad-leaved vegetation was important in the development of argillic (Bt) horizons, and coniferous vegetation was associated with the development of spodic (Bs) horizons. The complex mosaic of forest cover types results in striking differences in soils in the upper Great Lakes region (Bockheim 1997). Individual tree species influenced base cation distribution and cycling in Typic Haplorthods in the Upper Great Lakes region (Fujinuma et al. 2005). Pastor et al. (1982) illustrated the importance of vegetation on the distribution of soil taxa on a small (70 ha) island (Blackhawk Island) in the Wisconsin River (Fig. 3.12). Blackhawk Island is a wooded island in the Wisconsin River, which has cut deep narrow gorges in the Cambrian sandstone. The soils are dominated by sands and sandy loams of eleven different soil series. The island has several community types, all relatively undisturbed, with white oak, red oak, white pine, sugar maple, basswood, hemlock, white cedar, yellow birch, river birch, cottonwood,

Fig. 3.12 Trees on shallow soils over Cambrian sandstone and wind-thrown trees at Blackhawk island

and red maple, each found dominating in specific areas. Especially common are ferns; nearly one-third of the taxa of Wisconsin are present.

In a southern Wisconsin mixed oak forest, the giant earthworm (*Lumbricus terrestris*), an introduced species that has replaced the native earthworms, formed a coprogenic A horizon in as little as 30 to 40 years (Nielsen and Hole 1964). In a southern Wisconsin prairie, ants (*Formica cinerea*) play an important role in the soil development by constructing mounds, creating channels, and transporting materials (Baxter and Hole 1967) (Fig. 3.13).

3.3.2 Pre-settlement Vegetation

When the first French settlers arrived in 1634, most of the state was still covered with forest (Campbell 1906). At the beginning of the twentieth century, most of the natural forest had been logged (Whitson 1927). At the time of settlement by Europeans in the 1830s, Wisconsin was 63% forested, with 27% of the state land area in savanna and 10% in prairie and sedge meadow (Fig. 3.14). This map is based on the original land survey conducted around 1850 and systematic studies by Curtis (1959). The dominant forests were temperate mixed deciduous and conifers in northern Wisconsin that were comprised of eastern hemlock (*Tsuga canadensis*), sugar maple (*Acer saccharum*), yellow birch (*Betula alleghaniensis*), red pine (*Pinus resinosa*), and white pine (*Pinus strobus*) (40%). In southern and parts of western Wisconsin, the mesic forests were comprised of sugar maple, basswood (*Tilia americana*), red oak (*Quercus rubra*), white oak (*Quercus alba*), and black oak (*Quercus velutina*) (10%). Oak savannas or openings contained bur oak (*Quercus macrocarpa*), white oak, and black oak (21%). Pine-oak barrens composed of jack pine (*Pinus banksiana*), and Hill's oak (*Quercus ellipsoidalis*) occupied 7% of the land area.

3.3.3 Present Vegetation

At the present time, agricultural land occupies 46% of Wisconsin's total area, followed by forest (38%), wetlands (10%), with lesser areas in water (3%), urban and residential development (2%), and barrens (1%) (Figs. 3.15 and 3.16). The forest cover of the state has been reduced by 25%, native prairies by 6%, and oak savannas and pine-oak barrens by 25%. As of 2013, 77% of the land area of Wisconsin is privately owned, followed by the state (10%), county (7%), and federal government (5%) (Fig. 3.17). Forest land comprises 48% of the state, with 62% in private, non-industrial ownership, followed by county forests (14%), federal forests (10%), state forests (7%), private industrial forests (5%), and tribal lands (2%).

3.3.4 Vegetation and Soil Development

Vegetation is an important factor closely related to climate that is important in the development of Wisconsin soils: Alfisols occur under temperate deciduous forests; Spodosols are present under temperate mixed deciduous and conifer forests (especially where hemlock is a key component); Entisols are present under pine-oak barrens; Mollisols represent former prairies, and Histosols and wet mineral soils occur in stream bottoms and major wetlands (Fig. 3.18).

3.4 Relief

The influence of relief, or topography, includes elevation, aspect, and position on slope. The weathering of layer silicates was influenced by slope position in a catena of soils in southeastern Wisconsin (Borchardt et al. 1968). The more intensely leached upland soil had less amount of mica and greater amount of kaolinite than soils in the lower topographic position. To display landform elements of a valley in southwestern Wisconsin, Irvin and others (1997) showed the utility of "fuzzy set" ISODATA and a digital terrain model. Park et al. (2001) used a geographic-information system and a digital elevation model to provide a three-dimensional extension of a nine-unit soil landscape model for southern Wisconsin.

Dense, subsurface horizons similar to fragipans have formed in depressions in loess-covered till plains of southern Wisconsin (Park et al. 2006). Soils containing red clays (*Terra Rossa*) occur throughout the Driftless Area of Wisconsin, except in valley bottoms where loess deposition created a surface mantle in excess of 2 m (Evans and Hartemink 2014) (Fig. 3.19). One study suggested there were

Fig. 3.13 Ant hill in a restored prairie in southern Wisconsin, an earthworm in the glossic horizon of an Alfisol in central Wisconsin, and filled in earthworm tunnel with casts (scale in cm and inches)

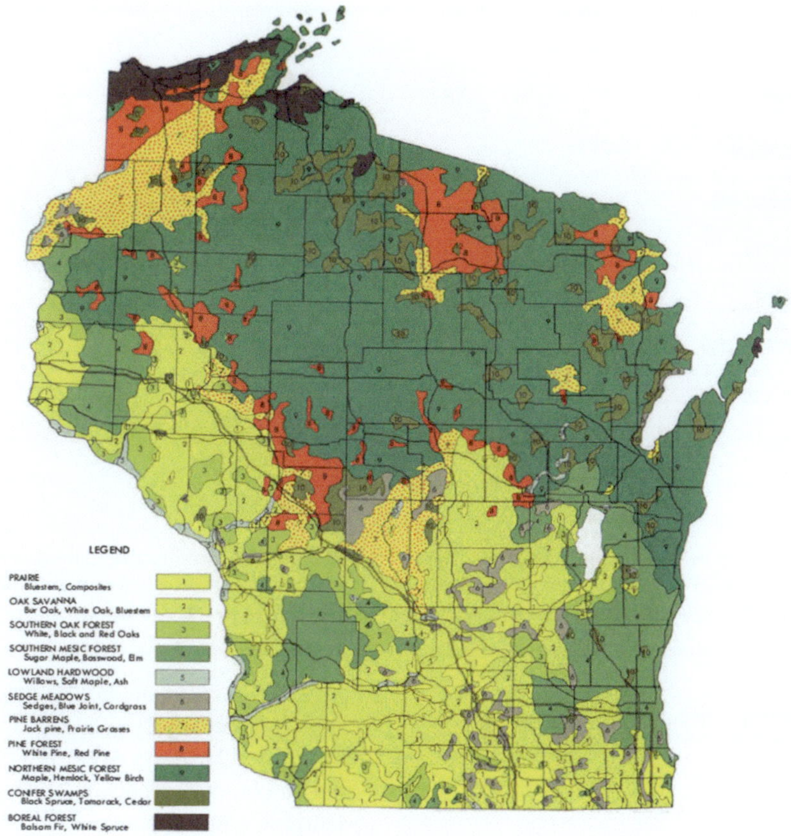

Fig. 3.14 Early vegetation of Wisconsin

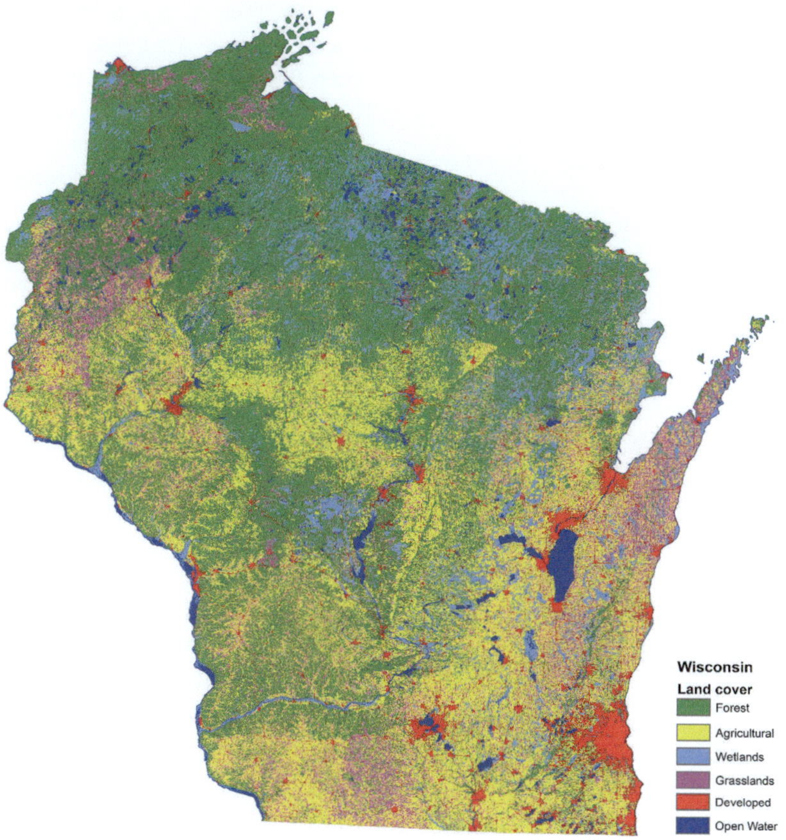

Fig. 3.15 Land cover in Wisconsin (compiled from National Land Cover Data 2006)

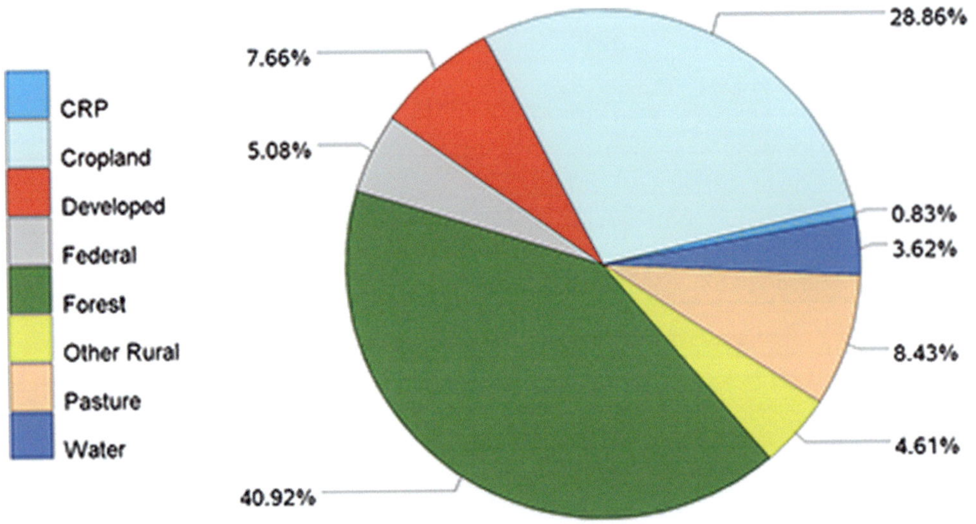

Fig. 3.16 Land cover in Wisconsin in 2014 (*Source* USDA Natural Resources Conservation Service). CRP is land in the Conservation Reserve Program—in exchange for a yearly rental payment, farmers enrolled in the program agree to remove environmentally sensitive land from agricultural production and plant species that will improve environmental health and quality

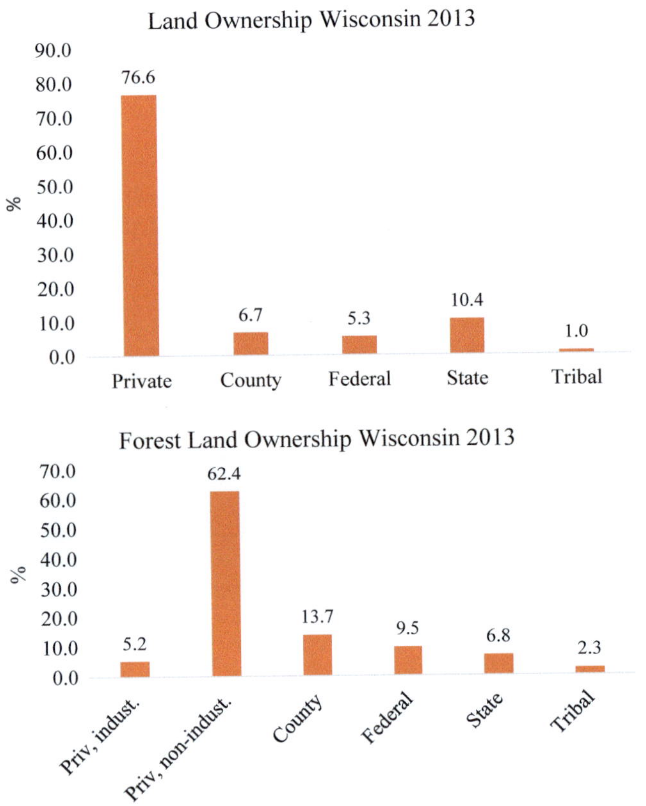

Fig. 3.17 Ownership of all land in Wisconsin (*top*) and forest land (*bottom graph*)

minimal differences in properties of the Fayette silt loam (Typic Hapludalfs), a forest soil common in the loess-mantled Driftless Area, as a function of slope (Frolking 1989). In contrast, Jacobs et al. (2012) showed considerable variation in soil horizonation and thickness in loess-mantled landscapes of southern Wisconsin.

Much of Wisconsin has a gently rolling topography. However, there is a considerable relief in the strongly dissected Driftless Area, and on the end moraines from the Green Bay, Lake Superior, and Chippewa Lobes (Fig. 3.20). The highest elevations are in the Northern Highlands Province (northcentral region), where elevations exceed 450 m a.s.l. The lowest elevations are adjacent to Lake Michigan (182 m).

3.5 Parent Materials

3.5.1 Previous Work

Parent materials are differentiated on the basis of chemical composition, particle size distribution, and their effects on susceptibility or resistance to breakdown by physical and chemical weathering (Wurman 1952). Madison and Lee (1965) delineated general areas of sandy soil in Wisconsin that are similar in mineralogy. Sandy soils derived from Cambrian or Ordivician sandstones contain primarily quartz, whereas sandy soils in northern Wisconsin have abundant feldspars, suggesting that they have not undergone substantial weathering. Lithologic discontinuities in soil parent materials were important in the occurrence of bisequal (Orthod sequum over an Udalf sequum) in northern

Fig. 3.18 Histosol augering and landscape in Jefferson County, Wisconsin

Wisconsin (Bockheim 2003). In his seminal paper "Wild soils of the Pine-Popple Rivers basin," Hole (1974) illustrated the effects of topography and parent material on the formation of "wild" (undisturbed) soils in northeastern Wisconsin through a series of block diagrams.

3.5.2 Geological Structure

Northeastern Wisconsin contains some of the oldest rocks in the world, including Lower Proterozoic and Upper Archean (Precambrian) granites and gneisses that are well over 1.5 billion years old (Fig. 3.21). These are overlain by Cambrian sandstones and Ordovian, Silurian, and Devonian formations of shale, dolomite, and other sedimentary rocks. In northern Wisconsin, the sedimentary rocks have been eroded following uplift and from repeated glaciations. In many parts of the state, glacial deposits (till, loess) cover the bedrock (Fig. 3.22). Most of the glacial deposits in Wisconsin are comprised of granites and gneisses from the Canadian Shield to the northeast.

3.5.3 Glacial Geology

Nearly 80% of the soils in Wisconsin are developed in glacial materials that largely differ in origin, composition, and thickness. The thickness of glacial cover is greater than 150 m on the Bayfield Peninsula (Fig. 3.22). In much of northern and eastern Wisconsin, the depth to bedrock ranges from 15 to 90 m. However, on the Door Peninsula, the thickness of drift cover is less than 15 m, whereas drift is lacking in the Driftless or nonglaciated area. The unconsolidated deposits overlying bedrock are comprised of two kinds of materials: glacial drift, including moraines and outwash; and post-glacial eolian materials, including silt (loess) and sand (Fig. 3.23). The loess thickness is about 0.5–5.0 m in southwestern Wisconsin but is largely absent in northeastern Wisconsin. About 10% of the soils have developed in glacial lake deposits.

Wisconsin has been subject to several glaciations for the past 2.1 million years. Glacial deposits fall into three age classes: early Holocene (ca. 9800 year BP), Late Wisconsinan 11,000–35,000 year BP), and pre-late Wisconsin (>35,000 year BP) (Attig et al. 2011) (Fig. 3.24). Glaciation in Wisconsin occurred during the advance of several lobes of ice that were part of the Laurentide Ice Sheet, including from west to east, the Des Moines Lobe, the Superior Lobe, the Chippewa Lobe, the Wisconsin Valley and Ontonagon Lobes, the Langlade Lobe, the Green Bay Lobe, and the Lake Michigan Lobe (Fig. 3.25). Whereas most of the glacial deposits are in the form of drift, including end moraines and ground moraine, there are areas of pitted outwash

Fig. 3.19 Freshly plowed fields in the Driftless Area, where *red* clay originating from the weathering of the underlying dolostone is brought to the surface

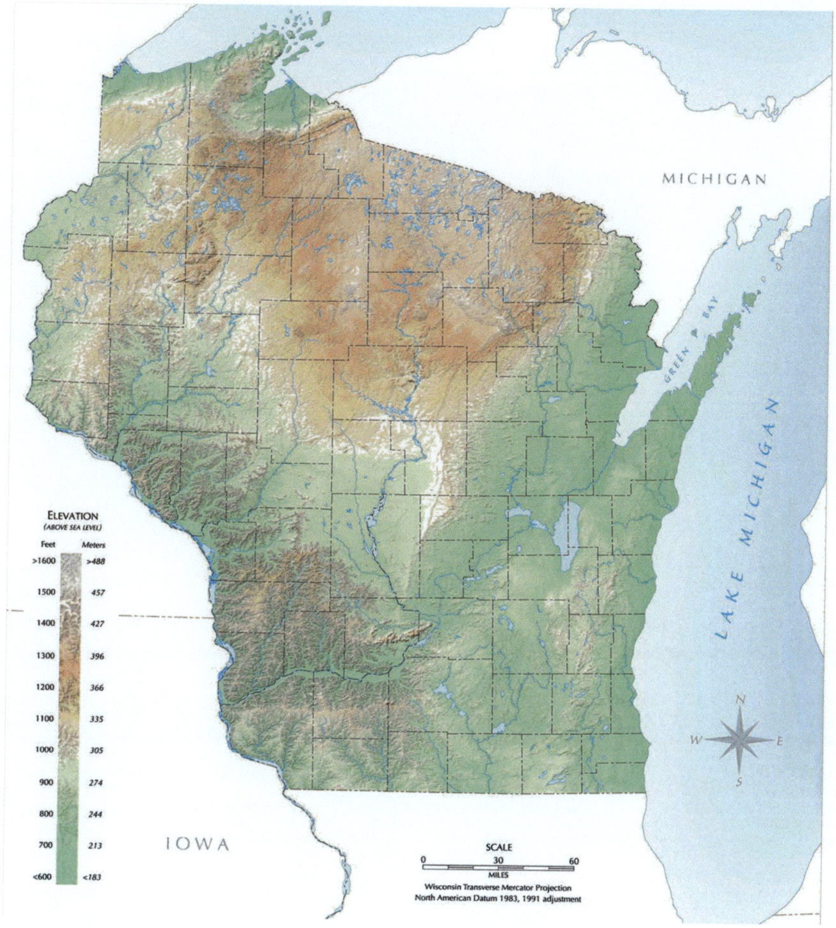

Fig. 3.20 Elevations of Wisconsin (m) (adapted from Wisconsin Geological and Natural History Survey 2004)

(northwestern and northeastern Wisconsin), unpitted outwash (central Wisconsin), and glacial lake basins (northwestern, central, and eastcentral Wisconsin) (Fig. 3.26). The Driftless Area covers about 35,700 km^2 in Wisconsin (25% of the state area) and has not been glaciated in the Quaternary. It is covered by a loess mantle, with outwash along the Wisconsin River.

Some of these features are evident in the relief map of Wisconsin, including the Driftless Area (A), Petenwill and Castle Rock Lakes in the former Glacial Lake Wisconsin basin (B), the Bariboo Hills to the south of the basin (C), the pronounced Almond Lateral Moraine (D) on the east side of the basin, and lakes in pitted outwash of northern Wisconsin (E) (Figs. 3.27 and 3.28). Figure 3.29 shows the Johnstown Moraine, which represents the southern limit of the Last Glacial Maximum. An aerial view, as shown in Fig. 3.30, shows the Johnstown Moraine and accompanying pitted outwash and kettle lakes along the distal margin. The Driftless Area is depicted in Fig. 3.31, an iconic image taken by the NRCS. Former Glacial Lake Wisconsin, which left 4700 km^2 of glaciolacustrine deposits in central Wisconsin approximately 19,000 to 15,000 years ago (Fig. 3.32). These deposits have been strongly reworked by wind.

Till in Wisconsin mostly has a sandy loam (40%) or loamy (38%) texture. However, clayey tills (10%) are present near Superior (Superior Lobe) and Michigan (Lake Michigan Lobe) lakes. Sandy tills (13%) occur in areas with abundant glaciolacustrine deposits or outwash. Tills from the Superior, Green Bay, and Lake Michigan Lobes are often calcareous. Pitted and unpitted outwash deposits are commonly sandy. The Driftless Area contains silty soils from loess deposition.

From descriptions and areas of the 740 soil series in Wisconsin, the primary parent material of the upper 1 m is loess (34%), or till (14%), alluvium (13%), outwash (12%), organic sediments (10%), glaciolacustrine (6%), and other materials such as colluvium, eolian, and glaciofluvial

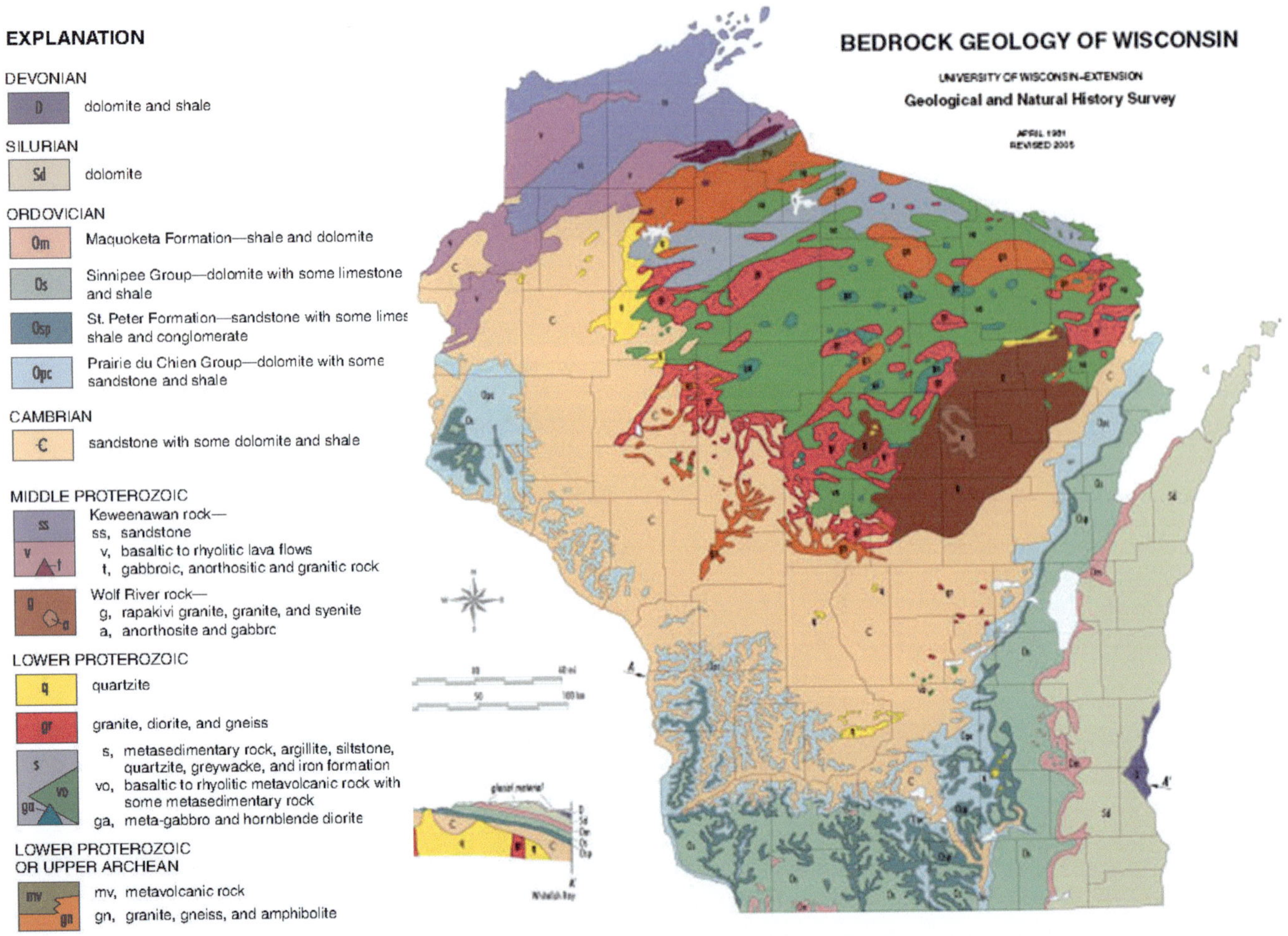

Fig. 3.21 Bedrock geology of Wisconsin (Wisconsin Geological and Natural History Survey 2005)

deposits (10%) (Fig. 3.33, upper). The second meter is mainly till (34%), or outwash (24%), residuum (13%), glaciolacustrine materials (8%), organic sediments and alluvium (7% each), and other materials such as colluvium, eolian, glaciofluvial, deep loess (7%) (Fig. 3.33, lower). Not many soils are derived solely from the weathering of the underlying bedrock.

3.6 Time

Time refers to stage of soil development and the susceptibility or resistance of different soil properties to change. Most of the soils of Wisconsin are derived from drift of Late Wisconsinan age (9500–30,000 year BP (Fig. 3.34). However, soils have been reported on pre-Late-Wisconsinan to

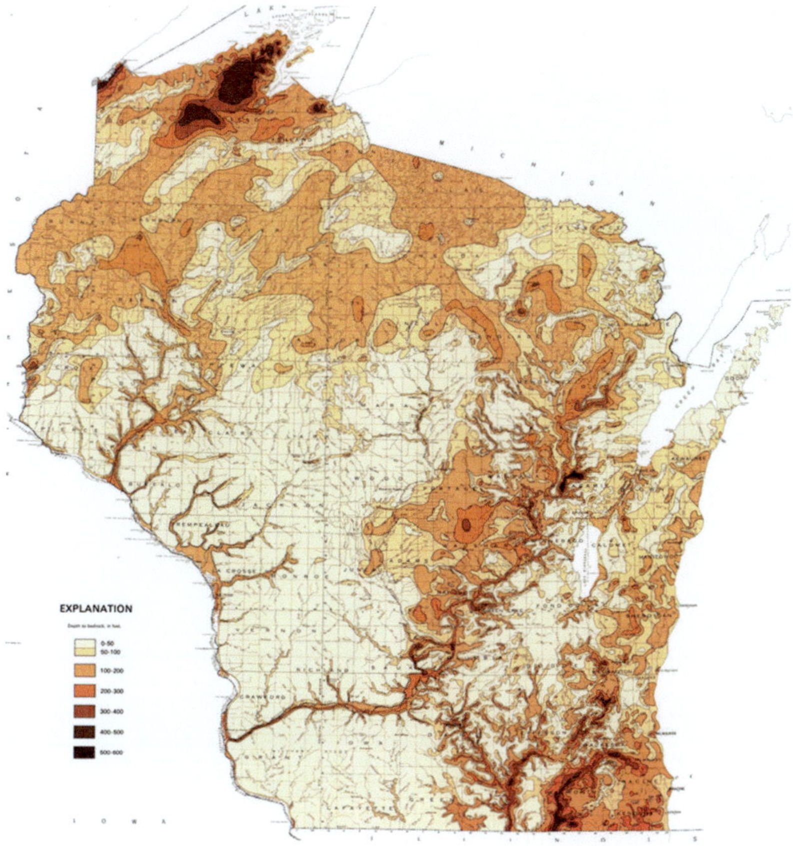

Fig. 3.22 Depth to bedrock in Wisconsin (Trotta and Cotter 1973)

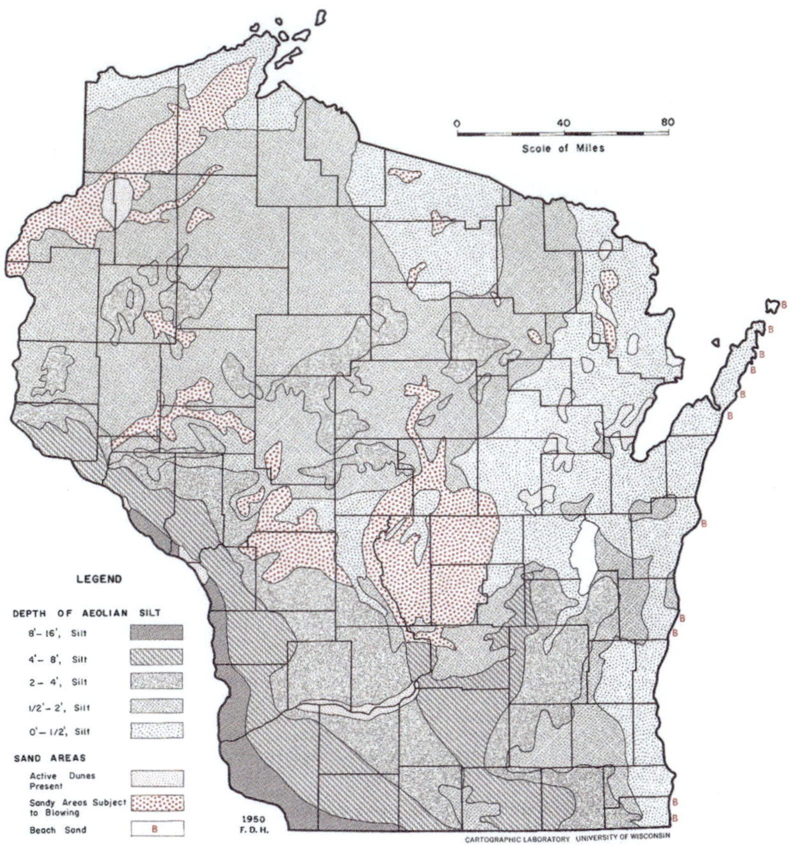

Fig. 3.23 Aolian silt and sand deposits in Wisconsin (Hole 1968)

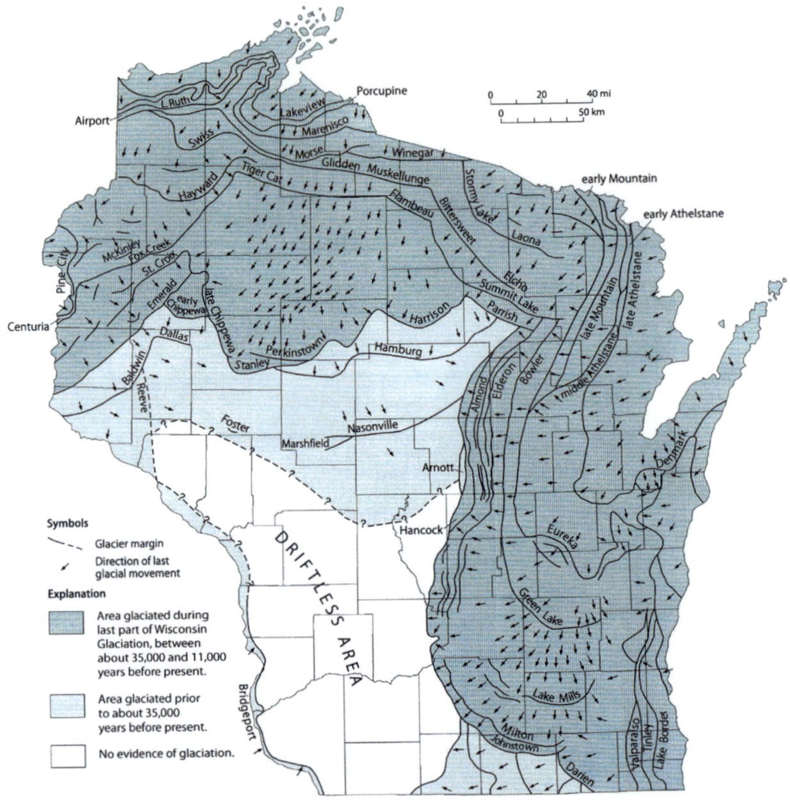

Fig. 3.24 Glacial deposits of Wisconsin (Attig et al. 2011)

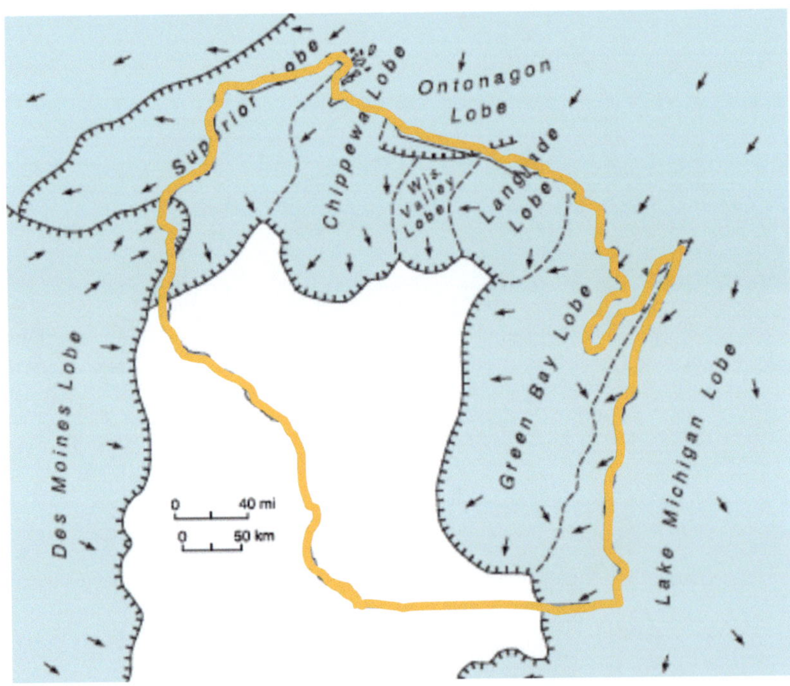

Fig. 3.25 Lobes of the Laurentide Ice Sheet (Attig et al. 2011). Arrows indicate the direction of ice movement

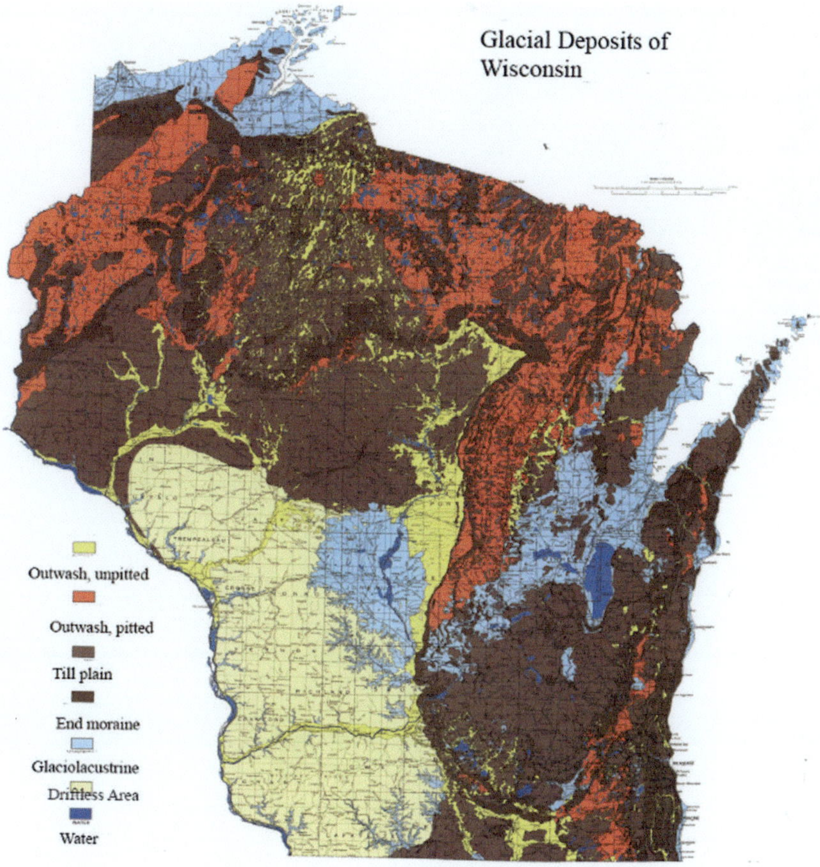

Fig. 3.26 Glacial deposits of Wisconsin (Wisconsin Geological and Natural History Survey 1976)

Illinoian drift (>30,000–300,000 year BP) and on pre-Illinoian drift between 780,000 and 2400,000 year BP (Syverson and Colgan 2011). The red clay pediment material in the Driftless Area is even older.

3.7 Humans

3.7.1 Previous Work

Humans have been in Wisconsin ever since the ice retreated about 13,500 years ago. They have altered the landscape by burning and changed the vegetation. Initially, this occurred at a slow pace but since the mid-1800s massive changes have occurred. All forests have been cleared, and agriculture has expanded. Below some examples are discussed how this has affected the soils.

Human cultivation erased the cradle-knoll micro-relief and replaced the forest floor and albic horizon with a plow layer, which resulted in slight reductions in nutrient contents of an Oxyaquic Haplorthods in northeastern Wisconsin (Gaikawad and Hole 1961). Grossman and Mladenoff (2008) concluded that agricultural cultivation results in more persistent changes than fire or clearcutting in forestry, particularly in the levels of P and Ca. Agriculturally induced erosion resulted in an "inverted horizon soilscape" in southcentral Wisconsin (Hartemink and Bockheim 2012).

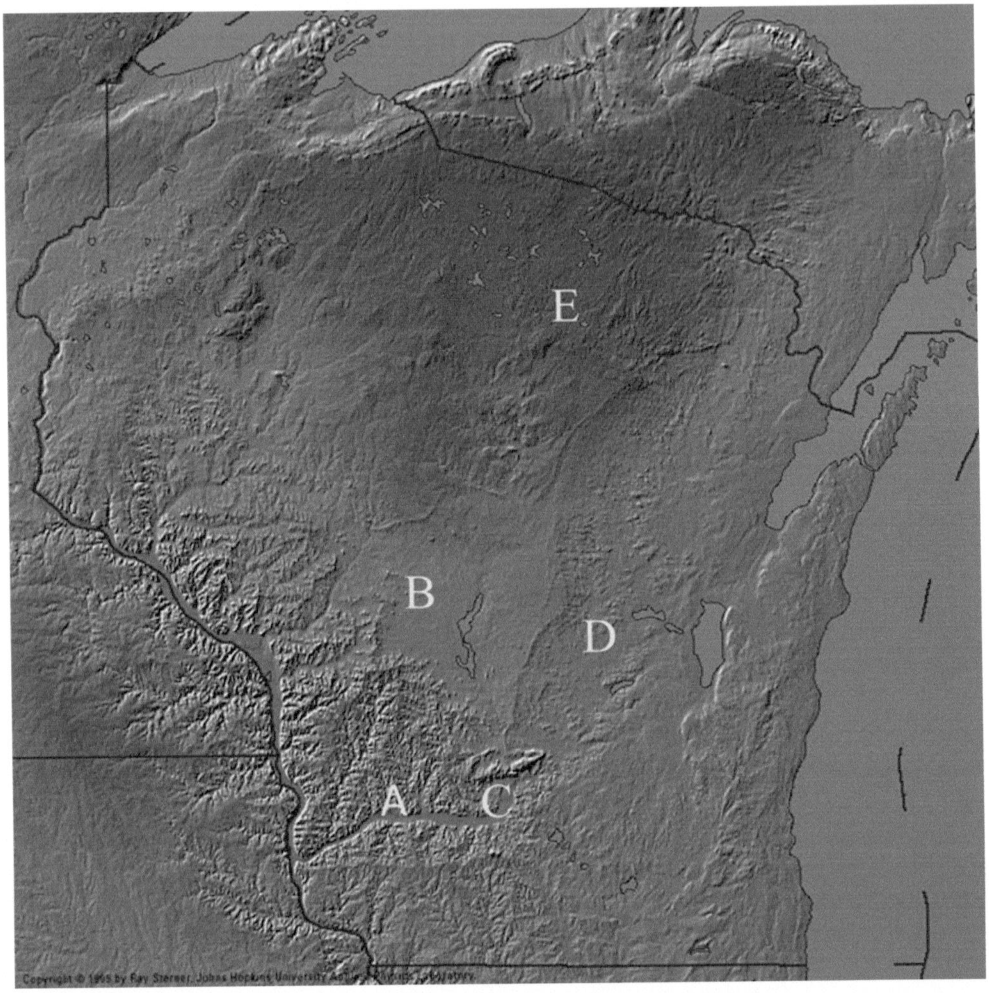

Fig. 3.27 Shaded relief map of Wisconsin, showing the Driftless Area (*A*), Petenwill and Castle Rock Lakes in the former Glacial Lake Wisconsin basin (*B*), the Bariboo Hills to the south of the basin (*C*), the pronounced Almond Lateral Moraine (*D*) on the east side of the basin, and lakes in pitted outwash of northern Wisconsin (*E*). Image by Ray Sterner, creator of the Color Landform Atlas of the United States

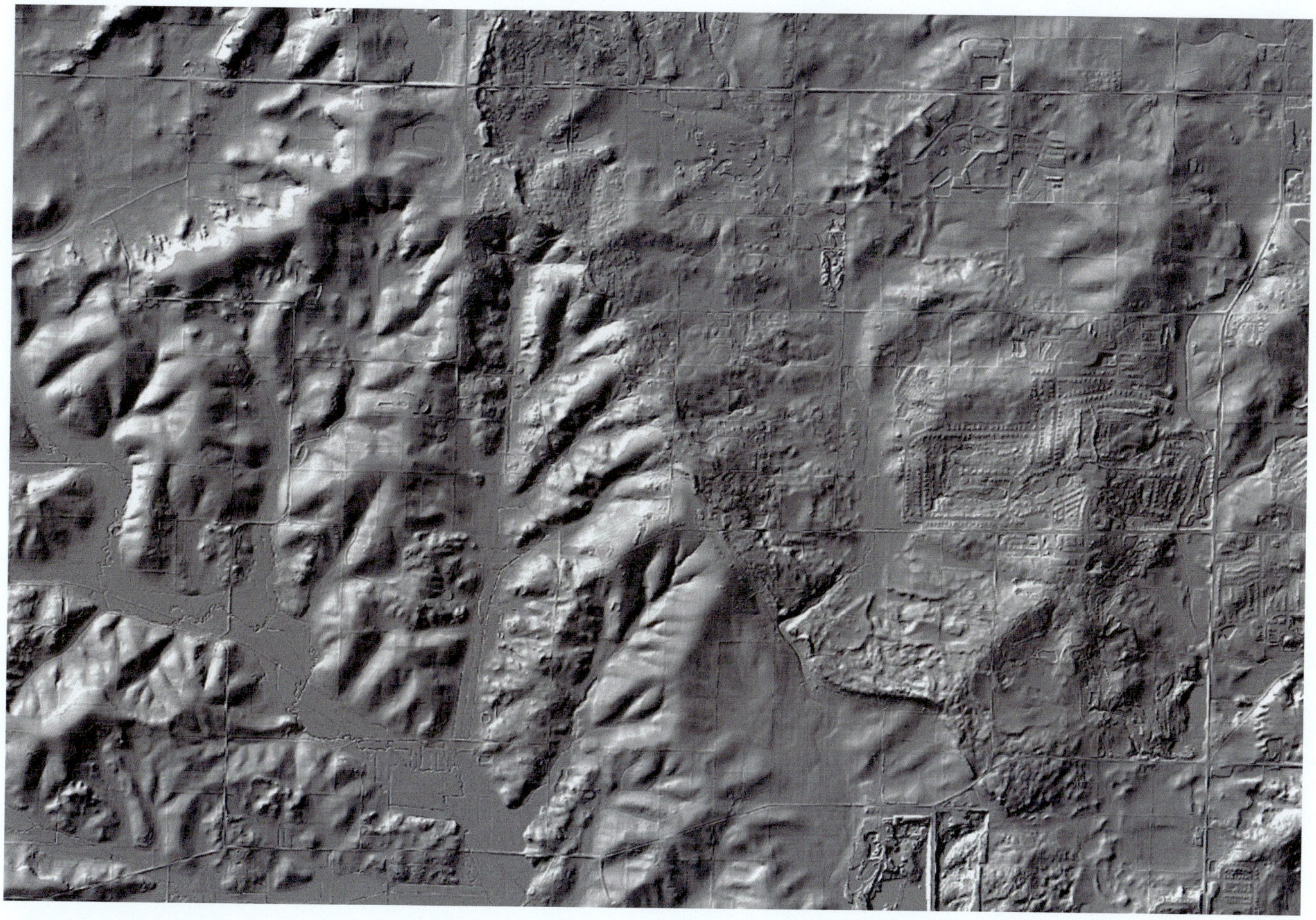

Fig. 3.28 Hillshade map of end moraine near Middleton with urban development on the pitted outwash. Sugar River and its tributaries in the left bottom part of the map

Fig. 3.29 Johnstown end moraine in southcentral Wisconsin

Fig. 3.30 Johnstown end moraine in southcentral Wisconsin with areas of pitted outwash and some kettle lakes (*lower picture*)

Fig. 3.31 Driftless Area (USDA Natural Resource Conservation Service (www.nrcs.usda.gov)

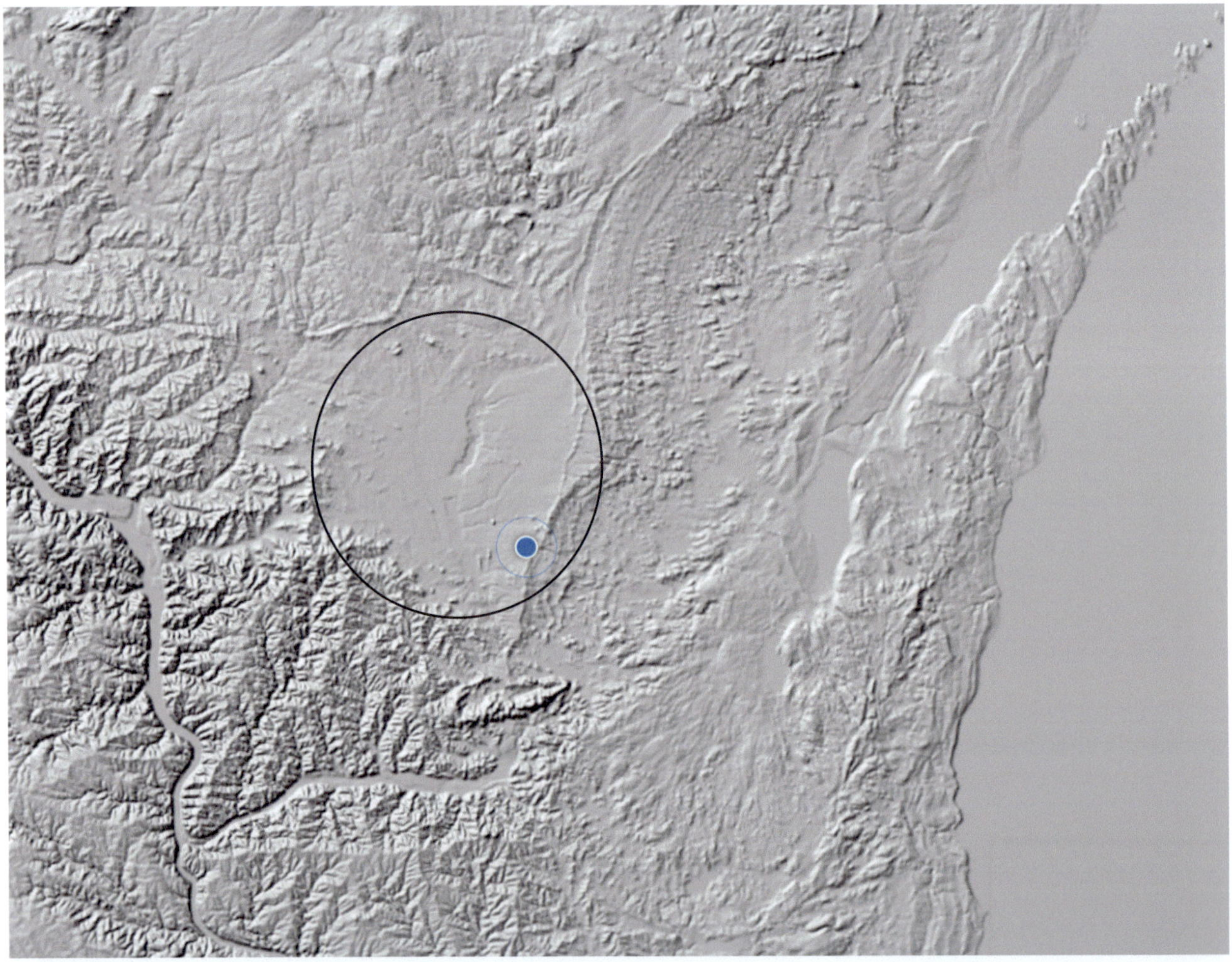

Fig. 3.32 Former Glacial Lake Wisconsin, which left 4700 km^2 of glaciolacustrine deposits in central Wisconsin approximately 19,000–15,000 years ago. Much of the former lake (also known as the central sands plains) is under irrigated agriculture (picture below with end moraine in the background)

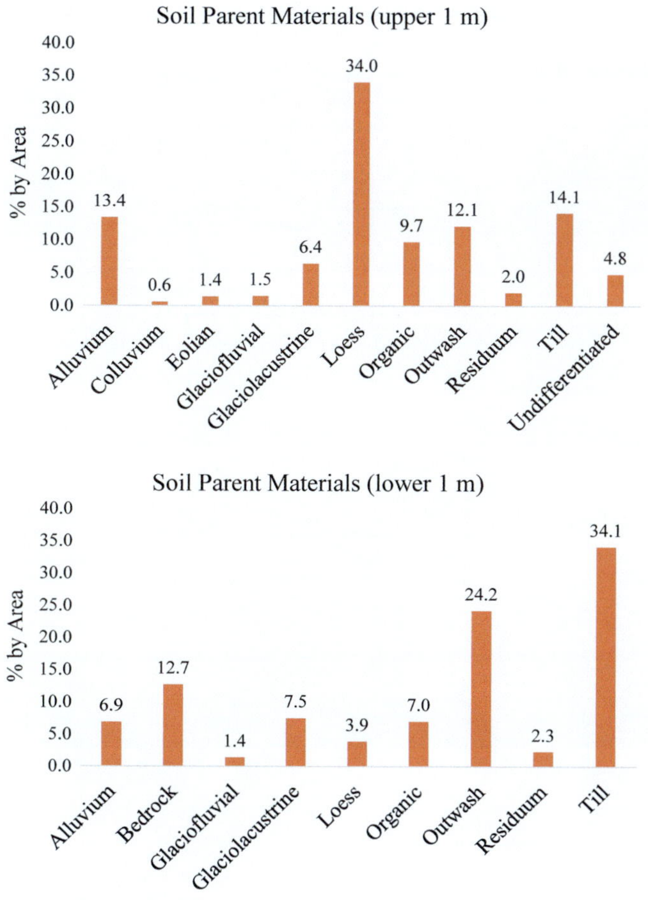

Fig. 3.33 Soil parent materials of the upper 1 m and lower 1 m from descriptions of 740 soil series

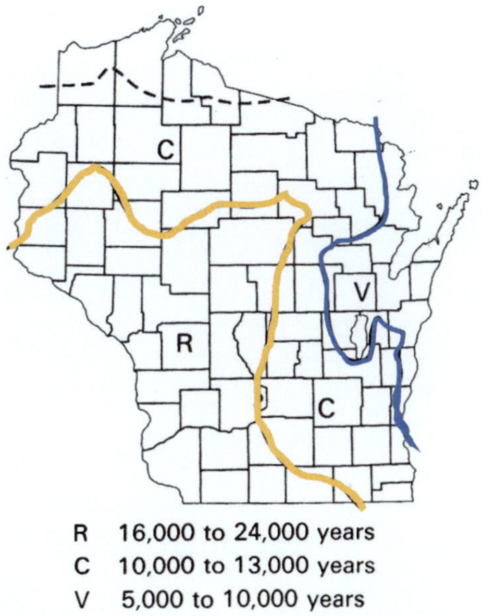

Fig. 3.34 Estimated age of the surface (after Hole 1976)

3.7.2 Paleo-Indians and Land Use

According to archeological studies, the earliest humans to enter Wisconsin were Paleo-Indians. One theory is that they crossed the Bering Land Bridge, a broad piece of land connecting the Eurasian and North American continents, during the waning stages of the Last Glacial Maximum approximately 13,500 years ago (Muñoz et al. 2010; Lambert and Loebel 2015). These early Americans hunted big game such as mastodon and gathered plants in what was then the southern Wisconsin tundra. Several mastodon butchering sites have been found in southern Wisconsin, the most famous one at Boaz.

During the Early Archaic or Late Paleo-Indian Tradition (11,250–8250 year BP), fishing and hunting of smaller mammals became more important as the glaciers retreated and the forest–tundra edge migrated northward. Forest fires increased, reflecting the shift toward a drier and warmer climate. During the Middle Archaic Tradition (8250–5250 year BP), the climate had warmed substantially and prairies became more evident. Ground stone tools and woodworking tools were developed during this time. During the Late Archaic Tradition (5250–3000 year BP), spear-throwing devices were developed. There was widespread trading of food and other resources among native peoples during this period. Copper trade items were introduced during this period.

Burial mounds were employed to interbodies and belongings during the Early Woodland Tradition (3000–2300 year BP. Cropping began during this period, especially the growing of corn, beans, and squash. Human-set forest fires became more common. Pottery was made from local clay deposits. These activities were expanded during the Middle Woodland Tradition (2300–1600 year BP), particularly by the Hopewell culture. Effigy mounds were constructed during the Late Woodland Tradition (1600–500 year BP). During the Mississippian Period (1100–400 year BP), stockade and permanent villages were established.

3.7.3 Modern Human Impacts

Soil erosion following clearing has been a problem throughout Wisconsin and has affected many of the soils. When the first settlers came in the Driftless Area in the 1850s, wheat was the primary crop grown and to lesser extent tobacco. Diseases and low wheat prices forced the settlers into dairy farming (Whitson 1927). As the land is sloping and much of the forest was logged, there was massive soil erosion in the Driftless Area and over 60% of the cropped land had lost 10–15 cm of its topsoil (Clark 1940).

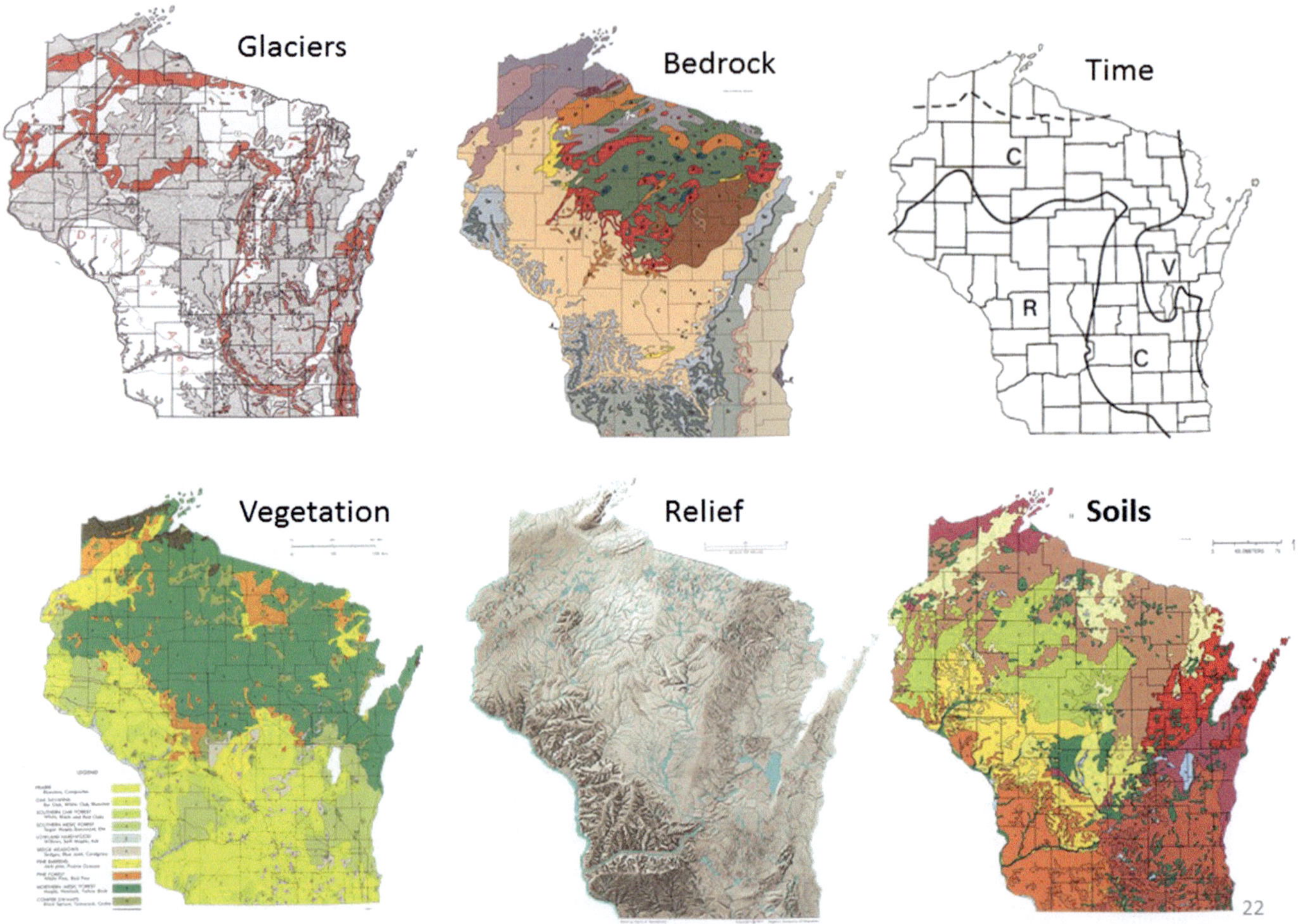

Fig. 3.35 Factors affecting the soil distribution in Wisconsin

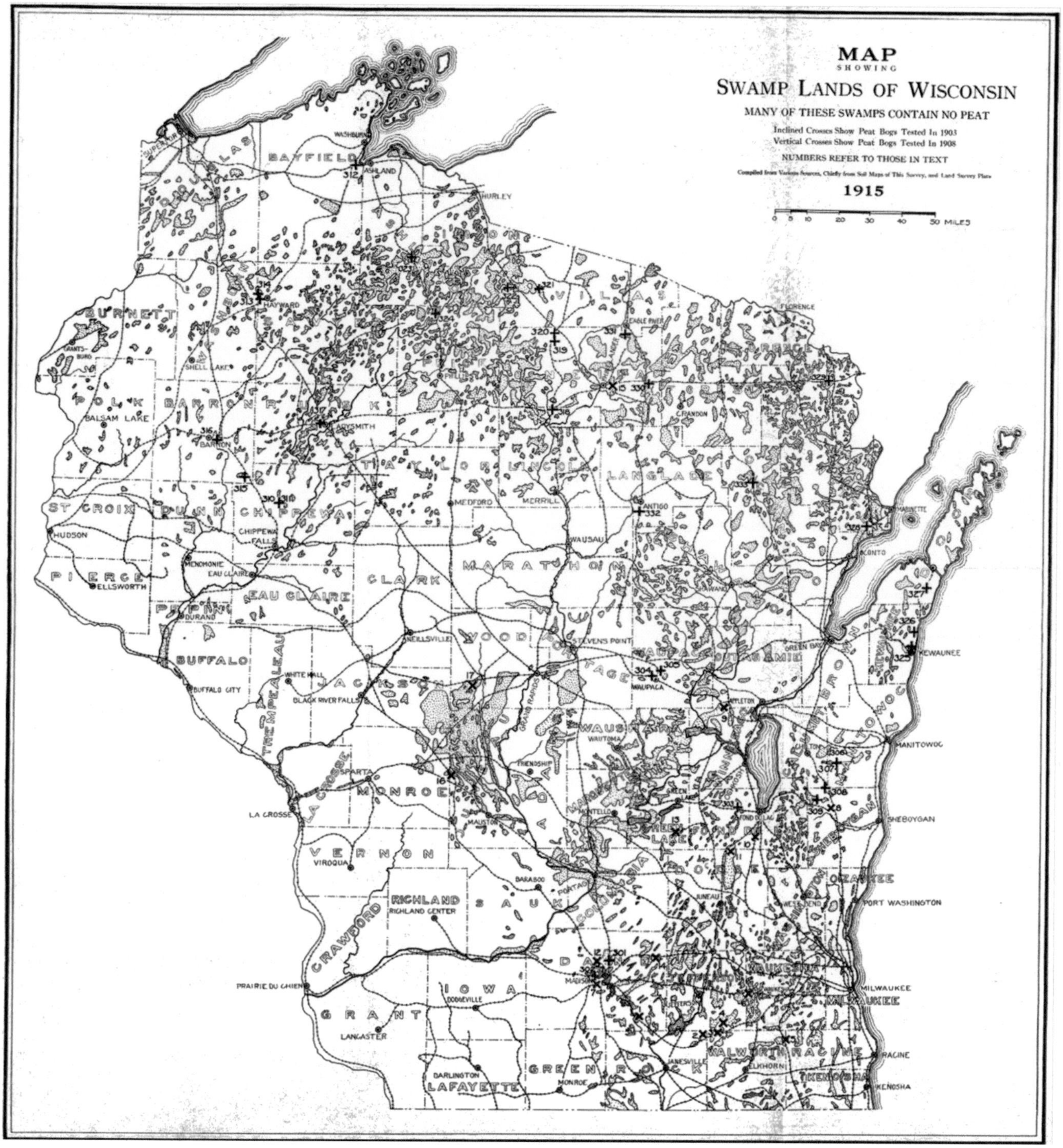

Fig. 3.36 Map of swamp lands in Wisconsin in 1915

According to Trimble and Lund (1982), average soil erosion rates in the 1930s were estimated to be about 34 Mg ha^{-1} year^{-1} but had decreased to about 8 Mg ha^{-1} year^{-1} in 1975. The reduction in erosion mainly resulted from improvements in land management and to a lesser degree by changes in land use; the total area under cropping had not changed much between the 1930s and 1975 but soil conservation management was greatly improved.

Other soil changes that were brought about by changes in land use include cultivation (mechanized, by hand), tillage, weeding, terracing, subsoiling, deep ploughing, manure, compost and fertilizer applications, liming, draining, and irrigation. Also the change in climate, hydrology, earth moving and paving has changed many of our soils.

3.8 Summary

The expression of a soil results from five factors operating collectively: climate, organisms, relief, parent material, and time; human activities are often considered a sixth factor (Fig. 3.35). The importance of climate as a soil-forming factor is evidenced by an ecotone, known as the "Wisconsin Tension Zone," that separates soils, dominantly Spodosols, with a frigid soil-temperature class (<8 °C at 50 cm) to the north and soils, dominantly Alfisols, with a mesic soil-temperature class (>8 °C at 50 cm) to the south. Vegetation is an important factor closely related to climate that is important in the development of Wisconsin soils: Alfisols occur under temperate deciduous forests; Spodosols are present under temperate mixed deciduous and conifer forests (especially where hemlock is a key component); Entisols are present under pine-oak barrens; Mollisols represent former prairies, and Histosols and wet mineral soils occur in stream bottoms and major wetlands (Fig. 3.36).

Topographic position along slopes exerts a strong control on soil development. About 80% of the soils are derived from glacial deposits. The Driftless Area in southwestern Wisconsin escaped the Quaternary glaciation. Most of the soils of Wisconsin are derived from drift of Late Wisconsinan age (9500–30,000 year BP). However, soils have been reported on pre-Late-Wisconsinan to Illinoian drift (>30,000–300,000 year BP) and on pre-Illinoian drift between 780,000 and 2,400,000 year BP. The red clay pediment material in the Driftless Area may be of Miocene (15 million year BP).

Although the Native Americans had an impact on soil formation through fires and occupation sites, European settlement in the mid-1850s caused the greatest change in soils. Clearing land for pastures, cultivation of land for agriculture, draining of wetlands, irrigation, and application of agrichemicals have altered the soils in the state.

4 Soil-Forming Processes

4.1 Introduction

In the previous chapter, we reviewed the soil-forming factors and how they affected the soils and their distribution in Wisconsin. In this chapter, we will discuss the soil-forming processes. Bockheim and Gennadiyev (2000) identified 17 generalized soil-forming processes and linked them to soil taxa and diagnostic horizons, properties, and materials in *Soil Taxonomy* (ST) (Tables 4.1 and 4.2) and illustrated them in simple diagrams (Fig. 4.1). They subsequently added two additional processes: cambisolization and pedoturbation. In Table 4.3, we show the dominant soil-forming processes in the seven orders and 16 suborders represented in Wisconsin. The importance of these processes can be seen by the number of soil series and the soil areas across the state. Below we describe each of the ten soil-forming processes that are operative in Wisconsin soils: argilluviation, biological enrichment of base cations, gleization, paludification, melanization, podsolization, base-cation leaching, and ferrallitization.

4.2 Soil-Forming Processes

4.2.1 Argilluviation

This process refers to the downward movement of clay in the soil profile. The argillic horizon must increase relative to the eluvial horizon or an underlying horizon and show evidence of clay movement (Buol and Hole 1959, 1961; Ranney and Beatty 1969). Argilluviation is a major process in Alfisols (> 35% base saturation) and Ultisols (< 35% base saturation) but may also occur in Mollisols (Argiudolls) and Alfic subgroups of Spodosols. Figure 4.2 shows a strongly developed argillic horizon in loess near Platteville.

4.2.2 Biological Enrichment of Base Cations

Vegetation plays an important role in maintaining the base cation (Ca, Mg, K, and Na) content of Alfisols and Mollisols (Fujinuma et al. 2005). Grasses and temperate deciduous forest types are effective in taking up and returning large amounts of base cations in litterfall, throughfall, stemflow, and belowground processes such as root exudation and fine-root turnover. Alfisols and Mollisols require that the base saturation be ≥ 35 and $\geq 50\%$, respectively, in a defined portion of the profile. A weaker version of biological enrichment of base cations is evident in Eutr-great groups of Inceptisols.

4.2.3 Gleization

This process refers to the presence of poorly drained (aquic) conditions evidenced by reductimorphic or redoximorphic features such as mottles and gleying(Richardson and Hole 1979). Reduction and oxidation processes have primarily affected iron and manganese compounds in soils and often result in visible morphological features that have been used for predicting soil moisture regimes. Gleization is recognized in Aqu-suborders in five of the seven orders identified in Wisconsin and may occur in the mineral within or below Histosols. Richardson and Hole (1979) studied the distribution of iron oxide and forms of redoximorphic features under various soil moisture regimes in northwestern Wisconsin. Well-drained soils either lacked redox features or they had a low chroma; somewhat poorly drained soils had pronounced redox features and Fe–Mn concretions; and very poorly drained soils were primarily gleyed. Figure 4.3 shows a moderately well-drained Entisol developed in alluvium in a valley near Plattevile. Gleying is common in 30-cm soil depth.

J.G. Bockheim and A.E. Hartemink, *The Soils of Wisconsin*,
World Soils Book Series, DOI 10.1007/978-3-319-52144-2_4

Table 4.1 Soil-forming processes in relation to diagnostic horizons, properties, and materials by order in *Soil Taxonomy*

Soil order	Diagnostic horizon, properties, material	Soil-forming processes	Representative horizon sequence
Alfisol	Argillic horizon (high base status)	Argilluviation biological enrichment of base cations	A/E/Bt/C
Andisol	Melanic epipedon Andic properties	Andisolization	A/Bw/C
Aridisol	Natric horizon Calcic, petrocalcic horizon Gypsic, petrogypsic horizon Argillic horizon Duripan Salic horizon [aridic soil moisture regime]	Solonization, Solodization Calcification Calcification Argilluviation Silicification Salinization	A/Eg/Btn/Bk/By/C A/Bkm/Ck A/Cym/Cy A/E/Bt/Ck A/B/Cqm Az/Cz
Histosol	Histic materials	Paludization	Oi/Oa/Oe
Mollisol	Mollic epipedon (high base status)	Melanization Biological enrichment of base cations	A/Bt/C
Oxisol	Oxic horizon	Ferrallitization	A/Bo/Cr
Spodosol	Spodic materials placic horizon Albic horizon	Podzolization base cation leaching	Oa/E/Bh/Bs/C
Ultisol	Argillic (low base status)	Argilluviation base cation leaching	E/Bt/C
Vertisol	(Slickensides, cracks)	Vertization	A/Css
Gelisol	Gelic materials	Cryoturbation	O/Bgjj/Cf
Inceptisol	Cambic (plus others)	Weak soil formation	A/Bw/C
Entisol	(None)	Very weak soil formation	A/C
All orders (except Aridisols)	Reductimorphic features [aquic soil moisture regime]	Gleization	A/Bg/Cg
(None)	Anthropic, plaggen horizons	Anthrosolization	Ap/Bw/C

From: Bockheim and Gennadiyev (2000)

Table 4.2 Occurrence of 16 secondary soil macro-processes in taxa of *Soil Taxonomy*[a]

Process	Generalized soil-forming process[b]	Soil taxa
Argilluviation	3	**Alfisols**; **Ultisols**; Aridisols (Argids); argi great groups of Aridisols, Mollisols; kandia great groups of Oxisols, alfic subgroups of Spodosols
Biological enrichment of bases	3	**Alfisols**: **Mollisols**; eutric great groups of Inceptisols
Andisolization	4	**Andisols**; andic subgroups of Spodosols
Paludization	1, 4	**Histisols**: histic great groups of Gelisols
Gleization	4	**Aqul-suborders** of all orders except Aridisols and Gelisols; Aqu great groups of Aridisols and Gelisols
Melanization	3	**Mollisols**; Inceptisols (Umbrepts); umbr great groups of Alfisols and Ultisols; hum great groups of Inceptisols
Ferrallitization	4	Oxisols
Podzolization	3, 4	**Spodosols**; spodic subgroups of Entisols and Andisols
Base cation leaching	2	**Spodosols**; **Ultisols**; dystr great groups of Inceptisols and Vertisols
Vertization	3	**Vertisols**; vertic subgroups of Alfisols, Aridisols, Entisols, Mollisols and Ultisols
Cryoturbation	3	**Gelisols**
Salinization	3	Aridisols (**Salids**); hal great groups of Inceptisols; sal great groups of Aridisols and Vertisols

(continued)

Table 4.2 (continued)

Process	Generalized soil-forming process[b]	Soil taxa
Calcification	3	Aridisols (**Calcids**, **Gypsids**); calcic great groups of Aridisols, Mollisols, and Vertisols; gypsic great groups of Aridisols and Vertisols
Solonization	3	**Natric great groups** of Alfisols, Aridisols, Mollisols, and Vertisols
Solodization	3	**Natric great groups** of Alfisols and Mollisols
Silicification	3, 4	Aridisols (**Durids**); dur great groups of Alfisols, Andisols, Inceptisols, Mollisols. Spodosols, and Vertisols
Anthrosolization	1	Entisols (**Arents**); Inceptisols (**Anthrepts**); anthropic subgroups of Aridisols, Inceptisols

[a]Bold face denotes primary occurence of a soil-forming process
[b]1 = Addition to soil; 2 = loss from soil; 3 = translocation within soil; 4 = transformation of material within soil (after Simonson 1959; Buol et al. 1997)
From: Bockheim and Gennadiyev (2000)

Table 4.3 Dominant soil-forming processes by soil order in Wisconsin[a]

Suborder	No. of series	Area (km^2)	Soil-forming processes
Aqualfs	46	3057	Argilluviation, biological enrichment of base cations, gleization pedoturbation
Udalfs	277	61277	Argilluviation, biological enrichment of base cations pedoturbation
Totals	**323**	**64335**	
Aquents	22	3219	Gleization
Fluvents	14	1574	Cambisolization (weak)
Psamments	43	11674	Cambisolization or podzolization (weak)
Orthents	4	546	Cambisolization (weak)
Totals	**83**	**17014**	
Saprists	23	12706	Paludization
Hemists	6	629	Paludization
Fibrists	1	6	Paludization
Totals	**30**	**13341**	
Aquepts	22	2441	Cambisolization, melanization (weak), gleization
Udepts	30	2609	Cambisolization, melanization (weak)
Totals	**52**	**5050**	
Aquolls	46	4342	Melanization, biological enrichment of base cations, gleization, Cambisolization or alluviation
Udolls	108	10156	Melanization, biological enrichment of base cations, cambisolization or argilluviation
Totals	**154**	**14499**	
Aquolls	35	3567	Podzolization, base cation leaching, gleization
Orthods	78	19702	Podzolization, base cation leaching, pedoturbation
Totals	**113**	**23269**	
Udults	1	3	Argilluviation, base cation leaching, ferrallitization (weak)

[a]All of the soils are subject to pedoturbation

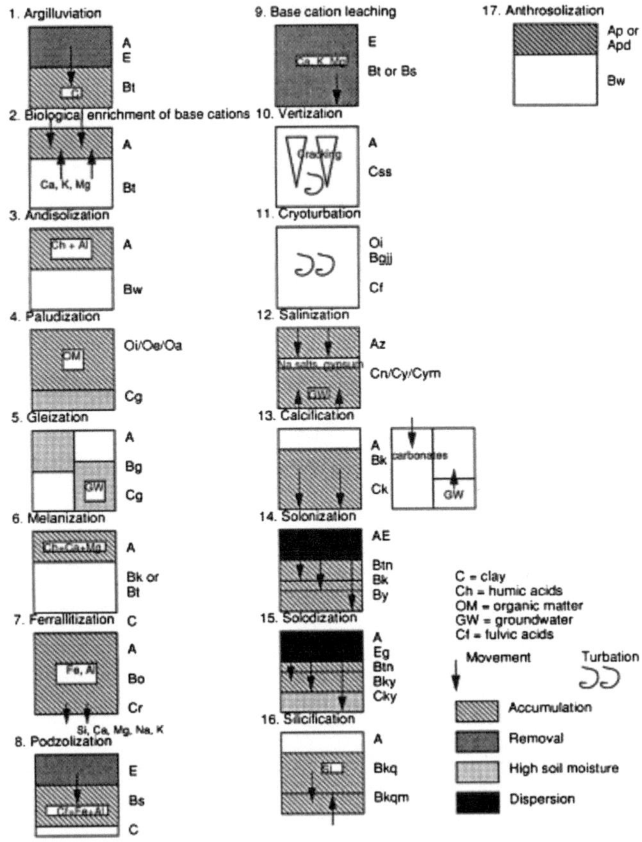

Fig. 4.1 Diagrams illustrating 17 key soil-forming processes as related to the 12 orders of Soil Taxonomy (Bockheim and Gennadiyev 2000)

Fig. 4.2 Afisol (Hapludalfs) with a strongly structured and developed Bt near Platteville

Fig. 4.3 Moderately well-drained Entisol (Fluvaquents) with gleying features with depth. Soil is in a valley and has developed in alluvial deposits. Oxidized Fe along larger pores and root channels

4.2.4 Cambisolization

This process was not recognized by Bockheim and Gennadiyev (2000) in their depiction of soil-forming processes. It was recommended to represent the early stages of soil formation, whereby the structure and color of the parent material are altered sufficiently to create a Bw horizon. Figure 4.4 illustrates a cambic or Bw horizon developed in coarse outwash.

4.2.5 Paludization and Ripening

This term pertains primarily to the deep (> 40 cm) accumulation of organic matter (histic materials) on the landscape usually in marshy areas. Most soils featuring paludization are in the Histosol order, but soils containing histic materials < 40 cm occur in Histic subgroups of Inceptisols, such as the Humaquepts. Ripening is a subprocess of paludization and refers to the chemical, physical, and biological changes following drainage and aeration of organic materials (Langton and Lee 1964).

4.2.6 Melanization

Some soils are characterized by the accumulation of well-humified organic matter within the upper mineral soil. In ST, these horizons are recognized as either a mollic or umbric epipedon and must have a dark color, at least 0.6% organic C, and be $\geq$ 18 cm in thickness (Soil Survey Staff 2014). Where soils subject to melanization are base-enriched, the humus accumulation is reflective of a mollic epipedon; where bases are depleted or where the parent materials contain low quantities of bases, the soils have an umbric epipedon. In addition to Mollisols, soils in Wisconsin in mollic subgroups include many Alfisols and some Inceptisols and Entisols. Umbric epipedons are present in some Inceptisols and Alfisols. Nielsen and Hole (1963) studied the natural processes of incorporation of organic matter into soils of the University of Wisconsin Arboretum. Storage of organic matter was largely aboveground at the forest sites and largely belowground at the prairie sites.

4.2.7 Podzolization

Podzolization is a collection of processes that includes eluviation of base cations, weathering transformation of Fe and Al compounds, mobilization of Fe and Al from surface horizons, and transport of these compounds to the spodic (Bs) horizons as Fe and Al complexes with fulvic acids and other complex polyaromatic compounds. Weaker degrees of podzolization occur in spodic subgroups of Entisols. In Wisconsin, podzolization occurs primarily in the northern third of the state, which is strongest under coniferous vegetation (Gaikawad and Hole 1961; Hole 1975) (Fig. 4.5).

4.2.8 Base Cation Leaching

This process is the opposite of biological enrichment of base cations and involves eluviation of Ca, Mg, K, and Na from the profile under leaching conditions, primarily in Spodosols and Dystr-great groups of Inceptisols.

4.2.9 Ferrallitization

Soils in the intertropical regions undergo a series of processes in which Al and Fe are concentrated and Si is lost from the profile as a result of primary and secondary mineral weathering. This process has occurred to a limited extent in a single soil series of northcentral Wisconsin, the Siouxcreek series (Aquic Hapludults). This series contains a red and brown sandy clay loam layer with quartz and chert gravel over Cambrian sandstone that was not glaciated (Mason et al. 1994). The soil contains a thick, well-developed paleo-argillic horizon. Many soils derived from loess or other sediments overlying clayey pedisediments and sandstone or limestone bedrock reflect the ferrallitization process.

4.2.10 Pedoturbation

Hole (1961) was the first to classify and describe nine types of soil mixing caused by plants, animals, frost, and other mechanisms. Baxter and Hole (1967) studied the influence of ants on pedoturbation, and Nielsen and Hole (1964) examined the role that earthworms played in forming coprogenous A horizons. Bockheim (2015) reported that 17% of the soil series in the western Great Lakes region contain a glossic horizon, which results from degradation of an argillic horizon, whereby clay and free iron oxides are redistributed, creating E/B and/or B/E genetic horizons (Fig. 4.6). Glossic soils occupy a belt to the north of the tension zone across Minnesota, Wisconsin, and Michigan. It is likely that the glossic soils result from a mid-Holocene warming of the climate that enabled eastern hemlock (*Tsuga canadensis*) to migrate southward, inducing pedoturbation.

Fig. 4.4 Gravelly and stony soil developed in outwash (Typic Udipsamments) with A horizon over weakly developed B over a C horizon

Fig. 4.5 Spodosol under mixed forest in Door County

Fig. 4.6 Alfisol with glossic horizon over dense red till near Stevens Point

4.3 Summary

Ten of the 19 generalized soil-forming processes are operative in Wisconsin soils, including clay migration (argilluviation) in the Alfisols; biological enrichment of base cations in Alfisols; gleization in Histosols and wet mineral soils (Aqualfs, Aquods, Aquents, Aquolls, and Aquepts); weak soil formation (cambisolization) in Inceptisols; paludization and ripening in Histosols; melanization in Mollisols; podsolization in Spodosols; base cation leaching in Spodosols; ferrallitization in Ultisols; and pedoturbation that occurs in most soils of Wisconsin.

5 The Soil Regions of Wisconsin

5.1 Introduction

Several efforts have been made to stratify the soils of Wisconsin into general regions based on location (northern, eastern, central, southwestern, southeastern, western) and on broad vegetation cover (forested, prairie) and texture of the soil parent materials (sandy, loamy, silty, silty over rock, etc.). In this chapter, we follow this approach, but in subsequent chapters, our approach examines the distribution of soil taxa (orders, suborders, great groups, subgroups, families, and series) and establishes soil regions based on great groups. Here, we will follow the soil regions from Hole (1976) and modified by Madison and Gundlach (1993) shown in Fig. 2.6.

5.2 General Soil Regions

5.2.1 Soils of Northern and Eastern Wisconsin

Forested, red, sandy, and loamy soils (soil region E)

This region encompasses deposits with a red color and sandy and loamy textures that are of Late Wisconsinan age (Fig. 2.6). Hole (1976) delineated 13 soil associations in this region (E) that were Alfisols, Entisols, Inceptisols, Mollisols, Spodosols, and Histosols. The dominant soil series recognized by Hole (1976) were the Emmet and Onaway (Hapludalfs), the Omega (Haplorthods), the Underhill (Eutroboralf; no longer recognized), the Solona (Argiudolls), the Angelica (Endoaquepts), and the Shiocton (Hapludolls). Figure 5.1 shows the distribution of soil orders along the Door Peninsula, and Alfisols and Spodosols are dominant. Figure 6.5 shows the distribution of great groups in Wisconsin; Haplorthods and Hapludalfs are dominant in soil region E. The Hortonville (Glossudalfs) and the Manawa and Onaway (Hapludalfs) series each occupy more than 400 km^2. Figure 5.2 shows plowed fields in Door County and common land use (apple orchards).

Forest, red, sandy, and loamy soils over dolomite (soil region Er)

This region only occurs in Door County, where thin layers of glaciolacustrine and beach deposits overlie dolomitic limestone. The dominant soil series of this region are the Longrie and Duel (Haplorthods), Bonduel (Hapludalfs), Kolberg (Glossudalfs), Summerville (Eutrudepts), and Namur (Hapludolls). These soils occur where dolomitic bedrock occurs within 100 cm of the surface (Fig. 5.2).

Forested, silty soils (soil region F)

Soil region F occurs in northcentral Wisconsin and includes large portions of Barron, Clark, Marathon, Rusk, Sawyer, St. Croix, and Taylor Counties (Fig. 5.3). The region features silty uplands and plains (Fig. 2.6). Hole (1976) identified 26 soil associations dominated by the Antigo, Brill, Freeon, Kennan, Milaca, Rozellville, Santiago, Spencer, and Withee series (Fig. 5.4)—all of which are Glossudalfs. Depressions in the landscape featured soils in Aqu-suborders, including the Adolph, Auburndale, Cable, Freer, and Marshfield series.

Soil region F contains mainly Glossudalfs (Fig. 5.4). Series occupying areas in excess of 400 km^2 include the Almena, Amery, Kennan, Loyal, Magnor, Rosholt, Santiago, and Withee (all Glossudalfs), the Chetek (Hapludalfs), and the Solona (Argiudolls).

Forested, loamy soils (soil region G)

This large region was identified by Hole (1976) as the loamy northern uplands and plains. Soil region G is strongly related to the occurrence of Late Wisconsinan end and lateral

J.G. Bockheim and A.E. Hartemink, *The Soils of Wisconsin*,
World Soils Book Series, DOI 10.1007/978-3-319-52144-2_5

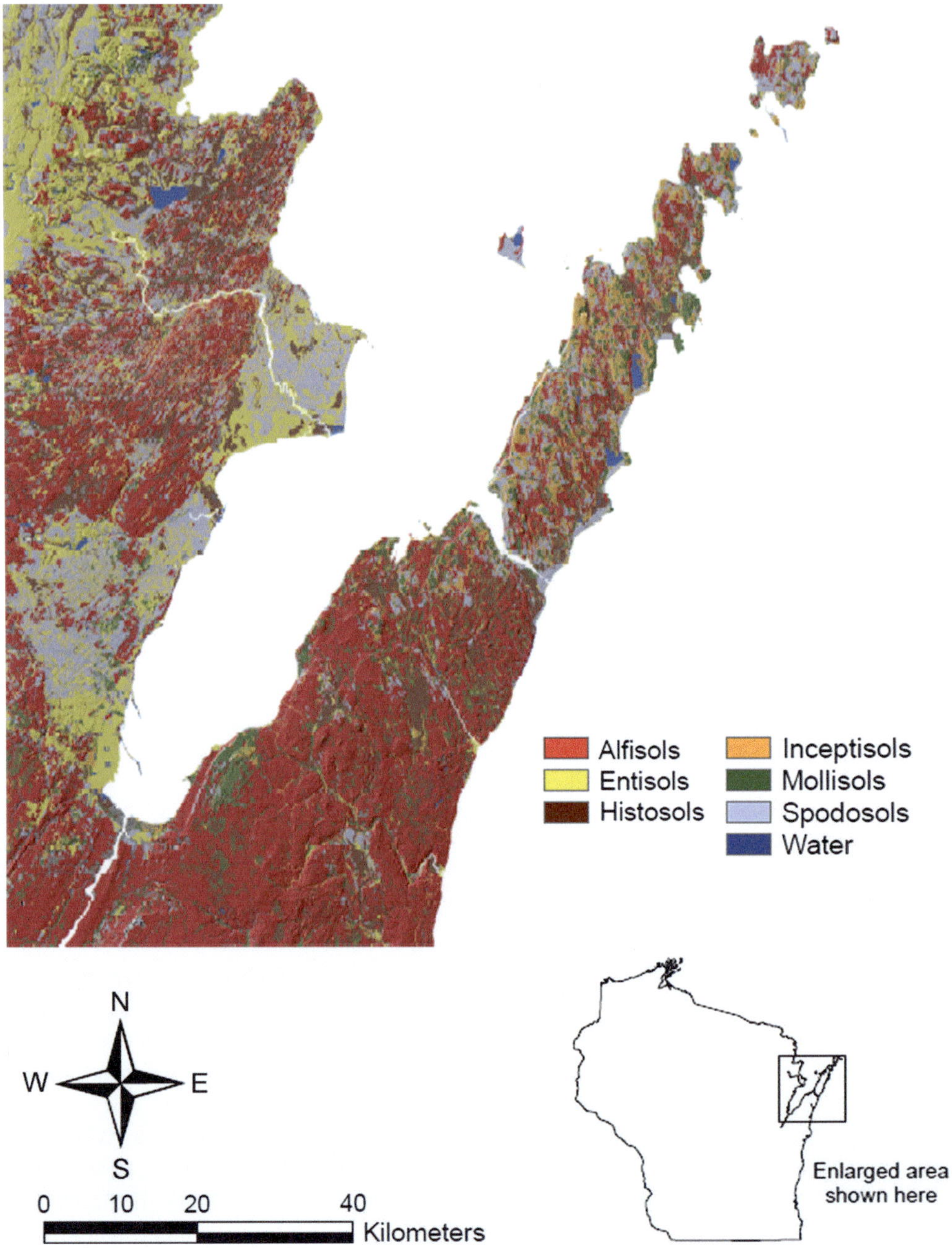

Fig. 5.1 Distribution of soil orders in soil region E, the Door Peninsula

moraines deposited by the Superior, Chippewa, Wisconsin Valley, Langlade, and Green Bay lobes (Fig. 5.3). Hole (1976) identified 28 soil associations comprised dominantly of Haplorthods (Goodman, Padus, Pence, Stambaugh, and Vilas series) and Fragiorthods (Gogebic series).

Soil region G is depicted on the great group map (Fig. 6.5) as predominantly Glossudalfs, Haplorthods, and Fragiorthods. Soil series with areas in excess of 400 km^2 include Glossudalfs (Amery, Antigo, Freeon, and Rosholt), Haplorthods (Newood, Karlin, Padus, Pence, and Vilas), Fragiorthods (Gogebic), and Hapludalfs (Chetek), and Endoaquepts (Minocqua).

Forested, sandy soils (soil region H)

Soil region H includes sandy soils in three large areas of pitted outwash, a smaller area of unpitted outwash, and isolated patches of outwash elsewhere in northern Wisconsin (Fig. 5.3). Hole (1976) identified seven soil associations comprised dominantly of the Omega, Pence, Vilas series—all of which are Haplorthods. In the great group map (Fig. 6.5), soil regions H and G appear as a single large unit, as both contain dominantly Haplorthods. Soil series occupying more than 400 km^2 each include Haplorthods (Croswell, Keweenaw, Karlin, Rubicon, Sarona, Sayner, and Vilas),

Fig. 5.2 Plowed fields (Alfisols) near Luxemburg in Door County that brought red clay to the surface. Fruitcrops (cherry, apples) are common and temperatures are moderated by Lake Michigan although much of the Door Peninsula has a frigid soil temperature regime

Udipsamments (Friendship, Grayling, Mahtomedi, Menahga), and Hapludalfs (Emmet).

Forested, red, clayey, or loamy soils (soil region I)

Soil region I, as recognized by Hole (1976), Madison and Gundlach (1993), includes two extensive areas of glaciolacustrine sediments reddened by iron oxides: the southern shore of Lake Superior and the eastern shore of Lake Michigan. The sediments were deposited by the Lake Superior and Green Bay and Lake Michigan lobes, respectively. Hole (1976) recognized 22 soil associations in soil region I that contained the Hibbing, Hortonville, Ontonagon, and Rudyard series (Glossudalfs), the Casco, Fox, Kewaunee, Manawa, Onaway, and Theresa series (Hapludalfs), the Superior (Haplorthods), the Pickford (Epiaquepts), the Gogebic (Fragiorthods), and the Poygan series (Epiaquolls).

Figure 5.3 shows northwestern Wisconsin and includes soil regions G, H, and I. Soil region I contains primarily Alfisols. The great group map identifies the Lake Superior component as Glossudalfs and the Lake Michigan component as Hapludalfs. Soil series with areas in excess of 400 km^2 include the Casco, Fox, Kewaunee, and Manawa (all Hapludalfs) and the Hortonville (Glossudalfs).

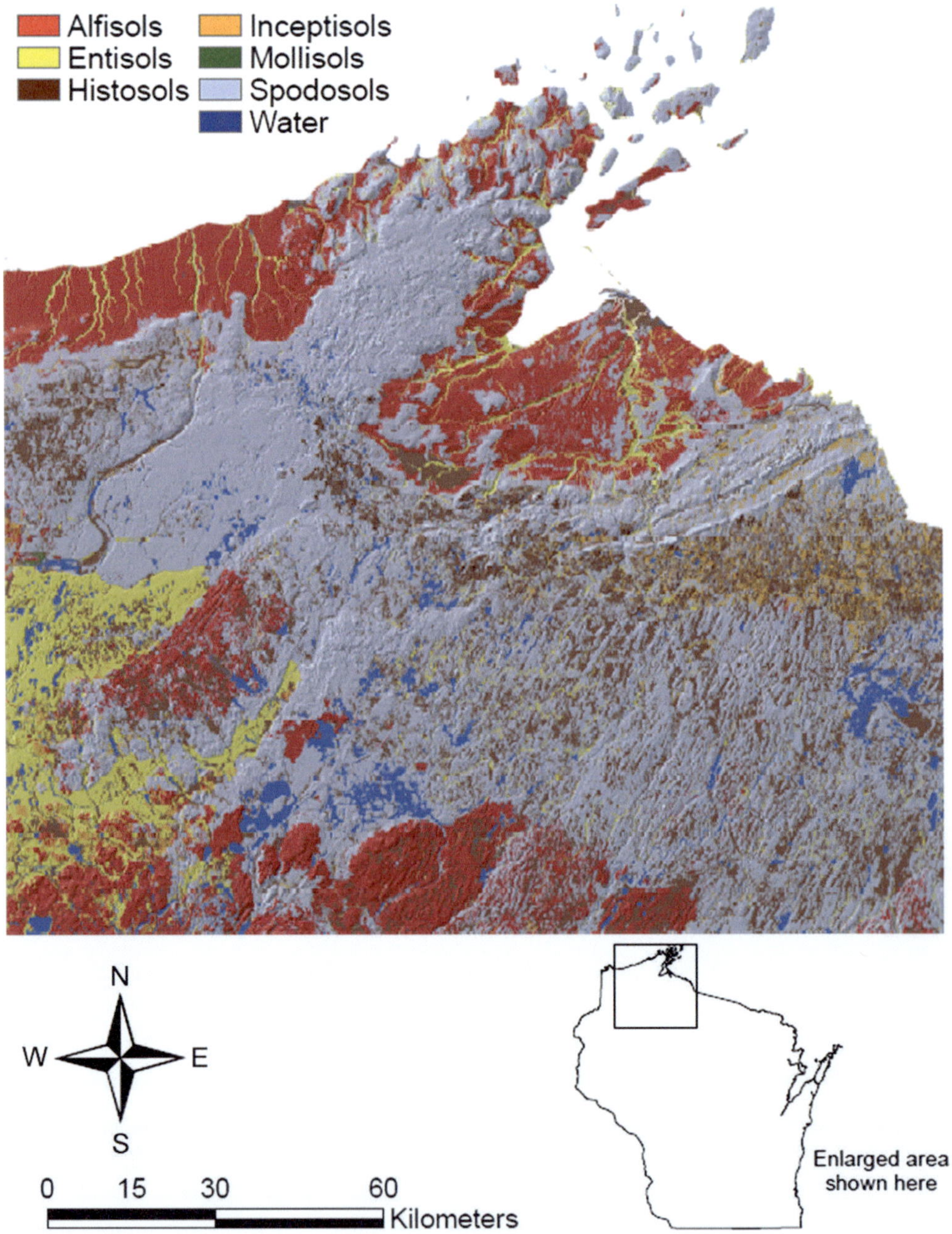

Fig. 5.3 Soil orders of northwestern Wisconsin. Soil region G contains primarily Spodosols, soil region H supports Entisols, and soil region I features primarily Alfisols

5.2.2 Soils of Central Wisconsin

Forested, sandy soils (soil region C)

Soil region C is comprised of the Central Sands region of Wisconsin that contains proglacial sands from Glacial Lake Wisconsin (Clayton and Attig 1989) (Fig. 5.5). Hole (1976) delineated 11 associations that contained forest soils and seven associations that contained some prairie soils (soil region Cm of Madison and Gundlach 1993). Of the 20 soil series identified in 11 associations, nine (45%) contain mainly Hapludalfs (Delton, Gotham, Meridian, Oshtemo, Richford, Shiffer, Wyocena, and Wyeville) (Fig. 5.6).

The STATSGO soil order map of central Wisconsin (Fig. 8.3) shows primarily Entisols. In the great group map (Fig. 9.5), soil region C contains dominantly Udipsamments. Soil series occupying more than 400 km^2 each in soil region C include the Friendship, Meehan, and Plainfield (all Udipsamments), the Tarr (Quartzipsamments), the Urne (Eutrudepts), and the Richford (Arenic Hapludalfs).

Prairie, sandy soils (soil region Cm)

Soil region Cm only occupies several small areas in southern Portage and Waushara. Hole (1976) recognized the Sparta (Hapludolls), Dakota (Argiudolls), and Granby series

Fig. 5.4 Pasture near Marshfield, these Alfisols are poorly drained and have gley at shallow depth and a glossic horizon. Withee soil series (frigid Aquic Glossudalfs). The soils are very deep, somewhat poorly drained and formed in loess or silty alluvium and in the underlying loamy till on ground moraines

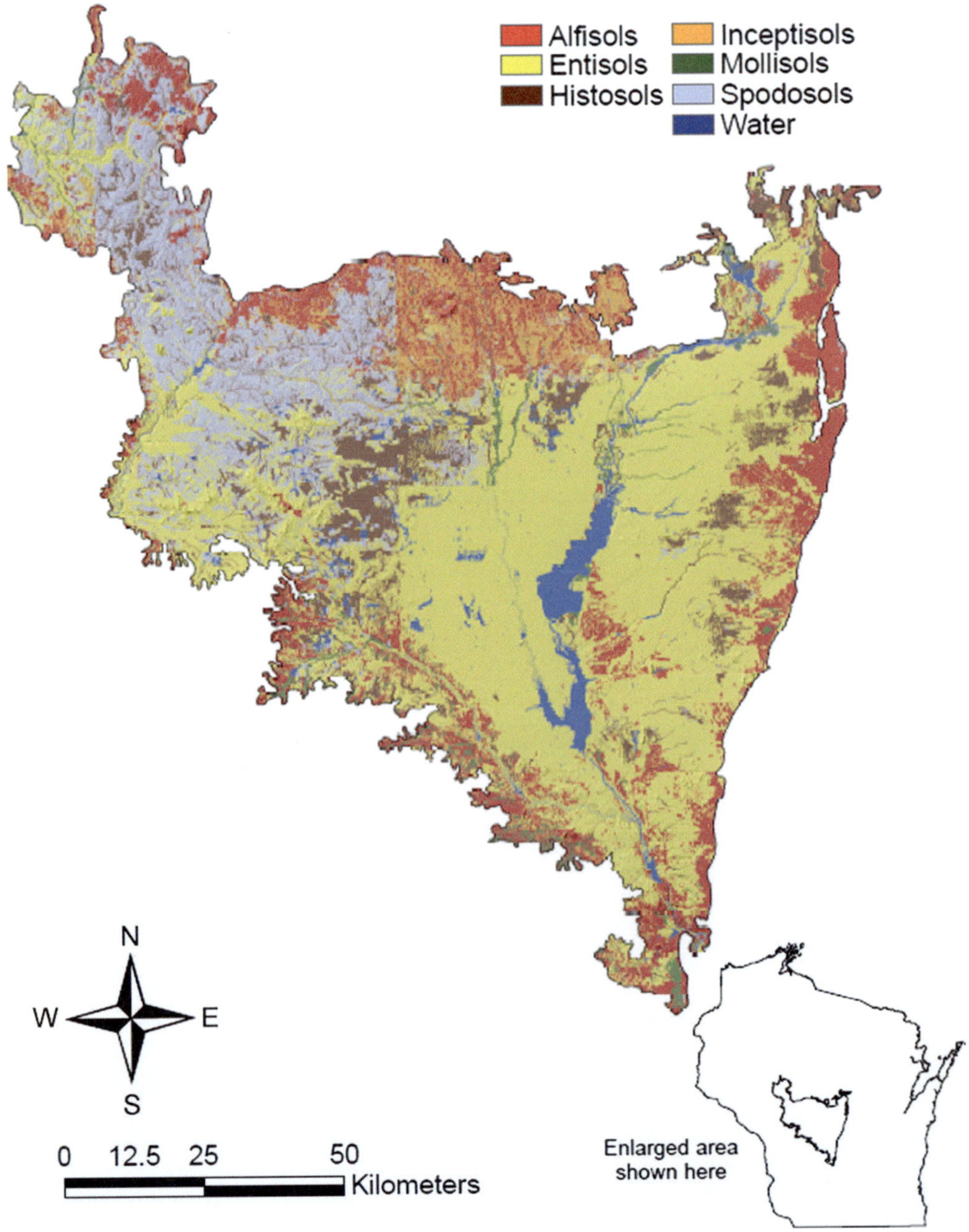

Fig. 5.5 Soil orders of Central Wisconsin, soil regions C and Cm

(Endoaquolls) in these areas. From the Web Soil Survey, the dominant series recognized in these areas today are the Billett (Mollic Hapludalfs), Gotham (Psammentic Hapludalfs), and Richford (Arenic Hapludalfs) series.

Forested, silty soils over igneous/metamorphic rock (soil region Fr)

This region is centered at Marathon County and has an area of about 2000 km^2. The dominant soils are Glossudalfs (Fenwood, Meadland, Rietbrock, and Rozellville series) and Endoaqualfs (Sherry series). This region was delineated by Hole as soil association F14, which includes Fenwood, Marathon, Rozellville (all Glossudalfs), and Cable (Epiaquepts). The great group map (Fig. 9.5) confirms that the soils are dominantly Glossudalfs. The Rietbrock (Glossudalfs) is also extensive in soil region Fr.

5.2.3 Soils of Southwestern and Western Wisconsin

Forested, silty soils (soil region A)

Soil region A was described by Hole (1976) as the soils of the southwestern ridges and valleys and is located almost

Fig. 5.6 The Central Sands plain of Wisconsin is intensively used for irrigated cropping. Some of these soils are moderately well or poorly drained but all of them are drought susceptible and irrigation is required for high crop yields

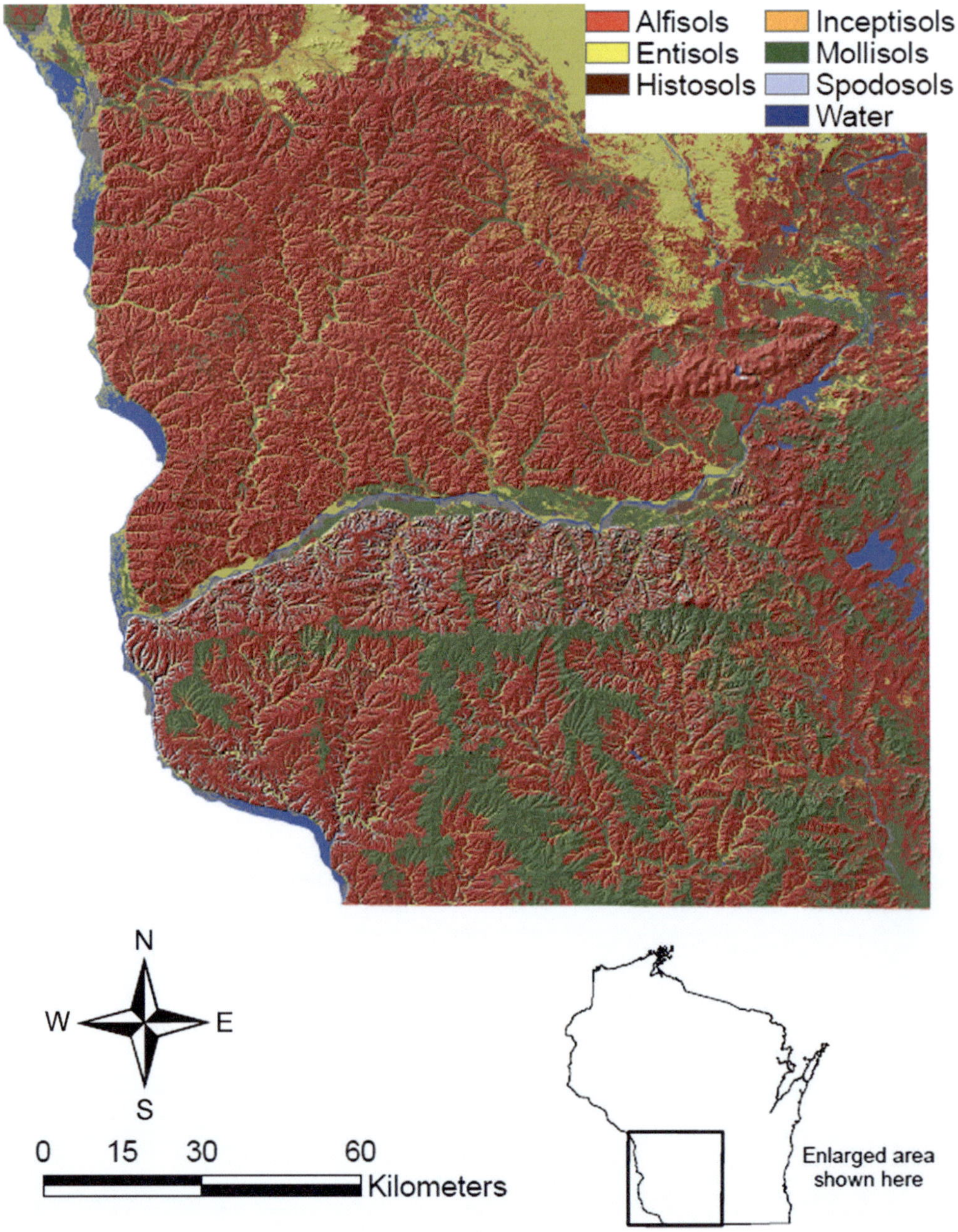

Fig. 5.7 Soil orders in the Driftless Area, soil regions A and Am

entirely in the Driftless Area (Fig. 5.7). The soils are derived primarily from loess. Hole (1976) identified seven associations of mainly forest soils and five of dominantly prairie soils. The forest soils in soil region A include primarily Hapludalfs (Bariboo, Bertrand, Derinda, Downs, Dubuque, Fayette, Meridian, Palsgrove, and Seaton series).

Figure 5.9 shows the abundance of Alfisols in the Driftless Area. On the great group map (Fig. 6.5), soil region A is comprised mainly of Hapludalfs (series listed above) and Paleudalfs. Soil series occurring in excess of 400 km^2 each include the Churchtown, Downs, Dubuque, Fayette, Hixton, La Farge, Newglarus, Palsgrove, and St. Charles (all Hapludalfs), the Arenzville (Udifluvents), and the Valton (Paleudalfs). Figure 5.8 shows an Alfisol of the Driftless Area.

Prairie, silty soils (soil region Am)

Prairie soils in the Driftless Area (soil region Am) are mainly Argiudolls (Ashdale, Dakota, Dodgeville, Richwood, Schapville, Tama, and Toddville series), with some Hapludolls (Muscatine) and Argiaquolls (Calamine). This region

Fig. 5.8 Alfisol near Dodgeville in the Driftless Area. Newglarus soil series (mesic Typic Hapludalfs). The Newglarus series consists of well-drained soils over to dolostone or other limestone. The soils formed in loess and clayey pedisediment and in loamy residuum weathered from the underlying dolostone

has extensive areas of Mollisols (Fig. 6.5). The Dodgeville and Tama series each have an area exceeding 400 km^2.

Forested soils over sandstone (soil region Dr)

This region is concentrated in Dunn, Jackson, Eau Claire, Monroe, Pepin, Trempeleau Counties and the parts of Clark and Wood Counties. Hole (1976) described this region as the western sandstone uplands, valley slopes, and plains and identified 13 soil associations. The soils were dominantly Hapludalfs (Fayette, Gale, Hixton, Norden, Northfield, and Seaton series). The Northfield is in the Lithic subgroup (bedrock within 50 cm of the surface), and the Gale, Hixton, and Norden are in the Typic subgroup but have sandstone bedrock within 1 m of the surface. Other common soils include several Glossudalfs (Arland, Kert, and Milaca series) and the Boone series (Quartzipsamments). Soils with restricted drainage in the region include the Merrillan (Epiaquods), Elm Lake (Epiaquents), and the Vesper and Veedum series (Epiaquepts).

The region has Quartzipsamments (Boone, Gosil, Tarr, and Twinmound series), Fluvaquents (Fordum and

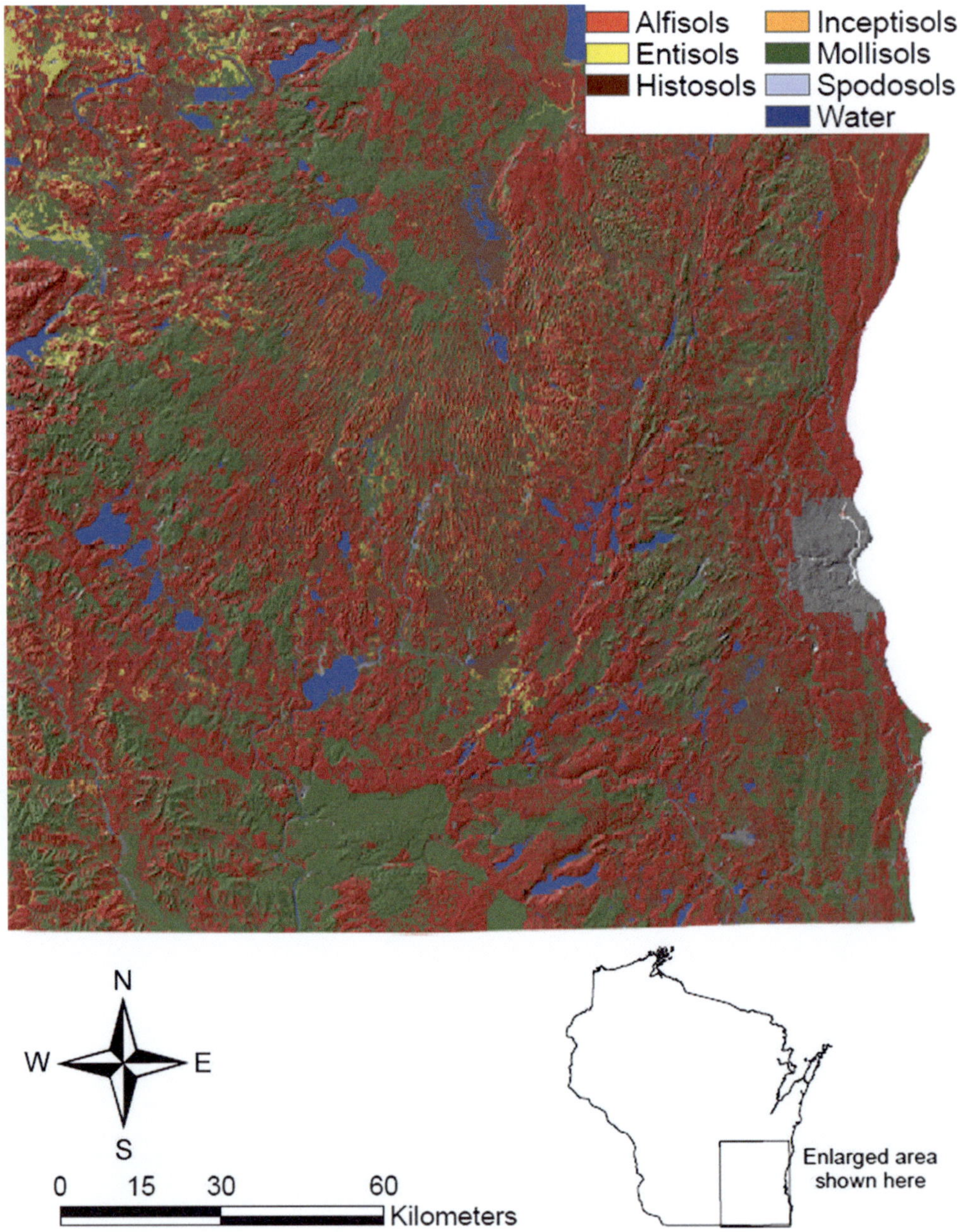

Fig. 5.9 Soil orders of southeastern Wisconsin, soil regions B and Bm

Fig. 5.10 Alfisol of the Arboretum in Madison (mesic Mollic Hapludalfs)

Fig. 5.11 Mollisol of the Arboretum in Madison (mesic Typic Endoaquolls)

Kalmarville series), Hapludalfs (soils listed above), Eutrudepts (Urne), and Dystrudepts (Elkmound, Impact, and Moppet series) (Fig. 6.5).

5.2.4 Soils of Southeastern Wisconsin

Forested, silty soils (soil region B)

Soil region B was identified by Hole (1976) as the southeastern uplands. Soils of this region are derived mainly from calcareous drift. Hole delineated 34 associations in this region, of which 24 were dominated by forest soils and 10 by prairie soils. The predominant forest soils in this region are Hapludalfs (Baraboo, Boyer, Casco, Dodge, Dubuque, Flagg, Fox, Hebron, Knowles, Lapeer, Leroy, Lomira, Matea, McHenry, Miami, Morley, Onaway, Ozaukee, Pardeeville, Pecatonica, Puchyan, Rodman, Sisson, St. Charles, Theresa, Whalan series), with some Epiaqualfs (Blount and Del Rey series) (Fig. 5.9).

Alfisols are very common in southeastern Wisconsin (Fig. 5.10) and Hapludalfs and Epiaqualfs in particular. Soil series with areas exceeding 400 km^2 each include the Casco, Eleva, Fox, Kidder, McHenry, Miami, Morley, Norden, and Theresa (all Hapludalfs).

Prairie, silty soils (soil region Bm)

The predominant prairie soils in this region are Argiudolls (Brookston, Durand, Elburn, Elliott, Hochheim, Nenno, Ogle, Plano, Ringwood, Ripon, Saybrook, Varna series), along with Endoaquolls (Ashkum, Pella, and Will series). Mollisols (Argiudolls) are common (Figs. 5.5 and 6.5). Soil series with areas in excess of 400 km^2 each include the Plano and Hochheim (Argiudolls) and Pella (Endoaquolls) (Fig. 5.11).

5.2.5 Statewide

Stream bottom and major wetland soils (soil region J)

Soil region J, soils of the stream bottoms and major wetlands, are scattered throughout the state. The three largest wetlands are the Necedah National Wildlife Refuge, the Sherwood–Camelot Lake area between Stevens Point and

Fig. 5.12 Sandy soils along the Wisconsin River near Arena

Wisconsin Rapids, and the Horicon Marsh Wildlife Area. Major river bottoms include those of the Wisconsin, Mississippi, Fox, St. Croix, Menominee, Chippewa, Rock, Milwaukee, and many smaller river basins (Fig. 5.12).

Hole (1976) recognized 15 soil associations in soil region J that included the Poygan series (Epiaquolls), the Granby and Pella series (Endoaquolls), the Au Gres and Monico series (Endoaquods), and other soil series in Aqu-suborders, including the Cable (Epiaquepts), Auburndale (Epiaqualfs), Freer (Glossaqualfs), and Brookston (Argiaquolls). The dominant great groups are Psammaquents (Newson, Ponycreek) and Haplosaprists (Adrian, Cathro, Dawson, Houghton, Loxley, Lupton, Markey, Palms, Seelyeville) (Fig. 6.5). Hydric soil series occupying more than 400 km^2 each include the Adrian, Beseman, Cathro, Dawson, Houghton, Loxley, Lupton, Markey, Palms, and Seelyeville (all Haplosaprists), Newson (Psammaquents), Capitola and Marshfield (Epiaqualfs), Minocqua (Endoaquepts), and Fordum (Fluvaquents). Hydric soils occupy 31,656 km^2 in Wisconsin, 58% of which are in Aqu-suborders and 42% are Histosols.

5.3 Summary

Previous analyses of the soils of Wisconsin have divided the state into regions based on geographic location, broad vegetation type (forest vs. prairie), and texture of parent materials. The analysis by Madison and Gundlach (1993) identified five geographic regions and 11 cover types of texture classes. Soil associations (associations of predominant soil series) are delineated within each of these regions. In this chapter, we follow this approach but update the list of dominant soil series and their classification in *Soil Taxonomy*.

6 Diagnostic Horizons and Soil Taxa

This chapter briefly examines the distribution of soils in Wisconsin by diagnostic horizons and taxa, beginning with the higher levels and continuing to the lower levels.

6.1 Diagnostic Horizons

The ochric epipedon (Fig. 6.1) is the most common diagnostic surface horizon of the soils in Wisconsin, accounting for 74% of the soil series and 76% of the state's land area. The average thickness of the ochric epipedon is 24 ± 14 cm in thickness, but it ranges from 3 to 109 cm (Table 6.1; Fig. 6.2). The second most common epipedon is the mollic horizon, which accounts for 21% of the soil series and 10% of the state's land area. The average thickness of the mollic epipedon is 39 ± 16 cm in thickness, but it ranges from 15 to 127 cm. Mollic horizons are only recognized in Mollisols; soils in mollic subgroups have an ochric epipedon. Mollic horizons are discussed in more detail in Chap. 11.

Histic epipedons are present in 5% of the soil series and cover 9% of the state area (Fig. 6.2). The average thickness of the histic epipedon is 95 ± 50 cm in thickness and it ranges from 20 to 203 cm. Most of the Histosols in Wisconsin are of Late Wisconsinan age, and the organic materials accumulated at a rate of about 6.3 mm/100 year (assumes age of soil to be 15,000 years), which is substantially less than rates reported in other Histosols of temperate regions (Frolking et al. 2001). Histic horizons are most common in Histosols but also occur in Humaquepts. Histic horizons are discussed in more detail in Chap. 12. Only six soil series in Wisconsin have an umbric epipedon, which averages 36 ± 7.6 cm in thickness. These soils are mostly Dystrudepts.

The argillic horizon is the most common diagnostic subsurface horizon in soils of Wisconsin, occurring in 61% of the soil series and on 60% of the state's land area. The argillic horizon averages 62 ± 37 cm in thickness, with a range of 10–272 cm (Table 6.1; Fig. 6.3). In Wisconsin, argillic horizons are most common (90% of soil series) in intermediate drainage classes (well drained, moderately well drained, and somewhat poorly drained) and are not evident in excessively somewhat excessively or poorly–very poorly drained soils. Dominant parent materials include loess (35%), till (18%), and alluvium (15%). Soils with argillic horizon tend to be in fine-silty and fine-loamy (33%) or fine and very-fine particle-size classes (29%). Argillic horizons

J.G. Bockheim and A.E. Hartemink, *The Soils of Wisconsin*,
World Soils Book Series, DOI 10.1007/978-3-319-52144-2_6

Fig. 6.1 A sandy soil from central Wisconsin with an ochric epipedon over a cambic subsurface horizon. Plainfield soil series (Typic Udispamments)

Table 6.1 Thickness and abundance of diagnostic horizons in Wisconsin

Horizon	Number of soil series	Avg. thickness (cm)	Std. dev. (cm)	Median (cm)	Maximum (cm)	Minimum (cm)
Epipedon						
Ochric	533	24	14	20	109	3
Mollic	149	39	16	36	127	15
Histic	33	95	50	88	203	20
Umbric	6	36	7.6	39	43	23
None	2	–	–	–	–	–
Subsurface						
Albic	220	13	8.6	10	48	3
Argillic	444	62	37	53	272	10
Cambic	120	48	28	43	145	8
Spodic	105	30	14	28	71	5
Glossic	148	33	22	28	124	8
Fragipan	8	60	36	56	124	15
Ortstein	0	–	–	–	–	–
Calcic	1	40	–	–	–	–
None	124	–	–	–	–	–

are most common in Udalfs (60% of all soil series), followed by Udolls (14%), Orthods (10%), and Aqualfs (9%).

The second most common diagnostic subsurface horizon is the cambic horizon, which occurs in 17% of the soil series but on only 9% of the state area. The cambic horizon averages 48 ± 28 cm in thickness and ranges from 8 to 145 cm (Fig. 6.3). The spodic horizon occurs in 15% of the soil series and on 18% of the state area, averages 30 ± 14 cm thickness, and ranges from 5 to 71 cm in thickness. The spodic horizon is treated in more detail in Chap. 9.

The glossic horizon occurs in conjunction with an argillic horizon in 100 soil series of Wisconsin and conjunction with an argillic and spodic horizons in an additional 47 soil series (Fig. 6.3). In the Wayka (Typic Epiaquods) series, the glossic horizon is accompanied only by a spodic horizon. The glossic horizon averages 33 ± 22 cm and ranges from 8 to 124 cm in thickness. The glossic horizon occurs almost exclusively (95% of soil series) in soils that are well drained, moderately well drained, or somewhat poorly drained. Soils containing glossic horizons are derived from till (32%), loess over till (19%), glaciolacustrine materials (13%), and alluvium over outwash (13%). The vegetation on soils with glossic horizons is either northern hardwoods with varying amounts of hemlock (58%) or mixed broad-leaved and coniferous forest (36%). Nearly all (97%) of the soil series with a glossic horizon have a mixed mineral class and a frigid soil-temperature class. Two-thirds of the soil series with a glossic horizon have a superactive cation-exchange capacity.

The albic horizon occurs in 227 soil series (Table 6.1), including 109 with an argillic horizon, 44 with a spodic horizon, 53 with an argillic and a spodic horizon, and six soil series with a cambic horizon. The albic horizon occurs at the surface of 19 Spodosols, two Alfisols (Sanborg and Neopit series), and one Inceptisol (Minocqua) in Wisconsin. Eighty-three percent of the soil series with an albic horizon are well drained, moderately well drained, or somewhat poorly drained. The dominant vegetation is northern

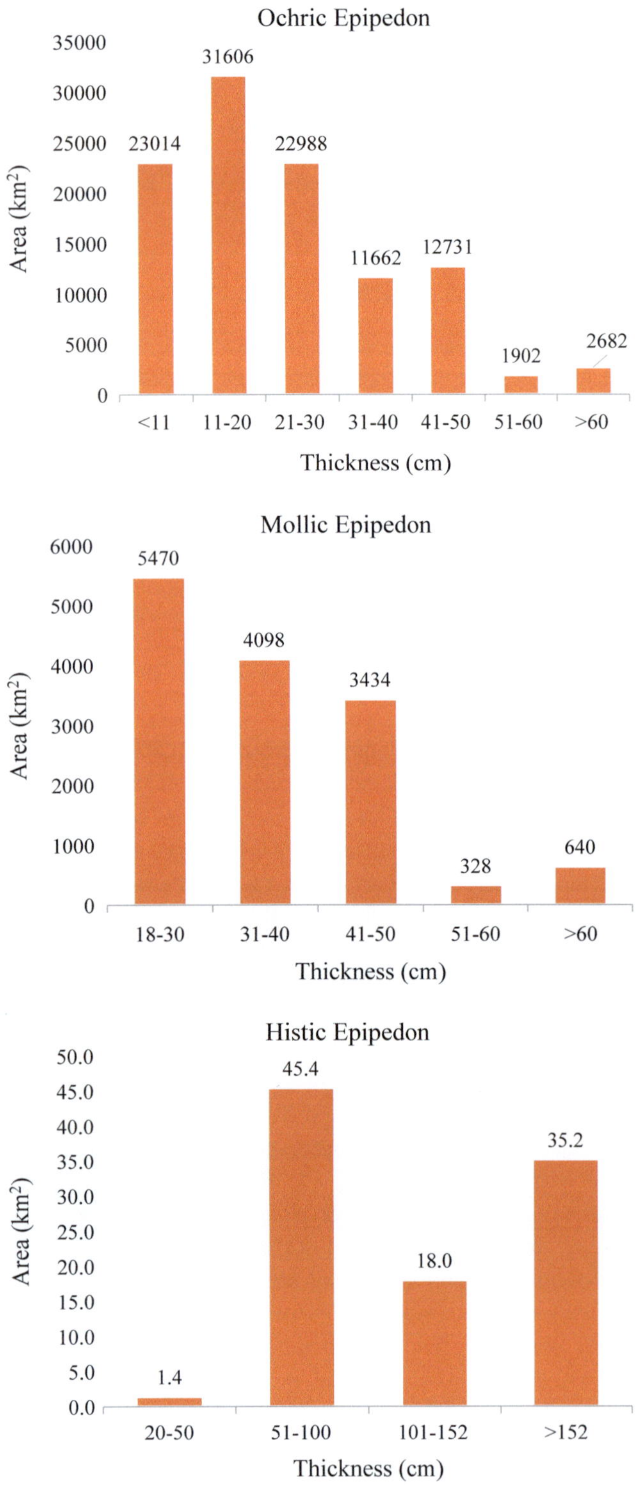

Fig. 6.2 Area (km^2) of diagnostic epipedons by depth class in the soils of Wisconsin

hardwoods and hemlock (39%) and mixed broad-leaved and coniferous forest (33%).

Only ten soil series in Wisconsin contain a fragipan; these soils account for a total area of 1659 km^2 (1.2% of the state's land area) and occur in soils along the northern tier of counties, especially Iron, Forest, Vilas, and Ashland counties. Fragipans are most common in bisequal soils on drumlins and end moraines that have received aeolian materials. A calcic horizon occurs only in the Cushing series, fine-loamy, mixed, superactive, frigid Haplic Glossudalfs that occur on calcareous loamy till in northwestern Wisconsin.

None of the soil series in Wisconsin are in the ortstein rupture-resistance class, i.e., has a layer 25 mm or more thick that is 50% or more cemented. Continuous ortstein is 90% or more cemented and has lateral continuity, restricting roots to vertical fractures with a horizontal spacing of 10 cm or more (Fig. 6.4). However, 21 Spodosol soil series in Wisconsin contain some reported ortstein (Table 6.2). These soils are primarily sandy Spodosols in the northernmost counties. A list of all soil series recognized in Wisconsin, along with the thicknesses of their diagnostic horizons, is given in Appendix C.

6.2 Higher Levels: Orders, Suborders, and Great Groups

Wisconsin has a diversity of soils and has soils representing seven of the 12 orders (58%) recognized in the USA, 15 of the 65 suborders (23%), and 32 of the 344 great groups (9%). Alfisols comprise 47% of the state's land area, followed by Spodosols (17%), Entisols (12%), Mollisols and Histosols (10% each), and Inceptisols (4%; Fig. 1.3). There is one Ultisol in WI, the Siouxcreek series, which comprises only 2.6 km^2. General soil maps at the order, suborder, and great group levels are shown in Figs. 6.5, 6.6, 6.7, and 6.8. A list of all soil series recognized in Wisconsin, as classified in *Soil Taxonomy*, is given in Appendix D.

6.3 Lower Levels: Subgroups, Series, and Families

The soils of Wisconsin are represented in 142 subgroups, 425 families, and 741 soil series. A complete list of soil series identified in Wisconsin is in Appendix A. Of these soil series, 432 (57%) were initially observed in Wisconsin, 305

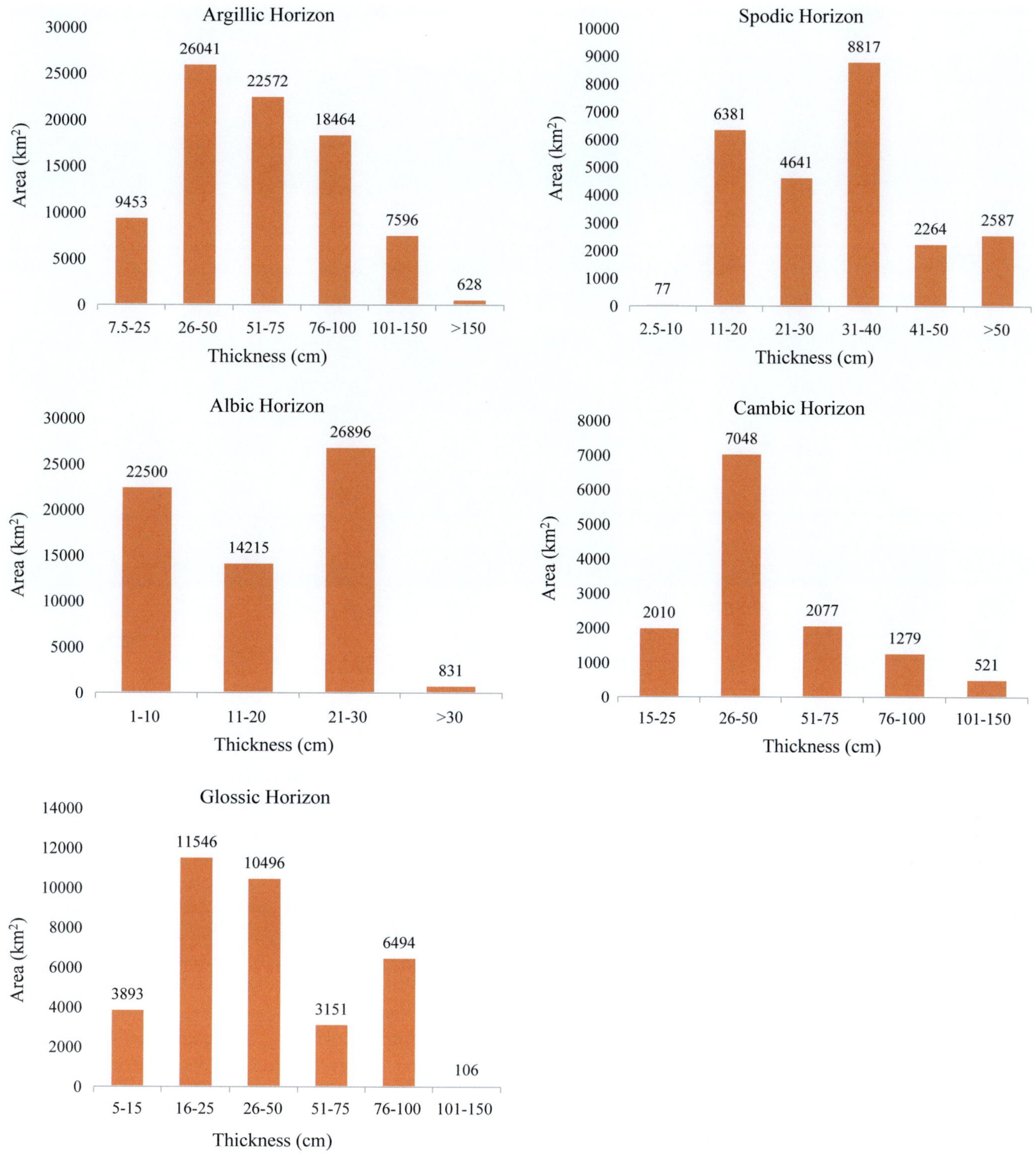

Fig. 6.3 Area (km^2) of diagnostic subsurface horizons by thickness class in soils of Wisconsin

Fig. 6.4 Ortstein in a Spodosol north of Sturgeon Bay in Door County

(40%) only occur in Wisconsin, and 132 (18%) are one of a kind and only occur in Wisconsin and an adjacent state, i.e., are endemic soils. There is a significant correlation between the number of soil series and area of these soil series according to great group (Fig. 6.9).

Whereas 52% of Wisconsin soils have a frigid soil-temperature regime, meaning that the mean annual soil temperature at 50 cm is 7 °C or colder, 48% have a mesic soil-temperature regime (MAST >7 °C) (Fig. 6.10). Eighty-seven percent of Wisconsin soil series have a mixed mineralogy, meaning that the soils contain a variety of minerals in the fine earth (<2 mm) fraction; other mineral classes represented in the state include isotic (25 series), siliceous (20), smectitic (15), and illitic (11), and minor occurrences of others; 23 series are derived from organic materials or sandy materials and are not placed in a soil mineral class. Nearly half (48%) of Wisconsin soils are in loamy particle-size classes, 22% are sandy, 18% are silty, 10% are clayey, and 2% are comprised of organic materials. Half (50%) of the soils in Wisconsin are in the superactive cation-exchange capacity class, meaning that they have clay-size minerals with a high-nutrient-holding capacity;

Table 6.2 Wisconsin soil series containing ortstein

Soil series	Area (km^2)	Soil classification	Depth (cm)	Ortstein (%)	Cementation
Abbaye	11	Coarse-loamy, mixed, active, frigid Alfic Oxyaquic Haplorthods	33–46	"few"	cs
Allendale	153	Sandy over clayey, mixed, semiactive, frigid Alfic Epiaquods	25–66	"few"	cs
AuGres	384	Sandy, mixed, frigid Typic Endoaquods	13–58	<50	–
Brownstone	6	Sandy-skeletal, mixed, frigid Typic Haplorthods	30–58	<50	–
Croswood	27	Sandy, mixed, frigid Oxyaquic Haplorthods	15–36	10–15	cw
Deerton	7	Sandy, mixed, frigid Typic Haplorthods	25–62	10	cw
Duel	3	Sandy, mixed, frigid Entic Haplorthods	23–53	10	cw
Ingalls	8	Sandy over loamy, mixed, active, frigid Typic Endoaquods	28–62	10–15	–
Iosco	185	Sandy over loamy, mixed, active, frigid Argic Endoaquods	28–71	<45	–
Ironrun	175	Sandy, siliceous, frigid Typic Endoaquods	30–62	<20	–
Ishpeming	87	Sandy, mixed, frigid Entic Haplorthods	15–33	10–40	cm
Keweenaw	1278	Sandy, mixed, frigid Alfic Haplorthods	28–69	35–75	cw
Lindquist	61	Sandy, mixed, frigid Lamellic Haplorthods	13–86	<35	–
Manistee	56	Sandy over clayey, mixed, active, frigid Alfic Haplorthods	43–71	"chunks"	cw-cs
Menominee	346	Sandy over loamy, mixed, active, frigid Alfic Haplorthods	46–58	"fragments"	–
Parkfalls	49	Coarse-loamy, mixed, superactive, frigid Alfic Epiaquods	43–71	5	–
Pequaming	34	Sandy, mixed, frigid Argic Endoaquods	28–38	30	cw
Rubicon	742	Sandy, mixed, frigid Entic Haplorthods	46–61	15	cw-cs
Sedgwick	97	Coarse-loamy over clayey, mixed, active, frigid Alfic Epiaquods	20–40	<50	–
Sultz	227	Sandy, mixed, frigid Entic Haplorthods	15–64	15–20	cw
Wainola	279	Sandy, mixed, frigid Typic Endoaquods	30–69	"fragments"	cw
	4215				

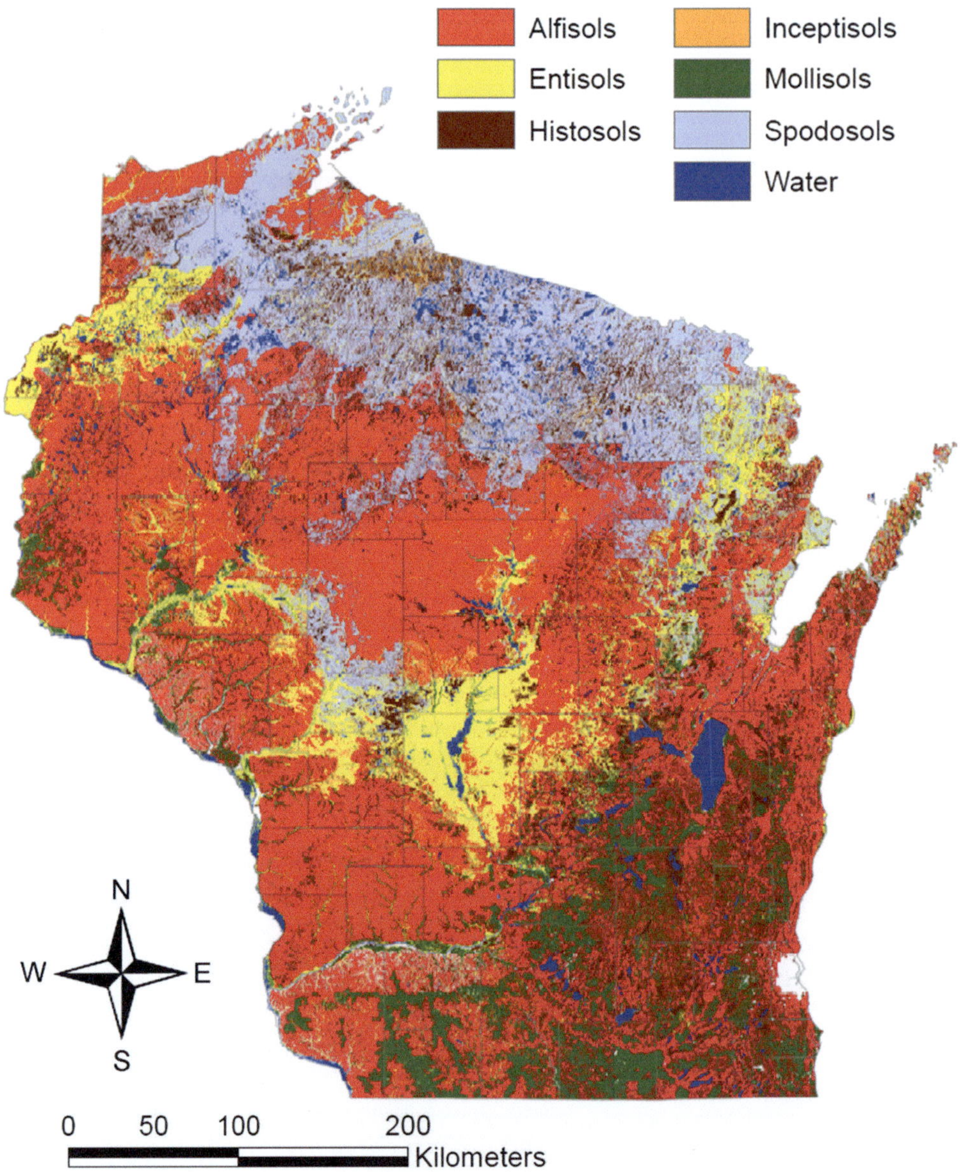

Fig. 6.5 Soil map of Wisconsin at the soil order level

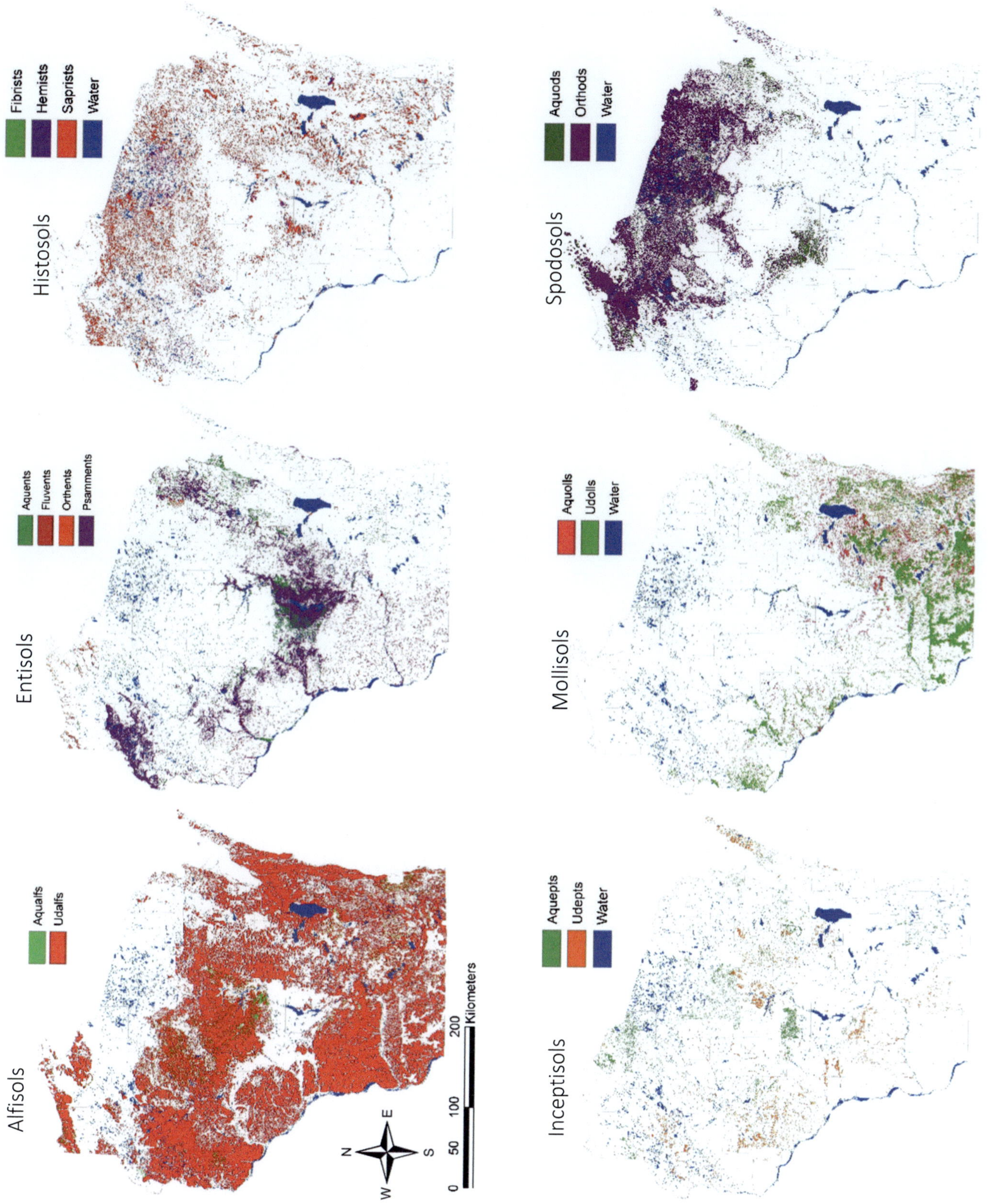

Fig. 6.6 Suborder soil maps of Wisconsin. Ultisols are not included here and cover less 260 ha

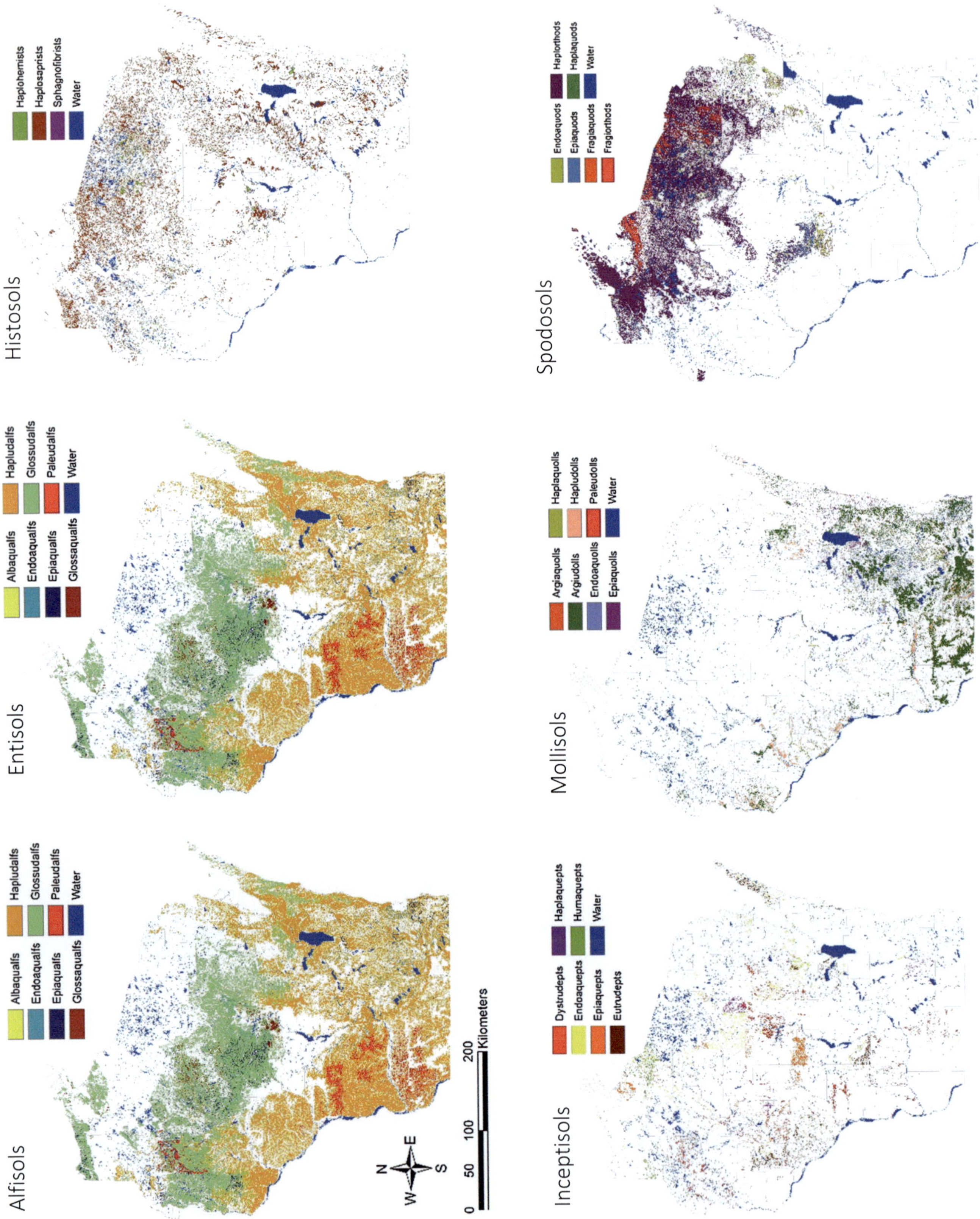

Fig. 6.7 Great group soil maps of Wisconsin. Ultisols are not included here and cover less 260 ha

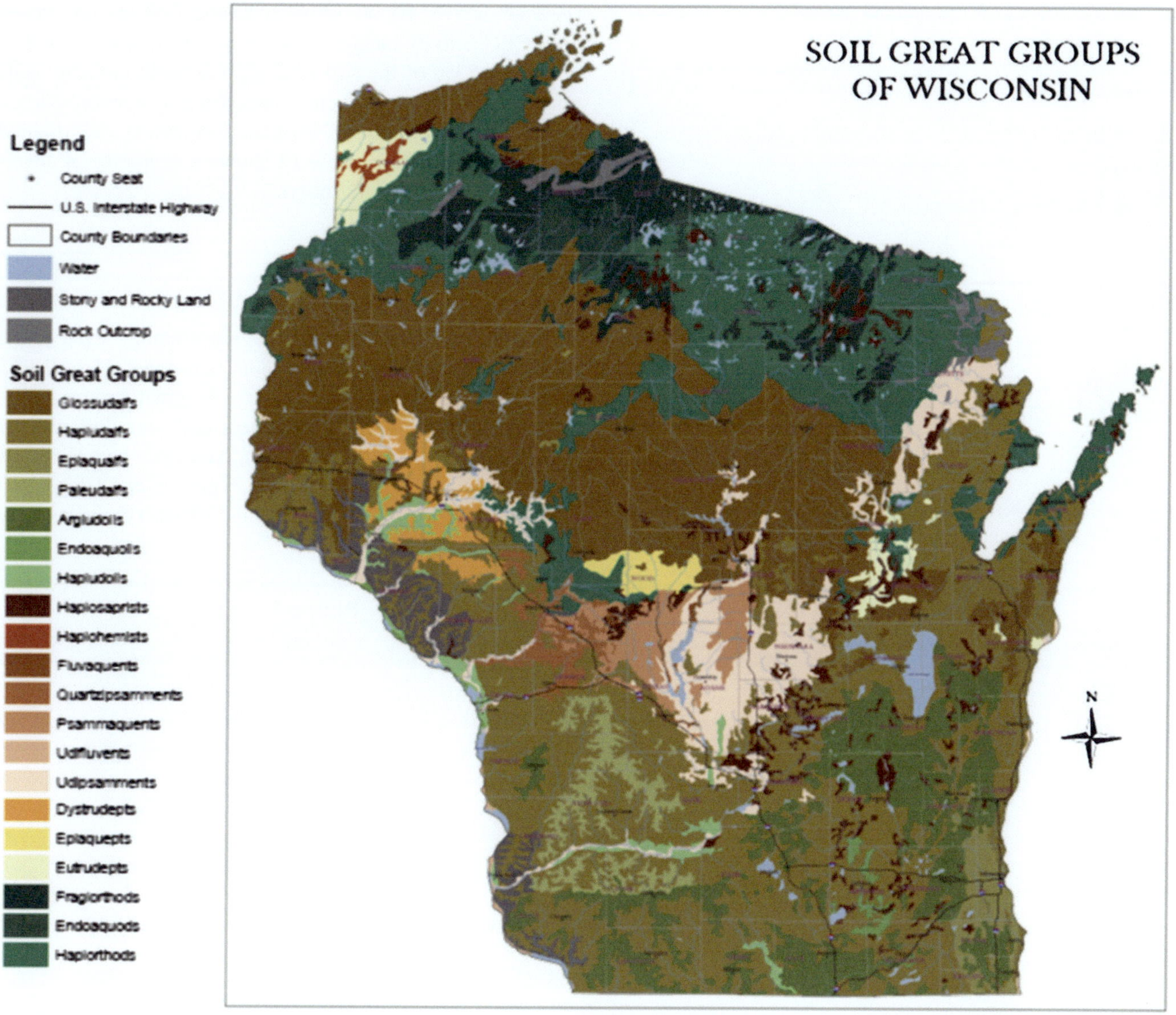

Fig. 6.8 Soil map of Wisconsin at the great group level (by J. Bockheim and L. Zamzow)

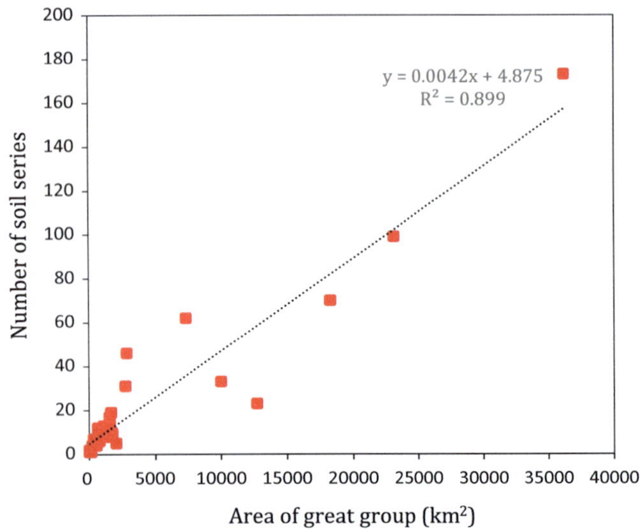

Fig. 6.9 Relation between number of soil series in a great group and area of the great group

19% are active, 4% are semiactive, and 27% are not placed in a CEC activity class because they are organic or sandy.

About 83% of the soil area of Wisconsin contains soils that are very deep (>150 cm) to a restricting layer or bedrock (Fig. 6.11, upper); and 40% of the soil area is well drained (Fig. 6.11, lower). Figure 6.12 shows a deep Alfisol developed under forest near Antigo.

6.4 Summary

Wisconsin has a diversity of soils representing seven of the 12 orders (58%) recognized in *Soil Taxonomy*, 15 of the 65 suborders (23%), and 32 of the 344 great groups (9%). Alfisols comprise 47% of the state's land area, followed by Spodosols (17%), Entisols (12%), Mollisols and Histosols (10% each), and Inceptisols (4%). There is one Ultisol soil series in WI, the Siouxcreek series, which comprises only

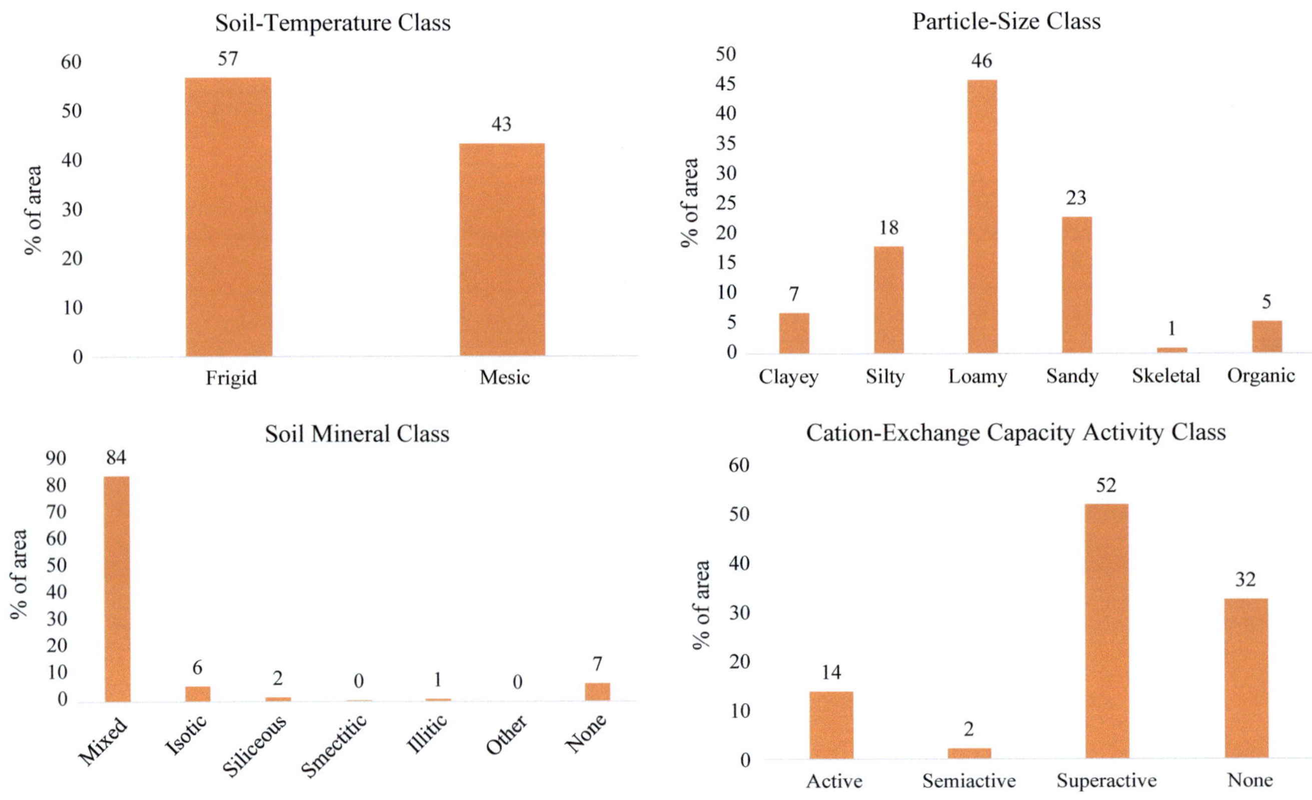

Fig. 6.10 Family classes of Wisconsin soils, including soil-temperature, particle-size, soil mineral, and cation-exchange capacity activity class

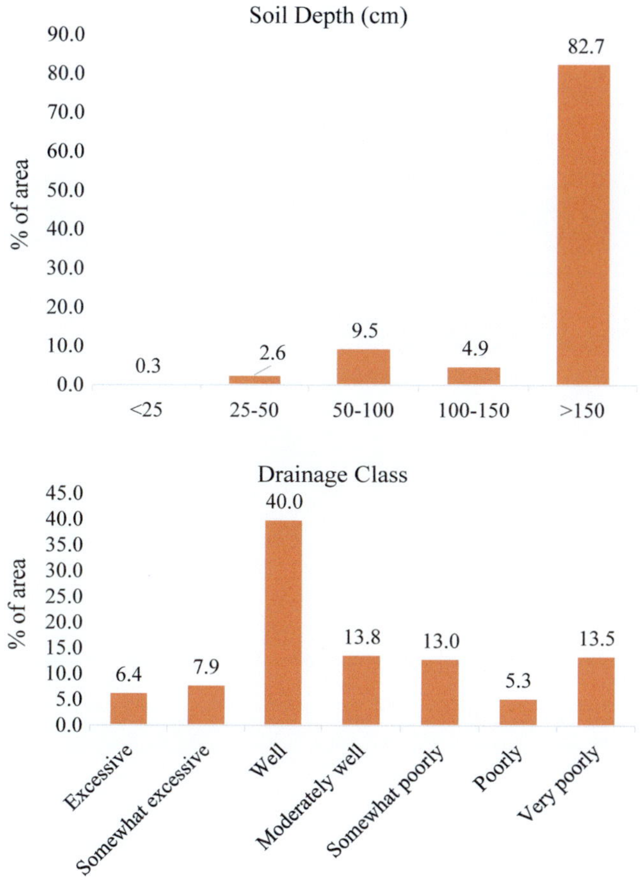

Fig. 6.11 Areal distribution of soils by depth class (*upper*) and drainage class (*lower*)

2.6 km^2. General soil maps at the order, suborder, and great group levels are provided. Soils of Wisconsin are represented in 142 subgroups, 425 families, and 723 soil series. Of these soil series, 432 (57%) were initially observed in Wisconsin, 305 (40%) only occur in Wisconsin, and 132 (18%) are endemic to the state. At the Family level, half of Wisconsin soils have a mesic soil-temperature regime and half have a frigid one; 87% of the soils have a mixed mineralogy; 48% of Wisconsin soils are in loamy particle-size classes; and 50% of the soils are in the superactive cation-exchange capacity class.

Fig. 6.12 Deep Alfisol developed under forest in the pitted outwash at the end moraine near Antigo

7 Taxonomic Soil Regions

7.1 Introduction

Kellogg (1930) divided the state into three broad soil taxonomic regions, including from south to north, the Gray-Brown Forest (Alfisols), Transitional (tension zone), and Podzols (Spodosols) (Fig. 2.9). The soil maps of Wisconsin from 1927 through 1968 (Chap. 2) were based on the associations of soil series. The 1993 map of Madison and Gundlach (Fig. 2.12) was similar to Hole's 1968 map (Fig. 2.10) except that the legend divided the soils by geographic region (northern, eastern, central, western, etc.), vegetation cover (forest vs. prairie), and texture of the soil materials (sandy, loamy, silty, clayey, etc.). This approach has been used in many other states (Bockheim and Gennadiyev 2015).

7.2 Taxonomic Soil Regions

Wisconsin is divided here into 16 soil regions based on the distribution of taxonomic (great groups) (Table 7.1; Fig. 7.1). The advantage of this approach is that it considers natural soil units that integrate the soil-forming factors and express the soil-forming processes. For example, the same soil taxa may exist on a variety of parent materials (e.g., Sect. 8.2).

7.2.1 Hapludalfs (Region 1)

Hapludalfs are the base—and clay-rich soils that support broad-leaved forest vegetation. The largest soil taxonomic region (region 1) includes the Hapludalfs developed in loess and till. These soils occupy 33,905 km^2 or 24.7% of the state's land area (Table 7.2). Soils of this region occur primarily in the southern one-third of the state and to the south of the Wisconsin Tension Zone (WTZ) (Fig. 3.2). The Hapludalfs occur in the Western Uplands and Eastern Ridges and Lowlands in soil regions A, B, E, and I (Lake Michigan red clays) (Table 7.1).

Hapludalfs are under mixed hardwoods, mixed oaks, oak-hickory forest, and oak-savanna. These soils occur in the Driftless Region and on end moraines and ground moraines outside the Driftless Region that are comprised of loess, till, and loess over till. The soils contain an ochric epipedon over an argillic that averages 60 cm in thickness horizon. About 17% of the Hapludalfs have an albic horizon. More than half (58%) of the Hapludalfs have a loamy particle-size class (fine-loamy, coarse-loamy, loamy, and loamy-skeletal); 91% have a mixed mineral class; 54% are in the superactive cation-exchange capacity activity class; 92% have a udic soil-moisture class; and 98% have a mesic soil-temperature regime.

Dominant (>400 km^2) Hapludalfs in soil region A include the Dubuque, Fayette, Seaton, St. Charles, Palsgrove, Kidder, Dodge, Churchtown, NewGlarus, and Downs series. In soil region B, the major Hapludalfs are the Casco, Theresa, McHenry, Miami, and Morley series. The dominant Hapludalfs in soil regions E and I (Lake Michigan red clays) include the Kewaunee, Manawa, and Fox series (mesic soil-temperature regime) and the Onaway and Emmet series (frigid soil-temperature regime).

Mesic Hapludalfs catenas from four general soil regions, A, B, D, and I, include the Fayette-La Farge-Eleva, Casco-Fox-Rodman, Kewaunee-Manawa-Poygon, and Downs-Boaz-Muscatine, and associations, respectively. The Fayette-La Farge-Eleva association occurs in Trempeleau County (soil region A) and contains well-drained soils that have an argillic horizon; they are all Typic Hapludalfs (Fig. 7.2). These soils developed under forest and are moderately deep to deep. Whereas the Fayette and Eleva series are underlain by sandstone at depths >2 m, the La Farge

J.G. Bockheim and A.E. Hartemink, *The Soils of Wisconsin*,
World Soils Book Series, DOI 10.1007/978-3-319-52144-2_7

Table 7.1 Relation of soil great groups in Wisconsin to soil-forming factors

Soil great group	Soil region[a]	Physiography[b]	Mid-1800s vegetation[c]	Landforms[d]	Parent materials[d]	Landform age[e]
Glossudalfs	F, F, I	Lake Superior Lowlands	Northern Hardwoods	Ground Moraine, end moraines	Till, Glaciolacustrine	Late WI, Pre–Late WI
		Northern Highland	Mixed hardwoods, mixed oaks	Lake plains	Loess cap	Pre-IL
Hapludalfs	A, B, E, I	Western Uplands, Eastern Ridges & Lowlands	Mixed pines, boreal forest Oak-hickory, oak-savanna	Ground moraines Driftless Area, end moraines	Loess, till, loess/till	Late WI
Epiaqualfs	B, F, J	Eastern Ridges & Lowlands	Mixed hardwoods, mixed oaks	End moraines	Till	Late WI
Paleudalfs	A, Dr	Western Uplands	Oak-savanna, mixed oaks	Driftless Area	Loess/clayey residuum/ limestone or sandstone bedrock	Late WI, Pre–Late WI Pre-IL
Argiudolls	Am, Bm	Western Uplands, Eastern Ridges and Lowlands	Prairie, mixed oaks	Driftless Area, end moraines Ground moraines	Loess, till, loess/till	Late WI
Endoaquolls	Bm, J	Eastern Ridges and Lowlands	Wet prairie, marsh	End moraines, ground moraine	Till, glaciolacustrine	Late WI
				Lake plains, outwash plains	Outwash	
Hapludolls	Cm, E, I	Western Uplands	Prairie, oak-savanna	Stream terraces, outwash plains	Alluvium, outwash	Late WI
Haplosaprists	J	All depressions	Wet mixed forest, marsh	Depressions outwash plains	Organic, organic/outwash	Late WI
				Ground moraine	Organic/till	
Haplohemists	J	All depressions	Wet conifers	Depressions outwash plains	Organic, organic/outwash	Late WI
				Ground moraine	Organic/till	
Fluvaquents	J	All river valleys	Wet hardwoods	Floodplains	Alluvium	Holocene
Quartzipsamments	Dr	Western Uplands–Central Plain	Oak-pine	Pediments	Residuum, alluvium/	Late WI and older
					Pedisediments	
Psammaquents	J	Central Plain	Marsh, oak-pine	Lake plains, outwash plains	Glaciolacustrine, outwash	Late WI
Udifluvents	A, B, J	Southern WI river valleys	Mixed forest, mixed hardwoods	Floodplains	Alluvium	Holocene
Udipsamments	C	Central Plain	Oak-pine	Lake plains, outwash plains	Glaciolacustrine, outwash	Late WI
Dystrudepts	Dr, Fr	Western Uplands	Oak-savanna, brush	River terrace	Alluvium	Late WI
Epiaquepts	J	Northern Highland	Mixed forest	Lake plains, drainageways	Loess/till, residuum	Pre-IL?
		Central Plain			Glaciolacustrine	
Eutrudepts	E, G, I	Lake Superior Lowlands	Mixed forest	Lake plains	Glaciolacustrine	Early Holocene
		Northern Highland				
Fragiorthods	G	Northern Highland	Northern hardwoods	End moraines	Till	Late WI

(continued)

Table 7.1 (continued)

Soil great group	Soil region[a]	Physiography[b]	Mid-1800s vegetation[c]	Landforms[d]	Parent materials[d]	Landform age[e]
Endoaquods	G, H, I, J	Northern Highland	Mixed forest	Stream terraces, outwash plains	Alluvium, outwash	Late WI
				Lake plains	Glaciolacustrine	
Haplorthods	G, H	Northern Highland	Northern hardwoods, mixed	Outwash plains, ground moraine	Outwash, till	Late WI
			Forest	End moraine		

[a]Madison and Gundlach (1993) and Soil regions of Wisconsin (map)
[b]WiscOnline: the Geographical Provinces of Wisconsin (map). http://www.wisconline.com
[c]Derived from vegetation delineations of Finley (1976) and vegetation of Wisconsin in the mid-1800s (map)
[d]Wisconsin Geological & Natural History Survey (1976) and Glacial Deposits of Wisconsin (map)
[e]Syverson and Colgan (2010). The Quaternary of Wisconsin: a review of stratigraphy and glaciation history. Wisc. Geol. and Natural History Survey Educ. Ser. 36

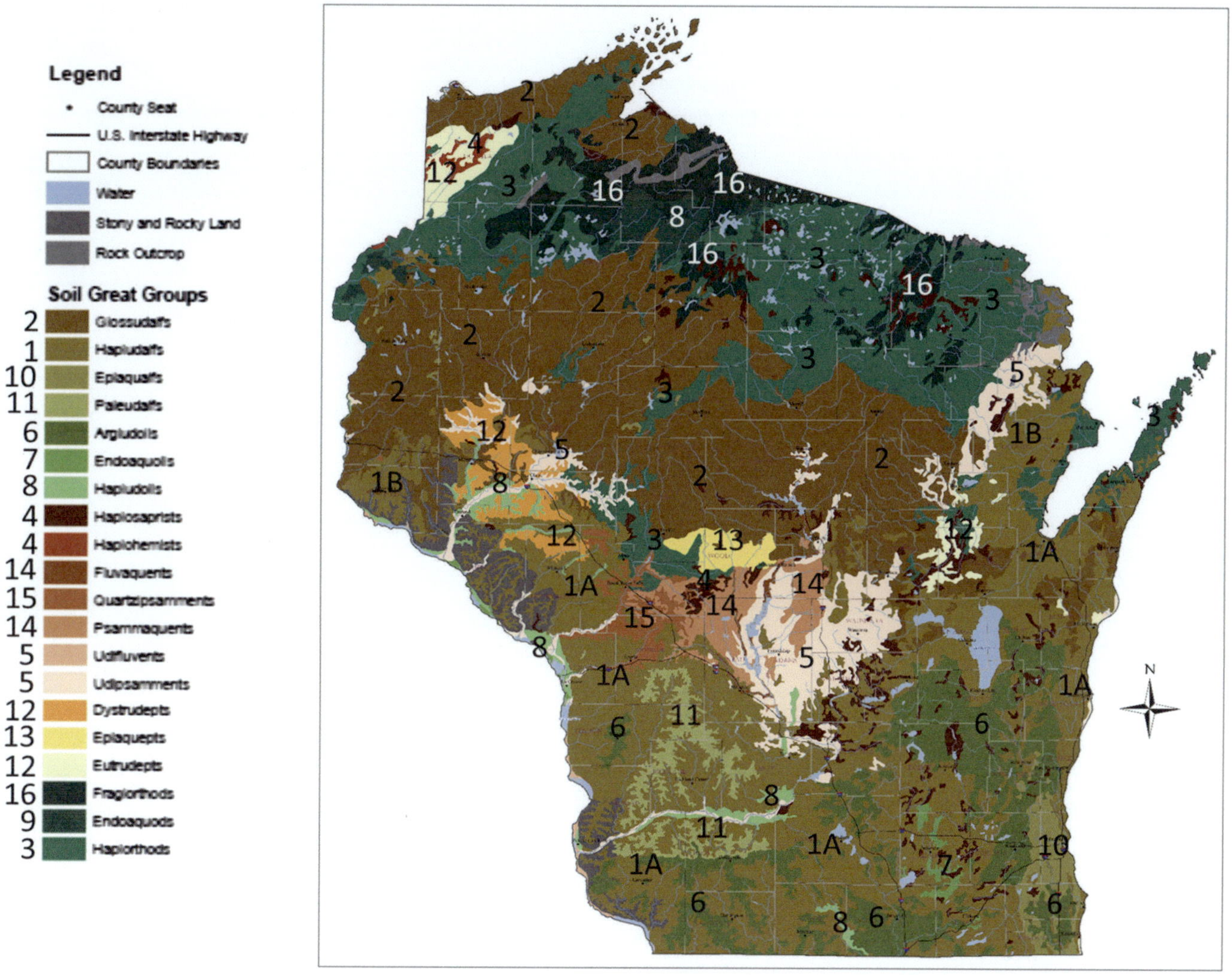

Fig. 7.1 Soil taxonomic regions of Wisconsin. The numbers refer to those given in the text in parentheses in this chapter

Table 7.2 Areas of soil great groups in Wisconsin

Region	Great group	No. of series	Area (km^2)	%
1	Hapludalfs	173	33,905	24.1
2	Glossudalfs	97	22,463	16.0
3	Haplorthods	68	20,133	14.3
4	Haplosaprists	22	12,551	8.9
5	Udipsamments	33	9665	6.9
6	Argiudolls	62	7524	5.3
8	Hapludolls	44	2696	1.9
7	Endoaquolls	30	2778	2.0
11	Paleudalfs	5	1110	0.8
14	Psammaquents	9	1819	1.3
15	Quartzipsamments	10	1620	1.2
9	Endoaquods	19	1772	1.3
5	Udifluvents	14	1704	1.2
9	Epiaquods	13	1360	1.0
12	Dystrudepts	17	1467	1.0
16	Fragiorthods	8	1361	1.0
10	Epiaqualfs	15	1526	1.1
10	Endoaqualfs	14	701	0.5
14	Fluvaquents	9	1087	0.8
12	Eutrudepts	11	1141	0.8
13	Endoaquepts	10	968	0.7
13	Epiaquepts	6	1025	0.7
10	Glossaqualfs	10	1538	1.1
7	Argiaquolls	8	643	0.5
4	Haplohemists	5	489	0.3
	Udorthents	4	96	0.1
7	Epiaquolls	6	632	0.4
13	Humaquepts	7	257	0.2
14	Epiaquents	4	363	0.3
9	Fragiaquods	1	179	0.1
14	Endoaquents	1	79	0.1
4	Sphagnofibrists	1	6	0.0
10	Albaqualfs	1	3	0.0
	Haplustepts	1	3	0.0
	Hapludults	1	3	0.0
	Haplustolls	1	284	0.2
	Other		5880	4.3
	Total	740	140,831	100

contains shaly sandstone beneath the loess. The Worthen series, a Cumulic Hapludolls, occur in lower portions of the landscape. This series has a thick (>60 cm) mollic horizon from upslope deposition and supports native prairie.

The Casco-Fox-Rodman association is common in Washington and Sheboygan counties (soil regions B and I) (Fig. 7.3). These soils are well drained to excessively drained, are calcareous, and have formed from stratified gravel and sand outwash. Whereas the Casco and Fox series are Hapludalfs, the Rodman is a Hapludolls and probably once was under prairie grasses.

The Kewaunee-Manawa-Poygon association is common in Calumet and Manitowoc alumet and Fond du Lac counties (soil regions B and I) (Fig. 7.4). The association occurs on

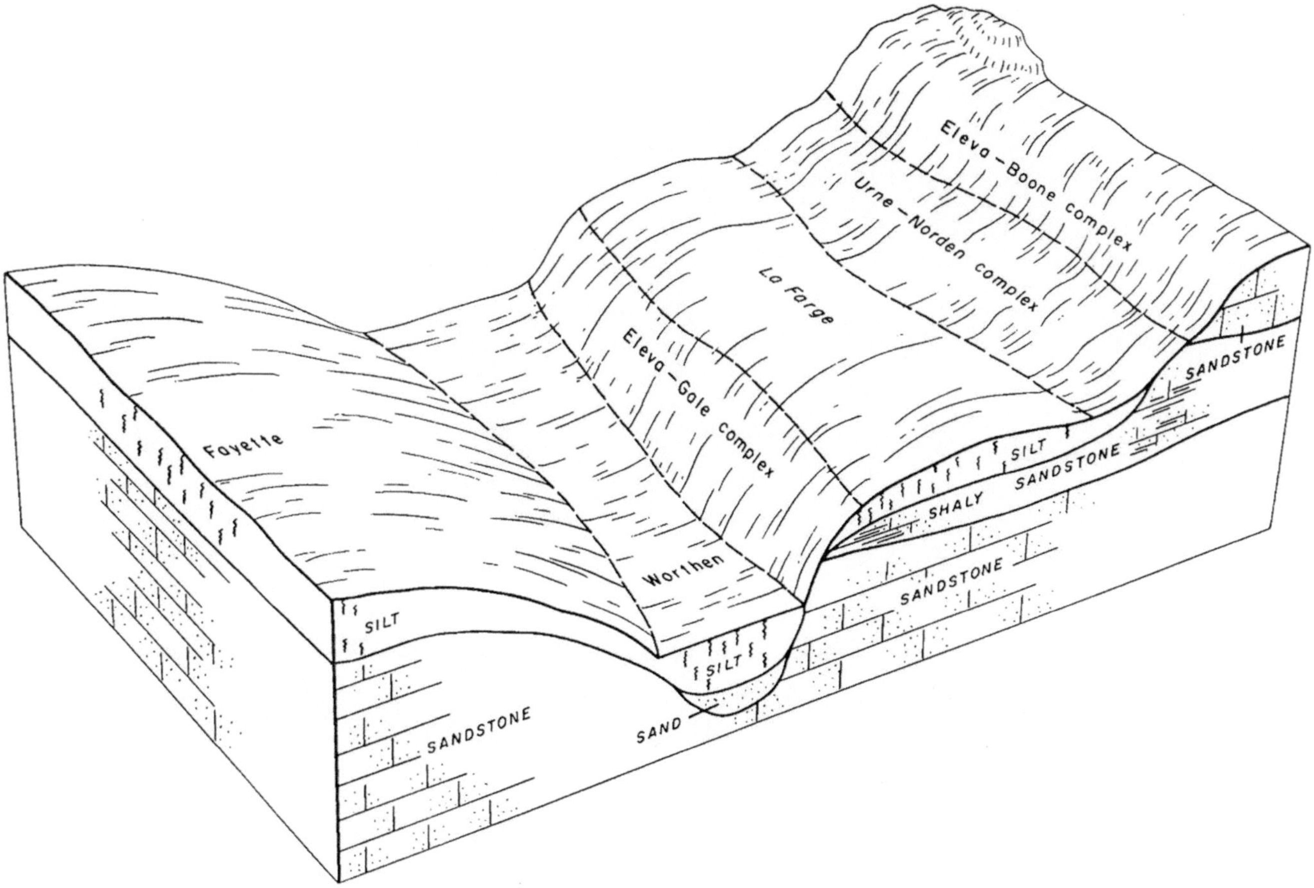

Fig. 7.2 Fayette-La Farge-Eleva association, Trempealeau County, Wisconsin

nearly level to sloping, well drained to poorly drained soils, and have formed from a thin layer of loess over clayey till. Whereas the Kewaunee and Manawa series have a clayey argillic horizon and are Hapludalfs, the Poygan series is a Typic Epiaquolls that was covered swamp deciduous forest.

The Downs-Boaz-Muscatine association is common in Trempeleau County (soil region D) and have formed from loess (>2 m) over sand or sandstone (Fig. 7.5). The Downs series is a well-drained upland soil (Mollic Hapludalfs) that originally supported oak-hickory savanna. The Boaz is a somewhat poorly drained Aeric Endoaquepts that featured wet savanna; and the Muscatine is a somewhat poorly drained Aquic Hapludolls that once supported tallgrass prairie.

About 2% of the Hapludalfs have a frigid soil-temperature regime, including three soil series with areas exceeding 400 km^2, the Chetek, Emmet, and Onaway series, which occur primarily in soil region I. The Onaway-Solona-Angelica association occurs in eastern Shawano County (soil region I) on nearly level to very steep, moderately well drained to poorly drained, loamy, upland soils (Fig. 7.6). The parent materials are loamy till on end and ground moraines. The Onaway series is an Inceptic Hapludalfs, the Solona is an Aquic Argiudolls, and the Angelica is an Aeric Endoaquepts. These soils supported mixed deciduous and conifers before clearing for agriculture and pasture (Fig. 7.7).

7.2.2 Glossudalfs (Region 2)

The Glossudalfs are forest soils in the northern half of Wisconsin that are undergoing a degradation of the upper part of the argillic horizon. The second largest taxonomic soil region (region 2) includes the Glossudalfs, which occur in two east–west bands, one just north of the tension zone in the Northern Highland and one further north in the Lake Superior Lowlands (Fig. 1.4). Glossudalfs support three vegetation types: boreal forest in the Lake Superior Lowlands and northern hardwoods and mixed pines in the Northern Highland. Predominant landforms include lake plains, end moraines, and ground moraine comprised of glaciolacustrine deposits and till with a thin loess cap. The soils occur on Greatlakean till of Early Holocene age. The Glossudalfs occupy 22,463 km^2, or 16.4% of the state's land area (Table 7.2). A Glossudalfs near Marshfield is shown in Fig. 7.8.

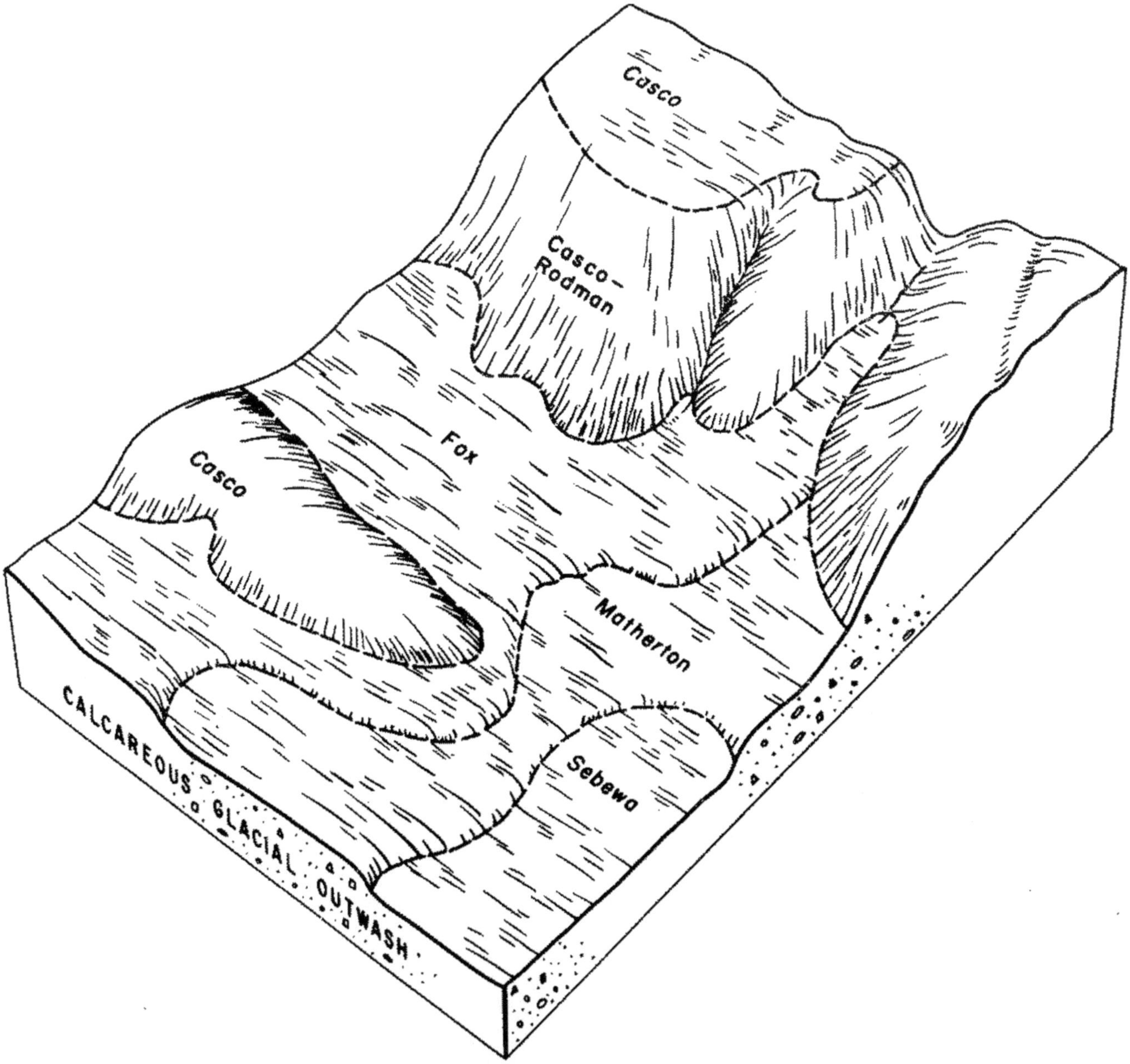

Fig. 7.3 Casco-Fox-Rodman association, Washington County, Wisconsin

Mean thicknesses of the albic, glossic, and argillic horizons in Glossudalfs are 17, 30, and 76 cm, respectively. More than two-thirds (68%) of the Glossudalf soil series are in loamy particle-size classes (coarse-loamy, fine-loamy, loamy, and loamy-skeletal); 99% have a mixed mineral class; 70% have a superactive cation-exchange capacity activity class; 96% have a frigid soil-temperature regime; and 92% have a udic soil-moisture regime.

Dominant soil series in the largest band include the Magnor, Freeon, Rosholt, Antigo, Withee, Loyal, Hortonville, Kennan, Amery, Santiago, and Almena. Of these series, only the Hortonville has a mesic soil-temperature regime. The northern band along Lake Superior is dominated by the Amnicon, Portwing, and Sanborn series, none of which occupies more than 365 km^2.

Two associations representing the general soil regions containing Glossudalfs include the Withee-Marshfield-Santiago (northern Wood County, soil region F) and the Hortonville-Symco-Manawa (Waushara County, soil region I). Soils of the Withee-Marshfield-Santiago association are nearly level to moderately steep, are well drained to poorly drained, and are derived from a thin layer of loess over loamy till (Fig. 7.7). The soils are arrayed along a drainage sequence: Santiago, Withee, and Marshfield. Whereas the

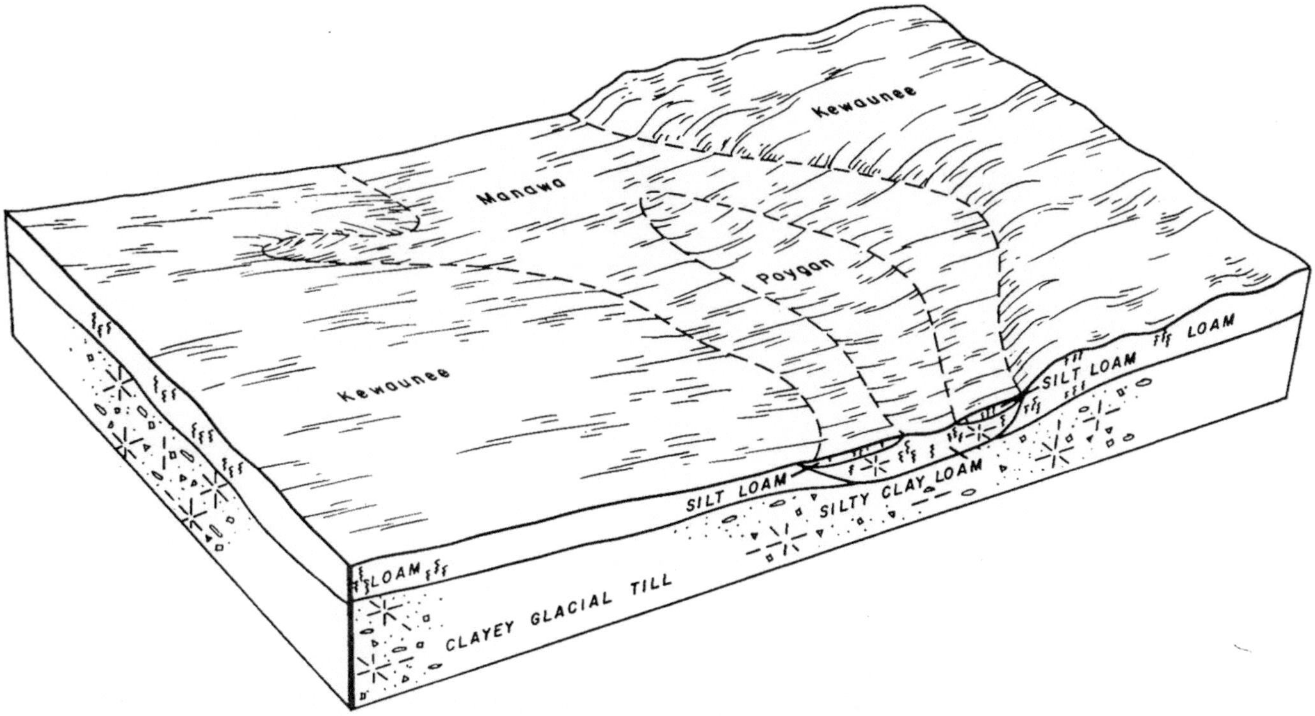

Fig. 7.4 Kewaunee-Manawa-Poygan association, Calumet and Manitowoc counties, Wisconsin

Santiago and Withee series are Glossudalfs with a degraded argillic horizon (E/B and B/E glossic horizon), the Marshfield is a Mollic Epiaqualfs that once was covered by wet deciduous-conifer forest. The cause of the degradation of the argillic horizon and formation of a glossic horizon is not known but may be related to changes in climate and vegetation (Hole 1975; Bockheim 2012).

The Hortonville-Symco-Manawa association is comprised of nearly level to sloping, well drained to somewhat poorly drained, silty, and loamy soils; they are found on moraines and in glacial lake basins (Fig. 7.9). The Hortonville series is a Haplic Glossudalfs; the Manawa and Symco series are Aquollic Hapludalfs. The soils often are calcareous and are under northern hardwoods vegetation.

7.2.3 Haplorthods (Region 3)

The Haplorthods are acid forest soils with iron-enriched B horizons. Haplorthods (soil taxonomic region 3) occur in soil regions D, H, and G, comprise the third largest taxonomic soil area, and occur in the northern one-third of the state. Haplorthods occupy 20,133 km^2, or 14.7% of the state's land area (Table 7.2). They occur on end moraines, ground moraine, and outwash plains comprised of till and outwash of Late Wisconsinan age. The vegetation is comprised of northern hardwoods and mixed broad-leaved and coniferous forests. The ochric, albic, and spodic horizons average 13, 8, and 32 cm in thickness, respectively. More than half (53%) of the Haplorthods have argillic and glossic horizons that average 43 and 40 cm in thickness, respectively. More than half (53%) of the Haplorthods are in the sandy particle-size class; 84% have a mixed mineral class; 46% are not in a CEC activity class (38% are superactive); all 68 of the soil series are frigid; and 81% are in the udic soil-moisture class.

Dominant soil series in soil region D include the Newot and Newood series; soil region G includes the Sarona and Padus series; and soil region H contains the Pence, Keweenaw, Vilas, Sayner, Rubicon, Croswell, and Karlin series.

The Rubicon-Sayner-Karlin association from Vilas County is representative of this soil taxonomic region (Fig. 7.10). These soils are nearly level to very steep, excessively drained and somewhat excessively drained, and derived from sandy outwash materials on uplands. The soils can be arrayed along a topographic sequence: Rubicon, Sayner-Karlin, Croswell, and Loxley-Dawson. Whereas the Rubicon, Sayner (Fig. 7.11), and Karlin series are Entic Haplorthods, the Croswell is an Oxyaquic Haplorthods, and the Loxley and Dawson are Haplosaprists. The mineral soils support pines, especially red pine and white pine, and the organic soils feature scattered swamp conifers such as black spruce and tamarack with an understory of blueberry, leatherleaf, sphagnum moss, and wintergreen (Fig. 7.12).

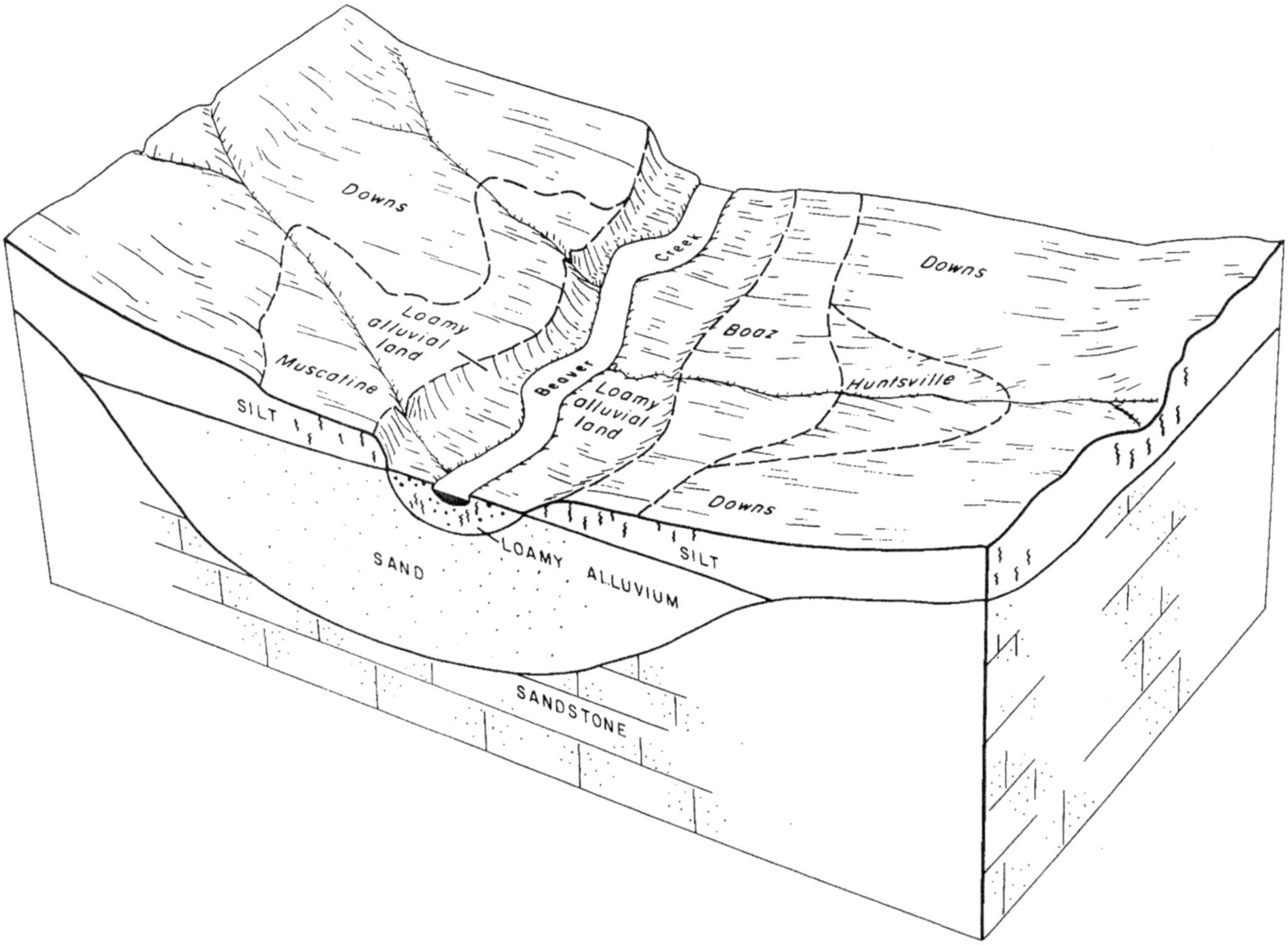

Fig. 7.5 Downs-Boaz-Muscatine association, Trempealeau County, Wisconsin

7.2.4 Haplosaprists-Haplohemists (Region 4)

Haplosaprists and Haplohemists are organic soils that have undergone substantial decomposition (ripening) and occur in wetlands. Haplosaprists (taxonomic soil region 4) and Haplohemists occupy 12,551 km^2 (9.2% of state's land area) and 489 km^2 (0.4%), respectively. They occur primarily in depressions in the glaciated region (soil region J). Haplosaprists support wet mixed forest and marsh sedges and grasses in areas underlain by till and outwash. The histic horizon in Wisconsin Histosols ranges from 20 to 203 cm and averages 95 cm in thickness.

More than three-quarters (77%) of the Histosols are euic, meaning that they have a pH >4.5 (in 0.01 M $CaCl_2$) in the control section. More than two-thirds (68%) of the Haplosaprists are in the frigid soil-temperature class.

The dominant series include the Cathro, Seelyeville, Lupton, Houghton, Markey, Loxley, Dawson, Palms, Adrian, and Beseman. All are frigid except the Houghton, Palms, and Adrian soil series. Haplohemists and Sphagnofibrists occupy 629 and 6 km^2 in the state, respectively. The Rifle and Greenwood peats are Haplohemists.

Two soil associations representative of the Haplosaprists are the Houghton-Palms-Adrian (mesic soil-temperature regime; Washington County) and the Haugen-Greenwood-Amery (frigid soil-temperature regime; Barron County). The three dominant series in the Houghton-Palms-Adrian association are all Saprists with organic materials derived from herbaceous materials (Fig. 7.13). The organic materials have accumulated in closed depressions on moraines. Whereas the organic materials are 40–130 cm thick in the Palms and Adrian series, it is more than 130 cm in the Houghton series.

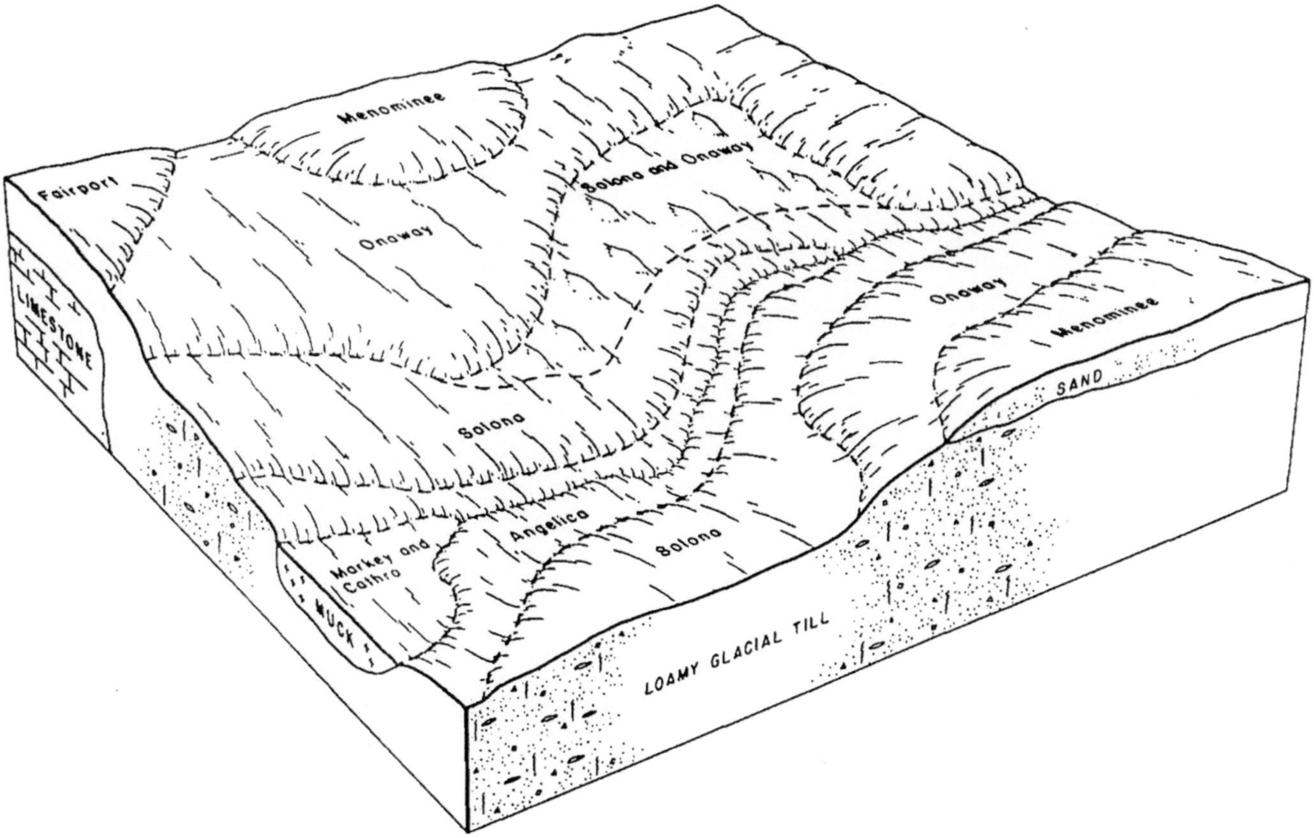

Fig. 7.6 Onaway-Solona-Angelica association, eastern Shawano County, Wisconsin

Only the Greenwood series is a Haplosaprist in the Haugen-Greenwood-Amery association (Fig. 7.14). This soil is derived from organic materials that are more than 150 cm thick on outwash plains, till-floored lake plains, and lake plains.

7.2.5 Udipsamments and Udifluvents (Region 5)

The Udsipamments and Udifluvents are poorly developed soils on sandy outwash or in floodplains. Udipsamments and Udifluvents (soil taxonomic region 5) constitute some of the least developed soils in Wisconsin. Udipsamments primarily are found in central Wisconsin in the basin of former Glacial Lake Wisconsin (soil region C) and in the outwash area of northeastern Wisconsin (soil region H). These soils comprise an area of 9665 km^2 (7.1% of state's land area). Udifluvents comprise 1704 km^2 (1.2%) and occur primarily in soil regions A and B. Udipsamments support scrub oak (*Quercus*)-jack pine (*Pinus banksiana*), as well as large areas under irrigated agriculture (Fig. 7.15). Udifluvents support mixed hardwoods or mixed broad-leaved and coniferous forest. The Entisols of Wisconsin often have only an ochric epipedon, which averages 20 cm in thickness.

Udipsamments are all in the sandy particle-size class, have a mixed mineralogy (except Graycalm and Grayling series, which are isotic), have a very low CEC, are either frigid (61%) or mesic (39%), and 85% have an udic soil-moisture regime (15% are aquic). The Udifluvents are in a variety of particle-size classes (57% are coarse-silty or fine-silty) and are in the mixed mineral class (except Absco series, which is siliceous); 79% are superactive; 71% are nonacid; 71% are mesic, and 29% are frigid; and all are udic (except the Absco series, which is aquic).

Dominant Udipsamments include the Plainfield, Menahga, Meehan, Grayling, Friendship, and Mahtomedi soil series; dominant Udifluvents are the Arenzville, Orion, and Chaseburg series.

Representative toposequences include the Plainfield-Brems association from Adams County, and the Menahga-Croswell-Mahtomedi association from Shawano County. The Plainfield-Brems association is nearly level to steep, excessively drained, and moderately well drained with sandy

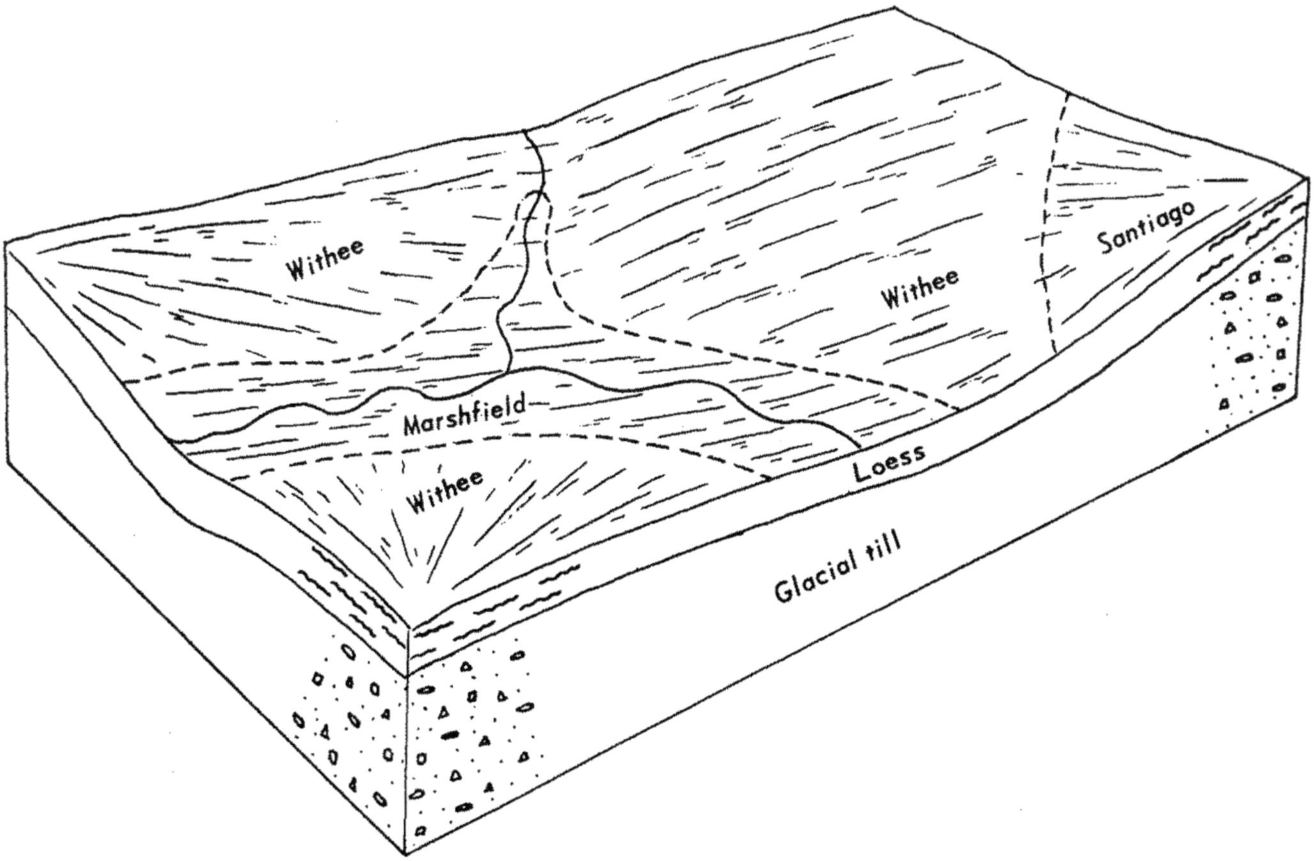

Fig. 7.7 Withee-Marshfield-Santiago association from Wood County

subsoil derived from sandy outwash deposits (Fig. 7.16). The soils occur along the slope from top to bottom: Plainfield, Brems, and Meehan. Whereas the Plainfield series is a Typic Udipsamments, the Brems and Meehan have a seasonally high water table at 60 cm and are classified as Aquic Udipsamments. These soils support jack pine and scrub oak. The Menahga-Croswell-Mahtomedi association is comprised of nearly level to steep soils that are excessively drained or moderately well drained, sandy, and occur on uplands (Fig. 7.17). The soils are derived from proglacial lake sediments and outwash. Whereas the Menahga and Mahtomedi series are Typic Udipsamments, the Croswell is an Oxyaquic Udipsamments. These soils supported jack pine and scrub oak.

The Arenzville-Alluvial association from Pierce County is representative of the Udifluvents associations. The Arenzville series is derived from post-settlement silty alluvium overlying buried soils with dark-colored A horizons (Fig. 7.18).

7.2.6 Argiudolls (Region 6)

The Argiudolls are the clay-enriched soils of the tallgrass prairie. Argiudolls (taxonomic soil region 6) represent former prairies in the Driftless Area (soil region A) and on drift in southeastern Wisconsin (soil region B). These soils cover 7524 km^2 (5.5% of land area). The sharp demarcation between Paleudalfs and Argiudolls on the great group map (Fig. 1.4) occurs along Military Ridge (Highways 151, 18), which represents the boundary between Prairie du Chien dolomite and sandstone to the north and the Sinnipee dolomite and limestone to the south (Fig. 7.19). Argiudolls support tallgrass prairie, which rarely occurs on the Cambrian sandstones of soil regions A and Dr. The Argiudolls contain a mollic (34 cm) and an argillic horizon (62 cm).

Nearly three-quarters (72%) of the Argiudolls are in the fine-loamy or fine-silty particle-size class; 92% have a mixed mineralogy; 73% are in the superactive CEC activity class; and 92% have a mesic soil-temperature regime.

Fig. 7.8 Glossudalfs near Marshfield Agricultural Research Station under forest. Marshfield soils series (Mollic Epiaqualfs)

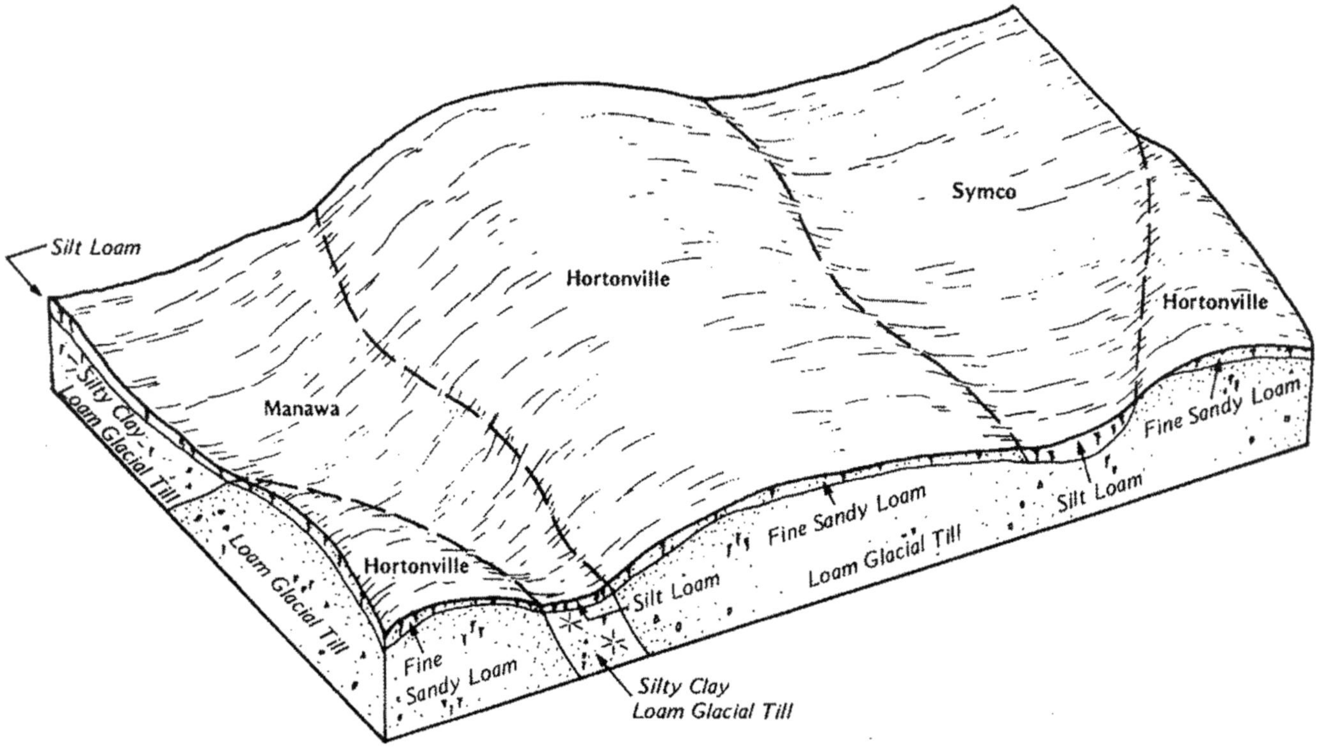

Fig. 7.9 Hortonville-Symco-Manawa association from Waushara County

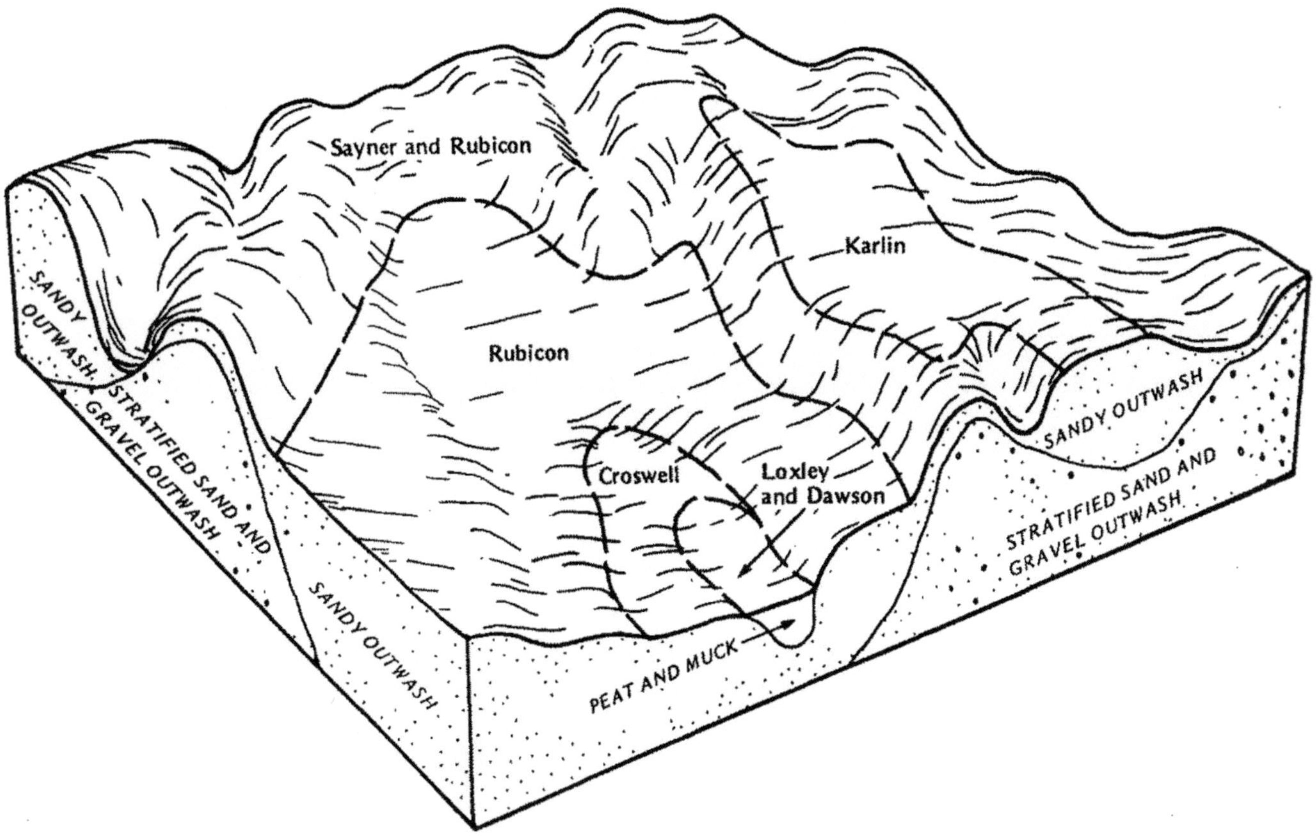

Fig. 7.10 Rubicon-Sayner-Karlin association from Vilas County

Fig. 7.11 Spodosol under hemlock in northern Wisconsin. Sayner soil series (Entic Haplorthods)

Fig. 7.12 Histosols surrounded by small lakes and Spodosols in northern Wisconsin

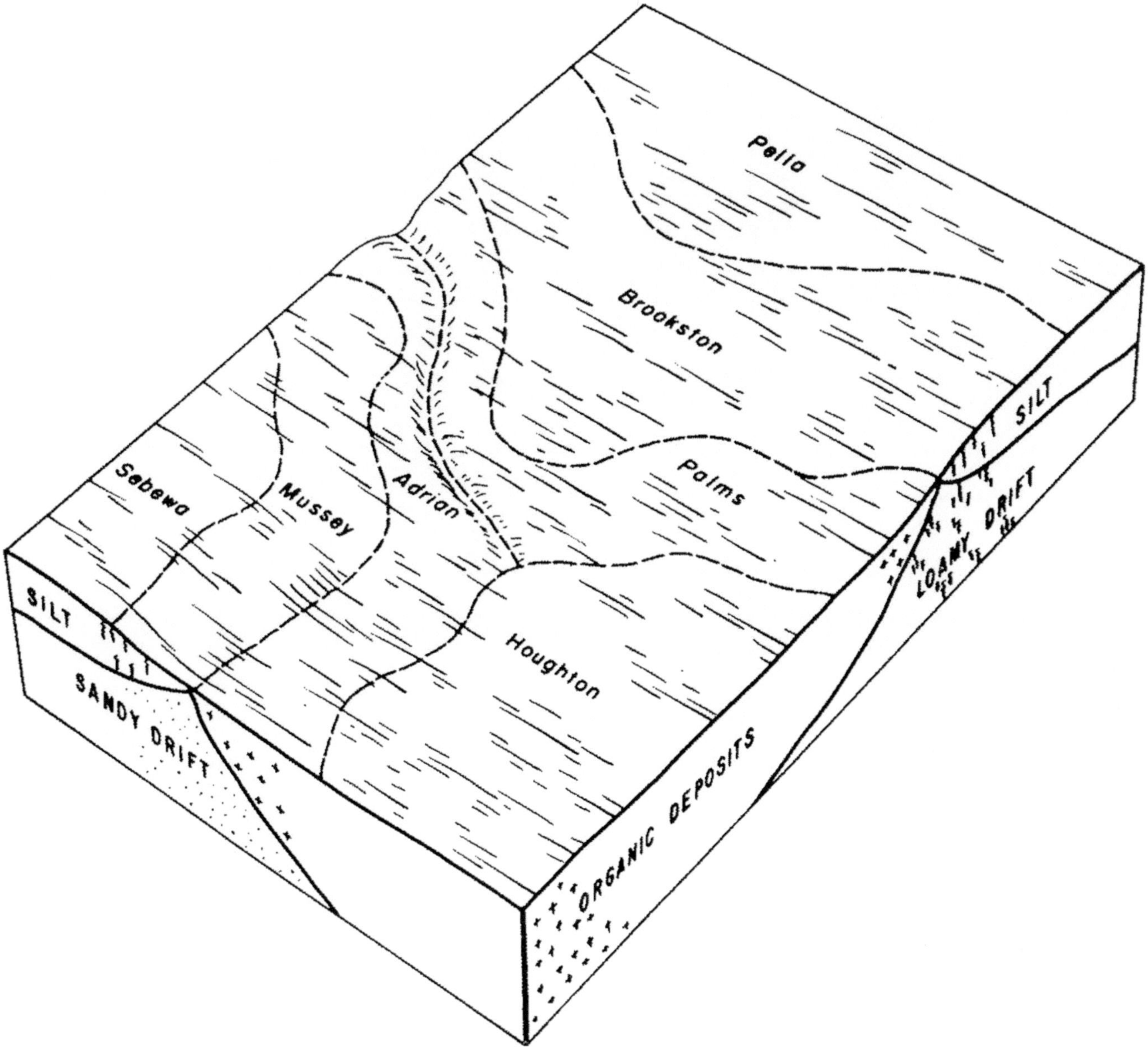

Fig. 7.13 Houghton-Palms-Adrian association from Washington County

Common Argiudolls include the Plano, Tama, and Dodgeville series in soil region A and the Hochheim, Markesan, and Rotamer series in soil region B. The Solona series, which occurs in proximity to Lakes Superior and Michigan, is an Argiudoll with a frigid soil-temperature regime.

Two catenas representative of this soil taxonomic region include the Plano-Warsaw-Dresden association from Rock County (soil region Am) and the Hochheim-Sisson-Casco association from Ozaukee County (soil region Bm). Soils of the Plano-Warsaw-Dresden association are deep and moderately deep, well drained and moderately well drained, nearly level to sloping, and derived from loess or loamy materials overlying sand and gravel outwash (Fig. 7.20). The Plano series is derived entirely from loess and the Warsaw is derived from fine-loamy materials over sandy outwash; both soils are classified as a Typic Argiudolls and had tallgrass prairie

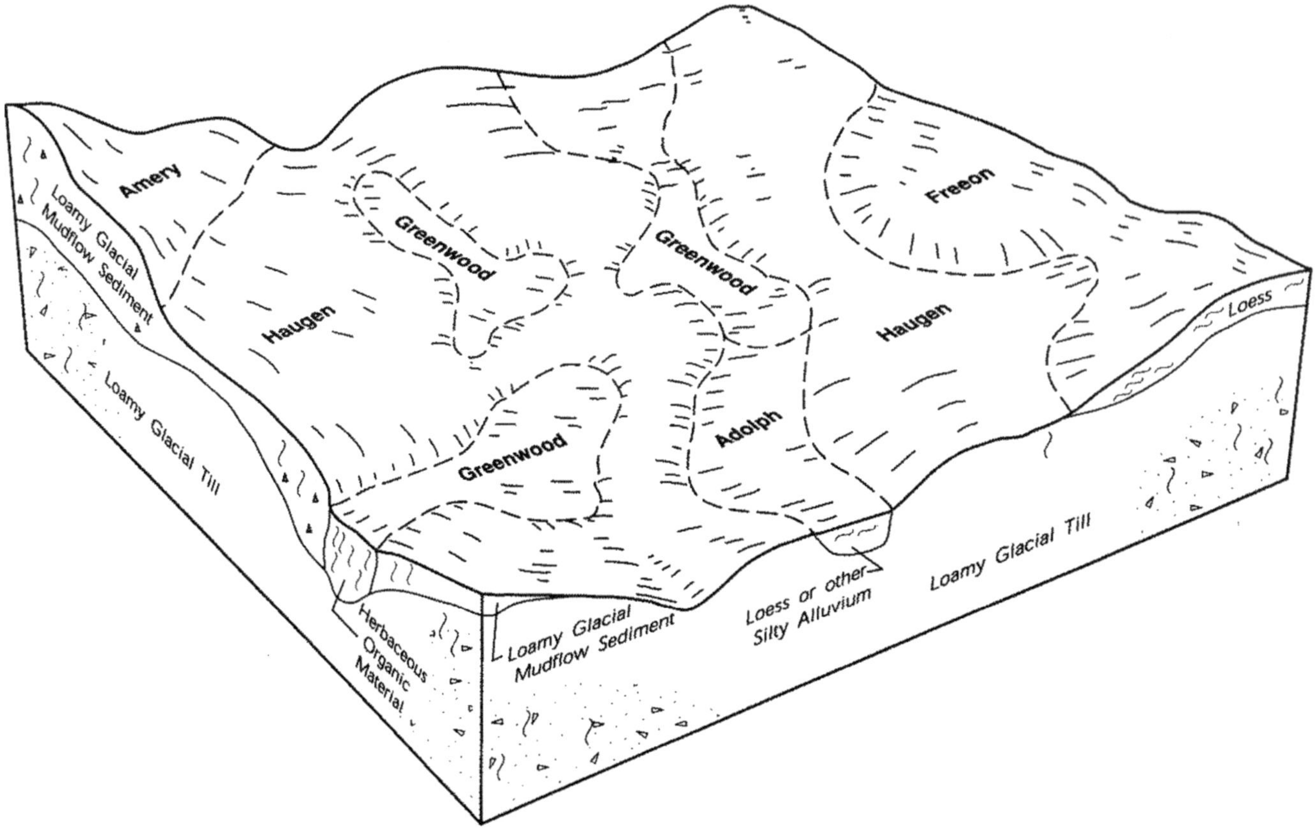

Fig. 7.14 The Haugen-Greenwood-Amery association from Barron County

vegetation prior to settlement; the Dresden is a Mollic Hapludalfs and once was under savanna vegetation.

The Hochheim-Sisson-Casco association consists of well-drained soils derived from calcareous loamy till, outwash, and glaciolacustrine deposits (Fig. 7.21). The Hochheim series occupies the upper landscape position, is derived from thin loess over till, and is classified as a Typic Argiudolls; the Sisson and Casco series are Hapludalfs derived from stratified loamy and silty materials and calcareous outwash, respectively. Whereas the Hochheim and Sisson soils originally support maple and basswood, the Casco soil had mixed oaks.

7.2.7 Aquolls (Region 7)

The Aquolls are the poorly drained soils with a mollic horizon. Aquolls (soil taxonomic region 7), which include the Argi-, Endo-, and Epiaquolls, cover 4,053 km^2 (3.0% of land area), primarily in the Eastern Ridges and Lowlands. This suborder supports wet prairie and marsh sedges and grasses on end moraines, ground moraine, lake plains, and outwash plains. The mollic and cambic horizons in Aquolls of Wisconsin average 43 and 53 cm, respectively. Eight of the Aquoll soil series have an argillic horizon (Argiaquolls) that is on average 53 cm thick.

The Aquolls represent a variety of particle-size classes, but 74% of the series occur in fine classes (clay, fine, fine-loamy, fine-silty, and very-fine); all 44 series are in the mixed mineral class; 75% are superactive; 75% have a mesic soil-temperature regime; and all have an aquic soil-moisture regime. Only the Pella (Endoaquolls) and Poygan (Epiaquolls) have areas in excess of 400 km^2.

The Waymor-Lamartine-Pella (Endoaquolls) association occurs in Kewaunee County a (Fig. 7.22). The soils are derived from loess over loamy till. The Pella series is a very deep, poorly drained soil formed from 80 cm of loess over stratified sandy loam till. The Kewaunee-Manawa-Poygon (Epiaquolls) association of Calumet and Manitowoc and Fond du Lac counties was described earlier (Fig. 7.4).

Fig. 7.15 Sandy soils (Typic Udipsamments) used for potato cultivation near Coloma

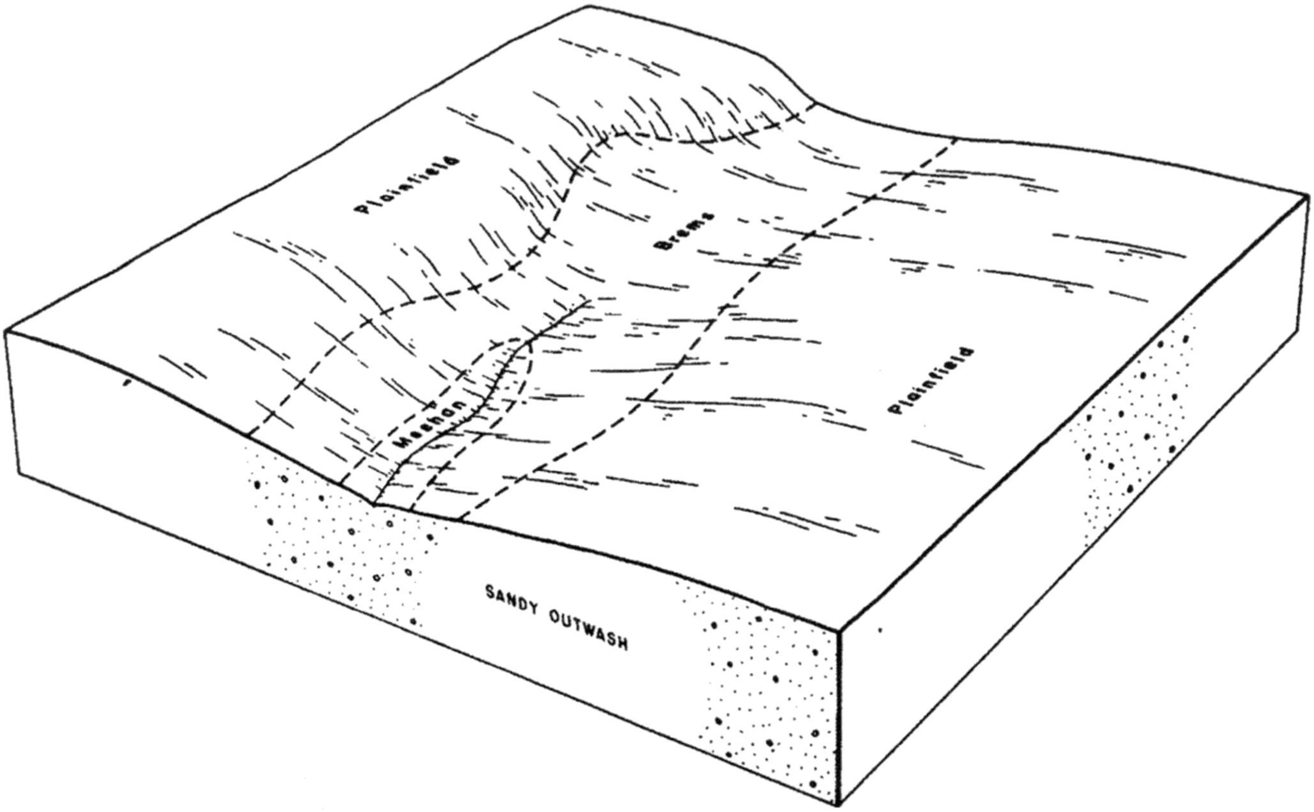

Fig. 7.16 The Plainfield-Brems association from Adams County

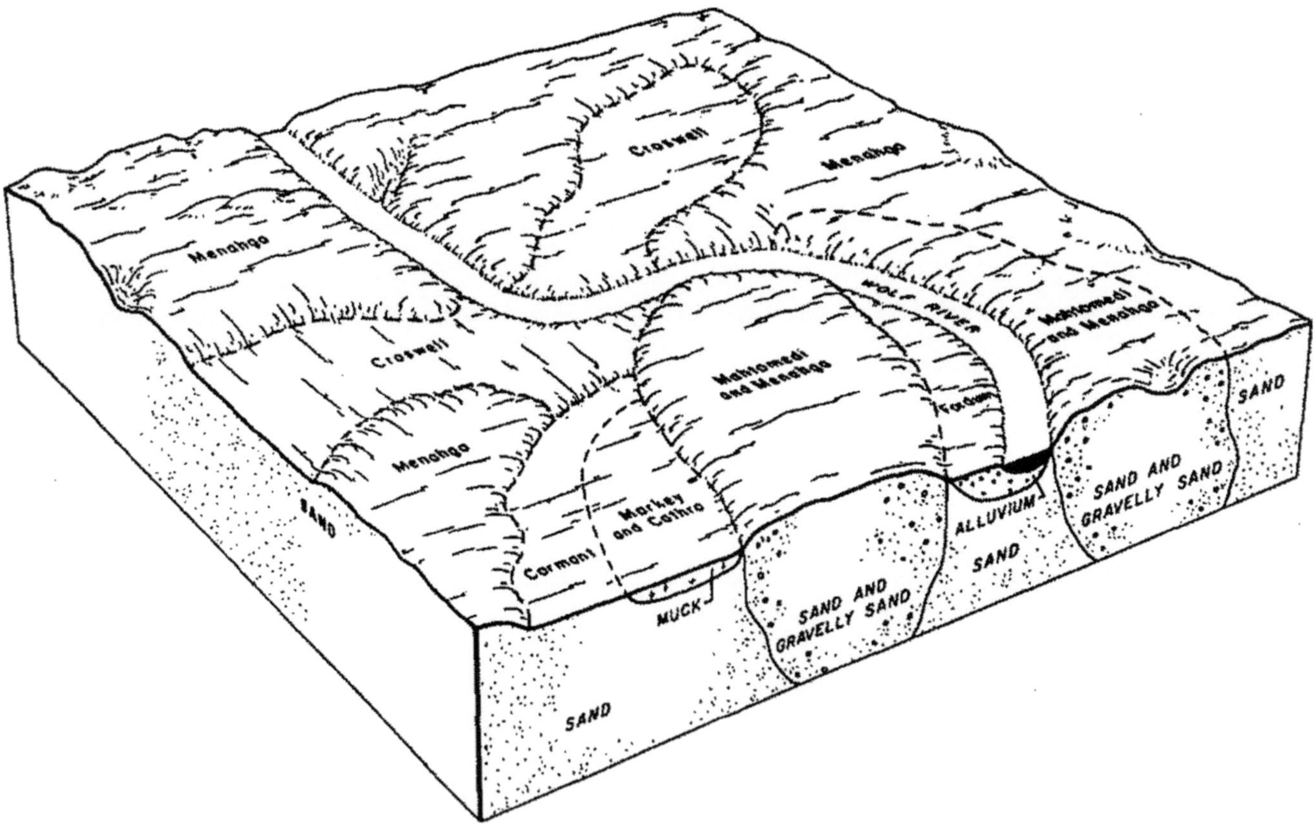

Fig. 7.17 The Menahga-Croswell-Mahtomedi association from Shawano County

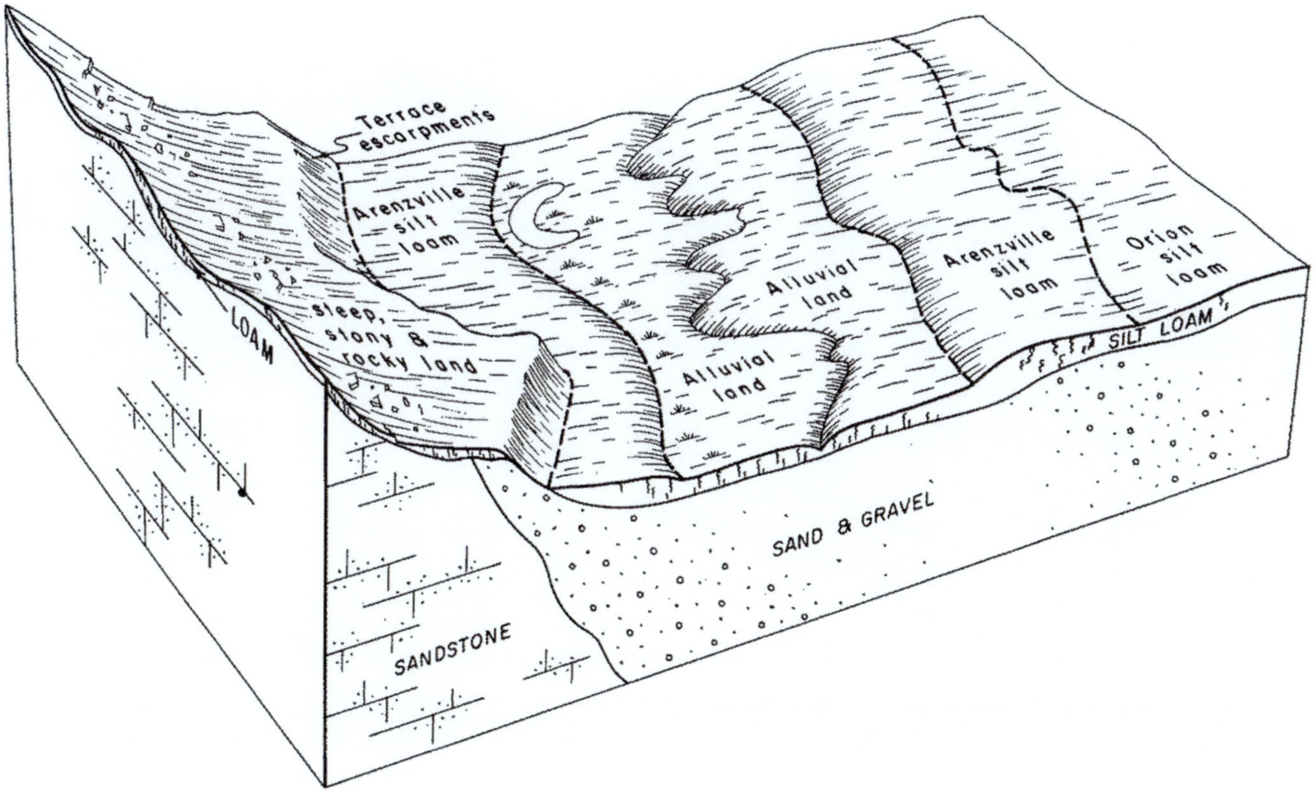

Fig. 7.18 Arenzville-Alluvial Land association from Pierce County

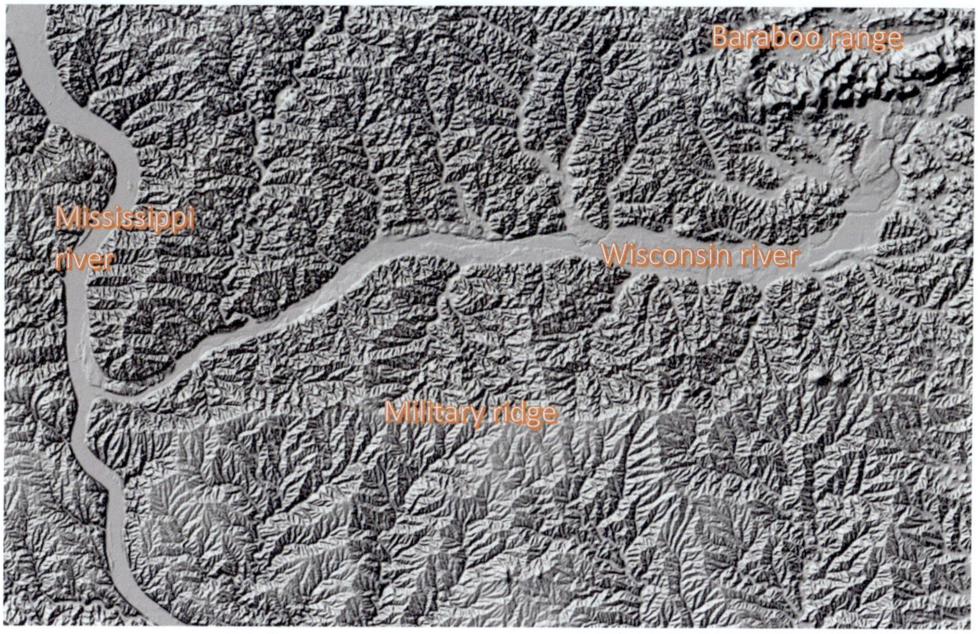

Fig. 7.19 Hillshade map of the Driftless Area showing the Wisconsin and Mississippi rivers, the Baraboo range (1.5 billion years old pink quartize and rhyolite) and the Military ridge which is the boundary between Prairie du Chien dolomite and sandstone to the north, and the Sinnipee dolomite and limestone to the south

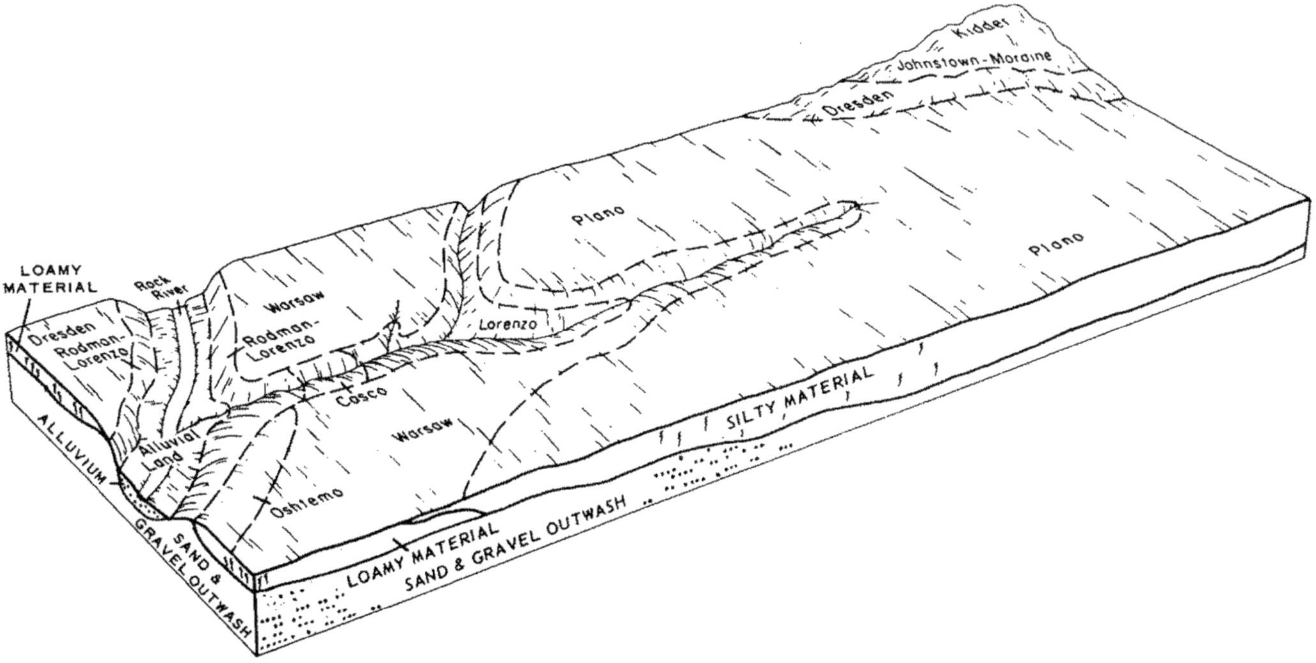

Fig. 7.20 Plano-Warsaw-Dresden association from Rock County

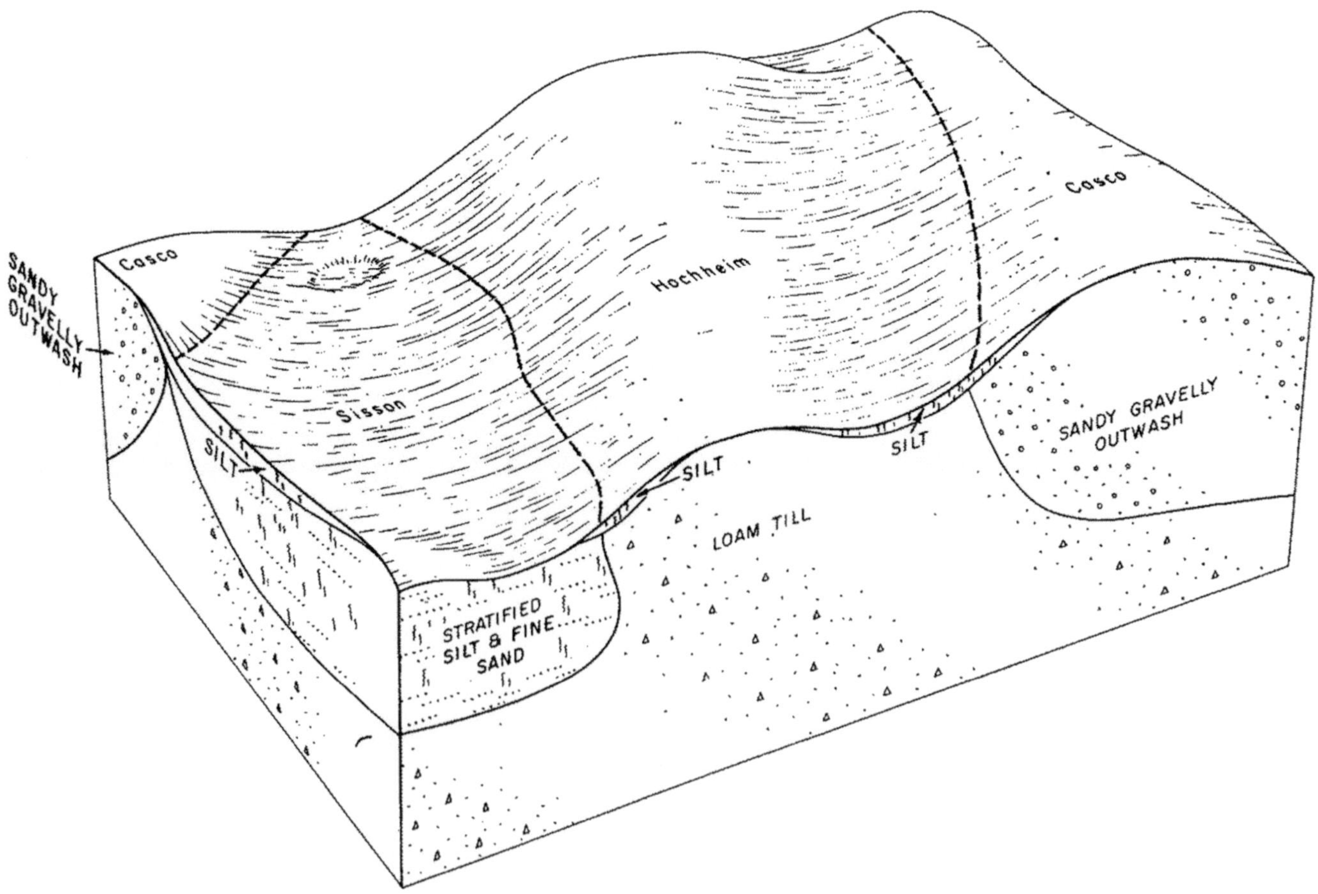

Fig. 7.21 Hochheim-Sisson-Casco association from Ozaukee County

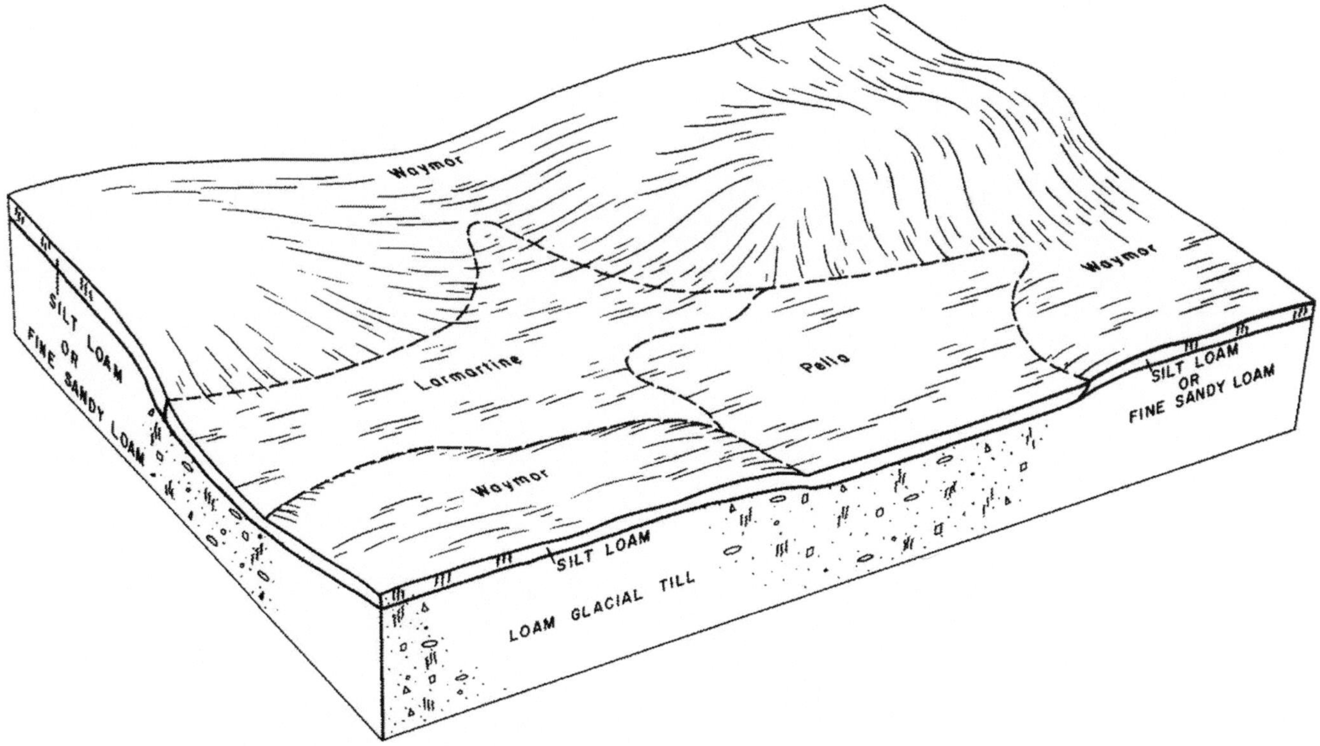

Fig. 7.22 Waymor-Lamartine-Pella association from Kewaunee County

7.2.8 Hapludolls (Region 8)

The Hapludolls are prairie and savanna soils that lack an underlying clay-rich horizon. The Hapludolls (soil taxonomic region 8) occupy 2696 km^2 or 2.0% of the land area in Wisconsin, primarily in soil regions Am, Bm, Cm, and Fr. Hapludolls occur on stream terraces and outwash plains comprised of alluvium and outwash and often form under prairie and oak-savanna vegetation. Hapludolls contain a mollic (average = 42 cm thick) and a cambic horizon (55 cm), respectively.

Hapludolls occur in a variety of particle-size classes ranging from fine-silty to sandy-skeletal and in a mixed mineral class (except the Brodale series which is carbonatic); more than half (54%) are in the superactive CEC activity class; 74% of the series have a mesic soil-temperature regime; and all except the Yahara series (aquic) have a udic soil-moisture regime.

Despite that there are 42 Hapludolls soil series, none of them have an area in excess of 400 km^2. The most extensive Hapludolls include the Sparta and Finchford series in soil region Dr and the Huntsville and Dickinson series in soil region Am.

The Sparta-Dakota association from Grant County (Fig. 7.23) is representative of this catena. The association contains deep, sandy, or loamy soils on terraces, underlain by sandy outwash and originally supporting tallgrass prairie-oak-hickory savanna. These soils occur on stream benches, or terraces, above the floodplains of major streams. The Sparta series (Entic Hapludolls) forms on the lower terraces and has E and Bt lamellae at depth (>175 cm). The Dakota series is a Typic Argiudolls on the more stable land surfaces.

7.2.9 Aquods (Region 9)

The Aquods are the poorly drained, podzolized soils that occur in wetlands. Aquods (soil taxonomic region 9), which include the Endo-, Epi-, and Fragiaquods, occupy 3611 km^2 or 2.4% of Wisconsin's land area in soil regions G, H, I, and J. They occur under mixed forest in depressions on stream

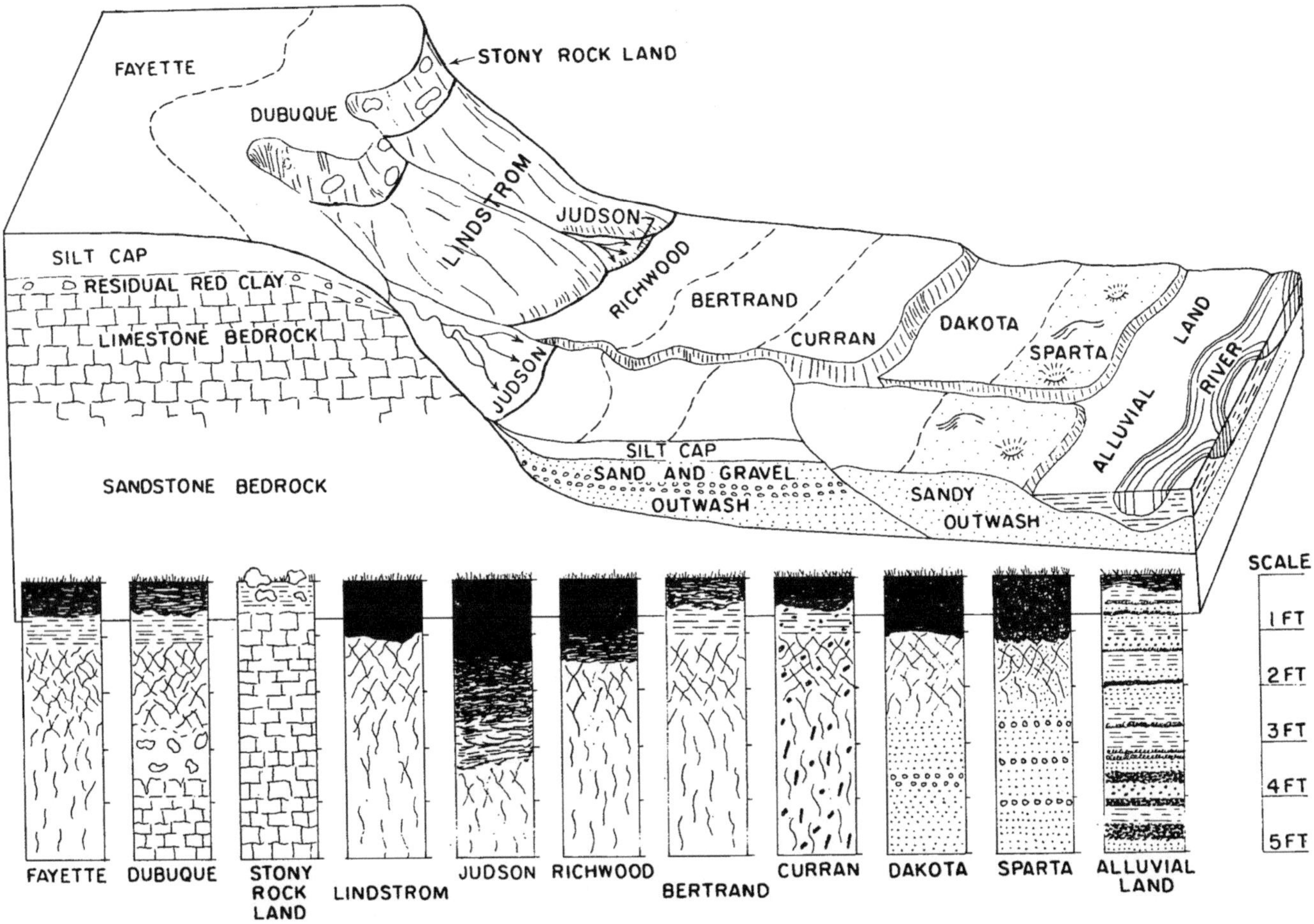

Fig. 7.23 Dakota-Sparta association on lower terraces in Grant County

terraces, outwash plains, and lake plains in the Northern Highland physiographic province. The ochric, albic, and spodic horizons average 16, 10, and 27 cm in thickness, respectively. Twenty-two of the 33 Aquod soil series are bisequal and have argillic and glossic horizons that average 44 and 35 cm in thickness.

Wisconsin Aquods are either coarse-loamy (61%) or sandy (36%); 94% of the series have a mixed mineralogy; 36% are superactive; all of them are frigid and aquic. Although none of Aquod soil series have an area in excess of 400 km^2, the dominant soil series are the Au Gres, Wainola, Worcester, and Iosco.

The Aquods generally occur the lower portion of the landscape. Representative associations include the Menahga-Croswell-Au Gres (Endoaquods) in Marinette County and the Fairchild (Epiaquods)-Elm Lake-Ludington from Clark County. The Menahga-Croswell-Au Gres catena is derived from pitted sandy outwash with the Menahga occupying the excessively drained upland position, the Croswell the moderately well-drained mid-slope position, and the Au Gres series and associated soils in the depressions. The Menahga soil supports jack pine, the Croswell has mixed deciduous and coniferous trees, and the Au Gres is covered by mixed swamp forest.

The Fairchild-Elm Lake-Ludington association (Fig. 7.24) is composed of moderately deep, nearly level to moderately steep, poorly drained to moderately well drained, and sandy and mucky soils on pediments. Although the Fairchild series, a somewhat poorly drained Ultic Epiaquods, occupies the upper landscape positions, the parent materials are strongly weathered, siliceous alluvium and residuum over shale. The Ludington also occupies the upper slopes

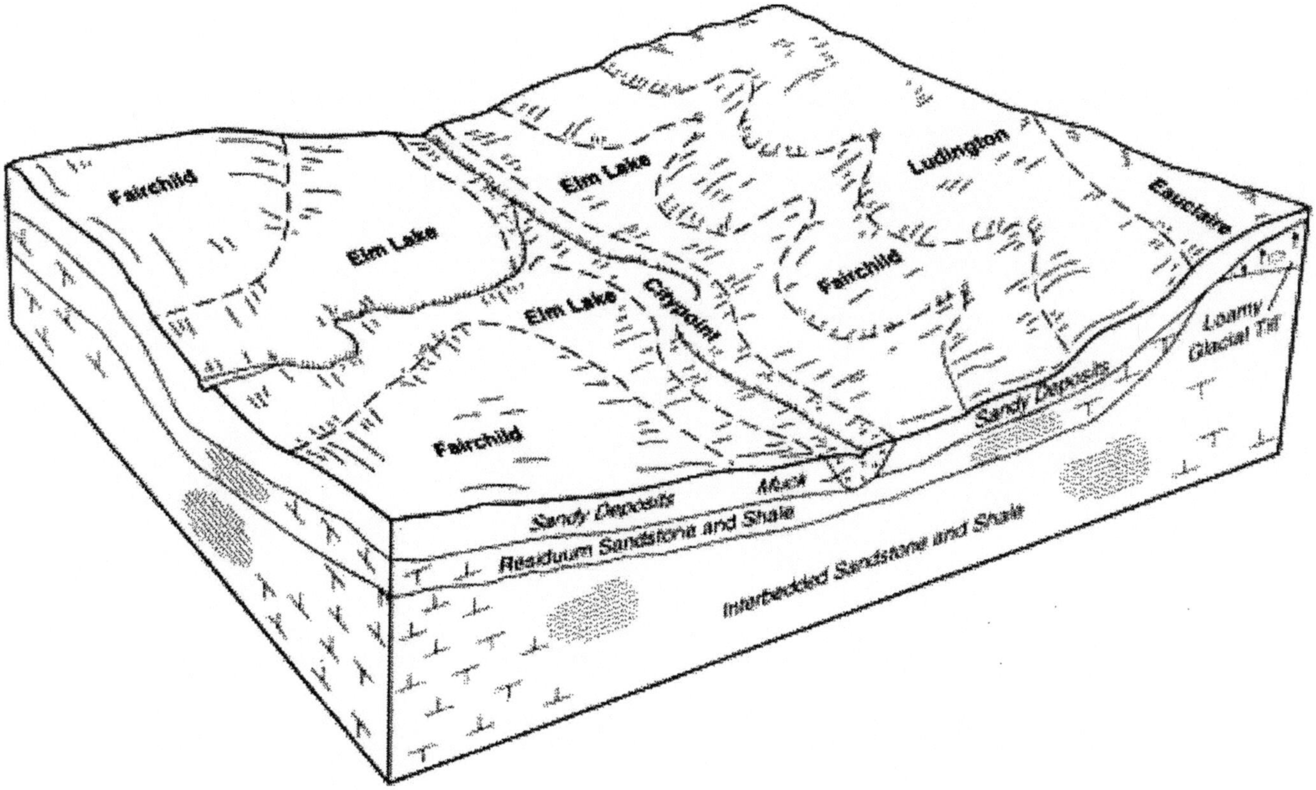

Fig. 7.24 Fairchild-Elm Lake-Ludington association from Clark County

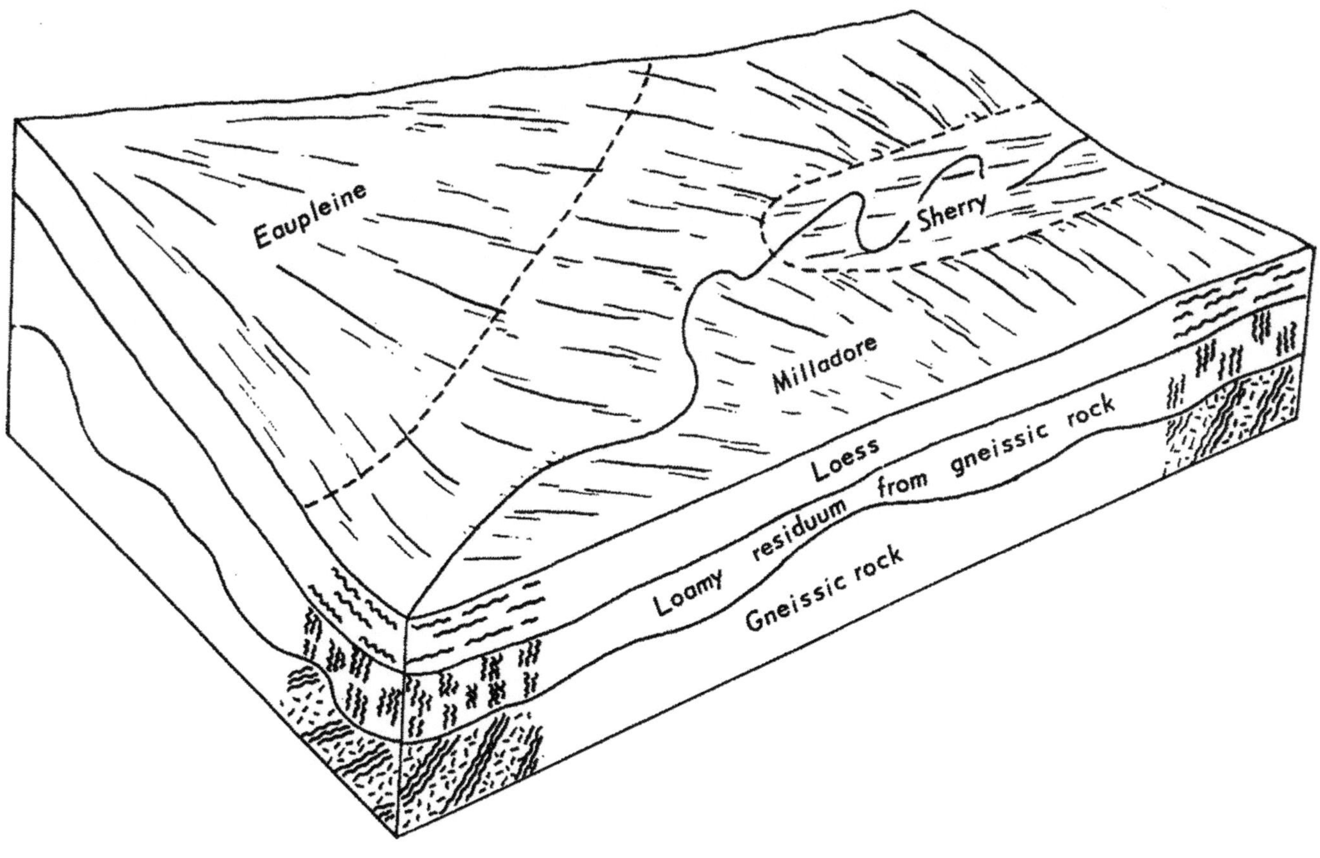

Fig. 7.25 The Milladore-Eaupleine-Sherry association in Wood County

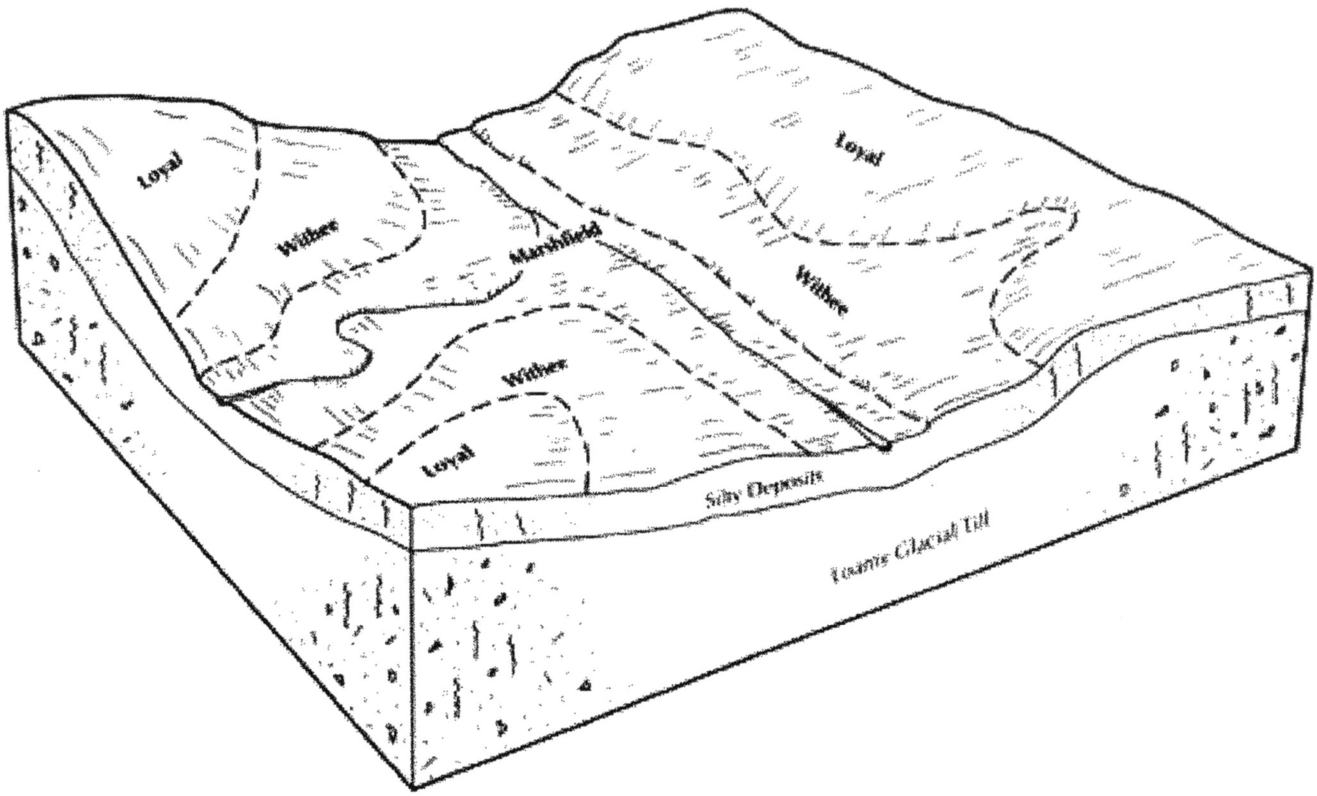

Fig. 7.26 The Loyal-Withee-Marshfield association in Clark County

and is a moderately well-drained Ultic Haplorthods. The lower slope positions contain the Elm Lake series, a poorly drained Humaqueptic Epiaquents.

7.2.10 Aqualfs (Region 10)

The Aqualfs are the poorly drained soils with a clay-enriched horizon. Aqualfs (taxonomic soil region 10), which include The Epi-, Endo-, Gloss-, and Argiaqualfs, occupy 3768 km^2 or 2.6% of the state's land area in soil regions B, F, and J. They occur under mixed hardwoods and mixed oaks in somewhat poorly drained to very poorly drained areas on end moraines. The ochric and argillic horizons average 29 and 64 cm in thickness, respectively. Albic and glossic horizons are common and average 18 and 28 cm in thickness.

Aqualfs tend to have fine textures with 82% of the series occurring in clayey, fine, fine-loamy, fine-silty, or very-fine particle-size classes; 83% have a mixed mineralogy; 55% are superactive; 68% are frigid.

In soil region F, the Capitola, Cebana, and Marshfield soil series each occupy an area in excess of 400 km^2, and all of them have a frigid soil-temperature regime. In soil region B, the Beecher and Blount series are dominant but do not occupy large areas and have a mesic soil-temperature regime.

Representative soil associations include the Milladore-Eaupleine-Sherry in Wood County and the Loyal-Withee-Marshfield in Clark and Marathon counties. Soils in the Milladore-Eaupleine-Sherry (Endoaqualfs) association are nearly level to sloping, and poorly drained to well drained that have a loam to silty clay loam argillic horizon (Fig. 7.25). These soils have formed in a thin layer of loess and loamy residuum from gneissic rocks on uplands. The Milladore and Eaupleine series are Glossudalfs; the Sherry contains a 65-cm-thick argillic horizon (Btg) partly in the loess and partly in the underlying loamy till.

Fig. 7.27 Poorly drained Alfisol developed in loess over dense till. Marshfield soil series (Mollic Epiaqualfs)

Fig. 7.28 Dolomitic limestone over sandstone in the Driftless Area of Wisconsin, soils are vegetated with oak

Soils of the Loyal-Withee-Marshfield (Epiaqualfs) association are very deep, nearly level to sloping, poorly drained to moderately well drained, silty soils on ground moraines (Fig. 7.26). The Loyal series, an Oxyaquic Glossudalfs, occurs on the summits, shoulders, and back slopes; the Withee, an Aquic Glossudalfs occurs on the footslopes and toeslopes; and the Marshfield, a Mollic Epiaqualfs, occurs in depressions and drainageways. The Marshfield series supports water-tolerant hardwoods and some conifers but is also used for cropping (Fig. 7.27).

7.2.11 Paleudalfs (Region 11)

The Paleudalfs are forest soils with continuous clay enrichment in the upper 150 cm. Paleudalfs (taxonomic soil region 11) occupy a triangular area in the Western Uplands (soil regions A and Dr) (Fig. 5.1). They occupy 1010 km^2 or 0.8% of the Wisconsin land area. Paleudalfs occupy the weathered uplands of Prairie du Chien dolomitic limestone (Fig. 7.28). The Haugen series is unique in that it formed from Pre-Illinoian till and mudflow deposits. The original vegetation (mid-1800s) was mixed oak and oak-savanna. Except for the Haugen series, the parent materials are loess of varying thickness over clayey residuum (pediment materials). Paleudalfs in Wisconsin contain ochric and argillic horizons averaging 35 and 140 cm in thickness, respectively.

Three of the five Paleudalfs are in the fine-silty particle size; all have a mixed mineralogy; all but the Wildale series are superactive; all but the Haugen are mesic; and all have an udic soil-moisture regime. The dominant Paleudalf soil series are the Valton (605 km^2) and the Haugen (320 km^2).

The Valton-Downs-Wildale association in Monroe County represents this soil taxonomic region (Fig. 7.29). The Valton (Mollic Paleudalfs) and Downs (Typic

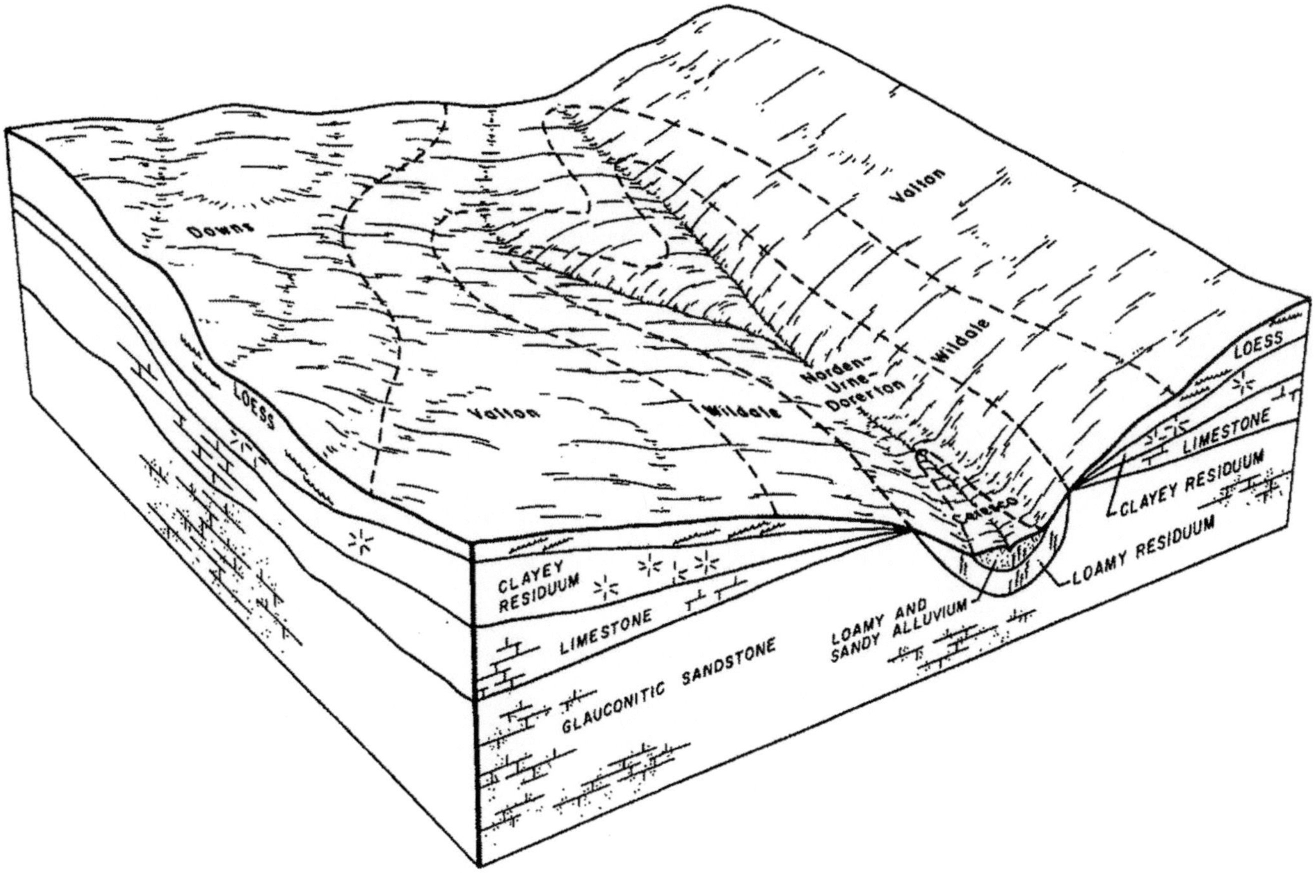

Fig. 7.29 The Valton-Downs-Wildale association in Monroe County

Hapludalfs) occur on summits and shoulders, and the Wildale (Mollic Paleudalfs) occurs on lower slopes.

7.2.12 Udepts (Region 12)

Udepts are moderately developed soils containing an umbric and/or cambic horizon. Udepts (soil taxonomic region 12) include the Dystrudepts that occupy soil regions Dr and Fr and the Eutrudepts that occur in soil regions E, G, and I. Collectively, they cover 2608 km^2 or 1.9% of the state's land area. Udepts support mixed forest, oak-savanna, and brush. They occur on lake plains and river terraces containing glaciolacustrine deposits and alluvium. The ochric and cambic horizons of Udepts in Wisconsin average 15 and 46 cm in thickness, respectively. Five of the six soils in Wisconsin containing an umbric horizon are Udepts, with the umbric horizon averaging 39 cm in thickness.

Udepts are in a variety of particle-size classes, but 54% are in loamy classes (coarse-loamy, fine-loamy, loamy, and loamy-skeletal) and 36% are in sandy classes (sandy, sandy-skeletal). Seventy-nine percent of the Udept series have a mixed mineralogy; 35% are superactive and 35% are not in a CEC activity class; 50% are frigid and 50% are mesic; and all have a udic soil-moisture regime. The predominant Eutrudepts are the Urne and Summerville soil series, and the major Dystrudepts are the Elkmound, Cromwell, and Cress series.

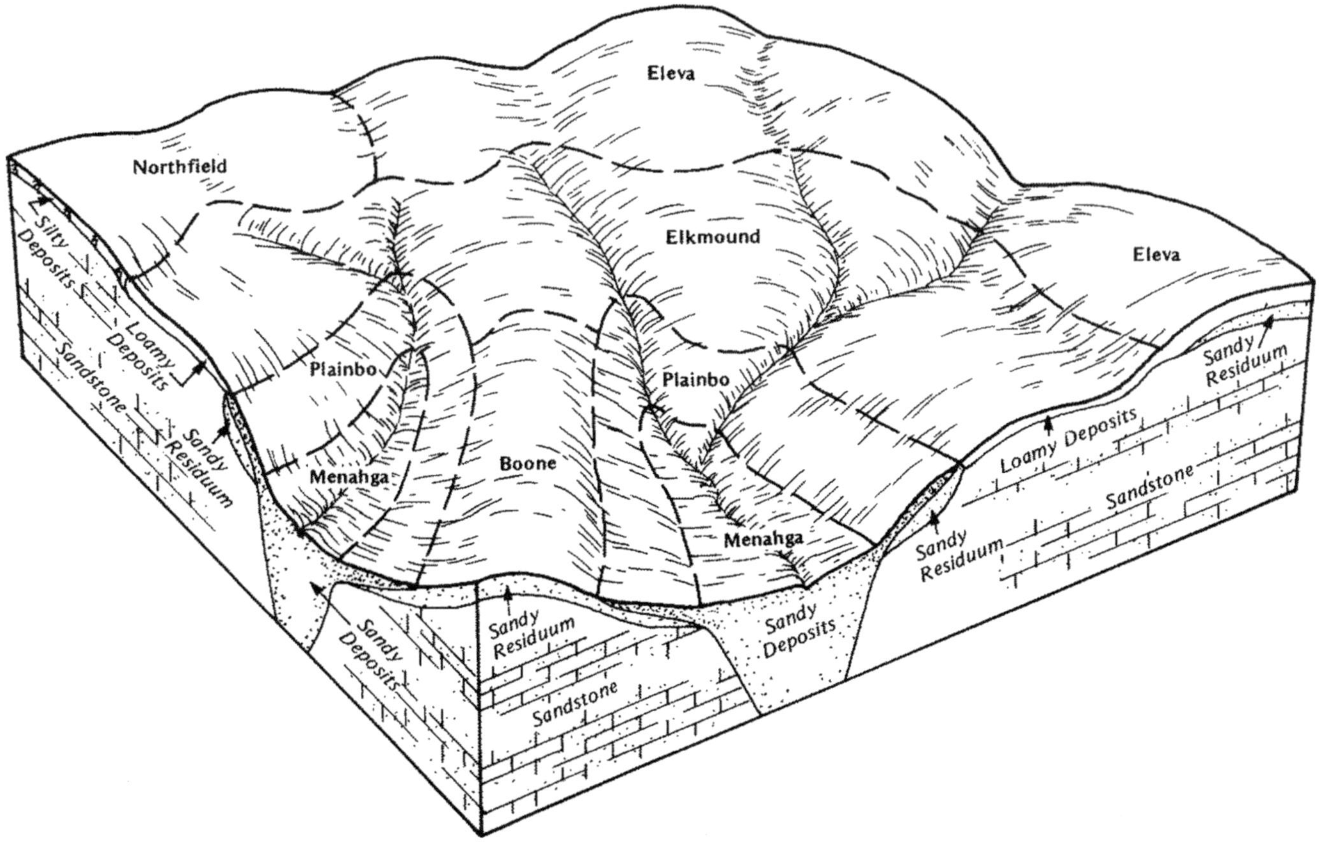

Fig. 7.30 The Elkmound-Plainbo-Eleva association in Chippewa County

The Urne series, a Dystric Eutrudepts, occurs on the mid-slope position and is derived from thin (70 cm) loess over glauconitic sandstone bedrock (Fig. 7.30). The Eutrudepts are represented by the Urne-La Farge-Rozetta association in Juneau County (Fig. 7.31). This association contains soils that are moderately deep, gently sloping to very steep, somewhat excessively drained to moderately well-drained composed of loamy and silty materials on uplands.

7.2.13 Aquepts (Region 13)

The Aquepts are the poorly drained soils with a cambic horizon. The Aquepts (soil taxonomic region 13), which include the Endo-, Epi-, and Humaquepts, occur primarily in soil region J and account for 2250 km^2 or 1.6% of Wisconsin's land area. Aquepts mostly support mixed forest and occur in lake plains and post-glacial drainage ways. The parent materials include loess over till, residuum, and glaciolacustrine deposits. The ochric and cambic horizons of Wisconsin Aquepts average 19 and 45 cm in thickness, respectively. Five of the series have a histic epipedon averaging 24 cm in thickness and four have an albic averaging 11 cm.

About 74% of the Aquepts series are in loamy classes (coarse-loamy, fine-loamy, and loamy-skeletal); 83% have a mixed mineralogy; 39% are superactive and 30% are active; 87% are nonacid; 83% are frigid; and all have an aquic soil-moisture regime. The most extensive Aquepts in Wisconsin are the Minocqua, Cable, Vesper, and Veedum soil series.

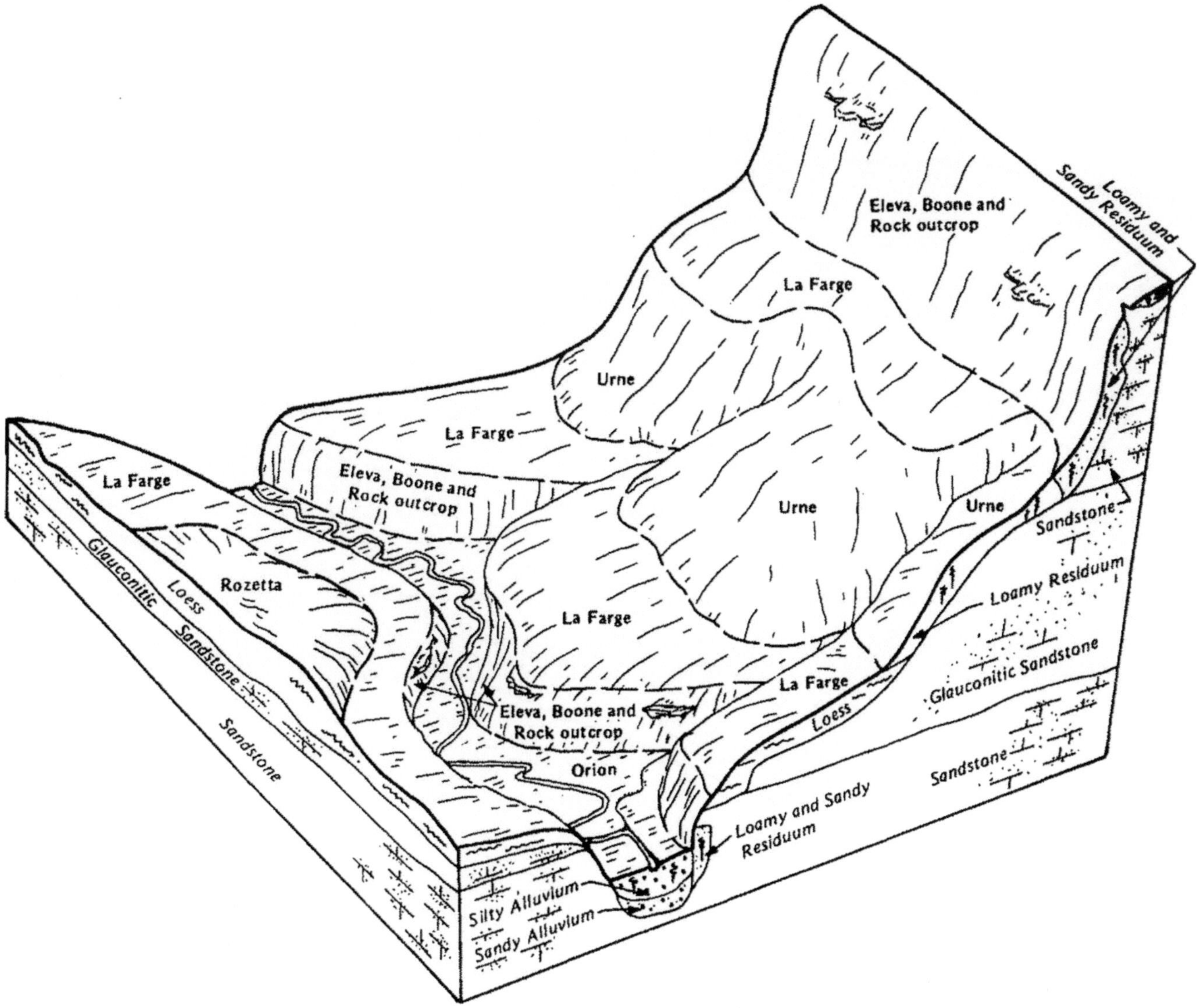

Fig. 7.31 The Urne-La Farge-Rozetta association in Juneau County

The Endoaquepts are represented by the Angelica series which is contained in the Onaway-Solona-Angelica association (Fig. 7.6), with the poorly drained Angelica series representing the Endoaquepts component. The Epiaquepts are represented by the Merrillan-Veedum-Humbird association in Jackson County (Fig. 7.32). This association contains soils that are moderately deep, nearly level to gently sloping, moderately well drained to poorly drained, loamy and mucky textures that occur on pediments. The Veedum series, a Humic Epiaquepts, is formed in silty alluvium and loamy residuum derived from the underlying interbedded sandstone and shale. The Veedum series is poorly drained and on nearly level terrain.

7.2.14 Aquents (Region 14)

Aquents are poorly drained soils lacking a diagnostic subsurface horizon. The Aquents (soil taxonomic region 14), which include the Psammaquents are composed primarily of

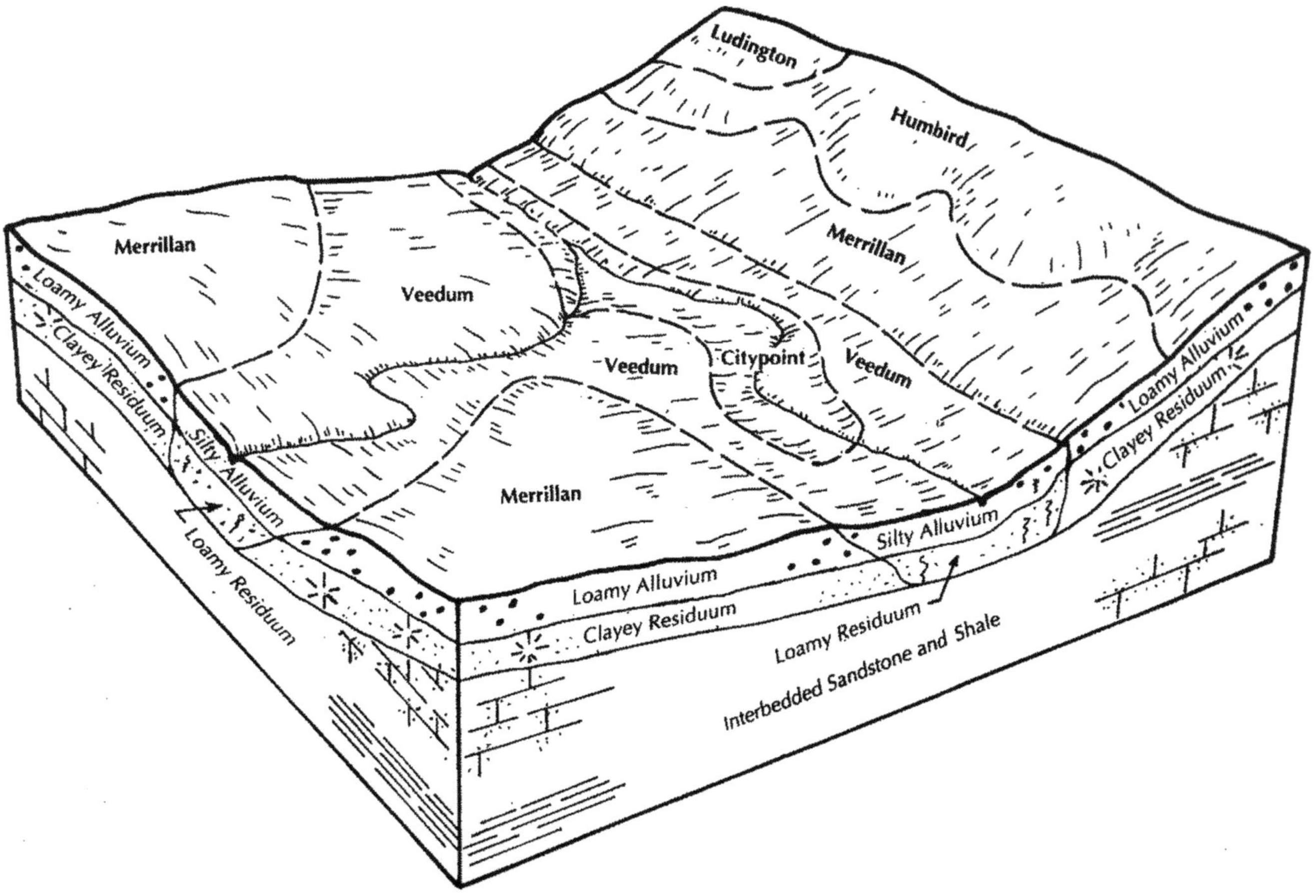

Fig. 7.32 The Merrillan-Veedum-Humbird association in Jackson County

Fluvaquents and Psamm-, Fluv-, Endo-, and Epiaquents, occupy 3348 km^2 or 2.5% of the state. The Fluvaquents occur in the floodplains of major river valleys throughout the state (0.8% of the land area). The Psammaquents occur primarily central Wisconsin within the former Glacial Lake Wisconsin basin (1.3% of land area), both of which are in soil region J. The vegetation includes wet hardwoods, marsh, and oak-pine, as well as irrigated agriculture. The Aquents contain only an ochric epipedon which averages 23 cm in thickness (Fig. 7.33).

Nearly two-thirds (65%) of the Aquents are sandy; 87% have a mixed mineralogy; 39% are in the semi- or superactive CEC activity classes; 52% are in the nonacid reaction class; 61% are frigid and 29% are mesic; and all are in the aquic soil-moisture class. The dominant Aquents include the Newson, Fordum, Roscommon, Elm Lake, and Arnheim soil series, each of which occupies more than 200 km^2.

The Newson-Dawson-Meehan association from Monroe County is representative of this soil taxonomic region (Fig. 7.34). This association contains soils on nearly level to gently sloping land; the soils are somewhat poorly drained to very poorly drained and are composed of peat and sand on floodplains, lake basins, and stream terraces. The Newson series, a Humaqueptic Psammaquents, is poorly drained and occupies depressions in sandy outwash.

Fig. 7.33 Poorly drained Entisol (Aquents) used for soybean cultivation near Grand Marsh

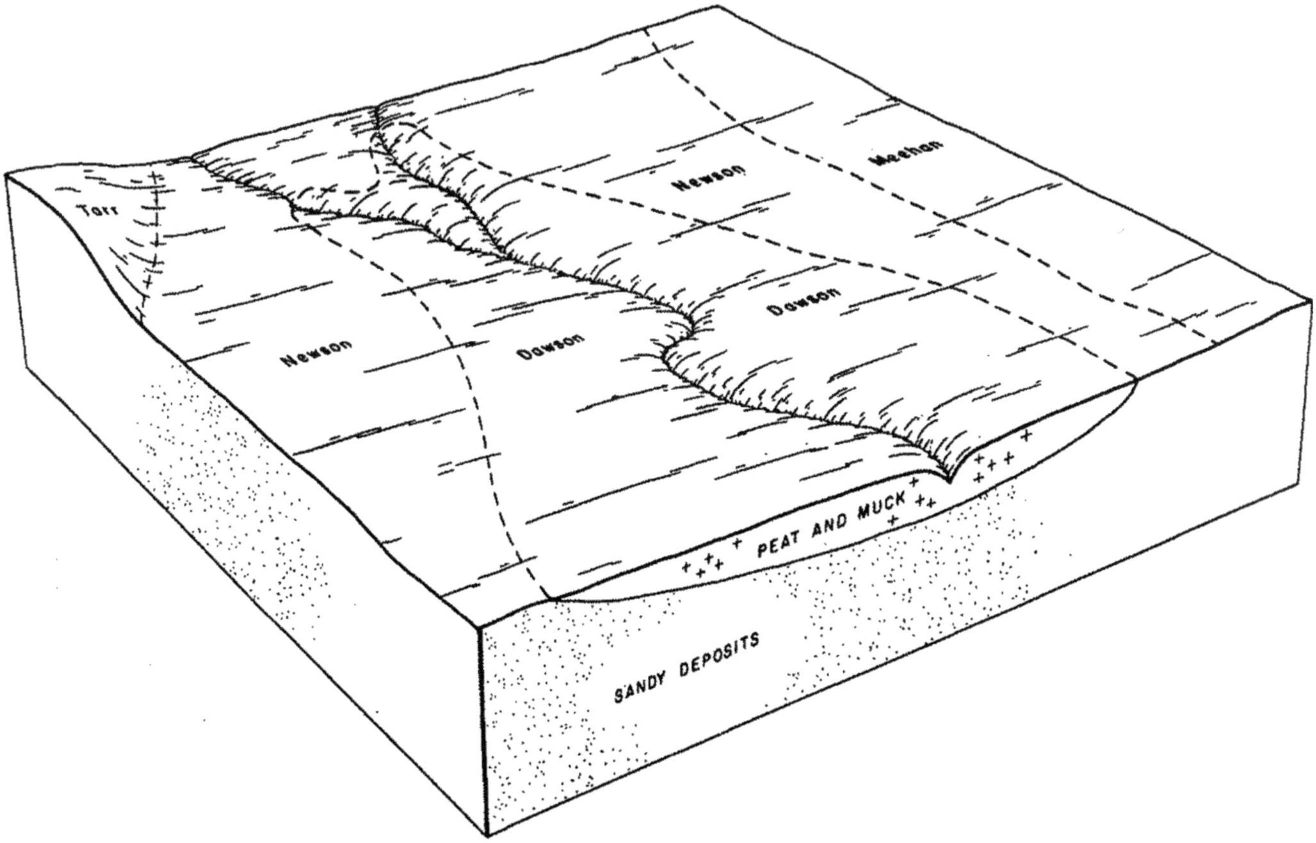

Fig. 7.34 The Newson-Dawson-Meehan association in Monroe County

7.2.15 Quartzipsamments (Region 15)

Quartzipsamments are quartz-rich, excessively drained, sandy soils. Quartzipsamments (soil taxonomic region 15) exist on sandy piedmonts in the Western Uplands (soil region Dr) of west-central Wisconsin and account for 1620 km^2 or 1.2% of Wisconsin's land area. These soils feature primarily oak-pine and are derived from alluvium, colluvium, and residuum over highly weathered sandstone. The ochric epipedon averages 16 cm in thickness. All of the soils are sandy; 8 of the 10 series have a mesic soil-temperature regime; and all are udic. The Tarr, Tint, and Tintson soil series have uncoated sand grains. The Tarr and Boone series each occupy >600 km^2.

The Boone-Elevasil-Tarr association occurs in Clark County and is representative of this soil taxonomic region (Fig. 7.35). This association contains soils that are moderately deep and very deep, nearly level to very steep terrain, well drained, and excessively drained composed of sandy and loamy materials on pediments, hills, and stream terraces. Although they are both Quartzipsamments, the Boone occupies the summits, shoulders, and backslopes, whereas the Tarr series is found on footslopes and toeslopes.

7.2.16 Fragiorthods (Region 16)

Fragiorthods are acid forest soils with iron enrichment (podzolized) and a "hardpan" (fragipan). Fragiorthods (soil taxonomic region 16) occur on end moraines in northernmost Wisconsin and comprise 1361 km^2 or 1.1% of the state's land area. The till commonly has a loamy eolian or

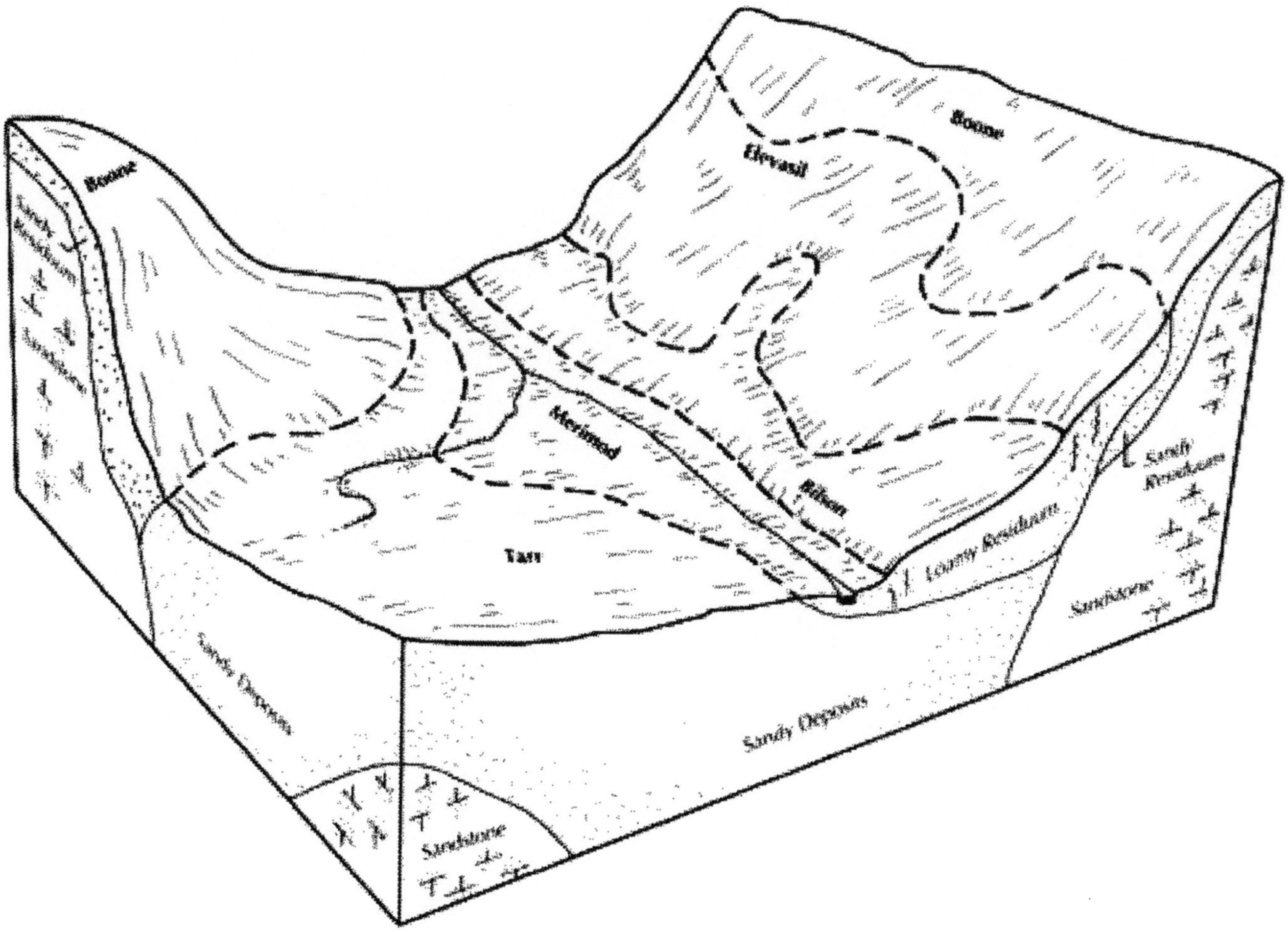

Fig. 7.35 The Boone-Elevasil-Tarr association in Clark County

loess cover. These soils support northern hardwood vegetation. Fragiorthods in Wisconsin are bisequal and have ochric, albic, and spodic horizons averaging 11, 7, and 34 cm in thickness, respectively, over argillic (64 cm), glossic (49 cm), and fragipan horizons (67 cm).

All of the Fragiorthods are coarse-loamy; 80% have a mixed mineralogy (20% are isotic); 60% are superactive; all are frigid; and all have a udic soil-moisture regime. The Gogebic and Wabeno series occupy >900 and 265 km^2, respectively.

The Champion association from Vilas County is representative for this soil taxonomic catena (Fig. 7.36). This association contains soils that are nearly level to moderately steep, moderately well drained, derived from silty materials, and occur on uplands. The Champion series is an Oxyaquic Fragiorthods that is derived from loamy loamy till covered

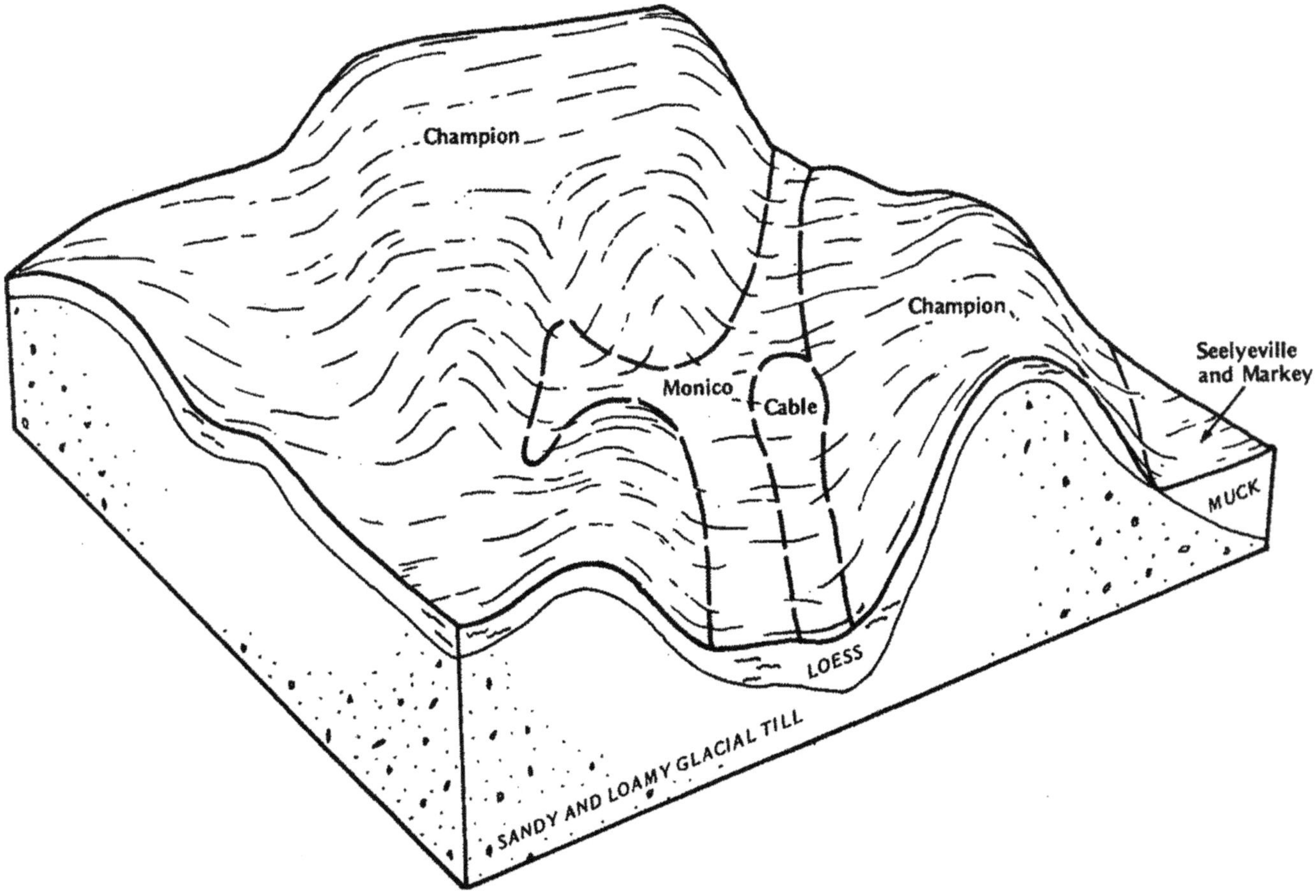

Fig. 7.36 The Champion association in Vilas County

by silty and loamy eolian sediments. The dominant vegetation on this soil is northern hardwoods and hemlock.

The remaining soil great groups not shown on the great group map include Udorthents (96 km^2), Haplustolls (284 km^2), and Hapludults (3 km^2). Land types such as alluvial land, dumps, gravel pits, landfills, rock land, sand pits, and urban areas compose 5,880 km^2 and are not shown on the map (Appendix E). A list of benchmark soil series, soils identified as importantly regionally with a state by the NRCS, is given in Appendix F.

7.3 Summary

Taxonomic soil regions more effectively capture the integrated effects of the soil-forming factors on soil distribution in Wisconsin. In this chapter, 16 soil taxonomic regions are identified from the great group soil map. The area distribution and dominant soil series are determined for each of the regions.

8 Alfisols

8.1 Distribution

Alfisols are the most abundant soil order in Wisconsin, accounting for 47% of the land area (Fig. 1.2) and 43% of the soil series in the state (Fig. 1.3). The distribution of Alfisols is shown at the order, suborder, and great group levels in Fig. 8.1.

In *Soil Taxonomy*, Alfisols are moderately leached soils that have relatively high fertility (Soil Survey Staff 1999, 2014). These soils form mainly under broad-leaved forest and have a subsurface horizon in which clays have accumulated. Alfisols occur in temperate humid and subhumid regions of the world. The broad-leaved vegetation plays an important role in cycling base cations that account for 35% or more of the CEC.

Alfisols in Wisconsin either have favorable seasonal distribution of precipitation and good drainage (Udalfs), or they are poorly drained (Aqualfs). The Hapludalfs and the Glossudalfs great groups compose 36,180 and 23,076 km^2 of the land area of Wisconsin, respectively, and account for 92% of the Alfisols in the state. Hapludalfs occur primarily in the southern mesic zone of Wisconsin (86% of Hapludalf soil series), and Glossudalfs occur mostly in the frigid northern zone (95% of Glossudalf soil series). Hapludalfs are dominant south of the tension zone (Fig. 3.4), whereas Haplorthods and Fragiorthods (Spodosols) are the most common north of the tension zone.

Wisconsin Alfisols are primarily in loamy particle-size classes (58%), followed by silty (24%) and clayey (19%) classes. Ninety-nine percent of Wisconsin Alfisols have a mixed mineralogy; 64% have a superactive CEC class; and 58% have a mesic soil-temperature class (42% frigid).

The Hapludalfs contain a number of soil series that occupy large areas (>400 km^2), including the Casco, Chetek, Churchtown, Dodge, Downs, Dubuque, Eleva, Fayette, Fox, Gale, Hixton, Kewaunee, Kidder, La Farge, Manawa, McHenry, Miami, Morley, Newglarus, Norden, Onaway, Palsgrove, Richford, Seaton, St. Charles, and Theresa series. Glossudalfs occupying areas in excess of 400 km^2 include the Almena, Amery, Antigo, Freeon, Hortonville, Kennan, Loyal, Magnor, Rosholt, Santiago, and Withee series. The Valton series, a Paleudalfs, has an area in excess of 400 km^2. Aqualfs with large areas include the Capitola, Cebana, and Marshfield soil series.

Aqualfs are shown in Fig. 8.3, including (a) the Curran series which is saturated from a seasonally high water table (Udollic Endoaqualfs) and (b) the Marshfield series which is saturated from the surface (Epiaqualfs). The Curran series has a mollic-like epipedon that is 35 cm thick, which is underlain by a Btg horizon with a moderate, coarse prismatic structure and masses of iron accumulation in the matrix (not readily visible). The water table is within a depth of 75 cm.

Udalfs are depicted in Fig. 8.2, including (a) the Kennan series which is a Haplic Glossudalfs, (b) the Fayette series which is a mesic Typic Hapludalfs, (c) the Dunbarton series which is a mesic Lithic Hapludalfs, and (d) the Valton series which is a Mollic Paleudalfs. The Kennan series has an A horizon (ochric epipdeon) that is 15 cm thick, which is underlain by B/E (15–35 cm depth) and E/B (35–50 cm depth) horizons, followed by the Bt horizons (50 cm to base of carpenter's rule at 75 cm). The B/E and E/B horizons comprise the glossic horizon, which represents the degradation of the argillic (Bt horizon). The Fayette series has the following horizons (depths) such as A (0–17 cm), E (17–43 cm), B/E (43–60 cm), and Bt (60–90+ cm).

The Chetek series has an ochric epipedon (A horizon) that is 12 cm thick and an argillic horizon that is 25 cm thick. The Bt horizons extend from the loamy alluvium through the

J.G. Bockheim and A.E. Hartemink, *The Soils of Wisconsin*,
World Soils Book Series, DOI 10.1007/978-3-319-52144-2_8

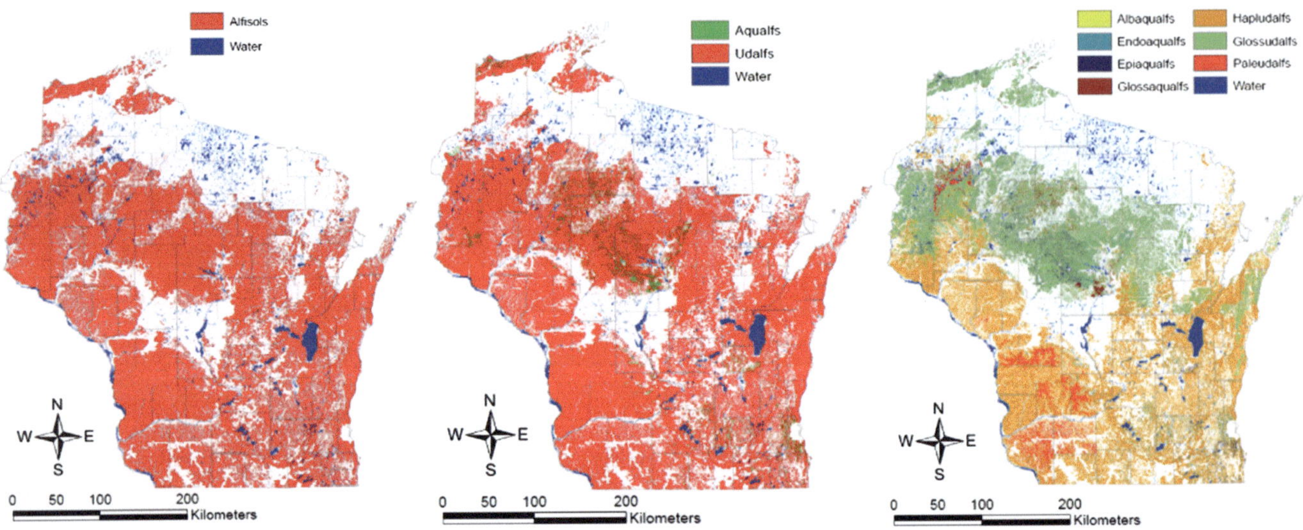

Fig. 8.1 Distribution of Alfisols orders, suborders, and great groups in Wisconsin

underlying sandy and gravelly outwash. As the Bt horizon is less than 35 cm thick, the Chetek soil is classified in the Inceptic subgroup.

The Valton series has a mollic epipedon (A horizon) that is 23 cm thick underlain by an argillic horizon (Bt) that extends beyond 150 cm in depth. The mollic epipedon was formed under oak-hickory savanna. The argillic horizon is formed in loess to a depth of 55 cm and in the underlying clayey residuum weathered from limestone.

8.2 Properties and Processes

The key properties of Alfisols are the presence of a clay-enriched (argillic) horizon and an abundance of base cations (Ca, Mg, and K) at the 25–100 cm depth. The pedons in Table 8.1 show that Alfisols may be formed from different parent materials, including clayey till (Badriver series), loess of varying thickness (the Dubuque 76 cm and Magnor, Sherry, Valton and Ozaukee series <56 cm), and sandy outwash (Richford and Rosholt series).

The series with a glossic horizon (Badriver, Magnor, and Rosholt) have E/B and/or B/E horizons reflecting the degradation of the argillic horizon. Two of the pedons have a mollic epipedon (Churchtown and Valton series) and developed under oak savanna. In about half of the pedons, the pH decreases with depth due to base cycling at the surface by the temperate deciduous forest.

The dominant processes in Alfisols are argilluviation and biological enrichment of base cations, with gleization and melanization occurring in some soils. Argilluviation refers to the movement of clay in the profile. The argillic horizon must contain a minimum clay increase relative to the eluvial horizon or an underlying horizon and show evidence of clay movement (Buol and Hole 1959, 1961; Allan and Hole 1968). Vegetation plays an important role in maintaining the base cation (Ca, Mg, and K) content of Alfisols (Bockheim 1997). Temperate deciduous forest types take up and return large amounts of base cations in litterfall, throughfall, stemflow, and belowground processes such as root exudation and fine root turnover (Fujinuma et al. 2005).

Gleization is a predominant process in the Aqualfs and refers to the presence of aquic conditions that are evidenced by reductimorphic or redoximrophic features such as mottles and gleying. The effects of reduction and oxidation particularly influence iron and manganese compounds, and these result in visible features that have been used for estimating soil-moisture regimes. Melanization refers to the accumulation of well-humified organic matter within the upper mineral soil horizons. These horizons have at least 0.6% organic C and are ≥18 cm in thickness.

Fig. 8.2 Examples of Udalfs in Wisconsin, including **a** Glossudalf, **b** the Fayette series, a mesic Hapludalfs, **c** Dunbarton series, mesic Lithic Hapludalfs, **d** the Valton series, a Mollic Paleudalfs

Fig. 8.3 Examples of Aqualfs in Wisconsin, including **a** the Curran series, an Udollic Endoaqualfs from southern Wisconsin (photograph by Northern Illinois University), and **b** the Marshfield series, an Epiaqualfs

Table 8.1 Analytical characteristics of some Alfisol soil series from Wisconsin

Horizon	Depth (cm)	Clay (%)	Silt (%)	Sand (%)	SOC (%)	CEC7 (cmol(+)/kg)	Base sat. (%)	pH H_2O
Badriver series; Aeric Glossaqualfs; Pedon No. 91P0371; Ashland, WI								
A	0–5	62.4	32.5	5.1	5.89	56.0	47	5.2
E	5–13	26.3	58.9	4.8	1.16	13.5	53	5.0
B/E	13–23	50.4	41.4	8.2	0.67	22.6	77	5.3
Bt1	12–33	59.3	33.7	7.0	0.57	28.8	84	5.6
Bt2	33–79	59.0	29.4	11.6	0.20	30.0	100	6.6
Btk	79–104	49.4	36.1	14.5	0.08	18.3	100	8.1
C	104–152	33.5	42.8	23.7	0.07	12.9	100	8.2
Capitola series; Aeric Epiaqualfs; Pedon 90P0057; Taylor, WI								
Eg	10–18	13.6	72.1	14.0	1.86	11.4	37	5.8
Btg1	18–30	16.7	66.1	17.2	0.27	12.0	55	5.0
Btg2	30–46	22.9	62.1	15.0	0.30	18.7	67	5.0
Btg3	46–61	20.4	56.9	22.7	0.16	17.7	79	5.2
2Bt	61–86	2.8	18.2	79.0	0.06	4.4	75	5.4
2C	86–152	4.4	21.7	73.9	0.03	5.2	92	5.9
Churchtown series; Mollic Hapludalfs; Pedon No. 01P0216; La Crosse, WI								
A	0–18	13.0	79.7	7.3	3.08			6.4
Bt1	18–58	10.5	80.6	8.9	0.52			4.7
Bt2	58	26.0	55.0	19.0	0.26			5.5
Dubuque series; Typic Hapludalfs; Pedon No. 11N0306								
Ap	0–25	15.1	73.2	11.7	1.28	11.0	88	5.3
Bt1	25–41	14.4	77.3	8.3	0.41	8.4	82	6.2
Bt2	41–76	23.2	70.2	6.6	0.26	11.9	87	6.5
2Bt3	76–109	36.2	11.3	52.5	0.18	18.3	73	5.2
2Bt4	109–163	12.5	2.5	85.0	0.05	5.8	78	5.2

(continued)

Table 8.1 (continued)

Horizon	Depth (cm)	Clay (%)	Silt (%)	Sand (%)	SOC (%)	CEC7 (cmol(+)/kg)	Base sat. (%)	pH H_2O
2Bt5	163–203	29.9	6.6	63.5	0.13	14.1	78	5.3
Magnor series; Aquic Glossudalfs; Pedon 88P0318; Barron, WI								
A	0–8	16.3	54.7	29.0	4.99	17.8	80	5.5
E	8–18	11.0	55.4	33.6	1.14	17.8	80	5.4
E/B	18–25	8.8	56.8	34.4	0.49	10.2	74	5.5
B/E	25–36	7.8	58.0	34.2	0.27	12.9	79	5.5
Bt	36–51	9.7	51.9	38.4	0.19	7.9	87	5.5
2Bt	51–64	11.9	34.5	53.6	0.16	9.4	94	4.8
2Cd	64–152	8.5	13.8	77.7	0.08	10.7	94	4.9
Ozaukee series; Oxyaquic Hapludalfs; Pedon 40A1561; Kenosha, WI								
Ap1	0–20	19.7	54.3	26.0	1.64	11.8	88	6.1
Ap2	20–28	21.8	53.4	24.8	1.48	12.0	90	6.2
Bt1	28–41	43.2	39.7	17.1	0.54	17.5	94	6.2
2Bt2	41–63	44.9	37.5	17.6	0.51	19.5		7.2
2BC	63–79	29.5	51.2	19.3	0.39	11.0		7.9
2C1	79–107	26.4	54.3	19.3	0.28	8.6		8.0
2C2	107–145	25.7	54.2	20.1	0.24	7.6		8.1
Richford series; Arenic Hapludalfs; Pedon 83P0844; Chippewa, WI								
Ap	0–23	5.1	18.0	76.9	0.98			6.5
E	23–56	4.5	19.1	76.4	0.22			6.6
Bw	56–66	7.2	19.8	73.0	0.20			6.5
Bt1	66–91	14.5	19.2	66.3	0.24			5.7
Bt2	91–102	13.7	12.6	73.7	0.18			5.3
Rosholt series; Haplic Glossudalfs; Pedon 89P0249; Barron, WI								
Ap	0–20	6.8	39.9	53.3	1.25	7.4		6.6
B/E	20–28	4.8	42.8	52.4	0.22	3.7		6.4
Bt	28–51	18.0	35.0	47.0	0.22	9.9		5.2
2Bt	51–71	7.9	20.1	72.0	0.13	6.1		5.2
Sherry series; Udollic Endoaqualfs; Pedon 40A1666; Wood, WI								
Ap1	0–10	25.1	63.9	10.6	4.21	25.5	53	4.4
Ap2	10–23	24.3	63.1	9.7	3.24	23.1	51	4.5
Bw	23–35	24.9	60.9	8.9	0.53	18.7	65	4.8
Bt1	35–52	27.2	55.5	11.4	0.20	20.0	83	5.1
2Bt2	52–83	14.6	25.4	6.8	0.07	11.2	96	6.1
2Bt3	83–124	9.5	16.3	5.1	0.04	6.2	98	6.5
2Bt4	124–141	16.8	50.3	7.5	0.08	19.8	100	6.6
2C1	141–171	18.2	48.5	8.6	0.07	14.8	100	6.8
Silverhill series; Ultic Hapludalfs; Pedon 92P0414; Jackson, WI								
Ap	0–20	7.1	31.9	61.0	1.18	6.5	80	5.3
Bt1	20–36	8.3	28.6	63.1	0.36	4.8	81	5.8
Bt2	36–66	11.0	24.6	64.4	0.23	5.8	72	5.5
2BC	66–81	4.3	8.0	87.7	0.06	2.5	52	5.0
2C	81–127	2.2	4.9	92.9	0.05	1.5	40	4.9
Valton series; Mollic Paleudalfs; Pedon 40A5208; Monroe, WI								
Ap	0–23	20.1	73.4	6.5	0.42	13.6	97	6.3
Bt1	23–36	26.8	68.1	5.1	0.19	15.0	87	5.8

(continued)

Table 8.1 (continued)

Horizon	Depth (cm)	Clay (%)	Silt (%)	Sand (%)	SOC (%)	CEC7 (cmol(+)/kg)	Base sat. (%)	pH H_2O
Bt2	36–56	31.3	60.9	7.8	0.21	17.0		5.3
2Bt11	56–81	41.5	48.4	10.1	0.17	23.5	78	4.9
2Bt12	81–107	44.8	42.0	13.2	0.14	23.9	82	5.1
2Bt21	107–140	50.4	40.3	9.3	0.14	25.9	80	5.1
2Bt22	140–168	70.6	19.8	9.6	0.12	37.1	85	5.0

Fig. 8.4 Sandy Alfisol used for soybean cultivation near Westfield. Richford soil series (Arenic Hapludalfs). Note shallow rooting of irrigated soybean

Fig. 8.5 Sandy Alfisol under pine forest. Richford soil series (Arenic Hapludalfs), with lamellae and redoximorphic features below 180 cm

Fig. 8.6 Sandy Alfisol under prairie grass. Richford soil series (Arenic Hapludalfs), with strong redoximorphic features below 120 cm

Fig. 8.7 Sandy Alfisol develop under forest vegetation near Westfield. Okee soil series (Arenic Hapludalfs)

Fig. 8.8 Alfisols of the West Madison Agricultural Research Station developed in loess over coarse outwash. Both soils are Typic Hapludalfs and about 50 m apart. Note iron accumulation at the textural discontinuity between the loamy loess and the sandy outwash

8.3 Use and Management

Wisconsin Alfisols are important for agriculture, grazing, forestry, and urbanization. Many of the most productive agricultural soils in the state are Alfisols, especially in the Hapludalfs great group (Fig. 8.4). Similarly, some of the most productive forest soils for hardwood sawlogs, veneer, and maple syrup production are Hapludalfs, including the Bertrand, Emmet, Fayette, Gale, Hixton, La Farge, McHenry, Seaton, and Theresa soil series (Johnson et al. 1993). In northern Wisconsin, some of the Glossudalfs, including the Amery, Antigo, Cushing, Kennan, Otterholt, and Withee soil series, are highly productive for forest products. Seven of the 10 most populated cities in Wisconsin, accounting for 1.2 million persons, occur within a Hapludalfs soil-map unit (Figs. 8.5 and 8.6). The Antigo silt loam (Haplic Glossudalfs) was selected as the state soil because of its importance in agriculture, grazing, and forestry (Figs. 8.7, 8.8, 8.9, 8.10, 8.11, 8.12, 8.13, 8.14, 8.15, 8.16).

Fig. 8.9 Alfisol developed in colluvium and loess near Platteville (Typic Hapludalfs)

Fig. 8.10 Alfisol under pine forest near Richland Center in the Driftless Area of Wisconsin (Typic Hapludalfs)

Fig. 8.11 Alfisol (Lithic Hapludalfs) in the Driftless area of Wisconsin with a thin layer of loess overlying red smectitic clays over Dolostone

Fig. 8.12 Alfisol developed in colluvium near Potosi in Wisconsin

Fig. 8.13 Alfisol under grassland and pine forest in Dane County, Wisconsin. Batavia soil series (Mollic Hapludalfs)

Fig. 8.14 Alfisol under forest near Platteville, note strongly developed clayfilms in the Bt horizon

Fig. 8.15 Deep Alfisol developed in loess near Lancaster in Wisconsin (Typic Hapludalfs)

Fig. 8.16 Alfisol under pasture near Belmont. Zones of strong reduction and oxidation below 80 cm soil depth. Mollic Hapludalfs

8.4 Summary

Alfisols are the dominant soil order in Wisconsin, accounting for 47% of the land area and 43% of the soil series in the state. Hapludalfs and Glossudalfs account for 92% of the Alfisols in Wisconsin. Whereas Hapludalfs primarily occur to the south of the tension zone, Glossudalfs are restricted to the north. The key properties of Alfisols are the presence of a clay-enriched (argillic) horizon and an abundance of base cations (Ca, Mg, and K). The dominant processes in Alfisols are argilluviation and biological enrichment of base cations, with gleization and melanization occurring in some soils. Alfisols are important for agriculture, grazing, forestry, and urbanization. Publ. G3452.

Spodosols

9

9.1 Distribution

Spodosols are the second most abundant soil order in Wisconsin, accounting for 17% of the land area (Fig. 1.2) and 15% of the soil series in the state (Fig. 1.3). The distribution of Spodosols is shown at the order, suborder, and great group levels in Fig. 1.9.

Spodosols (from Greek *spodos*, "wood ash") are acid soils characterized by a subsurface accumulation of humus that is complexed with Al and Fe. These soils usually have a light-colored albic or E horizon overlying a reddish brown spodic (Bhs, Bs) horizon.

Spodosols mainly occur to the north of the tension zone (Fig. 3.4) but also within the tension zone. The southernmost occurrence of Spodosols in Wisconsin is a 2500 km^2 "island" along the Clark-Jackson-Eau Clair county borders (soil region Dr) that features primarily Aquods (Fig. 9.1). These soils (Fairchild and Merrillan series) are derived from sandy and loamy alluvium overlying sandstone bedrock. There are several of southwest trending "bands" of Spodosols that occur on outwash in the Chippewa, Black, and Wisconsin River drainages that link with the Mississippi River. The Fragiorthods along the northern tier of counties (Fig. 9.1) occur on Late Wisconsinan end moraines and drumlins.

Spodosols are the most common on glacial drift, including outwash (45% of soil series) and till (29%). The vegetation is often mixed broad-leaved and coniferous forest (48%) or northern hardwoods and NHW-hemlock (44%). Spodosols can form very rapidly and occur on mid-Holocene beach ridges at the Ridges Sanctuary on Door Peninsula (Hole, unpublished). Most of the Spodosols in Wisconsin occur on drift of Early Holocene or Late Wisconsinan age.

Wisconsin Spodosols have either aquic conditions (Aquods suborder), or they more or less follow the central concept of the order (Orthods). Of the 105 Spodosol soil series, 69% are Orthods and 31% are Aquods. Of the Orthods soil series, 90% are Haplorthods and 10% are Fragiorthods, i.e., have a fragipan. Of the Aquods soil series, 54% are Endoaquods (saturated by a seasonally high water table), 43% are Epiaquods (saturated from the surface by rainfall and snowmelt), and 3% are Fragiaquods (have a fragipan as well as restricted drainage). The Haplorthods account for an area of 18,285 km^2 and are exceeded in area only by the Hapludalfs and Glossudalfs great groups (Table 9.1).

There is a common notion that Spodosols have a sandy texture. However, a larger proportion of Spodosol soil series in Wisconsin are loamy (53%) rather than sandy (45%). Eighty-six percent of Wisconsin Spodosols have a mixed mineralogy and 100% are in frigid soil-temperature class. Thirty-seven percent of Wisconsin Spodosols are in the superactive CEC class, 17% are in the active class, 7% are in the semiactive class, and 39% are subactive. The Au Gres, Ironrun, and Wainola soil series contain up to 50% ortstein (iron pan), which is less than the required 90% for them to be classified in the ortstein rupture-resistance class (Durorthods and Duraquods).

Many of the more common soil series are classified as Haplorthods, including the Sarona, Padus, Pence,, Keweenaw, Vilas, Sayner, Newood, Rubicon, Croswell, and Karlin. The Gogebic series, a Fragiorthods, is the only non-Haplorthods, with an area of >400 km^2.

The Ironrun series which is saturated from a seasonally high water table (Endoquods) is shown in Fig. 9.2. The Ironrun series contains an O horizon (0–5 cm), an A horizon from 5 to 10 cm, an E (albic) from 10 to 30 cm, a Bhs from 30 to 40 cm, a Bs1 from 40 to 60 cm, a Bs2 from 60 to 75 cm, and a C horizon from 75 cm to the bottom of the exposure. Masses of iron accumulation are evident in the Bs2 and C horizons.

Orthods are depicted in Fig. 9.3, including (a) the Gogebic series, a coarse-loamy, isotic, frigid Alfic Oxyaquic Fragiorthods (left) and the Rousseau series, a sandy, mixed, frigid Entic Haplorthods (right). The Gogebic series contains

J.G. Bockheim and A.E. Hartemink, *The Soils of Wisconsin*,
World Soils Book Series, DOI 10.1007/978-3-319-52144-2_9

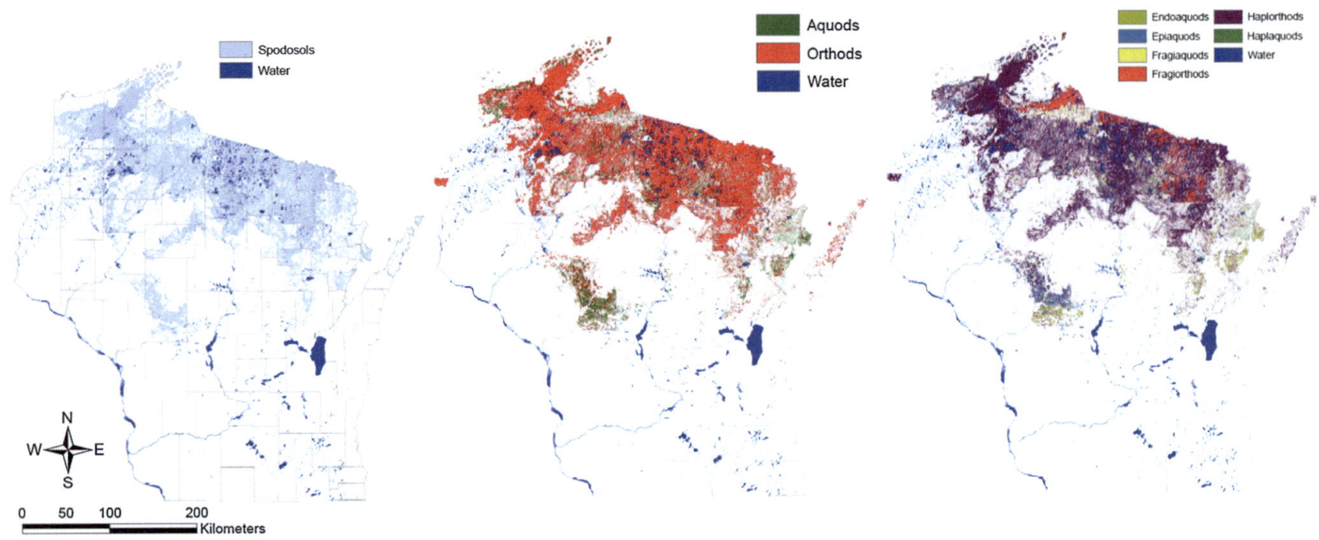

Fig. 9.1 Distribution of Spodosols (order, suborders and great groups) in Wisconsin

Table 9.1 Analytical characteristics of some Spodosol soil series from Wisconsin

Horizon	Depth (cm)	Clay (%)	Silt (%)	Sand (%)	SOC (%)	CEC7 [cmol(+)/kg]	Base sat. (%)	pH H_2O	Fed (%)	Ald (%)	Feo (%)	Alo (%)	Fep (%)	Alp (%)
Au Gres series; Typic Endoaquods; Pedon No. 83P0043; Vilas, WI														
O	0–5				9.50	27.6		4.3	0.3	0.1	0.14	0.13	0.1	0.1
E	5–15	0.6	12.3	87.1	0.57	3.0	20	4.3	0.2	tr	0.03	0.03		0.1
Bhs	15–20	5.4	11.4	83.2	1.95	13.0	11	4.6	1.7	0.6	0.96	0.59	0.8	0.5
Bs1	20–40	1.3	6.6	92.1	0.47	2.6	15	5.4	0.6	0.4	0.24	0.58	0.1	0.2
Bs21	40–63	0.1	2.2	97.7	0.15	0.6	17	5.4	0.2	tr	0.07	0.16		0.1
Bs22	63–86	0.1	2.1	97.8	0.26	0.9	22	5.4	0.2	0.1	0.03	0.15		0.1
C1	86–107	0.1	1.4	98.5	0.22	0.8	25	5.4	0.3	0.1	0.04	0.12		0.1
C2	107–152	0.0	0.8	99.2	0.09	0.5	40	5.5	0.2	tr	0.12	0.06	0.1	0.1
Fairchild series; Ultic Epiaquods; Pedon No. 92P0601; Clark, WI														
A	0–8	0.0	9.3	90.7	2.07	8.2	10	3.7	0.1	0.1	0.03	0.04		tr
E	8–28	0.0	4.0	96.0	0.16	0.8	25	4.0	tr	tr	tr	tr		0.1
Bhs	28–33	4.8	8.1	87.1	1.74	11.6	4	4.3	1.0	0.7	0.43	0.65	0.4	0.7
Bs	33–46	3.5	7.5	89.0	1.26	7.3	7	4.4	0.7	0.5	0.25	0.50	0.3	0.6
Bw	46–66	0.8	3.3	95.9	0.06	1.0	30	4.4	0.1	tr	0.04	0.03	tr	0.2
2Bt	66–81	30.4	22.2	47.4	0.18	9.7	8	4.2	0.4	0.1	0.08	0.15		0.1
Gogebic series; Alfic Fragiorthods; Pedon No. 85P0907; Ashland, WI														
E	0–20													
Bhs	20–28	7.5	65.0	27.5	3.12	22.0	18	4.5	2.2	0.7			1.2	0.6
Bs	28–53	1.2	34.6	64.2	1.21	10.5	12	4.9	1.3	0.6			0.4	0.4
B/Ex	53–89	0.8	32.7	66.5	0.23	4.0	20	5.2	0.8	0.1			0.1	0.2
C	89–193	3.3	31.3	65.4	0.07	4.2	93	5.7	0.9	tr			tr	tr

(continued)

Table 9.1 (continued)

Horizon	Depth (cm)	Clay (%)	Silt (%)	Sand (%)	SOC (%)	CEC7 [cmol(+)/kg]	Base sat. (%)	pH H_2O	Fed (%)	Ald (%)	Feo (%)	Alo (%)	Fep (%)	Alp (%)
Humbird series; Oxyaquic Ultic Haplorthods; Pedon No. 94P0155; Clark, WI														
A	0–3	4.9	31.3	63.8	4.40	15.5	11	4.2			0.11	0.07	0.1	0.1
E	3–13	3.5	30.3	66.2	0.69	3.5	6	4.3			0.05	0.03	tr	tr
Bs	13–33	10.3	30.0	59.7	1.05	8.2	7	4.5			0.30	0.16	0.3	0.2
Bw	33–51	7.4	25.3	67.3	0.29	3.7	5	4.7			0.13	0.07	0.1	0.1
2Bt1	51–64	31.5	39.2	29.3	0.30	11.0	16	4.7						
2Bt2	64–74	34.6	34.0	31.4	0.20	10.4	11	4.5						
Manitowish series; Oxyaquic Haplorthods; Pedon No. 81P0186; Vilas, WI														
A	0–8	9.0	39.6	51.4	2.82			5.4	0.1	0.3			0.5	0.3
Bhs	8–15	5.3	36.7	58.0	0.82	5.8	33	5.6	1.0	0.3			0.2	0.2
Bs	15–41	5.3	36.9	57.8	0.68	4.6	22	5.4	1.0	0.3			0.2	0.2
2Bs	41–48	2.0	14.5	83.5	0.35	3.7	27	5.7	0.4	0.3			0.1	0.2
2C	48–89	0.4	0.9	98.7	0.15			6.0	0.2	0.1			0.1	0.1
Newood series; Alfic Oxyaquic Haplorthods; Pedon 88P0546; Lincoln, WI														
A	0–3	7.9	62.7	29.4	6.38	21.5	45	4.7	0.9	0.1	0.24	0.14	0.3	0.2
E	3–8	6.0	45.4	48.6	1.11	7.5	35	4.8	0.8	0.1	0.20	0.08	0.1	0.1
Bs1	8–18	8.0	29.4	62.6	1.03	8.6	27	5.2	1.3	0.3	0.58	0.38	0.3	0.3
Bs2	18–30	6.0	31.0	63.0	0.62	6.4	31	5.4	1.0	0.2	0.22	0.14	0.2	0.3
E	30–43	4.5	29.6	65.9	0.27	4.1	44	5.3	0.9	0.1	0.24	0.14	0.1	0.2
E/B	43–84	5.9	24.4	69.7	0.11	4.4	66	5.5	1.0	0.1	0.18	0.06	0.1	0.1
Bt1	84–104	13.2	23.3	63.5	0.05	7.9	94	6.3	1.1	0.1	0.14	0.05	tr	0.1
Bt2	104–145	10.9	22.6	66.5	0.03	7.4	91	6.4	1.6	0.1	0.14	0.07		0.1
Cd	145–183	10.0	23.3	66.7	0.01	6.9	94	6.7	1.0	0.1	0.15	0.05		0.1
Pence series; Typic Haplorthods; Pedon No. 85P0914; Price, WI														
A	0–5													
Bhs	5–20	6.2	23.1	70.7	1.89	10.9	35	1.2	0.3				0.3	0.3
Bs	20–46	6.6	20.7	72.7	0.84	7.8	15	1.1	0.3				0.2	0.2
BC	46–71	2.3	0.3	97.4	0.16	2.2	27	0.4	0.1				tr	0.1
C	71–152	1.9	4.0	94.1	0.24	3.3	58	0.8	0.1				0.1	0.1
Vilas series; Entic Haplorthods; Pedon No. 79P0388; Oneida, WI														
E	3–8	3.3	11.5	85.2	1.27	6.6	12	0.4	0.1					
Bhs	8–18	5.8	8.7	85.5	0.98	6.2	10	0.6	0.3				0.3	0.3
Bs	18–48	3.6	6.1	90.3	0.44	2.8	25	0.5	0.2				0.1	0.2
BC	48–76	2.0	0.6	97.4	0.11	0.7	29	0.2	0.1					
C	76–152	0.8	0.0	99.2	0.03	0.3	33	0.2	0.1					

an albic horizon (7–15 cm), a spodic horizon (15–53 cm), a fragipan (64–112 cm), a glossic horizon (94–112 cm), and an argillic horizon (112–142 cm). The Rousseau series has a thin A horizon (0–2.4 cm), an albic horizon (2.5–20 cm), a spodic horizon (20–43 cm), a BC horizon (43–64 cm), and the underlying, lighter colored C horizon.

9.2 Properties and Processes

Key properties of Spodosols are the accumulation of soil organic C in the upper spodic (Bhs) horizon, the low base saturation, the acid conditions, and especially the accumulation of iron and aluminum in the spodic (Bhs, Bs) horizon

Fig. 9.2 The Ironrun series is a sandy, siliceous, frigid Typic Endoaquods

(Table 9.1). The pyrophosphate extraction (Fe_p, Al_p) yields organic-bound forms of iron and aluminum; the oxalate extraction (Fe_o, Al_o) yields organic and amorphous forms of Fe and Al; and the dithionite extraction (Fe_d, Al_d) yields all free forms of these elements, including crystalline. The data in Table 9.1 show that amorphous forms of Fe and Al are dominant which relate to the isotic mineral class in which many Wisconsin Spodosols occur.

Spodosols have been studied for over 100 years in Wisconsin. Kellogg (1930) recognized what later was classified as Spodosols only in the northernmost counties of Wisconsin within 35 km of Lake Superior of the border with Michigan. He classified four of the 30 principal soil types of Wisconsin as "Podsols": the Saugatuck, Orienta, Cornucopia, and Superior. The Saugatuck is an Aquod that is no longer recognized to occur in Wisconsin; the Orienta is no longer recognized as a soil series; the Cornucopia is now classified as a Haplic Glossudalfs; and the Superior is an Alfic Oxyaquic Haplorthods that has been mapped in northwestern Wisconsin. Nygard et al. (1952) related the distribution of Podzol soils in the upper Great Lakes region to comparatively high levels of precipitation, mixed deciduous–coniferous forests, and sandy loam till and sandy outwash parent materials.

Madison and Lee (1965) examined the mineralogy of ten sandy soil series in Wisconsin, two of which were Spodosols, including the Omega (Typic Haplorthods) and the Vilas (Entic Haplorthods) series. The sand fraction of these series had the lowest amounts of quartz and the greatest amounts of orthoclase feldspars and heavy minerals of the soils investigated.

Olson and Hole (1967) investigated the fragipan of a number of soils in northeastern Wisconsin, including the Goodman series, a bisequal Alfic Haplorthods. They proposed a genetic sequence whereby the fragipan developed from desiccation during the Climate Optimum 6000–4000 years ago. Over the past 2000 years, an increase in the distribution of hemlock caused podzolization in the upper part of the profile, and the fragipan provided favorable conditions for the development of an argillic horizon below the spodic sequum. Hole (1975) elucidated this concept in a case study from the Menominee Tribal Lands, a now-classic article published in *Soviet Soil Science* (*Pochvovedenie*). Bockheim (2003) examined 17 bisequal pedons on Greatlakean drift (ca. 12,000 year BP) in the upper Great Lakes region. In contrast to Hole, he argued that lithologic discontinuities—such as till over outwash—rather than climate and vegetation change led to the development of Spodosol/Alfisol sequa.

Fig. 9.3 Examples of Orthods in Wisconsin, including the Gogebic series [coarse-loamy, isotic, frigid Alfic Oxyaquic Fragiorthods (*left*)] and the Rousseau series [sandy, mixed, frigid Entic Haplorthods (*right*)]

In the soil survey of Florence County in northern Wisconsin, 19 of the 26 soil series were classified as Spodosols (Hole 1974). Gaikawad and Hole (1961) examined the Au Train series (Oxyaquic Haplorthods) and related an ortstein horizon to the presence of "cradle-knoll" micro-relief from windthrow of trees.

The dominant soil-forming processes in Spodosols are podzolization and base cation leaching, with gleization occurring in Aquods. Podzolization is a complex collection of processes that includes weathering transformation of Fe and Al compounds, mobilization of Fe and Al in surface horizons, and transport of these compounds to the spodic Bs horizon as Fe and Al complexes with fulvic acids and other complex polyaromatic compounds. Base cation leaching is the opposite of biological enrichment of base cations and involves the eluviation of Ca, Mg, K, and Na from the solum under extreme leaching conditions. Gleization was discussed earlier and refers to soil processes occurring under hydromorphic conditions.

9.3 Use and Management

Spodosols are used primarily for forestry, particularly for pulpwood and firewood production. The more productive series for forestry include Haplorthods (Croswell, Rousseau, Padus, and Pence series), Fragiorthods (Gogebic), and Endoaquods (Au Gres). Some Spodosols are used for pastures but their use for agricultural production requires liming and fertilization. It is likely that tillage and liming will result

Fig. 9.4 Spodosol under hemlock forest in northern Wisconsin. Sayner series (Entic Haplorthods)

in the dissolution of the Bhs horizon and the soil may change to an Entisol. The frequency of frosts also limits Spodosols for agricultural cropping. With temperature and rainfall further increasing in the coming decades, it may be that some of the northern Spodosols will be cleared of their forest vegetation and cropped.

9.4 Summary

Spodosols are the second most abundant soil order in Wisconsin, accounting for 17% of the land area and 15% of the soil series in the state. Spodosols occur to the north of the tension zone. Of the 105 Spodosol soil series, 69% are

Orthods and 31% are Aquods. Key properties of Spodosols are the sandy parent materials and textures, the accumulation of soil organic C in the upper spodic (Bhs) horizon, the low cation-exchange capacities and base saturation, the acid conditions, and especially the accumulation of oxalate- and pyrophosphate-extractable iron and aluminum in the spodic (Bhs, Bs) horizon. The dominant soil-forming processes in Spodosols are podzolization, base cation leaching, with gleization occurring in Aquods. Spodosols primarily are used for growing pulpwood and firewood (Fig. 9.4).

10 Entisols

10.1 Distribution

Entisols account for 12% of the land area (17,014 km^2) (Fig. 1.2) and 11% of the soil series of Wisconsin (Fig. 1.3). Soil maps for Entisols are shown for Wisconsin at the order, suborder, and great group levels in Fig. 10.1.

Entisols are recent soils. The central concept is soils that are developed in unconsolidated parent material with usually no genetic horizons except for an A horizon and a Bw (development of color and structure) horizon. All soils that do not fit into one of the other 11 orders are classified as Entisols. Thus, they have considerable diversity, both in environmental setting and in land use.

In Wisconsin, Entisols are contained in four suborders: Aquents (aquic soil-moisture class), Fluvents (derived from alluvium), Psamments (derived from sandy materials), and Orthents (others). The Udipsamments are the most extensive great group of Wisconsin's Entisols, occupying 9987 km^2 (59% of Entisol area) in the state. There are also moderately large areas of Psammaquents (1736 km^2) and Quartzipsamments (1687 km^2). Udipsamments occur on glaciolacustrine sands in soil region C and on pitted outwash deposits in northwestern and northeastern Wisconsin. Quartzipsamments occur on alluvial and lacustrine deposits overlying sandy pediment materials and sandstones in soil region Dr. The Psammaquents occur on glaciolacustrine deposits in the western soil region C and on outwash in the eastern portion of soil region C. Seventy-five percent of the Wisconsin Entisols are in sandy particle-size classes, 81% have a mixed mineralogy, 70% are in unreactive CEC classes, and 50% of the Entisol series have a frigid soil-temperature class (50% mesic).

Some of the most common soil series by great group include the Coloma, Friendship, Grayling, Graycalm, Plainfield (Udipsamments), the Deford and Roscommon (Psammaquents), the Boone and Tarr (Quartzipsamments), and the Arenzville, Chaseville, Juneau, and Emmert (Udifluvents).

The Waupaca series (coarse-silty, mixed, superactive, nonacid, and frigid Mollic Fluvaquents) is an example of an Aquent that occurs in Wisconsin (Fig. 10.2). This series contains a mollic-like epipedon (0–23 cm) over gleyed water-laid deposits in the basins of glacial lakes and flood plains.

The Plainfield soil series (Typic Udipsamments) is a common Entisol in the Central Sands plain of Wisconsin (Fig. 10.3). The soil contains an ochric epipedon over a weakly developed Bw horizon from 15 to 70 cm. The Bw horizons have a stronger hue and chroma than the C horizons and a weak coarse subangular blocky structure but do not qualify as a cambic horizon. Although it is lightly stained with iron oxides, the Bw horizon has a similar color to the C horizon and a single grain structure. Not all Entisols in the Central Sands plain are well or excessively drained (Fig. 10.4) but large areas are used for irrigated agriculture. Crop yields are high and in some parts two crops per year are grown. Aldo Leopold referred to these as the "golden sands."

10.2 Properties and Processes

The properties of and processes in Wisconsin Entisols are extremely varied. Analytical data of four common Entisols in Wisconsin from the NRCS are given in Table 10.1. Each of the soils has an ochric epipedon but lacks a diagnostic subsurface horizon. The Aquents have gleying (Bg and Cg horizons) at depth. The Psamments have 88% sand or more and

J.G. Bockheim and A.E. Hartemink, *The Soils of Wisconsin*,
World Soils Book Series, DOI 10.1007/978-3-319-52144-2_10

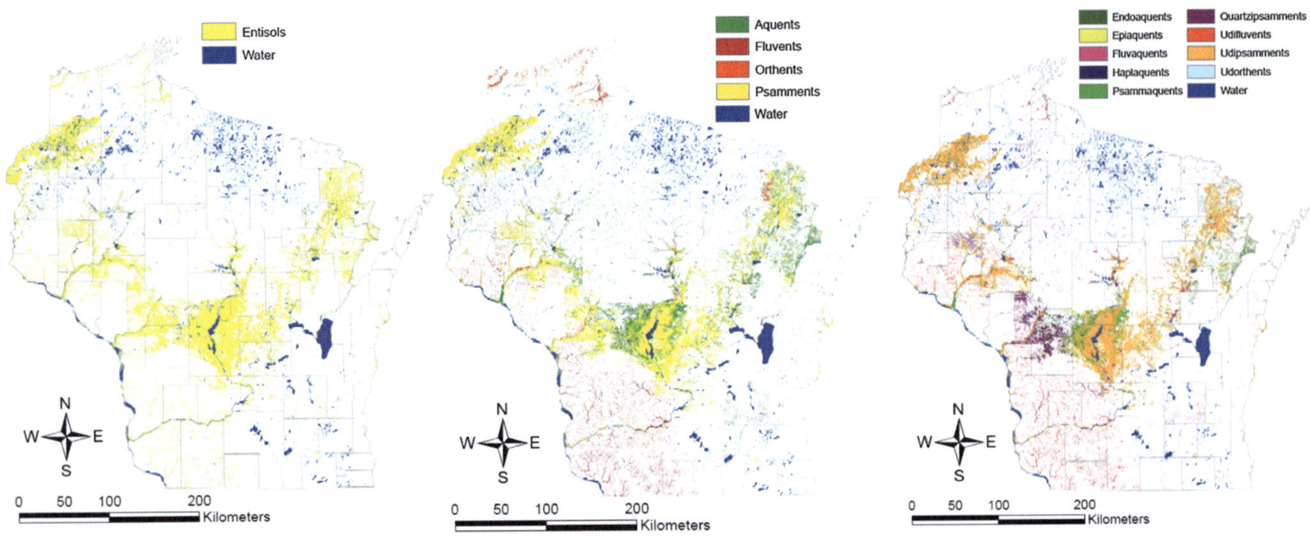

Fig. 10.1 Distribution of Entisols (order, suborders, and great groups) in Wisconsin

have very low CEC values and base saturation. Soil OC concentrations generally are low. The Fluvent has the typical buried A horizons and irregular depth distribution of SOC because of past flooding.

Madison and Lee (1965) studied some mineralogical characteristics of sandy soils in Wisconsin, including Entisols in soil regions C and H. A key finding was that soils derived from Ordivician sandstones in west-central Wisconsin can be characterized by their siliceous nature and grouped on the basis of their mineralogy. Figures 10.5 and 10.6 show weakly developed soils over sandstone in the southern part of the state—these soils are often less deep compared to the Entisols developed from sandy drift and in outwash.

10.3 Use and Management

Entisols are used in Wisconsin for irrigated vegetable crops in the central region, cranberry production in the west-central part of the state (Fig. 10.7), and for forestry, especially pulpwood and Christmas tree production, throughout the state. For red pine plantations, the Gotham (sandy Alfisols: Psammentic Hapludalfs) is exceptionally productive (site index 82) and the Brems (Aquic Udipsamments), Menagha and Plainfield series (both Typic Udipsamments) are moderately productive (site index 61–65) (Johnson et al. 1993). Continuously irrigated and cultivated Psamments may develop thick topsoils (up to 50 cm) (Watson and Hartemink 2015).

Fig. 10.2 The Waupaca series (coarse-silty, mixed, superactive, nonacid, and frigid Mollic Fluvaquents) is an example of Aquents in Wisconsin

Fig. 10.3 Examples of Typic Udipsamments (Plainfield soil series) in Wisconsin

Fig. 10.4 Sandy soils in the Central sands plain of Wisconsin, most of these soils are well drained (e.g., Plainfield and Richford) but some are moderately well drained as a result of their position in the landscape (e.g., Brems soil series)

Table 10.1 Analytical characteristics of some Entisol soil series from Wisconsin

Horizon	Depth (cm)	Clay (%)	Silt (%)	Sand (%)	SOC (%)	CEC7 (cmol(+)/kg)	Base sat. (%)	pH H_2O
Totagatic series; Typic Fluvaquents; Pedon No. 98P0250; Washburn, WI								
Oa	3–10	18.7	47.6	33.7	16.46	53.7	23	4.9
C1	10–20	6.2	9.7	84.1	0.85	6.1	59	4.6
C2	20–43	3.1	3.8	93.1	0.18	2.4	100	4.7
Cg1	43–71	3.2	2.1	94.7	0.58	3.6	100	4.1
Cg2	71–117	0.8	1.5	97.7	0.05	1.3	100	4.3
Cg3	117–152	1.3	2.8	95.9	0.33	2.1	71	3.7
Ponycreek series; Humaqueptic Psammaquents; Pedon No. 91P0486; Clark, WI								
A	0–15	1.4	11.0	87.6	3.03	9.1	4	3.6
Bg	15–71	0.0	2.8	97.2	0.11	0.5	20	4.1
Cg	71–152	0.0	0.9	99.1	0.08	0.5	20	4.5
Arenzville series; Typic Udifluvents; Pedon No. 40A5202; Iowa, WI								
Ap	0–30	14.5	82.5	2.0	1.70	17.0	100	6.8
C1	30–59	14.5	81.5	4.0	1.00	13.7	95	7.1
C2	59–79	17.5	77.2	5.3	1.70	17.7	99	7.0
C3	79–96	24.8	73.3	1.9	1.93	22.7	91	6.9
Ab1	96–112	26.2	69.2	4.6	3.20	37.3	76	6.3
Ab2	112–132	19.5	74.9	5.6	1.42	19.7	64	5.7
C4	132–152	16.1	77.5	6.4	0.92	15.8	61	5.6
Tarr series; Typic Quartzipsamments; Pedon No. 91P0481; Jackson, WI								
A	0–10	4.0	4.2	91.8	1.27	4.9	8	4.7
Bw1	10–23	3.9	3.6	92.5	0.37	2.1	5	5.0
Bw2	23–56	1.7	3.9	94.3	0.12	1.5	7	4.9
BC	56–69	0.2	2.3	97.5	0.07	0.7	3	4.9
C	69–152	0.0	0.8	99.2	0.03	0.3	3	5.1
Plainfield series; Typic Udipsamments; Pedon No. 40A1583; Waushara, WI								
Ap	0–18	4.3	7.2	88.5	0.46	1.4	51	6.9
Bw1	18–41	4.8	7.6	87.6	0.14	0.6	43	6.4
Bw2	41–71	2.2	1.2	96.6	0.09	0.5	39	5.7
BC	71–91	1.2	0.4	98.4	0.08	0.5	33	5.6
C	91–152	0.9	0.4	98.7	0.07	0.5	40	5.8

Fig. 10.5 Weakly developed soil over sandstone under oak trees in south central Wisconsin

Fig. 10.6 Weakly developed soil over sandstone under oak trees in south central Wisconsin

Fig. 10.7 Cranberry fields and harvest near Nekoosa, Wisconsin

10.4 Summary

Entisols account for 12% of the land area and 11% of the soil series of Wisconsin. Entisols are young soils. In Wisconsin, Entisols are contained in four suborders: Aquents, Fluvents, Psamments, and Orthents, with the Psamments being most dominant. The properties and processes in Entisols are weak and varied. Entisols are used in Wisconsin for irrigated vegetable crops in the central region, cranberry production in the west-central part of the state, and for forestry, especially pulpwood and Christmas tree production, throughout the state.

11 Mollisols

11.1 Distribution

Mollisols account for 10% of the land area (14,498 km^2) (Fig. 1.2) and 20% of the soil series of Wisconsin (Fig. 1.3). General soil maps at the order, suborder, and great group level are shown for Wisconsin in Fig. 11.1. Most Mollisols occur in the southern part of the state.

Mollisols (from Latin *mollis*, "soft") are the soils of grassland ecosystems. They are characterized by a thick, dark surface horizon. This fertile surface horizon, known as a mollic epipedon, results from the long-term addition of organic materials derived from turnover of plant roots.

In northcentral USA, the Udolls are distributed in a band that extends from Minnesota to southern Oklahoma but that is centered in Iowa and northern Illinois (Fig. 11.2). In Wisconsin, there is a break in the distribution of Mollisols with Iowa and southeastern Minnesota; the only linkage between Mollisols of Wisconsin is with Illinois. This suggests that Wisconsin represents the northeasternmost extension of Mollisols. Curtis (1959) suggested that some prairies in Wisconsin were initiated by Native Americans through burning.

Nearly three-quarters (60%) of the Mollisol soil series are well drained, moderately well drained, or somewhat poorly drained. Mollisols are most common on silt-rich parent materials, including loess (25% of soil series) and alluvium (25%), but they may occur on coarser materials, such as sandy outwash (16%). The native vegetation of 53% of the Mollisols was tallgrass prairie, but Mollisols are also formed under oak-savanna, marsh, and forest.

Mollisols are contained in two suborders: the Aquolls (Fig. 11.3) that have imperfect drainage and the Udolls that have good drainage and receive moisture throughout the growing season (Fig. 11.4). The Udolls are subdivided into two great groups: the Argiudolls, which contain an argillic horizon, and the Hapludolls, which follow the central concept of Udolls and generally have a cambic horizon. The Aquolls are subdivided into three great groups in the state: The Epiaquolls have restricted drainage because they are unable to divert rainfall through the profile; the Endoaquolls have a seasonally high water table; and the Argiaquolls have a clay-enriched horizon. The Argiudolls, Hapludolls, and Endoaquolls occupy 7823, 2826, and 2,721 km^2, respectively, which rank sixth through eighth among the great group extent in Wisconsin.

Endoaquolls include the Granby and Poy series; Epiaquolls include the Adolph and Poygan series; Argiudolls include the Ashdale, Dodgeville, Elburn, Hochheim, Plano, Richwood, Tama, and Warsaw series; and Hapludolls include the Port Byron and Sparta series.

Eighty-five percent of the Mollisol soil series in Wisconsin occur in the southern mesic zone; 69% are in the superactive CEC class; 98% are in the mixed mineralogy class; and 45 and 31% of the Mollisol series are in loamy and silty particle-size classes, respectively.

11.2 Properties and Processes

The key properties of Mollisols are the presence of a mollic epipedon over an argillic or cambic diagnostic subsurface horizon. The mollic epipedon is thick (>18 cm), dark-colored, contains organic C >0.6%, and has a base saturation >50%. The Marshan series, a fine-loamy over sandy or sandy-skeletal, mixed, superactive, mesic Typic Endoaquolls (tape measure in feet), is an example of an Aquoll in Wisconsin (Fig. 11.3). The soil contains a 25-cm-thick (the tape is in feet) mollic epipedon over a gleyed C. The following figures show Argiudolls derived from different parent materials, including red clay over dolostone at Platteville (Fig. 11.4), loess over till over dolostone at Arlington (Fig. 11.5), and loess at the West Madison Agricultural Research Station (Fig. 11.6). The Argiudolls are some of the highest producing soils in Wisconsin and many of these soils are used for growing corn and soybean (Fig. 11.7).

The Argiudolls in West Madison (Fig. 11.6) and the soil at OJ Noer Research Station (Fig. 11.8) show a thick A

J.G. Bockheim and A.E. Hartemink, *The Soils of Wisconsin*,
World Soils Book Series, DOI 10.1007/978-3-319-52144-2_11

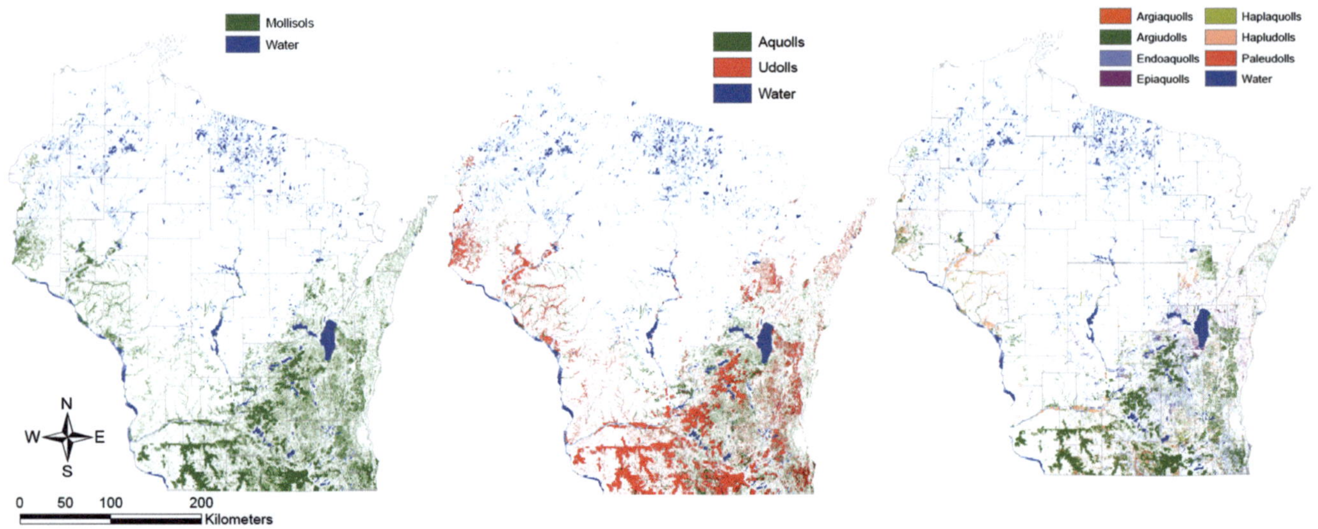

Fig. 11.1 Distribution of Mollisols (orders, suborders, great groups) in Wisconsin. *Note* Haplaquolls are no longer recognized in Soil Taxonomy. These soils would be in Endo- or Epi- great groups

Fig. 11.2 Distribution of Mollisols in northcentral USA

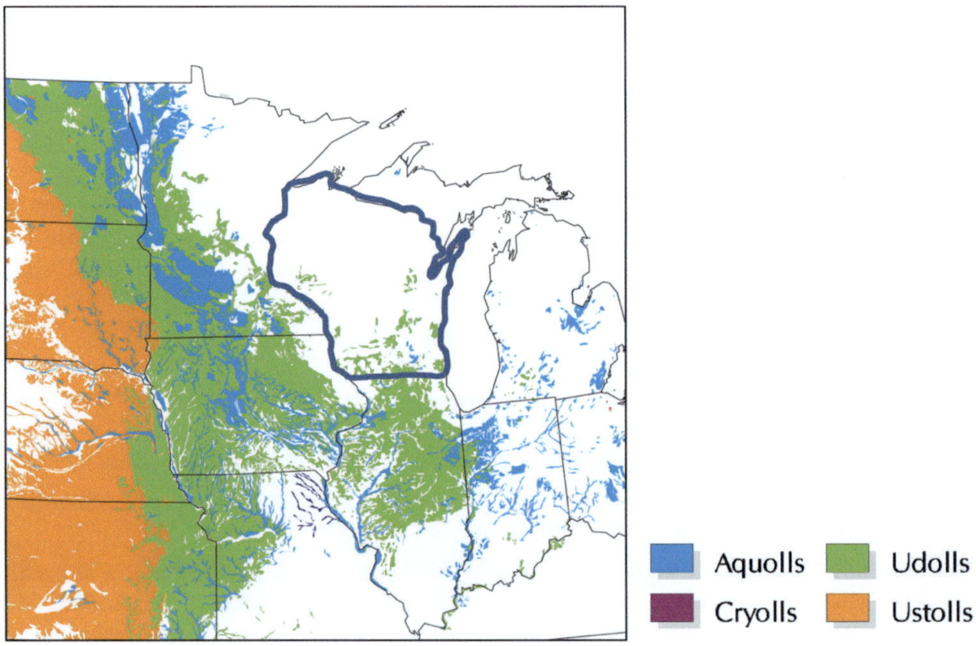

horizon over a buried A over an argillic horizon with some redox features. Buried soils in the lower parts of soilscapes are not uncommon and the result of sedimentation following erosion of soils on the slopes. No stratification is observed in these soils and any layering is these soils is destroyed by tillage. It is also possible that the sedimentation followed one or a few catastrophic rainfall events inducing massive erosion. Buried A horizons are often much darker than the overlying A horizons possibly as the soil organic matter is somewhat protected from further decomposition. The organic matter in such horizons is derived from different vegetation and may have formed when the soils were less well drained.

Many Argiudolls are formed in loess that has been deposited over coarse till or outwash. Figure 11.9 shows a Typic Argiudolls near Nelsonville that was cropped for many decades before it was reforested. Stones are commonly found in the loess part of the soil profile. They rose to the

Fig. 11.3 The Marshan series, a fine-loamy over sandy or sandy-skeletal, mixed, superactive, mesic Typic Endoaquolls (tape measure in feet), is an example of an Aquolls in Wisconsin

surface from frost heaving. As the stones affected tillage and harvesting equipment, they were collected each spring by the farmers and piled on the edge of the fields (Fig. 11.9). Stones are lifted upward by frost at rates of 1.5–5.5 cm/year (French 1996).

Analytical properties of several Mollisols are provided in Table 11.1. The soils are derived from clayey till (Poygan series—Typic Epiaquolls), loess or other silty sediments (Plano and Pella series), and sandy outwash (Sparta series). The Pella and Poygan (Typic Endoaquolls), and Plano series (Typic Argiudolls) contain primary (nonpedogenic) carbonates in the subsoil. The mollic epipedon ranges from 25 to 51 cm in thickness, 0.6–10.6% soil organic carbon, and 60–100% base saturation.

The dominant soil-forming processes in Mollisols are melanization and biological enrichment of bases; ancillary processes include argilluviation, gleization, and cambisolization. Melanization refers to the accumulation of organic matter that darkens the mineral soil. This process may be fostered by pedoturbation, including the activities of ants (Baxter and Hole 1967) and earthworms (Nielsen and Hole 1963, 1964).

It can be assumed that Mollisols have formed as a result of regular burning (natural or by humans) so that grassland vegetation dominates and not woody vegetation (Fig. 11.10). Burning is not so common and only takes placed controlled in prairie restoration projects. Many soils in Wisconsin have traces of burning such as charcoal and other black carbon.

11.3 Use and Management

Mollisols are highly fertile soils and used almost exclusively for crop production. They include soil series that are among the productive series in Wisconsin and the Midwest. These series are included in the "prime" and "unique" farmland groups and include the Judson (Cumulic Hapludolls); Plano, Dakota, and Tama (Typic Argiudolls); Ashkum (Typic Endoaquolls); Brookston (Typic Argiaquolls) and others.

Fig. 11.4 An Argiudolls near Platteville developed in loess over *red* clay and dolostone

Fig. 11.5 An Argiudolls near Arlington developed in loess over coarse till (*top*). At 150 cm, the coarse till covers striated dolostone

Fig. 11.6 A Troxel soil series (Pachic Argiudolls) with a buried A horizon at 70 cm soil depth developed in loess at West Madison Agricultural Research Station. The soil is located at the footslope

Fig. 11.7 Mollisol under corn at Arlington. Plano soil series (Typic Argiudolls)

Fig. 11.8 Buried soil that has developed into a Mollisol at OJ Noer Research Station

Fig. 11.9 A Mollisol under forest near Nelsonville (Typic Argiudolls). Field was cropped for many decades before reforested. Stones and boulders piled up along the edge of the field (lower picture). Many of these travelled upward through the soil profile over time (frost heaving). Every spring farmers removed the stones from the field and piled them up at the edges

Table 11.1 Analytical characteristics of some Mollisol soil series from Wisconsin

Horizon	Depth (cm)	Clay (%)	Silt (%)	Sand (%)	SOC (%)	CEC7 (cmol(+)/kg)	Base sat. (%)	pH H_2O
Pella series; Typic Endoaquolls; Pedon No. 40A1530; Dodge, WI								
						CEC8		
A1	0–10				10.6	72.0	85	6.5
A2	10–23	35.5	58.7	5.8	4.8	56.1	87	6.7
Ag	23–36	35.0	59.6	5.4	0.97	37.0	93	7.1
Cg1	36–61	30.4	64.9	4.7	0.32	26.6	94	7.3
Cg2	61–76	38.6	57.8	3.6	0.39	29.5	93	7.4
Cg3	76–117	39.4	57.5	3.1	0.39	28.5	93	7.3
2C	117	5.7	32.6	61.7	0.12			8.0
Poygan series; Typic Epiaquolls; Pedon No. 84P0923; Fond du Lac, WI								
Ap	0–25	39.3	49.0	11.7	2.86	38.6	100	7.6
Bg1	25–37	27.1	44.0	28.9	0.45	19.9	100	7.7
Bg2	37–50	33.1	44.0	22.9	0.28	23.1	100	7.8
Bg3	50–65	33.1	37.3	29.6	0.2	21.4	100	7.9
C1	65–100	47.3	41.3	11.1	0.27	17.3	100	8.1
C2	100–152	49.2	39.4	11.1	0.28	15.7	100	8.3
Plano series; Typic Argiudolls; Pedon No. 40A1585; Columbia, WI								
Ap	0–20	24.4	71.7	3.9	3.36	22.2	89	6.1
A1	20–38	23.0	73.6	3.4	2.04	16.6	60	5.3
A2	38–51	19.2	77.0	3.8	0.8	11.9	63	5.2
B1	51–66	25.8	70.6	3.6	0.48	15.1	79	5.3
Bt1	66–94	30.9	65.9	3.2	0.42	19.1	85	5.2
Bt2	94–112	28.1	68.4	3.5	0.2	18.8	89	5.3
Bt3	112–150	22.8	64.6	12.6	0.13	15.0	97	5.8
2Bt4	150–163	14.5	35.1	50.4	0.11	9.2	100	6.0
2Bt5	163–178	13.0	24.6	62.4	0.08	8.5	100	7.0
2C1	178–190	6.9	20.7	72.4	0.06	3.9	100	7.7
Sparta series; Entic Hapludolls; Pedon No. 90P0556; Grant, WI								
Ap	0–23	2.6	4.0	93.4	0.64	3.4	85	5.7
BA	23–33	2.2	4.4	93.4	0.62	3.1	65	5.5
Bw1	33–46	1.9	3.6	94.5	0.48	2.1	52	5.4
Bw2	46–69	0.1	3.5	96.4	0.28	1.3	62	5.3
BC	69–114	0.0	3.4	96.6	0.12	0.8	25	5.1
C	114–152	0.0	0.5	99.5	0.03	0.3	33	5.0

Fig. 11.10 Burned prairie as part of a restoration project near Paoli

11.4 Summary

Mollisols are the soils of grassland ecosystems and are characterized by a thick, dark surface mineral horizon. Mollisols account for 10% of the land area but 20% of the soil series of Wisconsin. Eighty-five percent of the Mollisol soil series in Wisconsin occur in the southern mesic zone; 69% are in the superactive CEC class; 98% are in the mixed mineralogy class; and 45 and 31% of the Mollisol series are in loamy and silty particle-size classes. The key properties of Mollisols are the presence of a mollic epipedon over an argillic or cambic subsurface horizon. The dominant soil-forming processes in Mollisols are melanization and biological enrichment of bases; ancillary processes include argilluviation, gleization, and cambisolization. Mollisols are used almost exclusively for crop production.

12 Histosols

12.1 Distribution

Histosols account for 10% of the land area (13,341 km^2) (Fig. 1.2) but only 4% of the soil series of Wisconsin (Fig. 1.3). There are ten Histosol soil series that occupy more than 400 km^2 each in Wisconsin. General soil maps at the order, suborder, and great group levels are given in Fig. 12.1.

Histosols (from Greek *histos*, "tissue") are soils that are composed mainly of organic materials. They contain at least 20–30% organic matter by weight and are more than 40 cm thick. Histosols are divided into three suborders based on the degree of decomposition of the organic matter, from least to most in decomposition: Fibrists, Hemists, and Saprists. Most Histosols in Wisconsin are Saprists (95% of Histosol soil area). In fact, Haplosaprists account for 12,706 km^2 and have the fourth largest area of soil great groups in the state. The average thickness of organic materials in Wisconsin Histosols is 101 cm (Fig. 12.2).

Seventy percent of Wisconsin's Histosols occur in the frigid northern zone; 73% of the Histosols are euic (i.e., having a pH of $\geq$4.5); and 47% are in terric subgroups (i.e., having 30-cm-thick mineral soil layers within the control section). Extensive (exceeds 400 km^2) Histosol soil series in Wisconsin include the Cathro, Seelyeville, Lupton, Houghton, Markey, Loxley, Dawson, Palms, Adrian, and Beseman, which are all Haplosaprists. The Greenwood and Rifle peats are the most abundant Haplohemists, each of which does not exceed 400 km^2.

An example of a common Histosol in Wisconsin is shown in Fig. 12.3. The Houghton series, a euic, mesic Typic Haplosaprists, has a series of sapric horizons (Oa) that extend to more than 130 cm. The water table in the image of the Houghton series is at 47 cm. Figure 12.4 shows Histosols (lower panel) occurring in depressions between drumlins (upper panel).

12.2 Properties and Processes

The key property of Histosols is the presence of large amounts of organic materials. In the two pedons in Table 12.1, the organic deposits range between 147 and 180 cm, and the organic carbon levels between 32 and 54%. The organic material in these soils has a high pH-dependent cation-exchange capacity; the CEC ranges between 93 and 109 cmol(+)/kg. Both pedons have low pH values (<4.7) and are dysic.

Some of the early work on Histosols was carried out by G.B. Lee and his graduate students at the UW-Madison. Frazier and Lee (1971) characterized three Wisconsin Histosols: a Fibrist, a Hemist, and a Saprist. They determined that fiber content was the single most useful characteristic in differentiating among the soils. Sodium pyrophosphate extract color was also a useful parameter for estimating the degree of decomposition. Organic C contents were highest in the Saprist and lowest in the Fibrist. Langton and Lee (1964) characterized Saprists (muck) from southern Wisconsin. The highly decomposed sapric materials were formed by faunal activity from arthropods and earthworms.

The dominant soil process in Histosols is paludization, the buildup of organic matter from conditions of restricted drainage, which leads to gleization (Bockheim and Gennadiyev 2000).

J.G. Bockheim and A.E. Hartemink, *The Soils of Wisconsin*,
World Soils Book Series, DOI 10.1007/978-3-319-52144-2_12

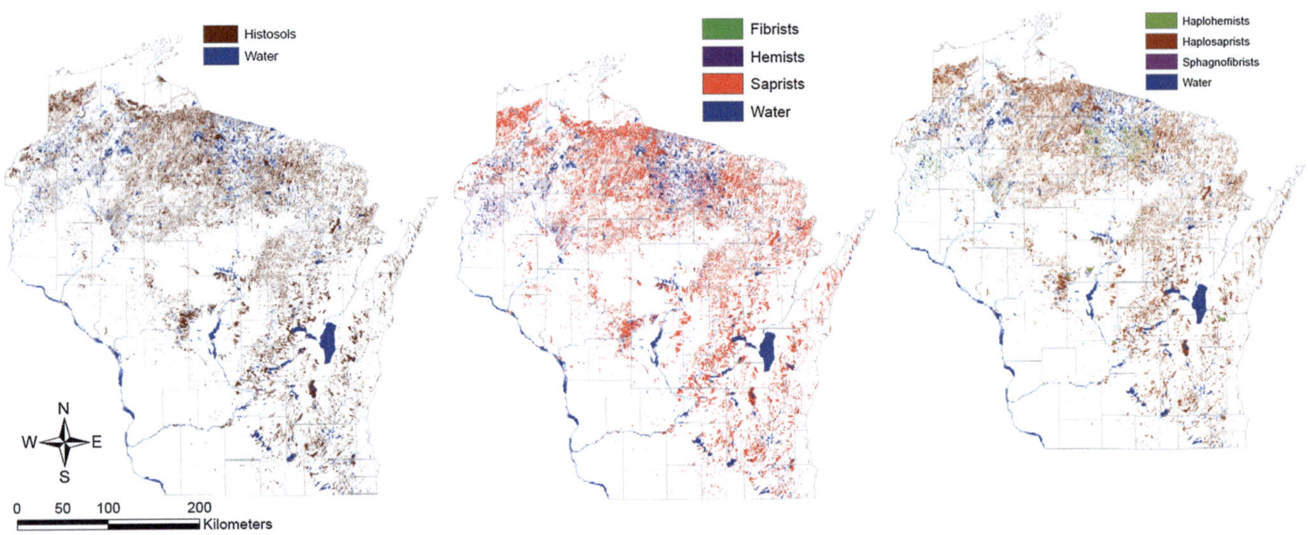

Fig. 12.1 Distribution of Histosols (orders, suborders, great groups) in Wisconsin

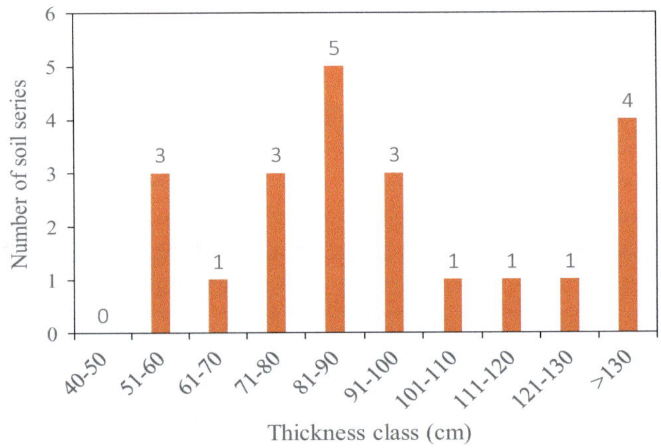

Fig. 12.2 Frequency distribution of Histosols thickness in Wisconsin

12.3 Use and Management

Histosols are used in Wisconsin for truck crops (onions, celery) and cranberries. Drainage will lead to subsidence as the organic material is decomposed (Fig. 12.5). Many Histosols are forested (Fig. 12.6), and some are productive for trees, including the Carbondale (Hemic Haplosaprists) and Palms (Terric Haplosaprists) for white ash and the Houghton (Typic Haplosaprists) for red maple (Johnson et al. 1993).

Fig. 12.3 Houghton series, a euic, mesic Typic Haplosaprists, is an example of a Histosol from Wisconsin (Photograph by J.W. Williams)

Fig. 12.4 Hillshade map over aerial photograph showing drumlins in Jefferson County. Soils on the drumlins are often Alfisols and Mollisols. Between the drumlins, there are large areas where drainage is impeded and Histosols have formed (lower picture)

Table 12.1 Analytical characteristics of some Histosol soil series from Wisconsin

Horizon	Depth (cm)	Clay (%)	Silt (%)	Sand (%)	SOC (%)	CEC7 (cmol(+)/kg)	Base sat. (%)	pH H_2O
Greenwood series; Typic Haplohemists; Pedon No. 40A1638; Oneida, WI								
Oi1	0–8				53.3	90.5		3.3
Oi2	8–13				43.2	95		3.5
Oe1	13–23				54.3	102		3.2
Oe2	23–50				47.9	189		3.8
Oe3	50–70				49.8	93.3		3.4
Oe4	70–100				53.2	102		3.6
Oe5	100–180							

(continued)

Table 12.1 (continued)

Horizon	Depth (cm)	Clay (%)	Silt (%)	Sand (%)	SOC (%)	CEC7 (cmol(+)/kg)	Base sat. (%)	pH H_2O
C	180–312							
2Cg	312–370							
Seelyeville series; Typic Haplosaprists; Pedon No. 40A1664; Oconto, WI								
Oap	0–15				34.1	85.9		5.2
Oa1	15–38				34.4	93		5.1
Oa2	38–82				32.5	104		4.4
Oa3	82–96				38.2	109		4.4
Oa4	96–147				43.8	109		4.6
2C	147–167	0.9	1.5	97.6				4.8

Fig. 12.5 Histosols (sapric soil material) used for truck crops (e.g., onions, celery) in Jefferson county. Special equipment is required on these soils; drainage required for sufficient aeration causes oxidation of the organic material and subsidence of the land

Fig. 12.6 Histosols in northern Wisconsin with fibric soil material, and dark-colored lakes as a result of high dissolved organic carbon levels. Vegetation is stunted due to adverse growing conditions

12.4 Summary

Histosols account for 10% of the land area but only 4% of the soil series of Wisconsin. Histosols are soils that are composed mainly of organic materials. They contain at least 20–30% organic matter by weight and are more than 40 cm thick. Most of Wisconsin's Histosols are Saprists (95% of Histosol soil area). Seventy percent of Histosols occur in the frigid northern zone; 73% are euic; and 47% are in terric subgroups. The dominant soil process in Histosols is paludization, the buildup of organic matter from conditions of restricted drainage, which leads to gleization. Histosols are used for truck crops and cranberries, whereas others are productive for trees.

13 Inceptisols

13.1 Distribution

Inceptisols cover 4% of the land area (5050 km^2) (Fig. 1.2) and 7% of the soil series of Wisconsin (Fig. 1.3). Maps of Inceptisols at the order, suborder, and great group levels are shown in Fig. 13.1. Inceptisols occur scattered throughout the state but tend to be absent in the southeast part.

Inceptisols (from Latin *inceptum*, "beginning") are soils that exhibit minimal horizon development. They are more developed than Entisols, but lack the features required for other soil orders. In Wisconsin, 52% of Inceptisols are Udepts and 48% are Aquepts. Inceptisols mainly contain ochric and cambic horizons which average 17 cm and 46 cm in thickness, respectively. Five of the 51 Inceptisol soil series in Wisconsin have an umbric epipedon, and five have a histic epipedon that average 39 and 24 cm in thickness, respectively.

Fifty-two percent of the Inceptisols in Wisconsin are in the frigid northern zone; 35% are nonacid; 37% are in the superactive CEC class; 79% have a mixed mineral class; and 79% are in loamy or silty particle-size classes. Common soil series of Inceptisols in Wisconsin include the Minocqua, Cable, Vesper, and Veedum soil series (Aquepts) and the Urne, Elk Mound, and Cromwell (Udepts).

Examples of Udepts and Aquepts derived from outwash over dolostone in southcentral Wisconsin are shown in Fig. 13.2. Profiles of a Typic Eutrudepts (left) and a Mollic Endoaquepts (right) are shown in the upper panel. These soils appear in the upper and mid-slope positions in the soilscape (old lake bottom) shown in the lower panel. The black soils shown in the upper right portion of the panel are poorly drained Mollisols (Typic Endoaquolls).

13.2 Properties and Processes

The five pedons in Table 13.1 represent each of the five great groups of Inceptisols identified in Wisconsin. Four of the pedons are derived from sandy deposits; the Cable series has formed from silty materials. Three of the five soils contain carbonates (Mosinee, Urne, and Makwa). One of the pedons (Makwa—Histic Humaquepts) has a mollic epipedon; one has a histic epipedon (Makwa); and the rest have an ochric (Mosinee—Typic Dystrudepts and Urne—Dystric Eutrudepts) or ochric and albic horizons (Minocqua—Typic Endoaquepts and Cable—Typic Epiaquepts). All five pedons have a cambic horizon.

There is a variety of soil-forming processes occurring in the Inceptisols, but cambisolization is the dominant process. Other processes include gleization, biological enrichment of bases, biological depletion of bases, and melanization.

13.3 Use and Management

Inceptisols have a variety of land uses in Wisconsin, including forestry (Lincoln and Iron counties), shiitake mushroom production (Marathon County), fruit-tree production (Door County), forestry and grazing (Eau Claire and Dunn counties), wildlife protection (Wood County), and crop production and grazing (Juneau County).

J.G. Bockheim and A.E. Hartemink, *The Soils of Wisconsin*,
World Soils Book Series, DOI 10.1007/978-3-319-52144-2_13

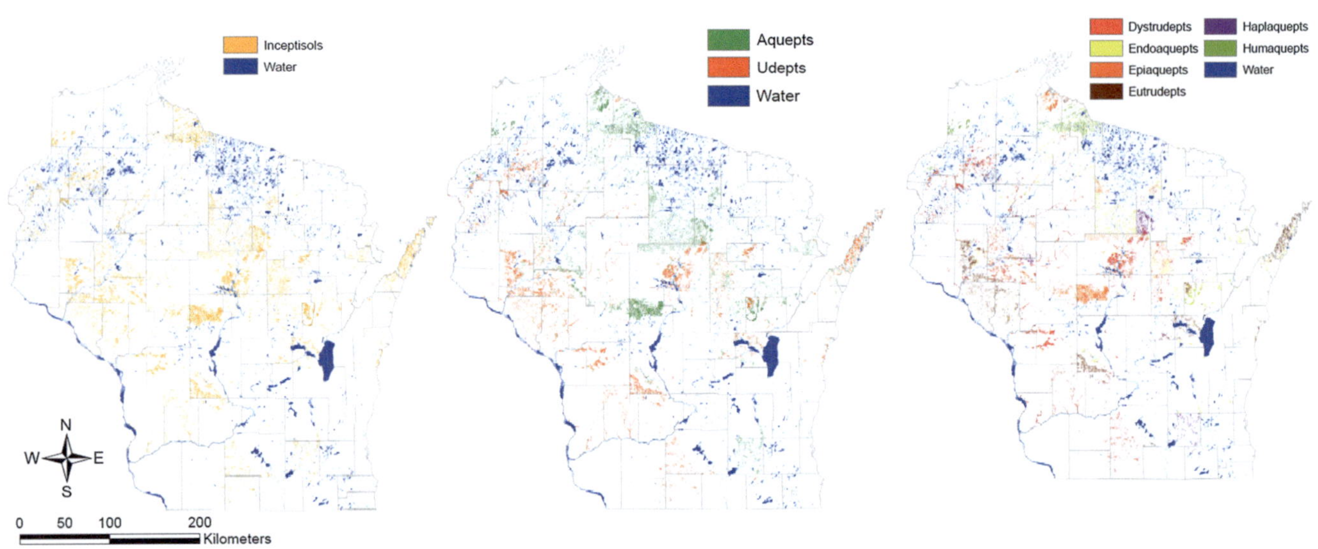

Fig. 13.1 Distribution of Inceptisols (soil orders, suborders, great groups) in Wisconsin

Fig. 13.2 Inceptisols in southcentral Wisconsin developed in outwash over dolostone, including Typic Eutrudepts (*upper*, *left*) and Mollic Endoaquepts (*upper*, *right*). In the lower part of the soilscape (old lake bottom), there are poorly drained Mollisols (Typic Endoaquolls)

Table 13.1 Analytical characteristics of some Inceptisol soil series from Wisconsin

Horizon	Depth (cm)	Clay (%)	Silt (%)	Sand (%)	SOC (%)	CEC7 (cmol(+)/kg)	Base sat. (%)	pH H_2O
Mosinee series; Typic Dystrudepts; Pedon No. 81P0826; Marathon, WI								
Ap	0–23	6.6	24.6	68.8	1.52	8.8	100	7.2
Bw1	23–38	4.5	26.0	69.5	0.39	5.4	96	7.0
Bw2	38–61	5.3	28.9	65.8	0.38	6.0	88	6.9
2Bw1	61–79	5.3	28.2	66.5	0.29	5.6	89	6.9
2Bw2	79–104	4.9	26.3	68.8	0.35	5.0	86	6.9
R	104–147	8.1	21.4	70.5	0.42	6.5	92	6.9
Urne series; Dystric Eutrudepts; Pedon No. 01P0227; LaCrosse, WI								
Ap	0–20	11.4	43.5	45.1	1.53	5.5	75	7.0
Bw1	20–36	9.1	29.4	61.5	0.24	4.3	88	7.0
Bw2	36–62	17.4	13.3	69.3	0.08	8.1	49	6.8
Minocqua series; Typic Endoaquepts; Pedon No. 99P0484; Waupaca, WI								
A	0–23	7.2	30.5	62.3				4.9
Bg1	23–48	2.8	28.4	68.8				5.5
Bg2	48–94	14.2	15.9	69.9				5.3
Cable series; Typic Epiaquepts; Pedon No. 82P0202; Vilas, WI								
A	0–10	16.3	66.3	17.4	5.51			5.2
E	10–15	12.8	69.0	18.2	2.96			5.0
Bg	15–28	6.9	73.7	19.4	0.50			5.3
Bw1	28–58	6.1	64.4	29.5	0.24			5.3
Bw2	58–81	9.7	43.6	46.7	0.14			5.5
Makwa series; Histic Humaquepts; Pedon No. 03N0888; Burnett, WI								
Oa	0–20	25.6	46.0	28.4	15.58	66.3	91	5.2
A	20–41	16.2	35.8	48.0	2.99	23.9	86	6.3
Bw	41–109	8.0	19.3	72.7	0.62	9.1	89	6.4
Cg	109–165	12.8	18.4	68.8	0.14	11.5	100	6.5
2C	165–203	13.6	82.7	3.7	0.19	9.1	100	7.6

13.4 Summary

Inceptisols account for 4% of the land area and 7% of the soil series of Wisconsin. Inceptisols have minimal horizon development. They are more developed than Entisols, but lack the features required for other soil orders. Fifty-two percent of the Inceptisols in Wisconsin occur in the Udepts suborder and 48% in the Aquepts suborder. Fifty-two percent of the Inceptisols in Wisconsin are in the frigid northern zone; 35% are nonacid; 37% are in the superactive CEC class; 79% have a mixed mineral class; and 79% are in loamy or silty particle-size classes. There is a variety of soil-forming processes in Inceptisols, but cambisolization is the dominant process. Other processes include gleization, biological enrichment of bases, biological depletion of bases, and melanization. Inceptisols have multiple land uses, including forestry, agriculture, and grazing.

14 Paleosols

14.1 Introduction

A paleosol may be defined as a soil that is no longer in equilibrium with the present environment (Ruhe 1965). Paleosols may be buried, relict (remain at the surface as ground-soils), or exhumed (previously buried material eroded from the surface) (Ruhe 1956). Numerous paleosols have been studied in Wisconsin. Early Holocene soils occur in northern Wisconsin on sediments deposited by an advance of ice into the Lake Superior basin 9.9 kyr BP (Clayton 1984). Most of the soils on the massive end moraines and drift contained within were deposited by late Wisconsinan ice during the Last Glacial Maximum (LGM) 10–30 kyr BP (Attig et al. 2011; Syverson and Colgan 2011). The oldest drifts in Wisconsin are of Illinoian to pre-late Wisconsinan in age (300–30 kyr) or from prior to the Illinoian Glaciation (2.4–0.78 Myr BP). Paleosols of pre-Quaternary age have been buried by a thin cover of loess in the Driftless Area or by thicker glacial sediments elsewhere. These will be discussed in a following section.

14.2 Paleosols of Wisconsin

Dury and Knox (1971) described duricrusts and deep-weathering profiles in southwestern Wisconsin attributed to a humid–tropical kind of pedogenesis from the mid-Miocene (15 Myr BP). Red clays associated with Prairie du Chien and Sinnipee (Galena) dolomites and loess were studied stratigraphically and geomorphically at ten locations in loess-covered Wisconsin Driftless Uplands by Frolking et al. (1983). The smectite-rich clay has coated dolomitic sands and was illuviated from the loess (Fig. 14.1). They proposed that quartz be used as an indicator of red-clay provenance.

Hole (1974) described relict bisequal (two-storied) soils on the Menominee Tribal Lands that recorded a distinct post-glacial climate change. Clays had migrated downward in the profile under temperate deciduous vegetation during the Hypsithermal Interval (ca. 8–4 kyr BP), forming Alfisols (base-rich forest soils with clay-enriched horizons). During the subsequent cooler period, Spodosols formed in the E horizon of the former Alfisols under the influence of temperate coniferous forest, resulting in bisequal (Spodosol over Alfisol) soils.

The soils investigated by Frolking et al. (1983) probably belong to the Valton series, a fine-silty, mixed, superactive mesic Mollic Paleudalfs. This soil is formed from loess and in the underlying clayey residuum from weathered limestone. The argillic horizon is more than 125 cm thick and is formed partially in the loess but mainly in the underlying clayey-limestone saprolite. The paleo-argillic horizon is yellowish red (5YR 4/6), is very strongly acidic, and contains up to 70% clay (Table 14.1). There are up to 15% resistant chert fragments in the saprolite.

The depth to paleo-argillic horizon or red clay is variable (Table 14.2, Fig. 14.2). Evans and Hartemink (2014) conducted detailed studies (144 pedons of 16 catenas) in a 70 km^2 area near Verona in the Driftless Area. Half of all the pedons had a red clay subsoil (mostly 5YR 3/4) and on average over 40% clay. The mean thickness of the loess above the red clay was 50 cm but it ranged from 7 to 130 cm. The thickness of the red clay ranged from 4 to 60 cm with an average thickness of 25 cm. The red clay was found in every part of the landscape except the valley bottoms: it was present in soils on summits, shoulders, backslopes, and footslopes (Fig. 14.3). The red clay is only found over dolostone, but not present on all dolostone ridges. It was concluded that only part of the variation can be explained by topography.

Mason et al. (1994) investigated a red and brown sandy clay layer containing quartz and chert gravel overlying Cambrian sandstone outside the limits of the most recent Pleistocene glaciation in Portage County. The layer was interpreted as a thick, well-developed paleo-argillic horizon

J.G. Bockheim and A.E. Hartemink, *The Soils of Wisconsin*,
World Soils Book Series, DOI 10.1007/978-3-319-52144-2_14

Fig. 14.1 Loess over red clay (Rountree formation) over dolostone in a mine pit near Richland Center, Driftless Area of Wisconsin

belonging to an Ultisol. Based on the degree of development, they suggested that the soil could be pre-Wisconsin, or even pre-Quaternary in age. The K-feldspar grains in the sandy clay loam layer were highly weathered.

The soil investigated by Mason et al. (1994) may be the Siouxcreek series, a coarse-loamy over sandy or sandy-skeletal, mixed, active, frigid Aquic Hapludults. This soil is derived from silty or loamy alluvium over sandy residuum and occupies only 2.4 km^2 primarily in Baron County. The soil contains a dark yellowish brown (10YR 4/4), 25-cm thick argillic horizon over a fine-grained sandstone paralithic contact at from 50 to 100 cm. The subsoil is strongly to very strongly acid (pH 4.9–5.1) and has a base saturation less than 60% (Table 17.1).

Fig. 14.2 Thick layers of red clay exposed in a dolostone quarry near Richland Center in the Driftless Area

Table 14.3 contains a provisional list of paleosol soil series in Wisconsin. These soils can be divided into five age groups:

(i) Holocene burials
(ii) relict soils on pre-late Wisconsinan to Illinoian till
(iii) relict soils on pre-Illinoian till
(iv) loess over clayey pedisediments containing a buried soil over sandstone bedrock
(v) loess over clayey pedisediments containing a buried soil over dolomitic or limestone bedrock (Fig. 14.4).

The first group includes the Juneau (Typic Udifluvents), Quarderer (Fluventic Hapludolls), Radford (Fluvaquentic Hapludolls), and Wallkill (Fluvaquentic Humaquepts) soil series, which contain young, poorly developed soils derived from alluvium. The second group includes soils derived from loess overlying drift of pre-late Wisconsinan to Illinoian age and includes the Durand (Typic Argiudolls), Flagg (Typic Hapludalfs), Myrtle (Mollic Hapludalfs), Ogle (Typic Argiudolls), Pecatonica (Typic Hapludalfs), and Winnebago (Typic Argiudolls) soil series. These soils contain strongly weathered Bt horizons. The third group contains soils derived from younger till, mudflow deposits, or loess overlying pre-Illinoian drift and includes the Haugen (Oxyaquic Paleudalfs) and Hersey (Mollic Hapludalfs) soil series. These soils contain deep, strongly weathered Bt horizons (Table 14.3).

The fourth and fifth groups include soils derived from loess overlying clayey pedisediments over sandstone or limestone. Group four includes the Basco (Mollic Hapludalfs), Dolph (Aeric Glossaqualfs), Elevasil (Ultic Hapludalfs), Fairchild (Ultic Epiaquods), Humbird (Ultic Haplorthods), La Farge (Typic Hapludalfs), Merrillan (Ultic Epiaquods), Silverhill (Ultic Hapludalfs), and Siouxcreek (Aquic Hapludults) soil series, all of which occur in the Western Sandstone Uplands (soil region Fr).

Soil series in group five occur mainly in the Driftless Area (soil regions A or Am) and include the Ashdale, Brinkman, Dodgeville, Dubuque, Dunbarton, Fivepoints, Milton, Newglarus (Fig. 5.5), Palsgrove, Pepin, Reedsburg, Rockton, Schapville, Sogn, Valton, Whalan, and Wildale series. The Brinkman, Reedsburg, Valton, and Wildale soil series are Paleudalfs with Bt horizons that extend to 150 cm or more.

Buried soils have been observed throughout Wisconsin, including the well-studied post-Two Creeks soil in Manitowoc County along Lake Michigan (Lee and Horn 1972) (Fig. 14.5). The Two Creekan substage of the Wisconsinan occurred from 12.5 to 11.0 ^{14}C kyr BP. The paleosol has a peaty layer containing spruce (*Picea glauca*) needles and a thin gleyed horizon with spruce wood fragments. Hogan and Beatty (1963) analyzed the properties of Peorian loess ($\sim$23 kyr BP) and buried in southwestern Wisconsin. The Sangamon paleosol, which formed approximately 100 kyr ago, displays strong colors and thick Bt horizons (Jacobs et al. 2009).

Table 14.1 Analytical characteristics of some paleosol soil series from Wisconsin

Horizon	Depth (cm)	Clay (%)	Silt (%)	Sand (%)	SOC (%)	Fed (%)	Ald (%)	CEC7 [cmol(+)/kg]	Base sat. (%)	pH H_2O
Valton series; Mollic Paleudalfs; Pedon No. 40A5208); Monroe, WI										
Ap	0–23	20.1	73.4	6.5	0.42	1.1	0.1	13.6	97	6.3
Bt1	23–36	26.8	68.1	5.1	0.19	1.3	0.2	15.0	87	5.8
Bt2	36–56	31.3	60.9	7.8	0.21	1.6	0.2	17.0		5.3
2Bt11	56–81	41.5	48.4	10.1	0.17	3.7	0.3	23.5	78	4.9
2Bt12	81–107	44.8	42.0	13.2	0.14	3.7	0.3	23.9	82	5.1
2Bt21	107–140	50.4	40.3	9.3	0.14	4.5	0.3	25.9	80	5.1
2Bt22	140–168	70.6	19.8	9.6	0.12	5	0.3	37.1	85	5.0
Siouxcreek series; Aquic Hapludults; Pedon No. 91P0448; Barron, WI										
Ap	0–20	10.7	55.3	34.0	2.08			13.0	58	6.3
E	20–25	8.4	54.4	37.2	0.37			5.8	57	6.0
E/B	25–36	9.6	54.0	36.4	0.25			5.7	68	6.1
B/E	36–56	11.7	51.3	37.0	0.24			6.7	55	5.2
Bt1	56–74	12.5	33.2	54.3	0.16			7.5	48	4.9
Bt2	74–81	9.2	19.2	71.6	0.09			5.3	45	4.9
2Cr	81–97	1.8	4.8	93.4	0.04			0.9	22	5.1
Silverhill series; Ultic Hapludalfs; Pedon 92P0414; Jackson, WI										
Ap	0–20	7.1	31.9	61.0	1.18			6.5	65	5.3
Bt1	20–36	8.3	28.6	63.1	0.36			4.8	72	5.8
Bt2	36–66	11.0	24.6	64.4	0.23			5.8	61	5.5
2BC	66–81	4.3	8.0	87.7	0.06			2.5	57	5.0
2C	81–127	2.3	4.9	92.9	0.05			1.5	43	4.9
Humbird series; Oxyaquic Ultic Haplorthods; Pedon No. 94P0155; Clark, WI										
						Feo	Alo			
A	0–3	4.9	31.3	63.8	4.40	0.03	0.07	15.5	11	4.2
E	3–13	3.5	30.3	66.2	0.69	0.02	0.03	3.5	6	4.3
Bs	13–33	10.3	30.0	59.7	1.05	0.13	0.16	8.2	7	4.5
Bw	33–51	7.4	25.3	67.3	0.29	0.12	0.07	3.7	5	4.7
2Bt1	51–64	31.5	39.2	29.3	0.30			11.0	16	4.7
2Bt2	64–74	34.6	34.0	31.4	0.20			10.4	11	4.5

Table 14.2 Published results on the depth and thickness of red clay subsoils in the Driftless Area of Wisconsin. After Evans and Hartemink (2014)

Reference	Depth (cm)	Thickness (cm)	Details
Chamberlin and Salisbury (1886)	na	215	Based on 1500 observations
		140	On the slopes
		245	On the ridges
		413	On broad uplands
Whitson (1927)	na	60–300	
Black (1970)	na	15–90	Common
		120–300	Rare
Ballagh and Runge (1970)	94	23	
Frolking (1978)	30–225	35–140	Quarries, roadcuts, and soil pits
Knox et al. (1990)	259	200	Dolostone quarry
SSURGO, NRCS (various years)	20–127	5–76	Based on 6 soil series in this study area
Evans and Hartemink (2014)	7–130	4–60	Based on 144 pedons

na = not available

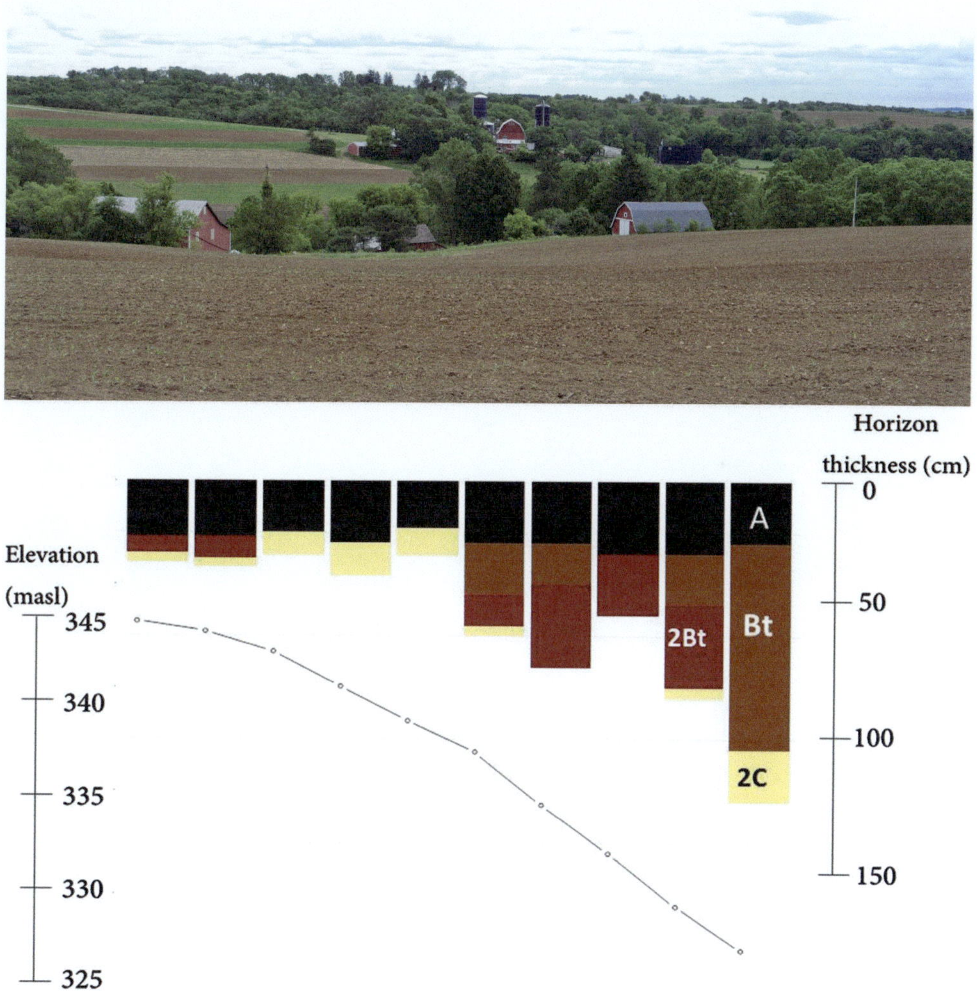

Fig. 14.3 Landscape and typical catena and soil profile diagrams (*A*, *Bt*, *2Bt* and *2C* horizons) in the Driftless Area near Verona, the 2Bt horizon is the paleo-argillic horizon or red clay (Terra Rossa). From Evans and Hartemink (2014)

Fig. 14.4 Red clay over weathered limestone near Verona

Table 14.3 Paleosols in Wisconsin

Soil Series	Soil region	Area (km^2)	Soil Taxonomy	Paleosol type[a]	Overlying material	Parent material	Approximate age
Ashdale	Am	210	Fine-silty, mixed, superactive, mesic Typic Argiudolls	Buried	Loess	SiC residuum from limestone	Rountree Formation?
Basco	Bm	53	Fine, mixed, superactive, mesic Mollic Paleudalfs	Buried	Loess	Clayey pedisediment over sandstone residuum	
Brinkman	A, Dr	44	Fine-silty, mixed, superactive, mesic Mollic Paleudalfs	Relict	–	Clayey pedisediment over dolostone bedrock	Rountree Formation?
Dodgeville	Am	585	Fine-silty over clayey, mixed, superactive, mesic Typic Argiudolls	Buried	Loess	Clayey pedisediment over dolostone bedrock	Rountree Formation?
Dolph	Fr, J	88	Fine, mixed, superactive, frigid Aeric Glossaqualfs	Buried	Loess	Clayey residuum from schist	?
Dubuque	A, B	2835	Fine-silty, mixed, superactive, mesic Typic Hapludalfs	Buried	Loess	Clayey pedisediment over dolostone bedrock	Rountree Formation?
Dunbarton	A	246	Clayey, smectitic, mesic Lithic Hapludalfs	Buried	Loess	Clayey pedisediment over dolostone bedrock	Rountree Formation?
Durand	Bm	69	Fine-loamy, mixed, superactive, mesic Typic Argiudolls	Buried	Loess	Strongly weathered till	Illinoian?
Elevasil	Dr	257	Coarse-loamy, siliceous, active, mesic Ultic Hapludalfs	Buried	Colluvium, alluvium	Sandy residuum over sandstone	Pre-Illinoian?
Fairchild	Dr	237	Sandy over loamy, siliceous, semiactive, frigid Ultic Epiaquods	Buried	Alluvium	Loamy residuum on pediments over sandstone	Pre-Illinoian?
Fivepoints	A	183	Clayey over loamy-skeletal, mixed, superactive, mesic Typic Hapludalfs	Buried	Loess	Clayey pedisediment over dolostone bedrock	Rountree Formation?
Flagg	B	44	Fine-silty, mixed, superactive, mesic Typic Hapludalfs	Buried	Loess	Strongly weathered till	Illinoian?
Haugen	G	322	Coarse-loamy, mixed, superactive, frigid Oxyaquic Paleudalfs	Relict	Till, mudflow	Densic till	Pre-Wisconsinan
Hersey	Dr	57	Fine-silty, mixed, superactive, mesic Mollic Hapludalfs	Buried	Loess	HERSEY till	Pre-Illinoian
Humbird	Dr	242	Coarse-loamy over clayey, mixed, semiactive, frigid Oxyaquic Ultic Haplorthods	Buried	Alluvium	Clayey residuum over sandstone	Pre-Illinoian?
Juneau	B, J	48	Coarse-silty, mixed, superative, nonacid, mesic Typic Udifluvents	Buried	Colluvium, alluvium	Till	Holocene?
La Farge	A, Dr	595	Fine-silty, mixed, superactive, mesic Typic Hapludalfs	Buried	Loess	Residuum glauconitic sandstone	?

(continued)

Table 14.3 (continued)

Soil Series	Soil region	Area (km^2)	Soil Taxonomy	Paleosol type[a]	Overlying material	Parent material	Approximate age
Merrillan	Dr	313	Coarse-loamy over clayey, mixed, semiactive, frigid Ultic Epiaquods	Buried	Alluvium	Clayey residuum over sandstone	Pre-Illinoian?
Milton	B	2	Fine, mixed, active, mesic Typic Hapludalfs	Buried	Loess	Till over limestone residuum	?
Myrtle	A	17	Fine-silty, mixed, superactive, mesic Mollic Hapludalfs	Buried (Sangamon)	Loess	Strongly weathered till	Illinoian
Newglarus	A	430	Fine-silty over clayey, mixed, superactive, mesic Typic Hapludalfs	Buried	Loess	Clayey pedisediment over dolostone bedrock	Rountree Formation?
Ogle	Bm	47	Fine-silty, mixed, superactive, mesic Typic Argiudolls	Buried (late Sangamon)	Loess	Strongly weathered till	Illinoian?
Palsgrove	A	659	Fine-silty, mixed, superactive, mesic Typic Hapludalfs	Buried	Loess	Clayey pedisediment over dolostone bedrock	Rountree Formation?
Pecatonica	B	92	Fine-loamy, mixed, superactive, mesic Typic Hapludalfs	Buried	Loess	Loamy till with paleosol	?
Pepin	A, Dr	191	Fine-silty, mixed, superactive, mesic Typic Hapludalfs	Buried	Loess	Clayey pedisediment over dolostone bedrock	Rountree Formation?
Quarderer	Cm	14	Coarse-silty, mixed, superactive, frigid Fluventic Hapludolls	Buried	Silty alluvium	Silty alluvium	?
Radford	Bm	80	Fine-silty, mixed, superactive, mesic Fluvaquentic Hapludolls	Buried	Silty alluvium	Silty alluvium	Holocene?
Reedsburg	A, Dr	34	Fine-silty, mixed, superactive, mesic Aquic Paleudalfs	Buried	Loess	Cherty, clayey residuum from limestone	
Rockton	Bm	101	Fine-loamy, mixed, superactive, mesic Typic Argiudolls	Buried	"Loamy sediments"	Clayey pedisediment over limestone bedrock	Rountree Formation?
Schapville	Am	1	Fine, mixed, active, mesic Oxyaquic Argiudolls	Buried	Loess	Clayey pedisediment over dolostone bedrock	Rountree Formation?
Silverhill	Dr	11	Coarse-loamy, siliceous, active, mesic Ultic Hapludalfs	Buried	Colluvium	Sandy residuum over sandstone	Pre-Illinoian?
Siouxcreek	Dr	2	Coarse-loamy over sandy or sandy-skeletal, mixed, active, frigid Aquic Hapludults	Relict	Silty residuum	Sandy residuum over sandstone	?
Sogn	Am, Bm	284	Loamy, mixed, superactive, mesic Lithic Haplustolls	Relict	SiCL residuum	Weathered limestone	
Valton	A	606	Fine-silty, mixed, superactive, mesic Mollic Paleudalfs	Buried	Loess	Clayey pedisediment over dolostone bedrock	Rountree Formation?

(continued)

Table 14.3 (continued)

Soil Series	Soil region	Area (km^2)	Soil Taxonomy	Paleosol type[a]	Overlying material	Parent material	Approximate age
Wallkill	J	30	Fine-loamy, mixed, superactive, nonacid, mesic Fluvaquentic Humaquepts	Buried	Alluvium	Histosols	?
Whalan	B, E, G	290	Fine-loamy, mixed, superactive, mesic Typic Hapludalfs	Buried	Loamy till	Clayey residuum from limestone	Rountree Formation?
Wildale	Dr	104	Fine, mixed, active, mesic Mollic Paleudalfs	Buried	Loess	Clayey residuum from limestone	Rountree Formation?
Winnebago	Bm	25	Fine-loamy, mixed, superactive, mesic Typic Argiudolls	Buried (Sangamon)	Loess	Strongly weathered till	Illinoian

[a]Buried–soil preserved by burial; relict–ancient soil preserved at surface; exhumed–ancient soil buried and re-exposed at surface by natural processes (Ruhe 1956)
Total km^2 = 9448
% Total soils = 6.7357

Fig. 14.5 Two Creeks soil near Two Creeks, Wisconsin. This soil is derived from till radiocarbon dated at 12.5–11.0 kyr BP. The buried soil occurs at the top of the shovel blade (Photo by T. Wilch)

14.3 Summary

Buried and relict paleosols occur throughout Wisconsin, particularly in the Driftless Area and in the northcentral region in areas of older drift. The paleosols range from mid-Holocene to Miocene in age. The paleosols are useful for reconstructing past climates and distinguishing among glacial stages and substages.

15 Endemic, Rare, and Endangered Soils

15.1 Introduction

Hole (1974) was a forerunner in recognizing the need to protect the "wild" soils surrounding our wild rivers. He identified "wild" soils as those which still function as integral parts of native ecosystems. He used the Pine-Popple Rivers basin in Florence County as an example of an area with wild soils. These soils were considered to be "wild" because of their importance in protecting riparian resources. Drohan and Farnham (2006) proposed that rare and threatened soils be recognized in the USA.

Endemism is a key concept in ecology and refers to plant or animal species that occur naturally and are confined to a particular geographic area (Magurran 1988). Endemism may occur at any scale and normally involves isolation of individual species, groups of species, or entire communities due to displacement by continental drift or island building. Guo et al. (2003) may have been the first to apply the term endemism to soils. They defined soil endemism as the "taxa richness of soil 'communities'" and suggested that soil endemism is predictable from soil formation theory: "The unique combination of state factor combinations from region to region would be expected to result in unique soils, and the number of state factor combinations would be expected to increase with increasing land area" (p. 113).

To validate that soils are endemic, Bockheim (2005) compared soil bodies occurring on different continents but having a similar combination of soil-forming factors using National Soil Survey Characterization Data. A key finding was that similar soil families occur within a given climate class (Köppen–Trewartha system) and vegetation type, reinforcing the concept of "zonal" soils. In a second study, soil series were compared within a soil family from randomly selected benchmark soil series using the State Soil Geographic (STATSGO) Database and evaluated in terms of the soil-forming factors contributing to endemism. This analysis showed that soil endemism is influenced by parent material (e.g., volcanic ash, clay, secondary silica) and topography (e.g., closed basins), i.e., "intrazonal" soils. Soils lacking a well-developed profile because of young parent material, or peculiar conditions of parent material, or relief ("azonal" soils) do not show endemism.

15.2 Endemic Soils of Wisconsin

Bockheim and Schliemann (2014) delineated soil richness on the basis of the number of soil taxa per county, and they identified endemic soils on the basis of having a centralized distribution, lacking competing soil series, and being unique to Wisconsin and some adjoining states. Counties within the transition zone contained a greater number of suborders, great groups, subgroups, families, and series than those without the tension zone (Fig. 15.1). Although the transition zone contains 13% of the state's land area, it has 40% of the endemic soils (Fig. 15.2). Endemic soils in Wisconsin occur in 7 orders, 14 suborders, 29 great groups, 83 subgroups, and 138 families; they include 159 soil series that cover 1.44 million ha (19% of state's land area). Spodosols occurred exclusively north of the tension zone, and Mollisols were most common south of the tension zone (Table 15.1). Alfisols were common to the north and south of the tension zone. The predominant soil properties related to soil endemism include the following: (i) a glossic horizon (34% of endemic soils), (ii) aquic conditions or oxyaquic subgroups (28%), and (iii) bisequal profiles (20%) (Table 15.2).

A re-examination of the NRCS database showed that 132 soils qualify as being endemic (Appendix G). Fifteen of

J.G. Bockheim and A.E. Hartemink, *The Soils of Wisconsin*,
World Soils Book Series, DOI 10.1007/978-3-319-52144-2_15

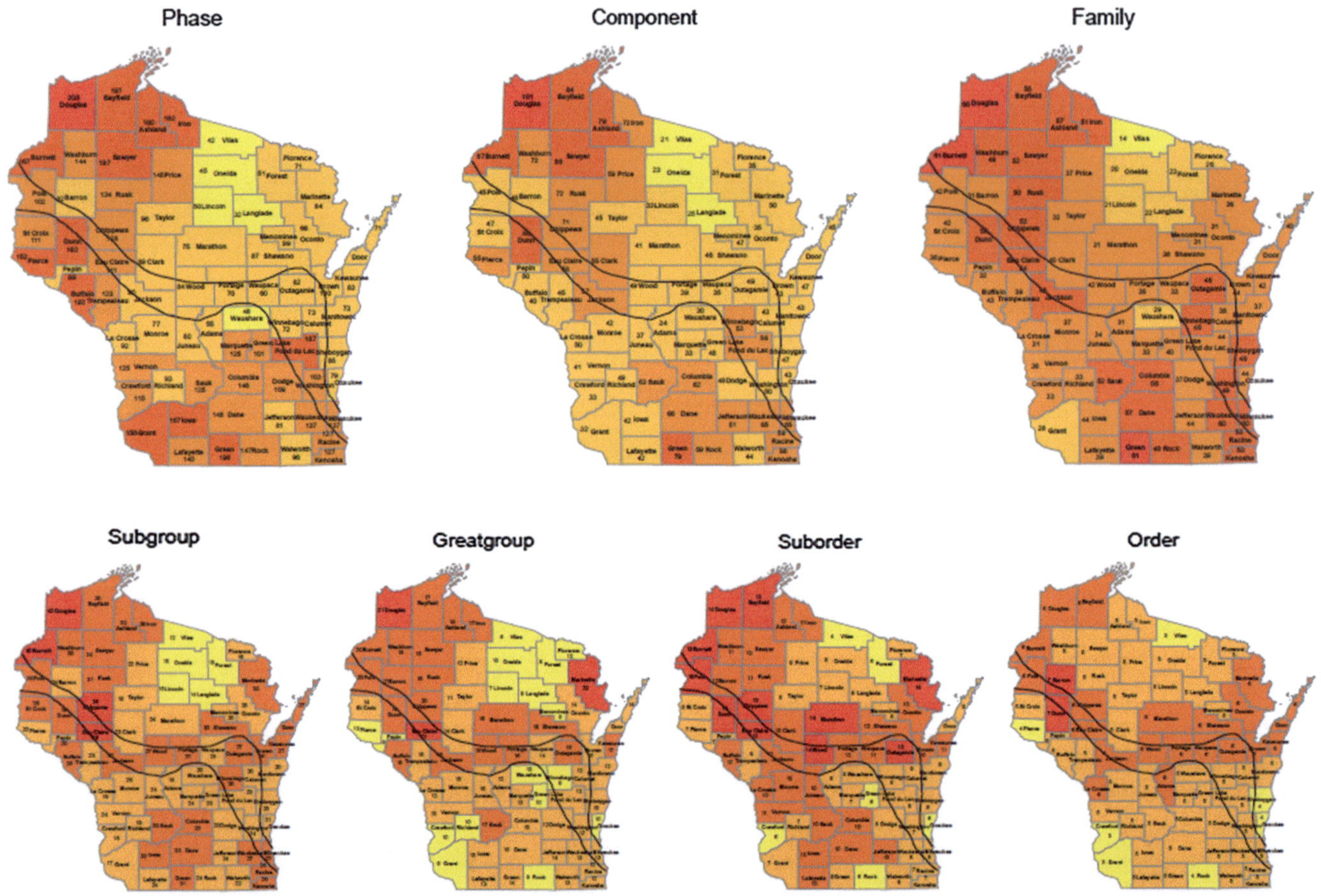

Fig. 15.1 Soil richness by county and level of Soil Taxonomy by phase, component (series), family, subgroup, great group, suborder, and order (data not normalized for differences in county areas)

these soils (11% of total endemic soils) occur in Bayfield County, which contains a diversity of parent materials, elevations, and proximity to Lake Superior. Two-thirds (67%) of these soils are considered rare, i.e., having an area of less than 10,000 ha (Ditzler 2003); 58% may be considered endangered, i.e., having an area of less than 10,000 ha and only occur in Wisconsin.

The ranking of the number of endemic soils by soil order generally follows the ranking of the number of soil series by soil order for the entire state, except that Inceptisols are overrepresented in the endemic soils and Mollisols and Alfisols are slightly underrepresented. Whereas 27% of the state's soil series are hydric, 34% of the endemic soils are hydric.

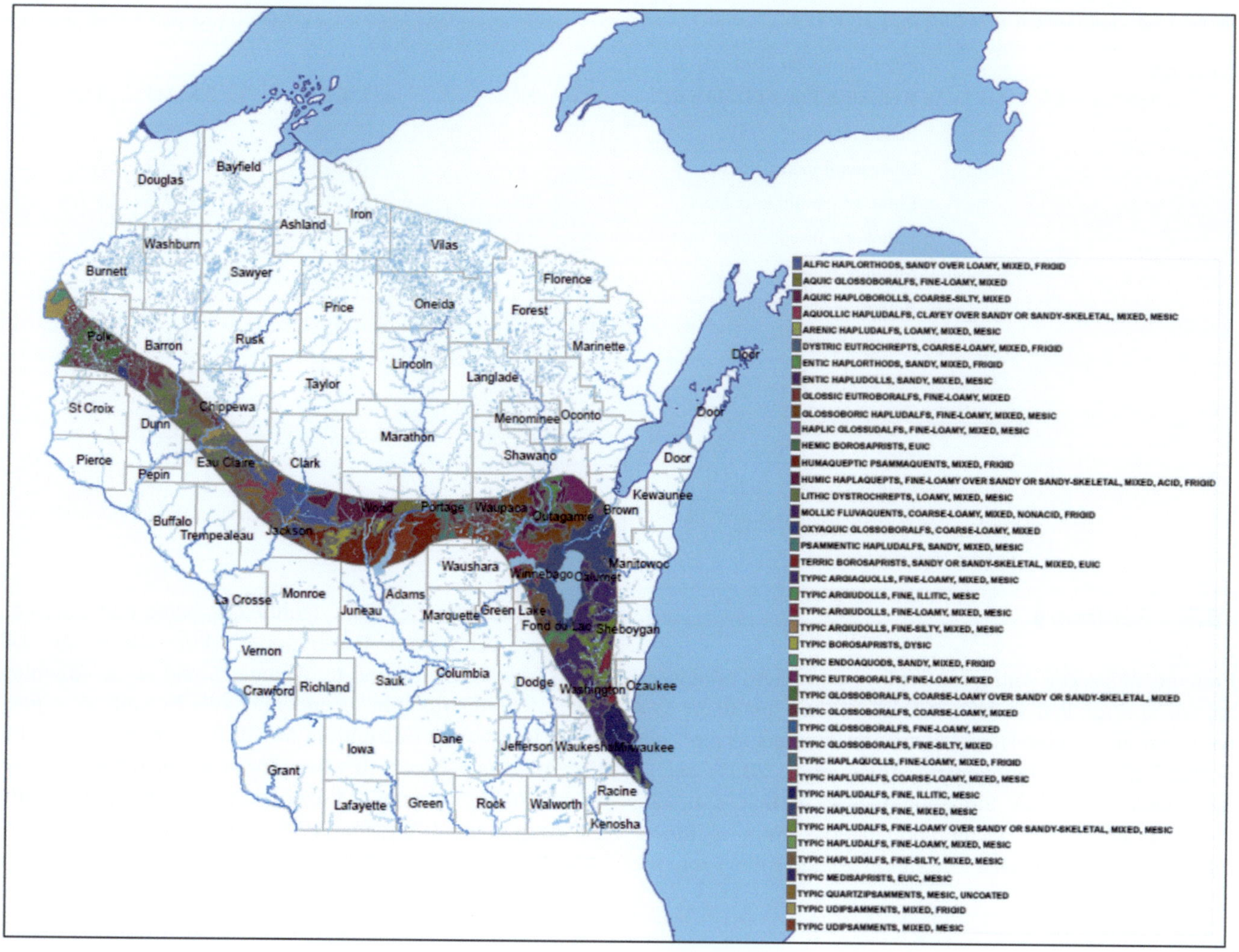

Fig. 15.2 Pedodiversity of the Wisconsin tension zone. *Note* Boralfs in Wisconsin are now recognized in Soil Taxonomy as Udalfs, Borolls as Udolls, Ochrepts as Udepts, Glossoboric subgroups as Glossic, Borosaprists as Glossoboric subgroups as Glossic, Borosaprists as, and Eutroboralfs as Hapludalfs

Table 15.1 Number of soil series in Wisconsin to the south and north of the tension zone by order and suborder

Order	South	North	Total
Spodosols	0	113	113
Alfisols	173	157	330
Mollisols	124	31	155
Inceptisols	19	35	54
Entisols	41	42	83
Histosols	10	23	33
Total	367	401	768
Suborder			
Hapludalfs	150	24	174
Glossudalfs	4	98	102

Table 15.2 Soil properties associated with endemic soils in Wisconsin

Factor	No. of soil series	% of total	Area (ha)	% of total
Relief (drainage class)				
Wet (SPD, PD, VPD, MWD)[a]	58	36.5	396,284	27.5
Parent material				
Siliceous mineral class	6	3.8	35,289	2.4
Lithic or paralithic contact	13	8.2	88,533	6.1
Lithologic discontinuity (not bisequal)	12	7.5	31,733	2.2
Multiple parent materials	8	5.0	39,434	2.7
Environmental change				
Glossic horizon	35	22.0	484,418	33.6
Bisequal	24	15.1	290,412	20.1
Time				
Base of Bt >150 cm deep	3	1.9	75,359	5.2
Total	159	100	1,441,462	100

[a]*SPD* Somewhat poorly drained; *PD* Poorly drained; *VPD* Very poorly drained; *MWD* Moderately well drained

15.3 Summary

Endemism is a key concept in ecology and refers to plant or animal species that occur naturally and are confined to a particular geographic area. Soil endemism refers to the "taxa richness of soil 'communities'" (Guo et al. 2003) and is predictable from soil formation theory in that a unique combination of state factor combinations from region to region would be expected to result in unique soils. From the NRCS database, 132 soils qualified as being endemic, i.e., having a centralized distribution, lacking competing soil series, and being unique to Wisconsin and some adjoining states. The distribution of endemic soils by soil order generally follows the distribution of soil series by soil order for the entire state, except that Inceptisols are overrepresented in the endemic soils and Mollisols and Alfisols are slightly underrepresented. Whereas 27% of the state's soil series are hydric, 34% of the endemic soils are hydric.

16 Wisconsin Soils in a Changing Climate

In the previous chapters, we have shown how much the soils of Wisconsin have been influenced by the past climates that carry a signature in many of the properties and soil profiles that we observe. The glacial periods stand out, but the tropical climate of the Silurian and even the Cambrian left traces in the soils of Wisconsin. In this chapter, we summarize the main climatic influences including climatic projections for the future.

16.1 Introduction

Climate is recognized as one of the five key soil-forming factors, with the others being parent material, topography, organisms, and time (Jenny 1941). Although short-term changes in soil are often driven by land use and human influences, most long-term changes in the soil are linked to changes in climate (Yaalon 1971). Climate controls the broad distribution of biomes on the Earth's surface, which in turn dictates the nature of the soils. Bockheim (2005) demonstrated that similar soil taxa formed on different continents when soil-forming factors, particularly climate, were comparable.

In the early soil classification systems used in the USA, soils were grouped into three orders: zonal, intrazonal, and azonal (Marbut 1927; Baldwin et al. 1938). Zonal soils were those with strong development, primarily in response to climate. Intrazonal soils had restricted development because of unique parent materials or other features such as restricted drainage, and azonal soils lacked development, because of a limited time for soil formation. Simonson (1989) reviewed the problems of the zonal soil classification systems, the main problems being that there were too many exceptions and soils were not recognized as natural bodies with measureable characteristics.

The current classification system, *Soil Taxonomy* (Soil Survey Staff 1999, 2014), recognizes two orders defined primarily on basis of soil climate: the Aridisols (soils of dry climates) and the Gelisols (soils of cold climates containing permafrost). Soil climate is recognized at the second hierarchical level (suborder) in seven of the 12 soil orders, including Alfisols, Andisols, Inceptisols, Mollisols, Ultisols, and Vertisols. A tenth order, the Spodosols, contains a cryic (mean annual soil temperature <8 °C) suborder (Cryods). Only the Histosols (organic soils) and Entisols (recent soils) are not subdivided at the order or suborder level based on soil climate.

In Chap. 14, we examined paleosols of Wisconsin and illustrated how they represented former climates. The objective of this chapter is to explore the relation of climate and soils in Wisconsin and to predict changes in soil patterns under climate warming. Evidence of past changes in the climate can be found in many soils (Fig. 16.2).

16.2 Distribution of the Soils in Relation to Climate

Climate plays a key role in the development of soils in Wisconsin. A sharp soil-temperature boundary separates frigid (mean annual soil temperature at 50 cm depth <8 °C) soils in the northern half of the state from mesic (mean annual soil temperature >8 °C) soils in the south (Hole 1976) (Fig. 16.3). Whereas Spodosols and Alfisols (Glossudalfs) are common in the frigid soil-temperature zone, Alfisols (light brown) and Mollisols (light green) are common in the mesic zone. Areas of Entisols and Histosols occur in both regions because they are comprised of sandy parent materials or have restricted drainage. The mesic-frigid boundary in Wisconsin corresponds rather closely to the

J.G. Bockheim and A.E. Hartemink, *The Soils of Wisconsin*,
World Soils Book Series, DOI 10.1007/978-3-319-52144-2_16

vegetation tension zone observed by Curtis (1959). The tension zone contains elements common to both the northern and southern flora and fauna.

16.3 Climate Change

In Wisconsin, USA, changes in soil cover have been identified from changes in climate and vegetation and from paleosols (Hole 1974, 1975, 1976; Olson and Hole 1967; Lee and Horn 1972; Mason et al. 1994; Bockheim 2003, 2015; Bockheim and Schliemann 2014). The buried Two Creeks paleosol, an Alfic Haplorthod (Lee and Horn 1972), in eastern Wisconsin, formed 11,800 years ago under spruce-fir vegetation (Maher and Mickelson 1996), when the climate was cooler and moister than today. Hole (1975) described relict bisequal (two-storied) soils (Fig. 16.1) on the Menominee tribal lands that recorded Holocene climate change. Clays migrated downward in the profile under temperate deciduous vegetation during the Holocene Climatic Optimum (ca. 8200–4200 cal. BP), forming Alfisols. During the subsequent cooling period (ca. 4200–2000 cal. BP), Spodosols formed in the E horizon of the Alfisols under the influence of temperate coniferous forest, resulting in bisequal (Spodosol over Alfisol) soils (Fig. 16.1).

A systematic effort to evaluate the effect of climate on soil distribution in Wisconsin has not been conducted. The pedological history of Wisconsin over the past 17,000 years can be reconstructed through (i) an understanding of the relation of the soil-forming factors to soil taxa, (ii) the times required to form diagnostic horizons and soil orders, and (iii) the evolution of soil-forming processes. The objectives of this chapter are to (i) reconstruct the distribution of soils in Wisconsin from past vegetation and the associated inferred climate during selected intervals of the Late Wisconsinan and Holocene (ca., 17,000 cal. BP to present) and (ii) predict changes in soil patterns in a climate-warming scenario Figs. 16.2 and 16.3.

Fig. 16.1 Bisequal soil of the Upper Peninsula in Michigan; these soils are also common in northern Wisconsin. Newood soil series (frigid Alfic Oxyaquic Haplorthods)

Fig. 16.2 Cryoturbated sandy sediments at about 1 m depth in a roadcut between Madison and Middleton

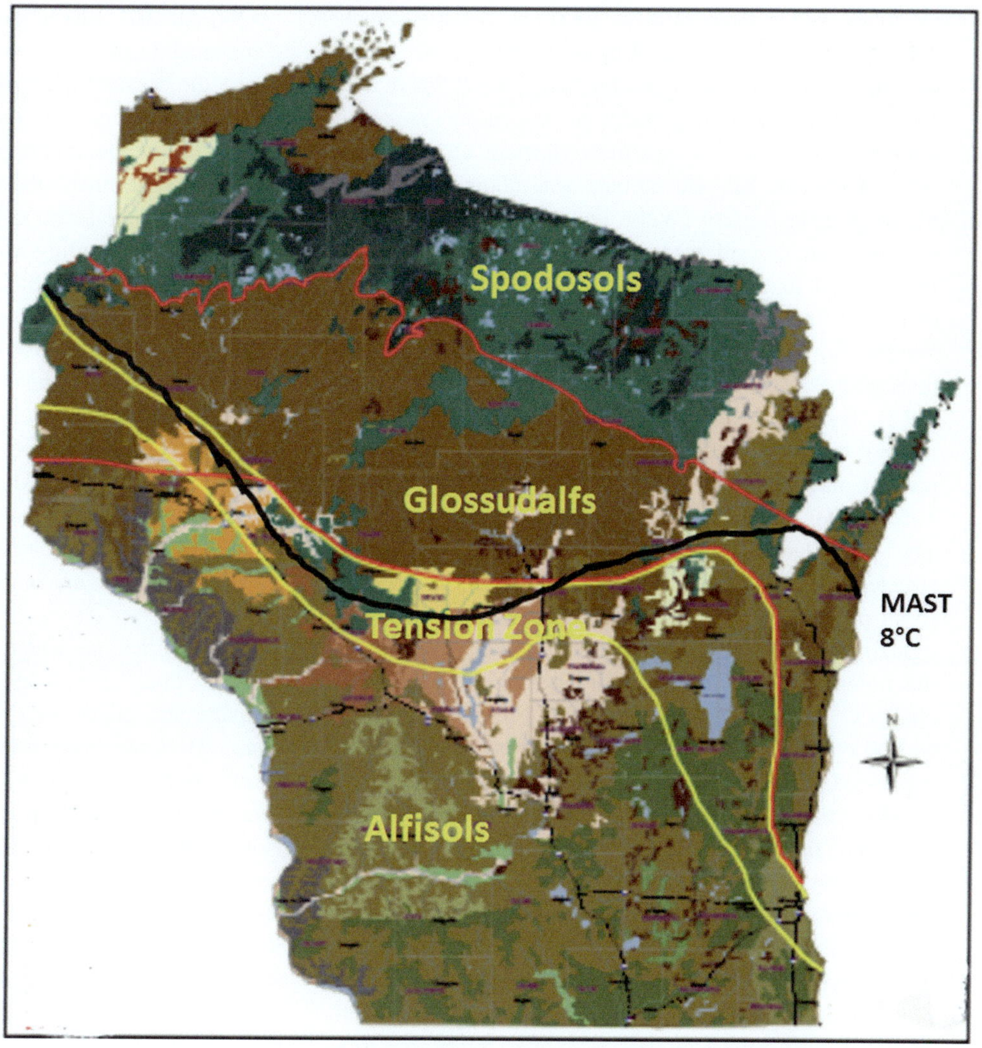

Fig. 16.3 The mesic-frigid (8 °C) soil-temperature boundary in Wisconsin is shown with the *black line*; the tension zone is delineated by the *yellow lines*; and the broad differentiation of Spodosols, Glossudalfs, and Alfisols (primarily Hapludalfs) is delineated by the *red lines*

16.4 Evaluating the Effects of Climate Change on Soils

We approximated the distribution of vegetation, ice, and soil order across Wisconsin based on the scientific literature and the development of a mental model. The following premises were made:

1. Glacial advances and retreat during the Late Wisconsinan and Early Holocene follow the chronology of Attig et al. (2011) and Syverson and Colgan (2011). By 14,000 cal. BP, ice from the Lake Superior-Chippewa, Green Bay, and Lake Michigan lobes persisted in northern and eastern Wisconsin. A small lobe (Des Moines lobe) extended from the west into westcentral Wisconsin. By 12,300 cal. BP, ice remained along the southwest shore of Lake Superior in northwestern Wisconsin and in the Green Bay area. By 11,800 cal. BP, the Laurentide glacier had retreated out of Wisconsin. Ice from the Greatlakean advance extended into northwestern Wisconsin and the Green Bay area and by 9800 cal. BP persisted only in what is now Lake Superior.
2. Permafrost existed beyond the margin of the Laurentide ice from 14,000 cal. BP (throughout most of the state) to 11,400 cal. BP (northern and eastern Wisconsin only) (Black 1954; Wayne and Guthrie 1993; Clayton et al. 2001). Relict periglacial features such as ice-wedge polygons and ice-wedge casts exist throughout much of the state, attesting to the former presence of permafrost (Black 1965; Clayton et al. 2001).
3. As derived primarily from fossil pollen data, the climate of Wisconsin was periglacial from 13,000 to 11,000 cal. BP (Bernabo and Webb 1977; Clayton et al. 2001), which would require a mean annual air temperature of

−2 °C or colder. From 8000 to 3000 cal. BP, the climate was drier and warmer than at present (Baker et al. 1992); the mean annual precipitation may have been 75–175 mm lower than at present and the mean annual air temperature 0.5–2.0 °C warmer (Bartlein et al. 1984). After 3000 cal. BP, the climate was about cooler and wetter than at present (Webb and Bryson 1972). There has been an overall warming over the last 200 years. From 1950 to 2006, the mean annual air temperature increased by about 0.6–1.1 °C, primarily in western Wisconsin, and the mean annual precipitation increased throughout most of the state by about 15% (WICCI 2011). By 2055, the mean annual air temperature over most of the state is expected to be 3.5 °C warmer than at present and the mean annual precipitation to be about 10% greater than at present (WICCI 2011).

4. The post-glacial vegetation history has been reconstructed for six regions of Wisconsin primarily from fossil pollen and can be divided into five intervals, recognizing the natural variations in climate of the state and the time-transgressive nature of vegetation response to changes in climate (Fig. 16.4). From 17,000 to 11,000 cal. BP, most of the ice-free areas of Wisconsin supported spruce (*Picea*)-sedge (*Carex*) tundra. From 11000 to 6800 cal. BP, there was a high species diversity (Bernabo and Webb 1977; Bartlein et al. 1984; Davis et al. 1986; Baker et al. 1992; Brugam et al. 2004). The spruce-dominated boreal forest was absent from the state; oak (*Quercus*) was dominant in the south; prairie (*Poaceae*) advanced into the western part of the state; and pines (*Pinus*) were abundant in the north. Peak warmth occurred during the Holocene Climatic Optimum 8200–4200 cal. BP. Mixed pines were dominant in the north and mixed oaks in the south. Prairie vegetation advanced into the western part of the state. After 3200 cal. BP, there was an increase in spruce pollen, coupled with the arrival of hemlock (*Tsuga*) and beech (*Fagus*) in northern Wisconsin. Since the arrival of Europeans in the mid-1800s, the forest cover has diminished by 25% from clearing for cropping and grazing.
5. The tension zone, which extends northwest to southeast across the central part of the state, contains plants and animals representative of northern mixed forests and eastern broadleaf forests (Curtis 1959). The tension zone is dynamic and responds to small-scale oscillations in climate over short time periods (Curtis 1959; Hupy and Yansa 2009a, b; Bockheim and Schliemann 2014).
6. Vegetation, climate, and soil order are strongly correlated in Wisconsin. Like elsewhere, climate and vegetation are the primary drivers of pedogenesis (Jenny 1941).

Six soil orders are found in Wisconsin, including Alfisols (47% of land area), Spodosols (17%), Entisols (12%), Mollisols and Histosols (10% each), and Inceptisols (4%) (Chap. 3). One relict Ultisol occurs to a limited extent (2 km^2) in westcentral Wisconsin. Three of the orders are widespread, including the Alfisols, Inceptisols, and Histosols. Although Spodosols and Mollisols occur in different climates, there is considerable overlap in climate between these orders and the more broadly distributed Alfisols, Inceptisols, and Entisols (Table 16.1). Vegetation is correlated with soil order, with Alfisols primarily occurring under broadleaf forests, Spodosols under mixed forest, Mollisols under prairie and oak savanna, Histosols under conifer

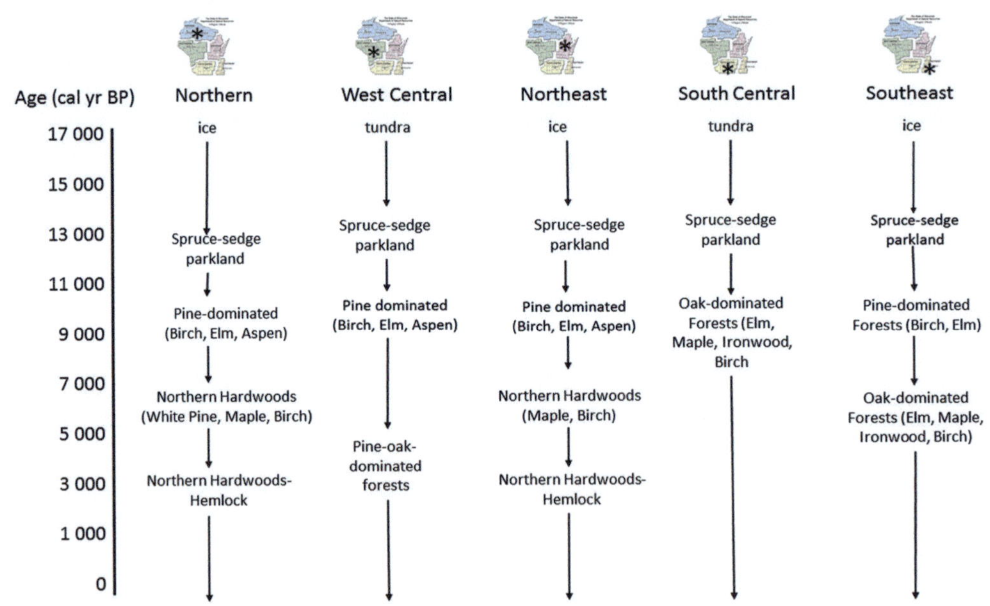

Fig. 16.4 Vegetation history of Wisconsin from the Late Wisconsinan to the present (derived from Wright et al. 1963; Webb 1974, 1987; Bernabo and Webb 1977; Bartlein et al. 1984; Heide 1984; Baker et al. 1992; Maher and Mickelson 1996; Brugham et al. 2004; Long et al. 2011)

Table 16.1 Past climates, vegetation, and soils in Wisconsin

Time (ky BP)[a]	Event	Proximity to current mesic-frigid boundary	Approx. MAAT (°C)	Approx. MAP (mm)	Dominant vegetation[b]	Dominant Soil orders[c]
135–80	Sangamon	North	14	800	Grassland, temperate deciduous forest	Mollisols and Alfisols
	Interglaciation	South	16	800	Grassland	Mollisols
22–12.3	Late Wisconsinan	North	−8	200	Glacial ice	–
	Glaciation	South	−3	200	Arctic tundra	Gelisols
12.3–9.4	Late Wisconsinan	North	−3	400	Arctic tundra	Gelisols
	Glaciation	South	0	400	Boreal	Spodosols
9.4–4	Hypsithermal	North	10			
	Interval	South	12			
4-present	Little Ice Age	North	>8	750	Temperate coniferous and deciduous forest	Spodosols and Alfisols
	(Neoglacial)	South	<8	750	Grassland and temperate deciduous forest	Alfisols and Mollisols

[a]Modified from Syverson and Colgan (2004)
[b]From palynological studies by Webb (1974); Davis (1977); Maher (1982); Heide (1984); Davis et al. (1986); Maher and Mickelson (1996)
[c]From studies of paleosols by Beaver (1966); Lee and Horn (1972); Hole (1974); Follmer (1983); Jacobs et al. (1997)

swamps and sedge meadows, and Entisols under pine barrens and lowland hardwood forests.

Landforms and parent material are important soil-forming factors in Wisconsin. Alfisols are most common in areas of loess and clayey glaciolacustrine deposits; Spodosols are dominant on pitted outwash and sandy and sandy loam till; and Entisols are prevalent on sandy lake deposits and unpitted outwash.

Based on experimental studies and the study of soil chronosequences, histic, mollic, and cambic horizons form in less than 400 years in the Great Lakes region (Fig. 16.5). Spodic subgroups and argillic horizons require longer periods to form, ca. 500 and 700 years, respectively. Histosols and Inceptisols can form in less than 1000 years, whereas Mollisols require about 3000 years to form. Spodosols and Alfisols generally from 3000 to 4000 years.

Gelisols were the common soils in Wisconsin from 29,000 to 18,000 cal. BP. With increasing temperatures, these soils changed into Inceptisols and subsequently into Spodosols. The conversion of Gelisols to Inceptisols has been reported in Swedish Lapland (Darmody et al. 2000) and the conversion of Gelisols to Spodosols in Wilkes Land, Antarctica (Beyer et al. 2000). The changes require a recession of the permafrost table and the development of soil color and/or structure (cambisolization) or podzolization (Bockheim and Gennadiyev 2000).

The evolution of Inceptisols to Spodosols, Alfisols, or Mollisols has been reported in Ontario, Canada (Mahaney 2015), Taiwan (Tsai et al. 2016), and Wisconsin (Hartemink and Bockheim 2013), respectively. These transformations require podzolization, clay illuviation, or melanization (Bockheim and Gennadiyev 2000). The E horizon of Alfisols may be converted to a Spodosol via podzolization (Hole 1975), and the spodic horizon may be converted to an argillic horizon by clay illuviation (Bockheim et al. 1996; Lindeburg et al. 2013). Mollisols may be converted to Entisols (Quideau and Bockheim 1996) or Alfisols under conditions of erosion and rapid decomposition of soil organic matter.

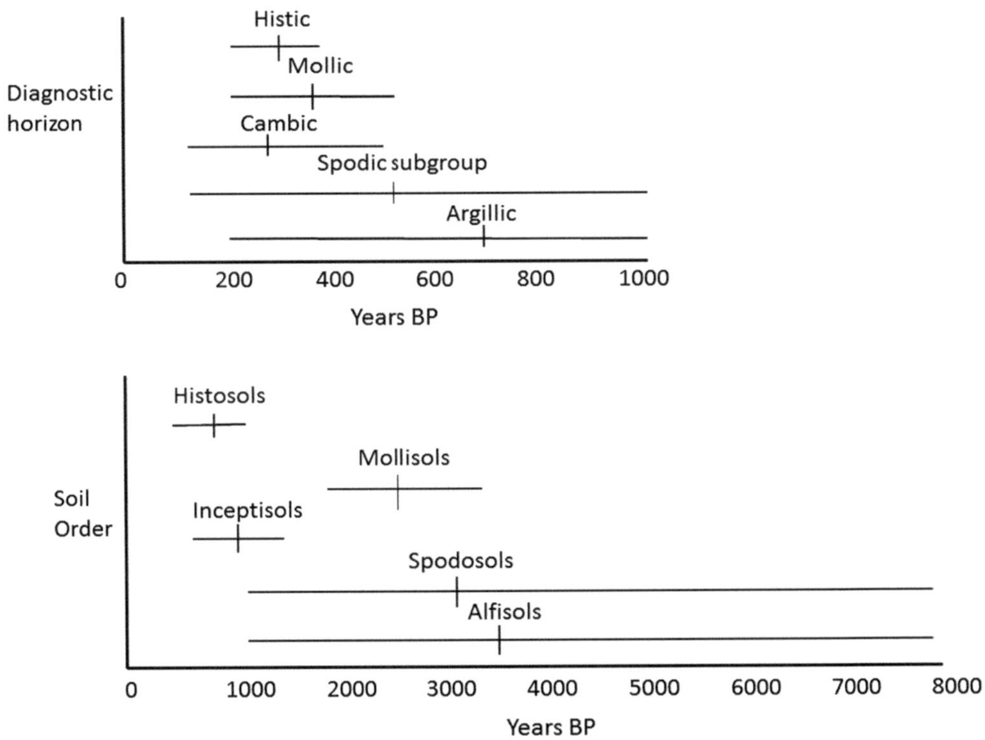

Fig. 16.5 Time required to form diagnostic horizons and soil orders in the western Great Lakes region (derived from soil chronosequence data summarized by Bockheim 2014)

16.5 Effects of Past Climate Change

From 29,000 to 21,000 cal. BP, Laurentide ice covered all but the southwestern 40–50% of the state. By 18,000 cal. BP, ice had retreated northward and eastward, exposing about 73% of the state. Between 29,000 and 18,000 cal. BP, the soils were covered by spruce-sedge tundra and were underlain by permafrost. Ice-wedge casts are present throughout this area (Black 1965). Gelisols were the dominant soil order (Fig. 16.6). From 14,000 to 12,800 cal. BP, the ice continued to retreat with spruce-sedge tundra, permafrost, and Gelisols occupying a narrow band to the south of the glacial margin (Table 16.3). Spruce-dominated boreal forest covered most of the state by 12,800 cal. BP, with Spodosols occurring to the north of the tension zone and Inceptisols to the south. Based on soil chronosequence data (Fig. 16.5), there have may have been insufficient time (1700 years) in a periglacial climate for Alfisols to form. Areas of sandy outwash and glacial lake deposits supported Entisols, which are still common soils in these areas today.

By 11,600 cal. BP, the Laurentide ice had melted, but a fringe of permafrost remained along the southwest shore of Lake Superior, as evidenced by former ice-walled lakes (Clayton et al. 2001). Spodosols were likely at their maximum extent during this time and may have occupied one-third of the land area (Fig. 16.6). The buried Two Creeks paleosol, an Alfic Haplorthod with spruce macrofossils, developed at this time (Lee and Horn 1972). Inceptisols or Alfisols persisted to the south of the tension zone. During 11,600–8,900 cal. BP, ice from the Greatlakean stage advanced into northernmost and easternmost Wisconsin (Clayton et al. 2001). Gelisols likely occupied a narrow fringe ($\sim$3% of the state's land area) along the ice front around 11,600 cal. BP, with Spodosols existing to the southern boundary of the current tension zone and Alfisols further to the south (Table 16.3).

The interval 8900–5700 cal. BP was a period of warmer and drier conditions, enabling prairie vegetation to prevail in western Wisconsin (Davis 1977; Heide 1984), resulting in Mollisols ($\sim$10% of state's land area) in the Driftless Area and allowing sufficient time for Alfisols to form throughout much of the southern part of the state (Fig. 16.6; Table 16.3). Spodosols likely retreated to the northern border of the tension zone.

From 5700 to 2600 cal. BP, the climate became cooler and wetter. Prairie vegetation migrated westward out of the state (Bernabo and Webb 1977), resulting in a fragmented distribution of Mollisols in southern Wisconsin (Table 16.2). Hemlock spreads rapidly over a large portion of northern Wisconsin (Davis et al. 1996). Hemlock enhances podzolization (Bockheim 1997). Spodosols migrated southward to the southern border of the tension zone, creating a zone of bisequal soils (Fig. 16.1). Beaver (1966) observed that bisequal soils were restricted to the frigid soil-temperature zone (Table 16.3).

Over the past 2600 years, the climate has become warmer in Wisconsin. Alfisols and Mollisols are probably at their greatest extent (Fig. 16.6). Spodosols occupy 17% of the land surface which is likely the smallest extent in the past 29,000 years.

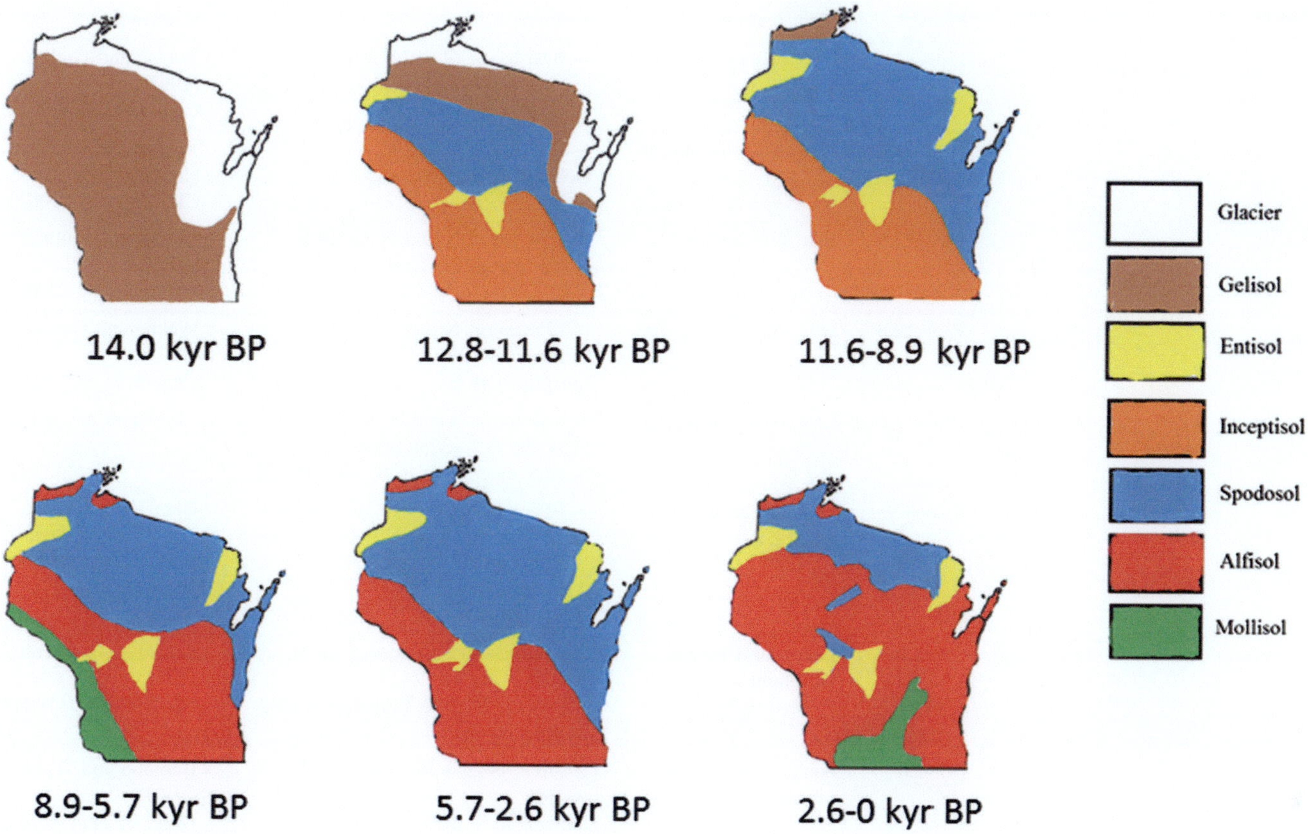

Fig. 16.6 Pedological history of Wisconsin from the Late Wisconsinan to the present (derived from vegetation history of Wisconsin; climate interpreted from fossil pollen; and current biome–soil relationships)

Table 16.2 Changes in soil order and diagnostic horizon from climate change, based on the literature

Soil Change		
Order	Diagnostic horizon	Reference
Gelisol → Spodosol	Recession of permafrosttable podzolization None → Spodic	Wilkes Land, Antarctica (Beyer et al. 2000)
Gelisol → Inceptisol	Recession of permafrost table development of color, structure None → Cambic	Swedish Lapland (Darmody et al. 2000)
Inceptisol → Spodosol	Podzolization Cambic → Spodic	Ontario, Canada (Mahaney 2015)
Inceptisol → Alfisol	Clay illuviation Cambic → Argillic	Taiwan (Tsai et al. 2016)
Inceptisol → Mollisol	Melanization Ochric → Mollic	Wisconsin, USA (Hartemink and Bockheim 2013)
Alfisol → Spodosol	Podzolization Albic → Spodic	Wisconsin, USA (Hole 1975)
Spodosol → Alfisol	Clay illuviation Spodic → Argillic	Oregon, USA (Lindeburg et al. 2013)

Table 16.3 Main features and approximate soil distribution by time period in Wisconsin

Period[a]	Main feature	Soil extent	Reference
15.0–14.6	Late Wisconsinan ice	~38% Ice in N and E; Gelisols in ice-free areas	Wayne and Guthrie (1993); Clayton et al. (2001)
14.6–14.0	Late Wisconsinan ice	~27% ice in N and E; Gelisols in ice-free areas	Wayne and Guthrie (1993); Clayton et al. (2001)
14.0–13.9	Late Wisconsinan ice	~27% ice in N and E; Gelisols in ice-free areas	Wayne and Guthrie (1993); Clayton et al. (2001)
13.9–12.8	Late Wisconsinan ice	~17% ice in N and E; ~80% Gelisols; ~10% Inceptisols in S	Wayne and Guthrie (1993); Clayton et al. (2001)
12.8–11.6	Late Wisconsinan ice	~10% ice in N and E; ~18% Gelisols in N and E; ~32% Spodosols in N; ~20% Inceptisols in S; 10% Entisols	Clayton et al. (2001)
11.6–8.9	Greatlakean advance	~10% ice and ~3% Gelisols in N and E; ~32% Spodosols in N; ~20% Alfisols in S; 10% Entisols	Bernabo and Webb (1977); Bartlein et al. (1984); Attig et al. (2011)
8.9–5.7	Holocene Climatic Optimum	Ice-free; ~20% Spodosols in N; ~40% Alfisols in S; 10% Entisols; ~10% Mollisols	Bernabo and Webb (1977); Bartlein et al. (1984)
5.7–2.6	cooler and moister climate	Ice-free; ~25% Spodosols in N; ~45% Alfisols in S; 10% Entisols	Bernabo and Webb (1977); Brugham et al. (2004)
2.6–0	warmer climate	Ice-free; 47% Alfisols; 17% Spodosols; 10% Entisols; 10% Mollisols	Bockheim and Schliemann (2014)

[a]These are identified as the following chronozones: Oldest Dryas, Bölling, Older Dryas, Allerød, Younger Dryas, Boreal, Atlantic, Subboreal, and Subatlantic, respectively (Walker et al. 2011)

16.6 Role of the Tension Zone in Soil Distribution

The floristic tension zone in Wisconsin follows the current mean July air temperature isotherm of 21 °C (Curtis 1959; Bockheim and Schliemann 2014). Based on the trajectory analysis of air masses, the Wisconsin tension zone represents the mean winter position of the Arctic frontal zone (Bryson 1966). Changes in vegetation over the past 17,000 years, in response to climate changes, were manifested by changes in the position of the tension zone (Hupy and Yansa 2009a, b). The current distribution of soils in Wisconsin reflects the changing positions of the tension zone.

Based on the reconstruction of the climate and vegetation, the Spodosols-Glossudalfs border in Fig. 16.3 may represent the tension zone 12,800–11,600 cal. BP, when spruce-sedge parkland supporting Spodosols bordered pine- and oak-dominated forests to the south with Inceptisols. During the Holocene Climatic Optimum (ca. 8900–5700 cal. BP), the tension zone would have been further north and positioned at the Glossudalfs-northern current tension zone border. Alfisols were the dominant soils with the shift in the tension zone to the north. During the cooler and moister period 5700–2600 cal. BP, the tension zone further shifted southward, accompanied by the degradation of the upper part of the argillic horizon in the Alfisols. It created Glossudalfs following an intensified podzolization process. Spodosols are currently restricted to a narrow band in northernmost Wisconsin.

The geographic shift in the tension zone changed the distribution of soils. Motew and Kucharik (2013) analyzed meteorological variables and suggested that the location of the Wisconsin tension zone had not migrated over the period 1947–2007 despite measurable changes in climate. Soil changes require much longer periods as summarized in Figs. 16.4 and 16.5.

16.7 Soil Development from Climate Change

From 1950 to 2006, the mean annual temperature increased by about 0.6–1.1 °C, primarily in western Wisconsin, and the mean annual precipitation increased throughout most of the state by about 15% (WICCI 2011). By 2055, the mean annual air temperature over most of the state is expected to be 3.51 °C warmer than at present (WICCI 2011). The amount of snowfall is a key climatic factor in how soils will respond to climate warming. Soils will cool as the climate warms in the Great Lakes region, because a thinner snowpack reduces the insulation from soil freezing (Isard et al. 2007; Pauli et al. 2013). The impact of warming on fire frequency will influence soil formation (Schulte et al. 2005). Increased fires may reduce the SOC contents of the soils, change mineralogy, and make the soils more prone to erosion. This may result in the loss of topsoil.

Based on past climate in Wisconsin, we anticipate the position of the tension zone to reflect changes in vegetation and soils over the next 50 years. Warming of the air temperature by 3.51 °C would initiate a northward shift in the tension zone so that frigid soils in the north would become mesic and mesic soils in the south would become even more mesic; but, they would not become thermic (MAST 15–221 °C).

Two additional factors pertain to the time required to form particular horizons, diagnostic epipedons and horizons, and soil orders and the susceptibility of a particular soil property or taxon to change. Most soil horizons require decades to centuries to form in soils with frigid and mesic soil-temperature regimes. The formation of diagnostic epipedons and subsurface horizons and soil orders requires centuries to millennia (Fig. 16.5). Therefore, under normal conditions, there will be a lag in the response of soils to a change in climate on the order of centuries to millennia.

Soil properties may change rapidly, slowly, or irreversibly in a threshold response to external forcing factors (Yaalon 1971; Chadwick and Chorover 2001). For example, it is unlikely that warming of the Central Sands Region of Wisconsin will result in the conversion of 14,000-year-old Entisols derived from glaciolacustrine sand to another soil order, because of pedological inertia. Some horizons are resistant to change. For example, ortstein layers in Spodosols of northern Wisconsin would be resistant to environmental change (Kaczorek et al. 2004).

South of the present tension zone, Alfisols transform Mollisols within hundreds of years from a warming event as the A horizon becomes thicker from the imprint of prairie vegetation to eventually form a mollic epipedon (Fig. 16.7). North of the present tension zone, Spodosols would be converted to bisequal soils (Orthod solum over Udalf solum) as bases are recycled by the deciduous vegetation replacing the coniferous vegetation and as clay is moved into the subsoil.

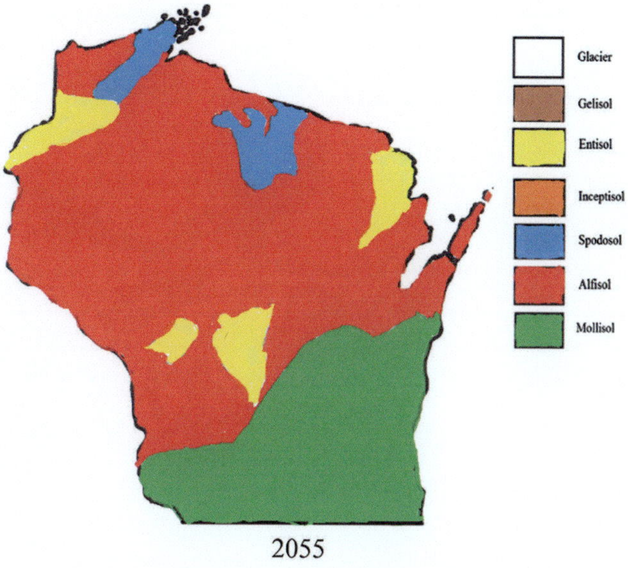

Fig. 16.7 Projected soil distribution in Wisconsin from the projected 3.5 °C increase in mean annual air temperature by 2055. These changes would require an additional ca. 1000 years to become evident

In view of the lag time in response of soils to climate change, we do not anticipate any major changes in soil geography over the next 50 years; these changes in soil distribution would have a lag time of at least 1000 years. Changes in soil development will occur primarily in the vicinity of the tension zone and along other ecotone areas, such as along forest/prairie transitions. However, changes in climate on land use could have considerable changes in soils over a relatively short time span (Watson and Hartemink 2015).

16.8 Summary

Throughout the Quaternary, Wisconsin has experienced major shifts in climate, accompanied by changes in the distribution of vegetation and soils. The most dramatic fluctuations in climate, vegetation, and soils have resulted from geographic shifts in the tension zone, where a sharp soil-temperature boundary separates northern forests from southern forests and Spodosols and bisequal soils from Alfisols (Fig. 16.1). Warming of the air temperature by 3.5 °C would initiate a northward shift in the tension zone so that frigid soils in the north would become mesic and mesic soils in the south would become even more mesic. The formation of diagnostic epipedons and subsurface horizons and soil orders requires centuries to millennia. There will be a lag in the response of soils to a change in climate on the order of centuries to millennia. Changes in soil development will occur primarily in the vicinity of the tension zone and along forest/prairie transitions. However, the effects of climate change on land use may initiate major changes in soils and should be the focus of future research.

Soils and Land Appraisal

17

17.1 Introduction

Rural land appraisal is based on (i) the productive capacity of the land, (ii) the potential appreciation of a land investment, and (iii) the market value of land (Appraisal Institute 2000). According to one appraiser: "Land holds a unique and pivotal position in social, political, environmental and economic theory. Land supports all life and stands at the center of human culture and institutions … It is nature's gift to mankind, which enables life to continue and prosper" (Gwartney 2015).

Soil cartography, which originated in Western Europe in the 1850s was based on ideas and classification approaches from agrogeology (Hartemink et al. 2013). Initial soil surveys were largely undertaken for land-tax purposes. Systematic soil surveys started in the early 1900s, and it has yielded important information for land appraisal by (i) providing soil maps (Phillips 1968), (ii) identifying land capability classes (Conrey 1935; Beek 1978; Gibbons and Haans 1976; Gersmehl 1980), and (iii) determining soil-yield potential (Klingelhoets and Westin 1954; Allgood and Gray 1978; Huddleston 1984; van Lanen et al. 1992).

Although soil survey information is essential for land use planning, not many studies exist in which the value of the information is tested. Here, we test the relationship between soil survey information (soil-classification level) and the value of agricultural land for Wisconsin, USA. All soils of the state of Wisconsin have been mapped in detail to the soil series level. The objective of this study is to (i) relate measures of agricultural land values to soil classes on a county basis and (ii) to develop a model for predicting the monetary value of agricultural land using the soil great group level and information on slope and stoniness of the soils.

17.2 Evaluating Monetary Data

Two parameters were used for land evaluation: (i) the agricultural land sale value in dollars per acre, and (ii) the average adjusted gross income in dollars. These data are available for each of the 72 counties for 2013 in the 2016 Wisconsin Blue Book (Wisconsin Legislative Reference Bureau 2015). Agricultural land sales includes land with buildings and other improvements, as opposed to bare agricultural land sales. The Wisconsin adjusted gross income is income reported to the Wisconsin Department of Revenue for income tax purposes and is based on the federal income tax definition of gross income as modified by certain additions and subtractions required by state law. The adjusted gross income includes off-farm income.

17.3 Evaluating Crop Yield Potential

The crop-yield potential is a relative ranking of a soil's ability to produce high-yielding crops. Laboski and Peters (2012) provided an overview of how to calculated the yield potential (YP) in Wisconsin (Table 17.1). Their analysis includes codes for soil group and soil yield potential, primarily for corn, which is the major crop in the state (Laboski and Peters 2012; Wisconsin Legislative Reference Bureau 2016). Three groups of soils based on soil texture were delineated: loamy (L), organic (O), and sandy (S). Soil yield potentials were categorized as high (H), medium (M), and sandy (S). Yield potential was based on soil texture group, drainage class, available water capacity, and depth to bedrock. Modifications are made based on soil-temperature regime and growing-degree days. The proportion of a county

J.G. Bockheim and A.E. Hartemink, *The Soils of Wisconsin*,
World Soils Book Series, DOI 10.1007/978-3-319-52144-2_17

Table 17.1 Soil-yield potential classes for Wisconsin (after Laboski and Peters 2012)

Soil-yield class	Numerical index	Soil group[a]	Drainage class[b]	AWC (cm)[c]	Bedrock depth (cm)[d]	STR[e]
High	3	L, O	W, MW, SP	M, H, VH	D	M, F
Medium[f]	2	L, O	E, SE, W, MW, SP, P, VP	VL, L, M, H, VH	S, D	M, F
Low	1	S				

[a]Soil group: L = loamy (sandy loam or finer); O = organic (Histosols); S = sandy (>75% sand in upper 20 cm or Psamm—suborder or great group)
[b]Drainage class: E = excessive; SE = somewhat excessive; W = well drained; MW = moderately well drained; SP = somewhat poorly drained; P = poorly drained; VP = very poorly drained
[c]Available water-holding capacity to 150 cm: VL = very low (<7.5 cm); L = low (7.5–15 cm); M = medium (15–23 cm); H = high (23–30 cm); VH = very high (>30 cm)
[d]Depth to bedrock: D = deep (>76 cm); S = shallow (<75 cm)
[e]Soil temperature regime: M = mesic (mean annual soil temperature at 50 cm 8–15 °C); F = frigid (MAST 0–8 °C)
[f]If any one of the properties, including drainage class, AWC, or bedrock depth, is not as listed under "H," the soil is assigned a yield class of "M"

that has been cropped was determined by using a GIS to overlay the soil great group map with the 2015 land cover map.

Soils in the sandy soil group (S) are generally considered to have a low yield potential (Table 17.1). Organic soils (Histosols) have a high yield potential if the soil temperature regime is mesic or a medium yield potential if it is frigid (and drainage is assumed). Mineral soils have a medium yield potential if one or more of the following three properties is limiting: (i) drainage classes other than well drained, moderately well drained, or somewhat poorly drained; (ii) available water-holding capacities that are less than 15 cm; and (iii) soil depths that are less than 75 cm. Soils without limitations are considered to have a high yield potential, unless they have a frigid soil-temperature class (mean annual soil temperature at 50 cm is <8 °C), in which case the crop yield potential is medium (M).

The number of soil classes by county was determined for use in the analysis (Table 17.2). The mean and standard deviations of the number of orders is 5 ± 0.6, suborders 9 ± 2, great groups 12 ± 3.1, and series 46 ± 18. The analysis suggested that orders were too broad at the county level, and soil series could not be used because they are often limited to one or several counties. We chose to conduct

Table 17.2 Numbers of soil taxa by county in Wisconsin

County	Year	Orders	Suborders	Great groups	Series
Adams	1984	6	8	11	31
Burnett	2006	6	12	18	102
Clark	2000	5	9	15	61
Dodge	1980	4	8	10	42
Florence	2004	5	7	10	40
Forest	2005	5	6	8	39
Jackson	2001	6	12	16	62
Lafayette	1966	5	9	11	36
Marathon	2003	6	12	15	42
Marquette	1975	5	6	8	31
Milwaukee	1971	5	8	10	60
Oneida	1993	5	8	9	27
Polk	1979	6	12	15	39
Rock	1974	5	8	10	54
Trempealeau	1977	5	8	10	35
Vernon	1969	5	8	9	34
Winnebago	1980	5	10	14	50
Average		5	9	12	46
Range		4–6	6–12	8–18	27–102
Stdev		0.6	2.0	3.1	18.1

Table 17.3 Yield potential of soil series >200 km^2 each by dominant great group in Wisconsin

Great group	Area (km^2)	Yield potential[a]			
		H	M	S	Soil series (>200 km^2 each)
Hapludalfs	33,806	30	11	3	Fayette, Kewaunee, Norden, Seaton, Manawa, St. Charles, Onaway, Fox, Palsgrove, Kidder, La Farge, Theresa, McHenry, Gale, Emmet, Dodge, Miami, Churchtown, Newglarus, Morley, Downs, Symco, Council, Lomira, Lamartine, Ozaukee, Lapeer, Dresden, Meridian, Sattre (H) Dubuque, chetek, Casco, Hixton, Eleva, Dorerton, Wyocena, Whalan, Elevasil, Dunbarton, Hayriver (M) Richford, Gotham, Boyer (S)
Glossudalfs	22,463	25	5	0	Magnor, Freeon, Antigo, Withee, Loyal, Hortonville, Kennan, Santiago, Almena, Amnicon, Spencer, Fenwood, Kert, Rietbrock, Ossmer, Anigon, Meadland, Sconsin, Tilleda, Waymor, Portwing, Sanborg, Flambeau, Vlasaty, Rozellville (H) Rosholt, Amery, Oesterle, Arland, Scott Lake (M)
Haplorthods	20,133	4	5	9	Sarona, Chequamegon, Stambaugh, Goodman (H) Padus, Newood, Karlin, Newot, Humbird (M) Pence, Keweenaw, Vilas, Sayner, Rubicon, Croswell, Menominee, Rousseau, Sultz (S)
Haplosaprists	12,551	3	9	0	Houghton, Palms, Adrian (H) Cathro, Sellyeville, Lupton, Markey, Loxley, Dawson, Beseman, Carbondale, Bowstring (M)
Udipsamments	9665	0	0	10	
Argiudolls	7524	7	2	0	Plano, Hochheim, Tama, Solona, Elburn, Warsaw, Ashdale (H) Dodgeville, Edmund (M)

[a]Yield potential for corn: H = high; M = medium; S = sandy (low) (Laboski and Peters 2012)

the analysis at the great group level, which have a fairly broad distribution in the state. Six great groups accounted for 79% of the soils of Wisconsin, including from greatest to least: Hapludalfs, Glossudalfs, Haplorthods, Haplosaprists, Udipsamments, and Argiudolls (Table 17.3).

Because there is variation in yield potential within a great group, we used a weighted ranking (Table 17.4). Soils with a high, medium, and sandy yield potential were assigned 3, 2, and 1 point, respectively. These were weighted by area and an overall weighted value was determined for each great group.

The distribution of soil great groups by county was determined using Web Soil Survey (Natural Resources Conservation Service 2016a) and Official Soil Series Descriptions, including the Soil Extent Mapping Tool (Natural Resources Conservation Service 2016b).

The agricultural land sale value and the average adjusted gross income were plotted against the proportion of individual or combined soil great groups. Regression coefficients were poorer for adjusted gross income so agricultural land sale value was used as the independent variable. All independent and dependent variables used in the analysis for the 72 counties in Wisconsin are given in Table 17.5. Step-wise multiple regression was used for predicting agricultural land sale value from county soil maps. One-third of the counties were randomly removed for testing the predictability of the model. All analyses were conducted using Minitab (Minitab, Inc., 2000).

17.4 Value of Agricultural Land Sales in Relation to Crop Yield

Nearly half (47% by area) of the soil series in Wisconsin are in the high potential yield class, with 31% in the medium class, and 22% in the sandy class. There was a highly significant, positive correlation between the value of agricultural land sales and the yield potential (Fig. 17.1). There is a $4066 increase in agricultural land value for each unit increase in yield potential.

17.5 Productivity of Soil Great Groups

Of the six dominant great groups (>2000 km^2), the Argiudolls and Glossudalfs (2.8) have the highest yield potential, followed by Hapludalfs (2.4) and Haplosaprists (2.3); the Haplorthods (1.8) and Udipsamments (1.0) have the lowest soil-yields (Table 17.6). These are confirmed by the proportion of the great group that is being cropped, with 95% of the Argiudolls under cropping and only 5% of the

Table 17.4 Number of soil series in a great group according to yield potential, along with weighted numerical rankings

Great group	Area (km^2)	High 3 pts.	Medium 2 pts.	Sandy 1 pt.	Total	Weighted points[a]
Hapludalfs	36,180	106	34	33	173	2.4
Glossudalfs	23,076	84	10	3	97	2.8
Haplorthods	18,285	15	22	31	68	1.8
Haplosaprists	12,706	7	15	0	22	2.3
Udipsamments	9987	0	0	33	33	1.0
Argiudolls	7323	51	11	0	62	2.8
Hapludolls	2826	23	7	14	44	2.2
Endoaquolls	2721	0	26	4	30	1.9
Paleudalfs	2021	5	0	0	5	3.0
Psammaquents	1736	0	0	9	9	1.0
Quartzipsamments	1687	0	0	10	10	1.0
Endoaquods	1642	6	5	8	19	1.9
Udifluvents	1574	12	0	2	14	2.7
Epiaquods	1539	5	3	5	13	2.0
Epiaqualfs	1526	5	10	0	15	2.3
Dystrudepts	1524	3	7	7	17	1.8
Fragiorthods	1417	3	5	0	8	2.4
Fluvaquents	1119	1	7	1	9	2.0
Eutrudepts	1085	6	3	2	11	2.4
Endoaquepts	1081	1	9	0	10	2.1
Epiaquepts	761	0	6	0	6	2.0
Endoaqualfs	701	8	6	0	14	2.6
Glossaqualfs	661	6	4	0	10	2.6
Argiaquolls	632	0	8	0	8	2.0
Haplohemists	629	1	4	0	5	2.2
Udorthents	546	0	1	3	4	1.2
Epiaquolls	526	0	5	1	6	1.8
Humaquepts	331	0	7	0	7	2.0
Haplustolls	284	0	1	0	1	2.0
Epiaquents	222	0	0	4	4	1.0
Fragiaquods	152	0	1	0	1	2.0
Endoaquents	120	0	0	1	1	1.0
Sphagnofibrists	6	0	1	0	1	2.0
Albaqualfs	3	0	0	1	1	1.0
Hapludults	3	0	1	0	1	2.0
Haplustepts	2	0	1	0	1	2.0
Sum	136,635	348	220	172	740	2.3

[a]Estimated by summing the product of number of soil series in a great group by yield potential points and dividing by the total soil series in a great group

Haplorthods. About half of the low-productivity Udipsamments are cropped, because of groundwater available for irrigation. The counties containing the most productive soils in Wisconsin include Rock, Dodge, Richland, Lafayette, Walworth, Pierce, St. Croix, Kewaunee, Dane, Ozaukee, Washington, Waukesha, Brown, Sheboygan, Fond du Lac, Green, Kenosha, Racine, Sauk, Barron, Langlade, Columbia, and Marathon (Table 17.3).

Table 17.5 Dominant soils in relation to agricultural land sales value and adjusted gross income by county in Wisconsin 2013

County	Agric. land sale value ($)[a]	Adjust. gross income ($)[b]	Year soil survey	Hapludalfs (%)	Glossudalfs (%)	Haplorthods (%)	Haplosaprists (%)	Udipsamments (%)	Argiudolls (%)	Steep, rocky (%)	Water (%)	Other (%)	Crop YP[c]
Adams county	3000	33,770	1984	16.5	0.0	0.0	5.3	62.2	0.0	0.6	0.0	15.4	1.5
Ashland county	1255	35,290	2006	0.0	22.2	23.5	17.0	0.2	0.0	2.8	1.7	32.6	2.1
Barron county	4132	41,420	2001	12.6	57.7	1.5	3.2	4.9	0.0	0.0	3.4	16.7	2.6
Bayfield county	1350	42,220	2006	0.0	19.0	54.4	5.3	0.0	0.0	0.9	2.6	17.8	2.1
Brown county	8120	55,140	1974	61.5	8.0	1.5	3.6	6.1	2.8	0.4	0.0	16.1	2.7
Buffalo county	4004	39,520	1962	46.6	0.0	0.0	2.4	1.1	2.5	10.3	0.6	36.5	2.3
Burnett county	2049	34,980	2006	9.1	8.7	0.1	12.3	44.8	1.0	0.0	7.0	17.0	1.6
Calumet county	6899	57,920	1980	52.7	8.4	0.0	6.5	2.4	7.8	0.0	7.3	14.9	2.5
Chippewa county	2710	44,880	1992	0.2	10.0	24.8	16.8	4.1	0.1	0.7	0.4	42.9	2.0
Clark county	3400	39,430	2000	0.8	55.8	11.5	4.8	0.5	0.0	0.1	0.5	26.0	2.5
Columbia county	6750	48,280	1978	45.8	0.0	0.0	10.0	7.5	19.2	0.6	0.0	16.9	2.6
Crawford county	2672	35,350	1961	57.9	0.0	0.0	0.0	0.1	0.4	25.4	0.0	16.2	2.1
Dane county	7434	63,890	1978	49.0	0.0	0.0	4.9	0.4	25.4	1.2	0.0	19.1	2.7
Dodge county	6289	47,640	1980	63.0	0.0	0.0	7.9	0.0	19.1	0.1	0.0	9.9	2.8
Door county	3994	48,040	1978	27.1	13.3	16.0	9.6	0.3	5.0	0.8	1.1	26.8	2.4
Douglas county	1146	41,830	2006	0.1	23.8	35.7	13.8	7.7	0.0	0.0	2.8	16.1	2.1
Dunn county	2853	44,090	2004	47.8	9.0	0.0	4.2	10.9	3.2	0.0	1.5	23.4	2.5
Eau Claire county	3078	68,660	1977	30.9	4.8	4.1	2.2	23.1	0.3	0.0	0.0	34.6	2.1
Florence county	1160	38,080	2004	0.0	2.9	67.2	14.1	0.0	0.0	3.3	0.0	12.5	2.0
Fond du Lac county	6305	49,620	1973	48.1	0.0	0.0	10.7	0.0	14.5	0.1	0.0	26.6	2.6
Forest county	1487	31,280	2005	0.0	0.0	45.9	24.6	0.0	0.0	0.0	3.3	26.2	1.9
Grant county	4727	39,520	1961	55.5	0.0	0.0	0.3	0.0	13.4	13.2	1.8	15.8	2.4
Green county	4742	49,770	1974	45.9	0.4	0.0	1.5	0.1	19.8	2.4	0.0	29.9	2.6
Green Lake county	5838	45,690	1977	37.5	0.0	0.0	8.5	7.2	20.7	0.7	0.0	25.4	2.5
Iowa county	4761	47,440	1962	46.7	0.0	0.0	0.9	0.0	21.1	11.7	1.1	18.5	2.4
Iron county	1005	34,980	2006	0.0	3.4	19.8	20.3	0.0	0.0	0.0	5.9	50.6	1.9
Jackson county	2972	40,680	2001	23.7	0.4	7.0	13.7	0.8	0.8	0.0	1.3	52.3	2.2
Jefferson county	6006	47,080	1979	49.6	0.0	0.0	3.7	0.5	12.2	0.0	3.6	30.4	2.5
Juneau county	3285	36,220	1991	31.7	0.0	0.0	0.7	37.4	0.0	0.0	3.7	26.5	1.9

(continued)

Table 17.5 (continued)

Kenosha county	5980	48,130	1970	41.0	0.0	0.0	6.6	0.0	19.3	0.0	0.0	33.1	2.6
Kewaunee county	4920	46,160	1980	33.5	37.5	1.8	9.1	0.7	2.9	0.0	0.6	13.9	2.7
La Crosse county	5080	51,070	2006	54.6	0.0	0.0	2.2	7.2	6.0	0.0	6.3	23.7	2.4
Lafayette county	5327	41,610	1966	50.5	0.0	0.0	0.1	0.0	30.9	1.6	0.0	16.9	2.8
Langlade county	3383	37,440	1986	0.0	58.3	23.7	10.7	0.0	0.0	0.0	0.0	7.3	2.6
Lincoln county	1780	42,490	1996	0.0	38.6	33.0	12.7	0.0	0.0	0.0	2.4	13.3	2.3
Manitowoc county	6421	46,460	1980	52.7	8.4	0.0	6.5	2.4	7.8	0.0	7.3	14.9	2.5
Marathon county	3231	51,040	2003	1.7	63.7	0.6	4.2	4.7	0.0	0.0	2.3	22.8	2.6
Marinette county	2953	38,700	1991	12.6	0.2	28.8	16.2	22.3	0.0	0.0	2.5	17.4	1.9
Marquette county	3800	38,050	1975	49.7	0.0	0.0	18.2	12.3	1.6	1.0	0.0	17.2	2.4
Menominee county		15,600	2004	0.0	27.2	32.1	13.5	15.2	0.0	0.0	2.2	9.8	2.1
Milwaukee county		45,620	1971	48.9	0.0	0.0	8.9	0.2	17.7	0.1	0.0	24.2	2.7
Monroe county	3276	39,850	1984	41.6	0.0	0.0	1.5	2.9	0.0	0.0	0.8	53.2	2.4
Oconto county	3594	44,030	1988	13.7	5.3	25.0	12.4	12.0	9.5	0.0	1.5	20.6	2.1
Oneida county	1699	43,460	1993	0.0	1.7	56.2	15.2	0.0	0.0	0.0	8.6	18.3	1.8
Outagamie county	8989	55,050	1978	32.4	20.5	4.9	9.3	0.7	3.4	0.1	1.1	27.6	2.5
Ozaukee county	5679	93,280	1970	52.6	0.0	0.0	6.7	0.0	19.9	0.5	1.0	19.3	2.7
Pepin county	4223	46,990	2002	54.3	0.0	0.0	2.4	9.9	1.3	0.0	6.7	25.4	2.3
Pierce county	4217	54,150	2006	62.4	13.8	0.0	0.0	2.7	6.4	0.0	3.0	11.7	2.7
Polk county	2466	43,190	1979	1.5	65.0	3.6	7.4	5.9	0.6	0.0	3.8	12.2	2.5
Portage county	2777	47,550	1978	28.8	19.4	0.0	11.7	22.0	0.0	0.1	0.0	18.0	2.3
Price county	1400	36,340	2006	0.0	28.1	30.6	22.3	0.0	0.0	0.0	2.3	16.7	2.2
Racine county	5939	49,890	1970	41.0	0.0	0.0	6.6	0.0	19.3	0.0	0.0	33.1	2.6
Richland county	2758	36,940	2006	77.9	0.0	0.0	0.6	1.2	4.6	0.0	0.9	14.8	2.8
Rock county	5883	45,690	1974	38.0	0.0	0.0	2.4	0.0	47.0	0.1	0.0	12.5	2.8
Rusk county	1990	33,620	2006	0.0	41.1	19.2	14.1	1.4	0.0	0.0	2.3	21.9	2.4
St. Croix county	4233	63,260	1978	20.9	48.2	0.0	1.0	1.6	7.9	0.0	1.2	19.2	2.7
Sauk county	3558	45,000	1980	60.9	0.0	0.0	1.8	3.2	4.8	0.2	1.2	27.9	2.6
Sawyer county	1982	35,780	2006	0.0	25.5	42.7	14.5	1.3	0.0	0.0	7.2	8.8	2.1
Shawano county	4803	39,300	1982	10.1	28.2	12.4	16.7	7.3	2.0	0.0	1.7	21.6	2.3

(continued)

Table 17.5 (continued)

Sheboygan county	6047	50,220	1978	45.3	5.0	0.0	8.3	0.9	16.9	0.5	0.0	23.1	2.7
Taylor county	2275	40,670	2005	0.0	49.7	15.4	14.6	0.0	0.0	0.0	1.3	19.0	2.5
Trempealeau county	4504	52,530	1977	58.9	0.0	0.0	2.5	0.1	1.7	3.4	0.0	33.4	2.5
Vernon county	4462	41,920	1969	66.2	0.0	0.0	0.3	0.0	1.0	23.0	0.0	9.5	2.2
Vilas county		37,280	1988	0.0	0.0	54.7	14.5	0.0	0.0	0.0	14.9	15.9	1.7
Walworth county	7399	49,330	1971	52.7	0.0	0.0	5.5	0.0	23.4	0.0	0.0	18.4	2.8
Washburn county	1932	37,050	2006	2.7	20.8	13.9	13.5	27.5	0.0	0.1	6.1	15.4	1.8
Washington county	7173	61,020	1971	47.6	0.0	0.0	10.5	0.0	19.9	0.0	0.0	22.0	2.7
Waukesha county	9112	78,080	1971	48.9	0.0	0.0	8.9	0.2	17.7	0.1	0.0	24.2	2.7
Waupaca county	3728	44,250	1984	17.7	42.3	0.7	9.0	15.3	0.0	0.0	2.0	13.0	2.4
Waushara county	2855	39,230	1989	40.0	2.1	0.0	12.1	36.9	0.0	0.0	1.5	7.4	2.0
Winnebago county	5029	50,090	1980	27.8	1.7	12.3	7.7	1.0	2.0	0.0	22.7	24.8	1.9
Wood county	2994	45,530	1977	1.8	31.1	1.0	6.9	20.3	0.0	0.0	0.9	38.0	2.1
	4480	50,670	Avg.	29.9	12.9	10.1	8.3	6.4	6.7	1.5	2.3	21.9	2.4

[a]Agricultural land sales 2013 (WI Blue Book 2015)
[b]Wisconsin adjusted gross income 2013 (WI Blue Book 2015)
[c]Weighted average of yield: Hapludalfs, Glossudalfs, and Argiudolls = high (3); Haplorthods, Haplosaprists, and "others" = medium (2); Udipsamments = sandy (1); steep & rocky = unproductive (0)

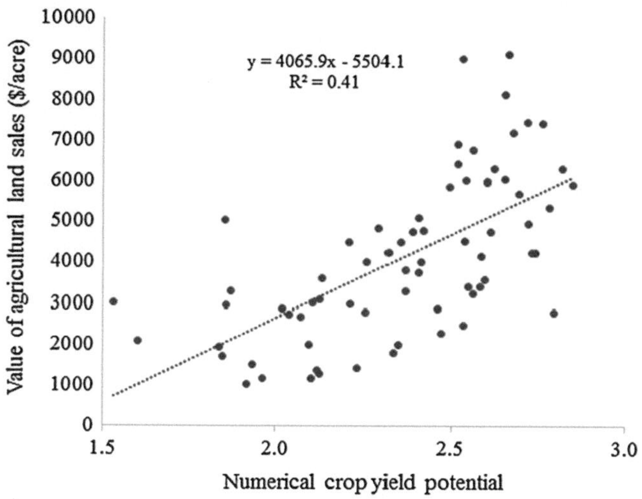

Fig. 17.1 Value of agricultural land sales ($ per acre in 2013) in relation to numerical yield potential

17.6 Soil Great Groups, Land Values, and Family Income

Agricultural land was more expensive when the proportion of Argiudolls + Hapludalfs increased in a county, but the correlation was negative for the less productive Haplorthods + Haplosaprists + Udipsamments (Fig. 17.2). Nearly half of the Haplorthods and all of the Udipsamments are sandy, and they have a low yield potential. Haplosaprists have a high yield potential, but drainage would be required for most crops. As wetlands, these soils are often protected.

The Glossudalfs, which have one of the highest weighted yield potentials (2.8) and occupy 16% of the land area in Wisconsin, correlate poorly with agricultural land sales (Fig. 17.3). The low value of Glossudalfs may pertain to their low growing degree days (< 2200) in northern Wisconsin and the fact that a large portion of the Glossudalf

Table 17.6 Area, number of soil series, yield potential, and proportion of area cropped for the six dominant soil great groups (of 32) in Wisconsin (rankings given in parentheses)

Great group	Area (km^2)	Soil series (no.)	Numerical yield pot.	Proportion cropped (%)
Hapludalfs	33,806 (1)	173(1)	2.4 ± 0.79 (7)	75(4)
Glossudalfs	22,463 (2)	97(2)	2.8 ± 0.47 (2)	60(5)
Haplorthods	20,133 (3)	68(3)	1.8 ± 0.79 (26)	5(14)
Haplosaprists	12,551 (4)	22(8)	2.3 ± 0.48 (11)	20(9)
Udispamments	9665 (5)	33(6)	1.0 ± 0 (32)	50(7)
Argiudolls	7524 (6)	62(4)	2.8 ± 0.52 (1)	95(1)
Others	34,688	285	1.9	10

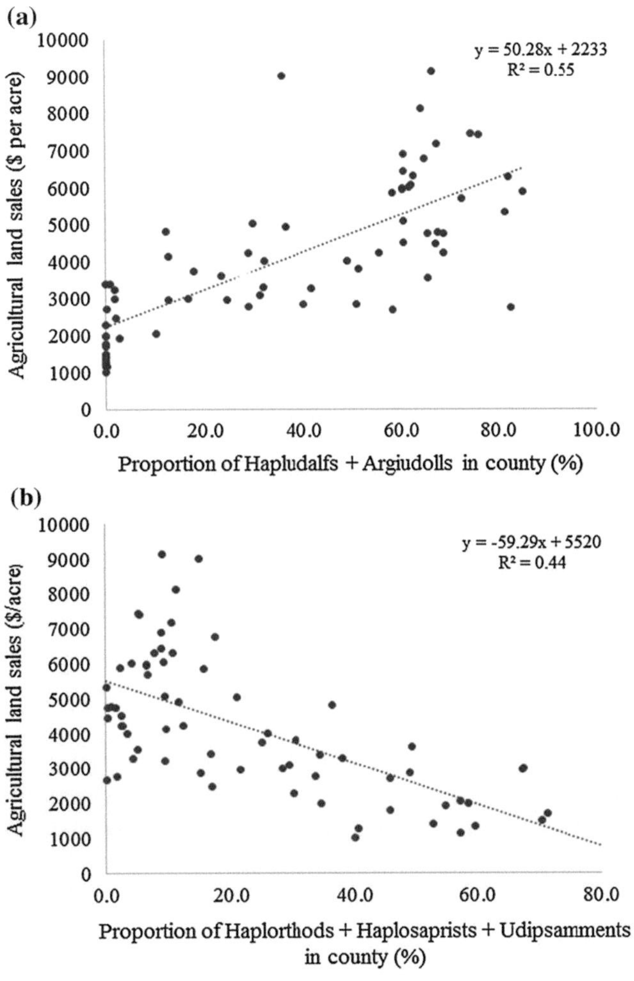

Fig. 17.2 Agricultural land sales ($ per acre in 2013) in relation to proportion (%) of Hapludalfs and Argiudoll great groups (**a**) and Haplorthods, Haplosaprists, and Udipsamments (**b**)

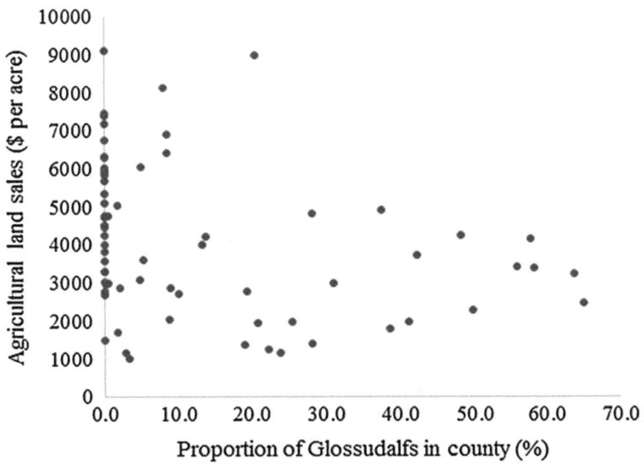

Fig. 17.3 Agricultural land sales ($ per acre in 2013) in relation to proportion (%) of soil series in the Glossudalf great group by county

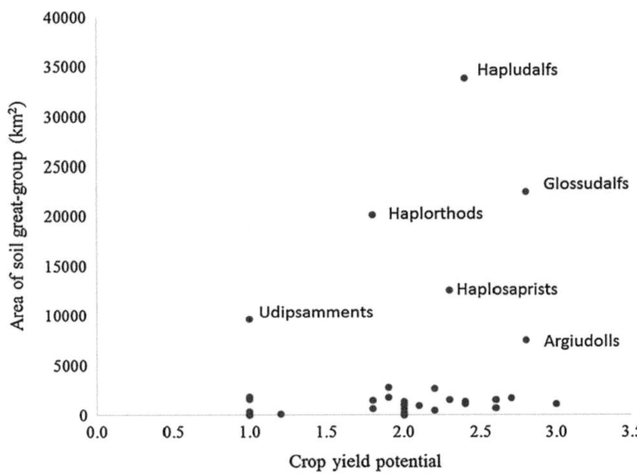

Fig. 17.4 Soil-yield potential in relation to great group in Wisconsin

area is in the Managed Forest Law program (Wisconsin Department of Natural Resources 2016). This program requires that the landowner give public use of the land in return for a lower tax rate. The relation between area of soil great group and crop yield potential is shown in Fig. 17.4, with the six major great groups identified.

The use of soil great groups in this state-wide study had the following advantages: (i) the six great groups accounted

Table 17.7 Step-wise (forward) multiple regression model for predicting agricultural land sales value from proportion of soil great groups and steep and stony land

Parameter[a]	%Argiudolls	%Hapludalfs	%Glossudalfs	%Steep and stony land
Constant = 1544				
Regression coefficients	99	54.7	18.9	−62
S	1341	1074	1043	1015
R^2	54.9	71.7	73.9	75.8
Adjusted R^2	53.9	70.4	72.2	73.6
Cp	36.8	8.4	6.4	4.9

[a]S = fraction by which the square of the standard error of regression is less than the adjusted coefficient of determination (R^2) of the model; Cp = Mallows' or Akraika information criterion is the fit of a regression model using ordinary least squares; a small value implies precision

for 79% of the land area (Tables 17.3 and 17.4); (ii) there were strong positive correlations with the soil great group level and the monetary value of agricultural land (Figs. 17.2 and 17.4). Soils series could not be used in this analysis because they often occur in one to several counties only rather than a state-wide basis.

There was a significant correlation between numerical crop yield and the value of agricultural land sales (Fig. 17.1), and the numerical soil yield evaluations of soil series within a great group had low standard deviations (Table 17.6). Whereas loamy Haplorthods have a potential of 2.40 ± 0.50, sandy Haplorthods have a yield potential of 1.00 ± 0. There appear to be three problems with the yield potentials. The northern half of the Glossudalfs is cropped to a lesser extent than the southern half, suggesting that growing degrees (GDDs) are critical for their use and possibly for their yield potential assessment. The proportion of Glossudalfs by county was poorly correlated with agricultural land sales value (Fig. 17.3). Glossudalfs with <2200 GDDs could be assigned a numerical soil yield index of 2 and those with >2200 GDDs.

In the system of Laboski and Peters (2012), Haplosaprists are assigned a high (H) or medium (M) soil yield value (Table 17.4), but these potential yields can only be obtained when these soils are drained. In Wisconsin, there is an effort to protect wetlands, which often contain Haplosaprists. Therefore, it may be more appropriate to assign Haplosaprists a low (L) or medium (M) value than a high (H) yield potential. The same argument could be made for soils in Epi- and Endo- great groups. These soils often occur in wetlands and they require tiling to improve. They also should be assigned yield values of medium (M) or low (L) rather than high (H)—unless they are drained.

17.7 A Numerical Model for Predicting Land Value

A step-wise (forward) multiple regression model using proportions of great groups and the percentage of steep and stony land predicted 74% of the variation in agricultural land sales value (Table 17.7). The proportion of Argiudolls exerted the greatest influence (54%), followed by Hapludalfs (16%); the proportion of Glossudalfs and steep and stony land little the variation (3.2%).

The model was also applied to 20 counties that were randomly withheld from the model preparation. The correlation coefficient between predicted and actual agricultural land sales value was 0.56, with a probability value of 0.01 (Fig. 17.5). Some differences in predicted and actual values

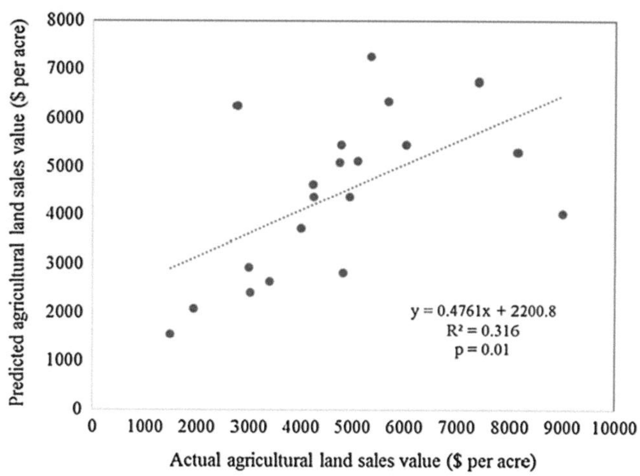

Fig. 17.5 Actual and predicted agricultural land sales value

are to be expected because (i) the land values were reported on a county-wide basis rather on a township or smaller-scale level; (ii) the land values include properties with structures and land improvements; and (iii) mean values were employed rather than median values, which may more accurately reflect differences in "bare" and developed land value.

On a county scale, the following technique is recommended for predicting agricultural land sales value. The first step is to obtain a map of the county, which can be derived from a digital version of a published soil survey or Web Soil Survey. The proportion (%) of soil map units, generally soil series, can be obtained in tabular form from the published soil surveys or from Web Soil Survey. The soil map units are converted to soil great groups from Official Soil Series Descriptions. The proportion of each great group is plugged into the equation:

Agricultural land sales value = 1544 + 99 (% Argiudolls) + 54.7 (% Hapludalfs) + 18.9 (% Glossudalfs) – 62 (% steep and stony land).

This approach could be used to develop equations for other great groups or other soil classes and possibly using soil series at the township level, if the data are available.

17.8 Summary

From this analysis, the following can be concluded as follows:

- Using great group level soil information and a semi-quantitative yield rating showed good correlations between the yield potential and the monetary value of agricultural land
- A clear positive relationship was established between the more fertile soils (Hapludalfs, Argiudolls) and the price of land
- When a county has a higher proportion of low fertility soils (e.g., Haplorthods, Haplosparists, and Udipsamments), the price of land is significantly lower
- A multiple regression model involving soil great group information, slope of the land, and stoniness is able to predict fairly accurately the price of land in some other neighboring states.

Soil taxa data in the NRCS databases is not only useful for exploring spatial soil information and land suitability, but also useful for predicting the monetary value of agricultural land.

18 Current and Future Soil Research

18.1 Introduction

The study of soils in Wisconsin is ongoing. All 72 counties of the state have been mapped at a resolution of 1:30,000 or finer. Official soil descriptions and extent maps are available for all 733 series identified in the state. Primary laboratory characterizations are available for more than 1000 pedons in the state. So the soil conditions have been well mapped and there is considerable information available. In this chapter, we will discuss the need to update the information and discuss some ways how this can be done. In addition, we will discuss some soil research areas that in our view need attention.

18.2 Soil Mapping and Databases

As evidenced by some of the maps, there are correlation issues that need to be resolved. County boundaries are readily visible between soil orders (Fig. 6.3), suborders (Fig. 6.4), and great groups (Fig. 6.5). This is not unique to Wisconsin and occurs in many other states as well. It is the effect of a county by county approach to soil mapping, and soil mismatches at the country boundaries as a result of differences in mapping approaches, concepts, techniques, soil surveyors and differences in the time that the counties were mapped. Various numerical approaches exist that harmonize the mapping units across county and state boundaries (Sun et al. 2010; Odgers et al. 2014).

There are taxonomic issues that will require attention not only for soils of Wisconsin, but also throughout the USA. The increased complexity of the system has been criticized by international soil taxonomists and by ecologists and geologists using the system (Hartemink 2015). The problems facing the use of soil classification are as follows: too many classification systems, changing too frequently, too many characteristics, data too difficult to obtain, too complex, too difficult terminology, and classification specialists also disagree (2001). There are efforts underway to change and simplify *Soil Taxonomy* (Stolt and Needelman 2015) as well as numerical classification systems (Hempel et al. 2013; Michéli et al. 2016).

For *Soil Taxonomy* (ST) to survive in the next several decades, the following changes may be necessary: (i) simplification and expansion of the number of diagnostic epipedons; (ii) reconsideration of the use of soil climate at the suborder level or the definition of soil-temperature and soil-moisture classes for more widespread application outside the USA; and (iii) addition of new orders based on new or existing diagnostic subsurface horizons (Bockheim 2014).

At the present time, four of the eight epipedons—the ochric, mollic, umbric, and histic—are used in 99% of the soil series identified in the USA. The definition of the mollic epipedon has become unnecessarily complex. In view of the interest in computing soil stores of organic C in response to global warming, changes will need to be made in the number and ranges in properties of diagnostic surface horizons. Some of the subsurface horizons are overutilized, and others are underused. The argillic horizon, which occurs in 42% of the soil series in the USA, could be subdivided into other clay-enriched horizons. The agric horizon should be used on a trial basis for heavily tilled soils.

A persistent criticism of ST is that climate is a soil-forming factor and should not be used to classify soils at the order and suborder level. The rejoinder to this criticism is that soil climate is used—and not atmospheric climate—and that it is a valid soil property. It is a transient soil property, but so are pH and soil-moisture content.

J.G. Bockheim and A.E. Hartemink, *The Soils of Wisconsin*,
World Soils Book Series, DOI 10.1007/978-3-319-52144-2_18

18.3 Current and Future Soil Assessments

The digital age has enabled improved usage of the existing data that was collected over many decades of field research, cartography, soil analysis, and report writing. This book could not have been written without the efforts of hundreds of soil surveyors and 130 years of soil research in Wisconsin. Detailed soil mapping in Wisconsin required an intensive level of field investigation and sampling. Soil scientists mapping in the field were supported by laboratory and correlation staff in Wisconsin and in the National Soil Survey Center in Lincoln, Nebraska. Four soil scientists could map an average county in about four years. Including state, national, and partner support staff, approximately 1500 staff years were needed to complete the initial detailed soil survey of Wisconsin. Most funding, leadership, and staff for the detailed soil survey of Wisconsin were provided by the USDA, NRCS. Significant funding and assistance was also provided by the University of Wisconsin; county, state and tribal governments, and other federal agencies. Much of the legacy soil data can be accessed through the soil portal of NRCS and through handheld devices (Beaudette and O'Geen 2010) and the ISEE project.

In an early stage, soil survey information in Wisconsin was used in small scale waste management and the development of innovative soil disposal systems (Bouma 1973). It

Fig. 18.1 Area predicted to contain red clay above various probability thresholds (30–80%) in a 7000 ha area in the Driftless Area of Wisconsin, USA. From Evans and Hartemink (2014)

has also been used in local land use ordinance control and farmland preservation legislation (Klingelhoets 1972, 1978) and forestry (Cain 1990). The soil maps of Wisconsin have been used for land evaluations for crop suitability (Ye et al. 1991), for predicting solute transport through the landscape (Macur et al. 2000), for assessing carbon stocks (Arriaga and Lowery 2005), and in spatial studies on human health issues related to blastomycosis or lyme disease (Guerra et al. 2002; Baumgardner et al. 2005). The digital use of existing data (also called legacy data) has yielded new insights in the soils of Wisconsin, its properties and distribution, but has also shown that there is an urgent need to update the information in several parts of the state. Below we discuss some of the developments in soil science and how they are being applied to study the soils of Wisconsin.

Scientists at the UW-Madison have been long engaged in quantitative techniques and digital soil mapping, and one of the first IUSS workshops on Pedometrics was held in Madison in August 1997. Many digital soil mapping studies have been conducted in Wisconsin. Digital Soil Mapping or predictive soil mapping is the computer-assisted production of soil type and soil property maps. It involves the creation and population of soil information by the use of field and laboratory observational methods coupled with spatial and nonspatial soil inference systems (McBratney et al. 2006). A digital soil-mapping system tested in several counties of Wisconsin is SoLIM (Soil Land Inference Model). SoLIM is a fuzzy inference scheme for estimating and representing the spatial distribution of soil types in a landscape (Zhu et al. 1997), and it has been used in various studies (Shi et al. 2004; Qi et al. 2006; Smith et al. 2006) across the state of Wisconsin. In the Central part of Wisconsin, prototype category theory has been used in soil mapping (Qi et al. 2006). A prototype-based approach was developed to acquire and represent knowledge of soil-landscape relationships. This knowledge was applied in digital soil mapping using a fuzzy logic system. The created maps seems more accurate in terms of both soil series prediction and soil texture estimation than the case-based reasoning approach (Qi et al. 2006), although a detailed comparison between traditional survey methods and SoLIM has not been made.

Numerical classification methods have been used to delineate landscape units in a study area in southwestern Wisconsin (Irvin et al. 1997). Park et al. (2001) used a process-based terrain characterization to identify the spatial distribution of soils in southern Wisconsin. Shi et al. (2004) used fuzzy soil mapping to test the SoLIM (Soil Land Inference Model). Zhu et al. (2004) used a multiscale tree-structured spatial model to map soil properties in southcentral Wisconsin. Evans and Hartemink (2014) mapped the distribution of red clay subsoil cover by loess in southwestern Wisconsin using digital soil-mapping techniques. They also showed the probability that the soils contain red clay subsoil (Fig. 18.1). The development of probability and uncertainty maps are one of the advantages of digital soil mapping. Among others, it allows for prioritizing the locations of new soil observations and samplings. Adhikari and Hartemink (2016) and Adhikari et al. (2017) used digital soil-mapping techniques to map soil organic carbon for a 5200 ha study area in the Driftless Area, and across Wisconsin using existing data (Fig. 18.2). For the predictions model, rule-based regression kriging was selected in which terrain parameters, soil, land use, and climate information were used as SOC predictors (Fig. 18.3).

Besides digital soil-mapping techniques, proximal sensors have been used to map soils at a fine scale resolution. Proximal sensors include the use of a human-informed mechanical device (cone penetrometer) that allows for

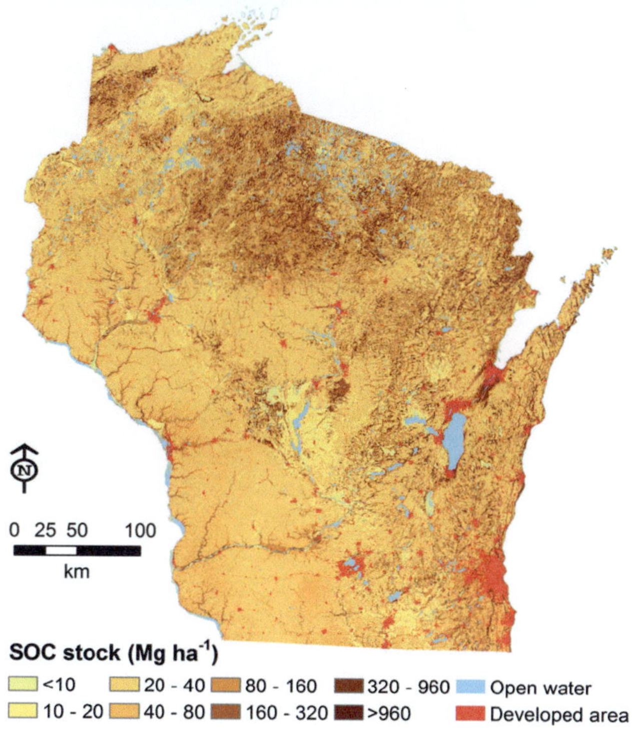

Fig. 18.2 Soil organic carbon stocks (t/ha for top 30 cm) for Wisconsin. After Adhikari et al. (2017)

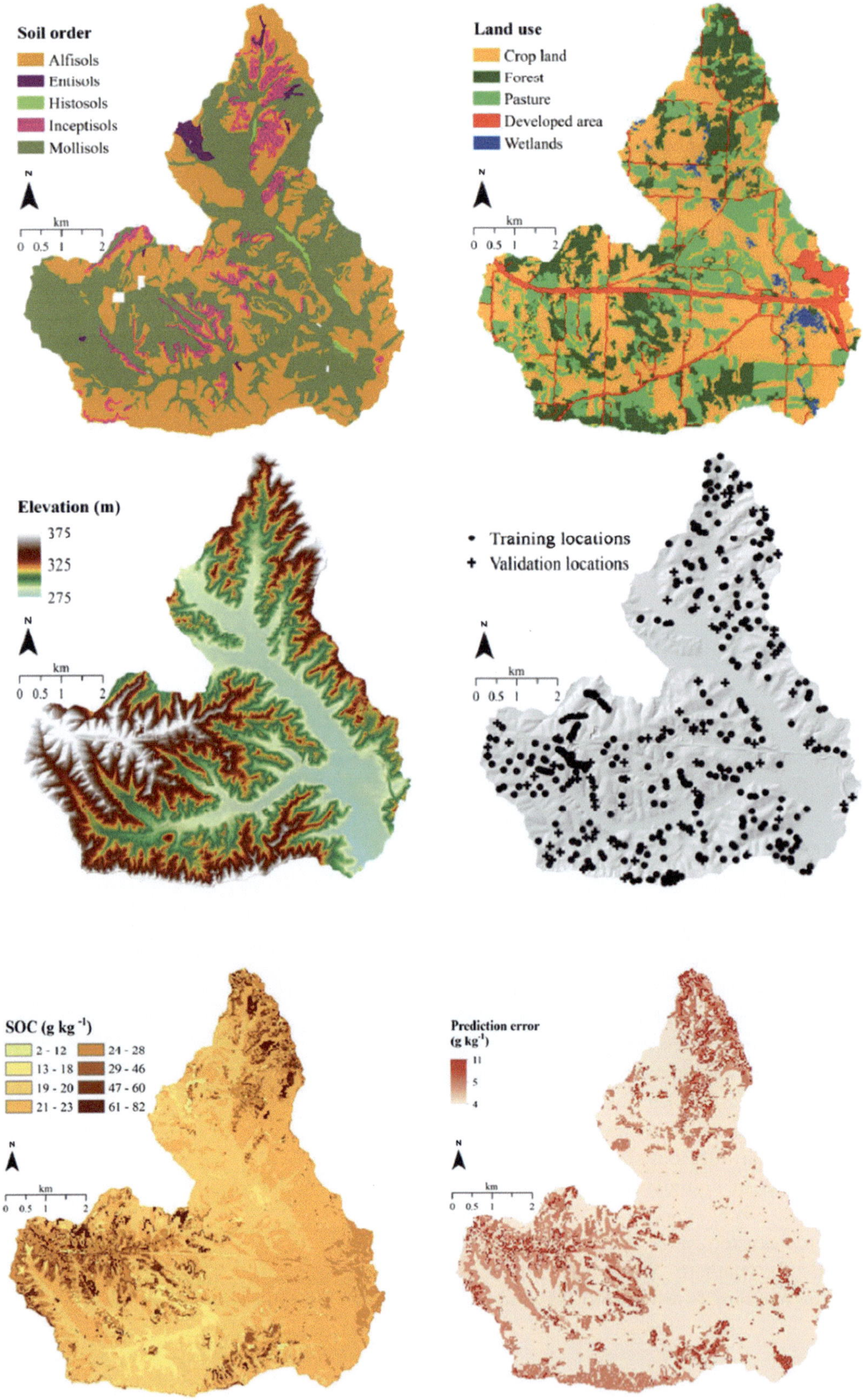

Fig. 18.3 Soil orders, land use, digital elevation model (9 × 9 m grid), sampling locations and predicted maps of soil organic carbon content, and the prediction error for a 5200 ha near Verona in Dane County. After Adhikari and Hartemink (2015)

developing three-dimensional (3D) soil maps (Grunwald et al. 2000, 2001; Rooney and Lowery 2000; Zhu et al. 2004; Arriaga and Lowery 2005). These maps are based on digital elevation models (DEM), thus they are largely physiographic based soil property maps. A cone penetrometer is calibrated for a given soil profile and mapping is completed using statistical applications to detect differences in soil properties with depth. Upon detection of a different soil profile, expert knowledge is introduced to recalibrate the new soil-map unit. This process is continued to produce 3D maps (Fig. 18.4). These and other mechanically developed maps are being used for site-specific (precision) farming. Also, ground penetrating radar (Kung and Donohue 1991; Lowry et al. 2009) and electromagnetic induction have been applied in mapping small areas of soils in Wisconsin (e.g. Morgan et al. 2000; Sudduth et al. 2005).

Recent work focuses on digital soil morphometrics which is defined as the application of instruments and techniques

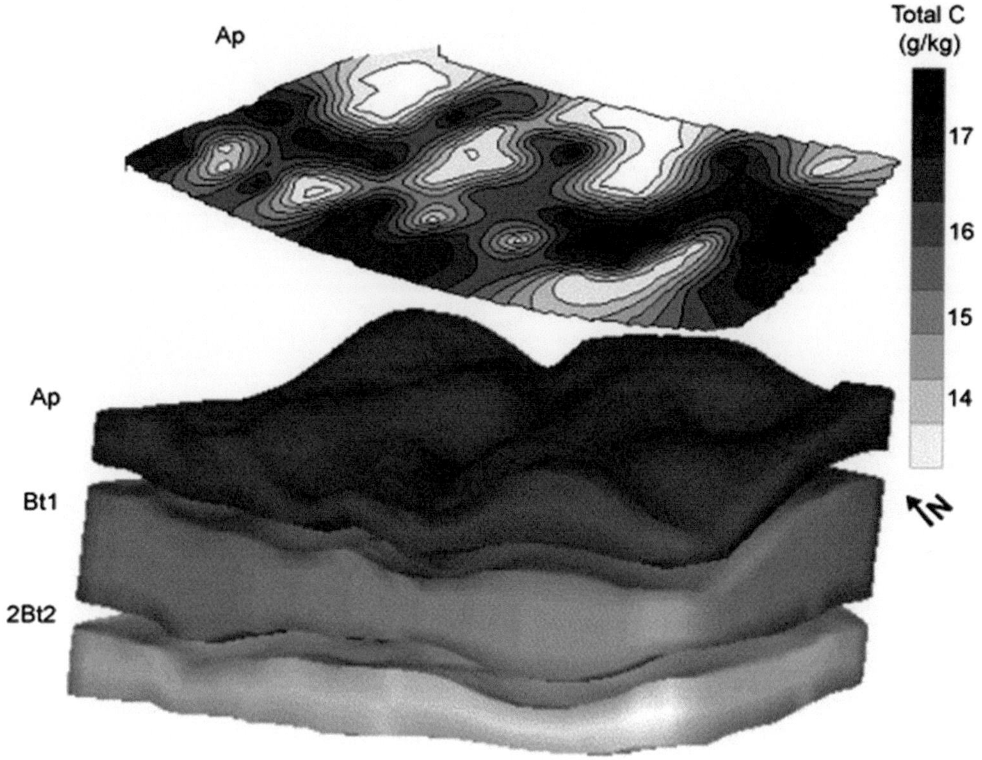

Fig. 18.4 3D Soil mapping of total carbon showing the effect of elevation and soil horizon thickness on soil C distribution of the Ap horizon. The mapping was conducted at Lancaster Agricultural Research Station and the soil was a Dubuque silt loam (fine-silty, mixed, mesic, Typic Hapludalfs). From Arriaga and Lowery (2005)

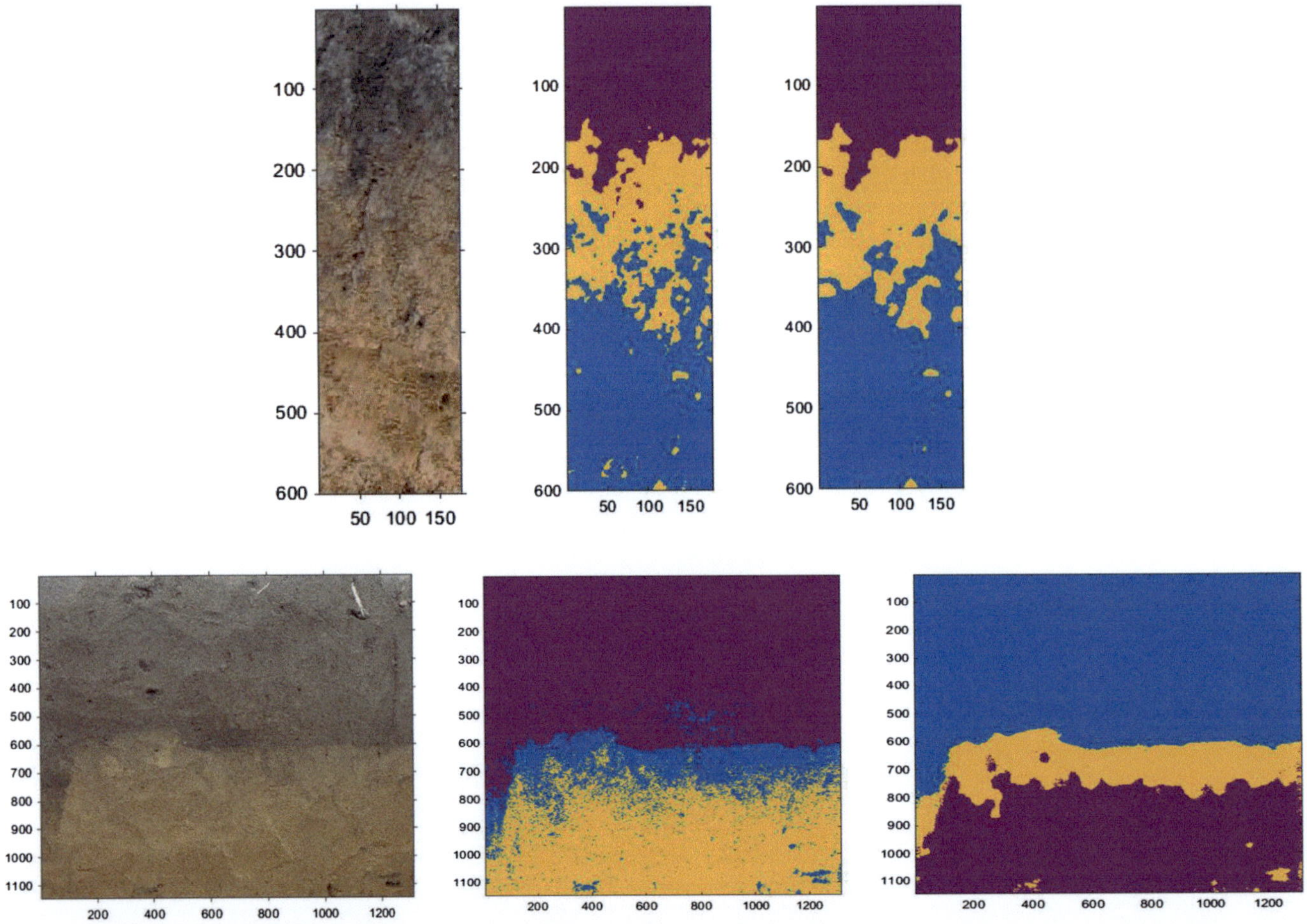

Fig. 18.5 Examples of image analysis of a Mollisol (*top*) and Entisol (*bottom*) using cluster analysis and pixel filtering to verify and obtain soil horizon boundaries and homogeneity (purity) of each horizon. MS graduate research of Jenna Grauer Gray (2016)

for measuring and quantifying soil profile attributes and deriving continuous depth functions (Hartemink and Minasny 2014). This largely focuses on the pedon scale, horizon boundary assessment, purity or homogeneity of the horizons (Fig. 18.5). Proximal soil sensing and other tools are used to enhance soil descriptions, including horizons, texture, color, structure, moisture, mottles, consistence, carbonates, rock fragments, pores, and roots. Tools for these analyses include vis-NIR, pXRF, GPR, electric resistivity, and range of emerging technologies.

Appendix A
Formative Elements for Wisconsin Soils in Soil Taxonomy

Taxonomic level	Formative element	Meaning
Orders		
	-alf	Alfisols
	-ent	Entisols
	-ist	Histosols
	-ept	Inceptisols
	-oll	Mollisols
	-od	Spodosols
	-ult	Ultisols
Suborders		
	aqu	Aquic conditiions (wet)
	fibr	Fibric organic materials (least decomposed)
	fluv	Floodplain
	hem	Hemic organic materials (intermediate in decomposition)
	orth	Central concept
	psamm	Sandy textures
	sapr	Sapric organic materials (most decomposed)
	ud	Udic soil-moisture class (never dry for >90 cumulative days or >45 consecutive days)
	ust	Ustic soil-moisture class (dry for >180 cumulative days or >90 consecutive days)
Great groups		
	Alb	Contains an albic € horizon
	Arg	Contains an argillic (Bt) horizon
	Dystr	Base saturation <60%
	Endo	Saturated from below
	Epi	Saturated from above
	Eutr	Base saturation >60%
	Fluv	Floodplain
	Fragi	Contains a fragipan (Bx) horizon
	Gloss	Contains a glossic (E/B, B/E) horizon
	Hapl	Minimum horizon development
	Hum	Presence of organic matter
	Pale	Excessive development
	Psamm	Sandy texture
	Quartzi	Quartz-rich
	Sphagno	From Sphagnum mosses
	Ud	Udic soil-moisture class (never dry for >90 cumulative days or >45 consecutive days)

(continued)

J.G. Bockheim and A.E. Hartemink, *The Soils of Wisconsin*,
World Soils Book Series, DOI 10.1007/978-3-319-52144-2

(continued)

Taxonomic level	Formative element	Meaning
Subgroup	Aeric	Aeration
	Alfic	Contains argillic (Bt) horizon
	Aquertic	Intergrade with Aquerts
	Aquic	Aquic conditions
	Aquollic	Intergrade with Aquolls
	Arenic	Sandy materials 50–100 cm thick
	Chromic	High chroma
	Cumulic	Thickened epipedon
	Dystric	Low base saturation
	Entic	Least developed of a great group
	Fluvaquentic	Intergrade with Fluvaquents
	Fluventic	Floodplain
	Fragic	Contains a fragipan (Bx) horizon
	Glossic	Contains a glossic (E/B, B/E) horizon
	Haplic	Sequentially the last subgroup within a great group
	Hemic	Hemic organic materials
	Histic	Histic properties
	Humaqueptic	Intergrade with Humaquepts
	Humic	Presence of organic matter
	Inceptic	Intergrade to Inceptisols
	Lamellic	Presence of lamellae (fine-textured bands)
	Limnic	Presence of a limnic layer
	Lithic	Bedrock with 50–100 cm of surface
	Mollic	Contains a mollic epipedon
	Oxyaquic	Seasonal high water table
	Pachic	A thick epipedon
	Psammentic	Intergrade with Psamments
	Terric	Mineral layer within 50 cm of surface
	Thapto	A buried soil
	Typic	Central concept
	Udollic	Intergrades with Udolls
	Ultic	An argillic horizon with 200 cm of the surface
	Umbric	Presence of an umbric epipedon
	Vertic	Presence of cracks

Appendix B
Soil-Forming Factors, Wisconsin Soil Series

Soil series	Area (km^2)	Soil depth class	Drainage class	Slope (%)	Parent material (surface)	Parent material (subsurface)	Vegetation
ABBAYE	11	MD	MWD	1–70	Till L/ss bedrock	Bedrock	NHW-H
ABSCO	236	VD	MWD	0–3	Alluvium, S, L	Alluvium	Mixed hardwood forest
ABSCOTA	2	VD	MWD	0–6	Alluvium S	Alluvium	Wet hardwoods
ACKMORE	64	VD	SPD	0–5	Alluvium Si	Alluvium	Prairie
ADDER	23	VD	VPD	0–1	Organic/alluvium S	Alluvium	Marsh grasses, sedges
ADOLPH	84	VD	VPD	0–2	Mantle L/till L	Till	Marsh grasses, sedges
ADRIAN	507	VD	VPD	0–1	Organic/outwash S	Outwash	Marsh grasses, sedges
AFTAD	53	VD	MWD	0–12	Glaciolacustrine L	Glaciolacustrine	NHW
ALANGO	0	VD	SPD	1–4	Glaciolacustrine C	Glaciolacustrine	Mixed forest
ALBAN	66	VD	WD	0–30	Glaciolacustrine L	Glaciolacustrine	NHW
ALCONA	47	VD	WD	0–70	Glaciofluvial L	Glaciofluvial	Mixed forest
ALDO	25	VD	MWD	0–3	Outwash S	Outwash	Mixed forest
ALGANSEE	166	VD	SPD	0–4	Alluvium S	Alluvium	Wet Mixed forest
ALLENDALE	153	VD	SPD	0–12	Outwash S/glaciolacustrine C	Glaciolacustrine	Mixed forest
ALMENA	435	VD	SPD	0–6	Loess/Till L	Till	Mixed forest
ALPENA	34	VD	SED	0–50	Glaciofluvial L	Glaciofluvial	Mixed forest
ALSTAD	2	VD	SPD	0–4	Till L calc	Till	NHW
ALTDORF	48	VD	PD	0–2	Loess/till C	Till	Wet hardwoods
AMASA	9	VD	WD	0–70	Outwash L/S	Outwash	Boreal
AMERY	991	VD	WD	1–45	Till SL	Till	Mixed forest
AMNICON	374	VD	MWD	0–15	Till C	Till	Boreal
ANGELICA	119	VD	VPD	0–2	Till L	Till	Boreal
ANIGON	259	VD	WD	0–25	Outwash S	Outwash	Mixed forest
ANNALAKE	152	VD	MWD	0–18	Glaciofluvial L	Glaciofluvial	Mixed forest
ANNRIVER	101	VD	VPD	0–2	Loess/alluvium L/outwash S	Outwash	Mixed forest
ANTIGO	1242	VD	WD	0–30	Loess/alluvium L/outwash S	Outwash	NHW
ANTON	14	VD	MWD	0–15	Till C/glaciolacustrine L, S	Glaciolacustrine	Mixed forest
ARBUTUS	1	MD	ED	2–6	Alluv S/igne and meta bedrock	Bedrock	Mixed forest
ARENZVILLE	464	VD	MWD	0–5	Alluvium Si	Alluvium	Oak-savanna
ARGONNE	23	MD	MWD	0–15	Till L	Till	NHW

(continued)

J.G. Bockheim and A.E. Hartemink, *The Soils of Wisconsin*,
World Soils Book Series, DOI 10.1007/978-3-319-52144-2

(continued)

Soil series	Area (km^2)	Soil depth class	Drainage class	Slope (%)	Parent material (surface)	Parent material (subsurface)	Vegetation
ARLAND	291	MD	WD	1–35	Till L/ss bedrock	Bedrock	NHW
ARNHEIM	203	D	PD	0–1	Alluvium L	Alluvium	Wet mixed forest
ASHDALE	210	D	WD	0–20	Loess/residuum LS	Residuum	Prairie
ASHIPPUN	12	VD	SPD	0–6	Loess/till L	Till	NHW
ASHKUM	223	VD	PD	0–3	Colluv/till SiCL	Till	Marsh grasses, sedges
ASHWABAY	94	VD	MWD	0–45	Outwash S/till C	Till	Mixed forest
ATTERBERRY	5	VD	SPD	0–6	Loess	Loess	Prairie
AU GRES	384	VD	SPD	0–6	Glaciofluvial S	Glaciofluvial	Wet mixed forest
AUBURNDALE	208	VD	PD	0–3	Loess/till L	Till	Wet hardwoods
AUGWOOD	13	VD	SPD	0–3	Outwash S/till GLS	Till	Mixed forest
AUSABLE	13	VD	VPD	0–2	Alluvium S	Alluvium	Wet hardwoods
AZTALAN	94	VD	SPD	0–6	Outwash L/lacust C, Si	Outwash	Prairie
BACH	26	VD	PD	0–2	Glaciolacustrine Si	Glaciolacustrine	Wet hardwoods
BADRIVER	181	VD	SPD	0–3	Till C	Till	Boreal
BALMORAL	1	VD	MWD	0–3	Alluvium Si, L/outwash S	Outwash	Prairie
BANAT	1	D	SPD	0–4	Mantle L/S calc	Mantle	Mixed forest
BARABOO	113	D	MWD	2–60	Loess/qtzite bedrock	Bedrock	Oak-hickory
BARREMILLS	43	VD	MWD	1–6	Loess	Loess	Prairie
BARRONETT	98	VD	PD	0–2	Glaciolacustrine Si	Glaciolacustrine	Wet hardwoods
BARRY	32	VD	PD	0–3	Till L	Till	NHW
BASCO	53	MD	WD	2–30	Loess/pedised. C/ss residuum	Residuum	Oak-savanna
BATAVIA	11	VD	WD	0–12	Loess/outwash L	Outwash	Oak-savanna
BEARPEN	43	VD	SPD	0–3	Loess/alluvium Si	Alluvium	Oak-hickory
BEARTREE	2	S	VPD	0–2	Alluvium L/mudst bedrock	Bedrock	Mixed forest
BEAVERBAY	46	VD	MWD	2–30	Loess/alluvium L	Alluvium	NHW
BEAVERCREEK	26	VD	WD	1–15	Alluvium KL	Alluvium	Oak-hickory
BEECHER	7	VD	SPD	0–6	Mantle L/till CL	Till	Oak-savanna
BELLECHESTER	3	VD	ED	12–90	Colluv, resid S/ss bedrock	Bedrock	Prairie
BELLEVILLE	0	VD	VPD	0–2	Glaciofluvial S/till L	Till	Wet hardwoods
BELLEVUE	336	VD	MWD	0–3	Alluvium Si	Alluvium	NHW
BERGLAND	8	VD	PD	0–2	Glaciolacustrine C	Glaciolacustrine	Boreal
BERTRAND	7	VD	WD	0–35	Alluv Si/alluv S	Alluvium	Oak-hickory
BESEMAN	472	VD	VPD	0–1	Organic	Organic	Marsh grasses, sedges
BIGISLAND	5	VD	SED	15–45	Alluvium S/till L	Till	Mixed forest
BILLETT	55	VD	WD	0–20	Alluvium L, S	Alluvium	Oak-savanna
BILLYBOY	97	VD	WD	0–3	Alluvium Si, L/outwash S	Outwash	NHW
BILMOD	32	VD	WD	0–3	Alluvium L/alluvium S	Alluvium	Oak-hickory
BILSON	133	VD	WD	0–20	Alluv L/alluv S	Alluvium	Oak-hickory
BJORKLAND	9	VD	VPD	0–2	Glaciolacustrine S/C	Glaciolacustrine	Marsh grasses, sedges
BLACKRIVER	15	VD	MWD	0–6	Alluv Si/outwash S	Outwash	NHW
BLOUNT	7	VD	SPD	0–6	Till CL	Till	Oak-hickory
BLUFFTON	13	VD	VPD	0–2	Alluv L/till L	Till	Wet mixed forest
BOAZ	52	VD	SPD	0–3	Alluvium Si	Alluvium	Wet hardwoods
BOGUSCREEK	1	VD	WD	0–3	Alluvium Si/outwash S	Outwash	NHW
BONDUEL	71	MD	SPD	0–3	Till L/dolo bedrock	Bedrock	Mixed forest

(continued)

(continued)

Soil series	Area (km^2)	Soil depth class	Drainage class	Slope (%)	Parent material (surface)	Parent material (subsurface)	Vegetation
BOONE	617	MD	ED	1–90	Residuum S/ss bedrock	Bedrock	Oak-pine
BOOTS	35	VD	VPD	0–2	Organic	Organic	Wet conifers
BOPLAIN	15	MD	ED	0–60	Outwash S	Outwash	Oak-pine
BOREA	11	VD	SPD	0–3	Till C/glaciolacustrine L, S	Glaciolacustrine	Mixed forest
BORTH	47	VD	MWD	0–6	Glaciolacustrine C/S	Glaciolacustrine	NHW
BOWSTRING	266	VD	VPD	0–1	Organic	Organic	Marsh grasses, sedges
BOYER	398	VD	WD	0–50	Till S, L/outwash S	Outwash	Oak-hickory
BRAHAM	0	VD	WD	6–30	Eolian S/till L	Till	Mixed forest
BRANDER	18	VD	MWD	0–3	Loess/outwash S	Outwash	NHW
BRANSTAD	4	VD	MWD	2–20	Till L calc	Till	NHW
BREMS	339	VD	MWD	0–8	Outwash S	Outwash	Mixed oaks
BRENNYVILLE	1	VD	SPD	0–6	Loess/till L	Till	Mixed forest
BREVORT	79	VD	VPD	0–2	Glaciolacustrine S/L	Glaciolacustrine	Wet mixed forest
BRICE	2	VD	WD	2–6	Eolian S, L	Eolian	Prairie
BRIGGSVILLE	102	VD	WD	0–20	Glaciolacustrine C, Si	Glaciolacustrine	Mixed hardwood forest
BRILL	102	VD	MWD	0–6	Loess/outwash S	Outwash	Mixed forest
BRIMLEY	3	D	SPD	0–6	Glaciofluvial L	Glaciofluvial	Mixed forest
BRINKMAN	44	VD	MWD	2–12	Loess/residuum C	Residuum	Oak-hickory
BRODALE	7	VD	ED	6–90	Residuum L/ls bedrock	Bedrock	Prairie
BROOKSTON	130	VD	PD	0–3	Till L	Till	Mixed hardwood forest
BROWNCHURCH	58	VD	WD	12–35	Alluvium L/Si	Alluvium	Oak-hickory
BROWNSTONE	6	MD	ED	0–15	Eolian S/ss bedrock	Bedrock	Mixed forest
BRUCE	4	D	PD	0–2	Glaciolacustrine L, S	Glaciolacustrine	Mixed forest
BURKHARDT	182	VD	SED	0–30	Alluvium L/outwash S	Outwash	Prairie
BUSHVILLE	1	VD	SPD	0–6	Outwash S/till L	Till	Mixed oaks
CABLE	340	VD	VPD	0–2	Loess/till L	Till	Wet mixed forest
CADIZ	38	VD	MWD	2–12	Loess/till SiCL	Till	Mixed hardwood forest
CALAMINE	9	D	PD	0–12	Loess/residuum calc shale	Bedrock	Marsh grasses, sedges
CAMPIA	57	VD	WD	0–25	Glaciolacustrine Si	Glaciolacustrine	NHW
CAPITOLA	707	VD	PD	0–2	Alluvium Si, L/till SL	Till	Mixed forest
CARBONDALE	369	VD	VPD	0–2	Organic	Organic	Wet conifers
CARYVILLE	12	D	WD	0–3	Alluvium S, L	Alluvium	NHW
CASCO	699	VD	SED	0–70	Outwash S calc	Outwash	Mixed oaks
CATHRO	1823	VD	VPD	0–2	Organic	Organic	Wet mixed forest
CEBANA	567	VD	VPD	0–2	Loess/till L	Till	Mixed forest
CERESCO	36	VD	SPD	0–3	Alluvium L	Alluvium	Wet hardwoods
CHABENEAU	11	VD	MWD	0–18	Eolian/outwash	Outwash	Boreal
CHAMPION	14	VD	WD	0–70	Eolian L/till L	Till	NHW-H
CHANNAHON	49	S	WD	0–25	Mantle L/ls bedrock	Bedrock	Prairie
CHANNING	9	VD	SPD	0–3	Outwash L/S	Outwash	Boreal
CHARLEVOIX	93	VD	SPD	0–6	Till SL	Till	NHW

(continued)

(continued)

Soil series	Area (km^2)	Soil depth class	Drainage class	Slope (%)	Parent material (surface)	Parent material (subsurface)	Vegetation
CHASEBURG	277	VD	WD	0–15	Alluvium Si, L	Alluvium	Mixed hardwood forest
CHELMO	3	VD	PD	0–2	Glaciolacustrine C/outwash S	Outwash	Mixed forest
CHELSEA	74	VD	ED	0–45	Eolian S	Eolian	Oak-hickory
CHEQUAMEGON	341	VD	MWD	2–30	Loess/till SL	Till	NHW
CHETEK	750	VD	SED	0–45	Outwash S	Outwash	Mixed forest
CHINWHISKER	29	VD	MWD	0–4	Glaciofluvial S	Glaciofluvial	Mixed forest
CHIPPENY	2	MD	VPD	0–12	Organic/ls bedrock	Alluvium	Wet mixed forest
CHURCHTOWN	455	VD	WD	6–30	Mantle L/loess	Loess	Oak-savanna
CITYPOINT	98	MD	VPD	0–1	Organic/residuum ss	Residuum	Marsh grasses, sedges
CLEMENS	3	VD	SPD	0–2	Alluvium L-sk/S-sk	Alluvium	Mixed forest
CLYDE	21	VD	VPD	0–4	Outwash L/till L	Till	Prairie
COFFEEN	37	VD	SPD	0–2	Alluvium Si	Alluvium	Oak-savanna
COFFTON	10	VD	SPD	0–3	Alluvium Si	Alluvium	Oak-savanna
COLAND	1	VD	PD	0–5	Alluvium	Alluvium	Prairie
COLOMA	303	VD	SED	0–70	Outwash S	Outwash	Mixed oaks
COLWOOD	168	VD	VPD	0–3	Glaciolacustrine Si, L	Glaciolacustrine	Wet hardwoods
COMSTOCK	143	VD	SPD	0–3	Glaciolacustrine Si	Glaciolacustrine	Mixed forest
CONOVER	1	VD	SPD	0–6	Till L	Till	Mixed hardwood forest
CORMANT	194	VD	VPD	0–3	Outwash S	Outwash	Mixed forest
CORNUCOPIA	103	VD	WD	2–45	Glaciolacustrine C/L, S	Glaciolacustrine	Boreal
COSAD	12	VD	SPD	0–8	Glaciolacustrine S/C	Glaciolacustrine	Wet hardwoods
COUNCIL	312	VD	WD	2–50	Alluvium L	Alluvium	Mixed hardwood forest
CRESS	195	VD	SED	0–35	Alluvium L/outwash S	Outwash	Mixed forest
CREX	33	VD	MWD	0–3	Outwash S	Outwash	Oak-pine
CROMWELL	238	VD	SED	0–40	Outwash S	Outwash	Mixed forest
CROSSETT	4	VD	SPD	0–6	Till L calc	Till	Mixed forest
CROSWELL	518	VD	MWD	0–12	Glaciofluvial S	Glaciofluvial	Mixed forest
CROSWOOD	27	VD	MWD	0–6	Outwash S/till L	Till	Mixed forest
CRYSTAL LAKE	145	VD	MWD	0–45	Glaciolacustrine Si	Glaciolacustrine	NHW
CUBLAKE	118	VD	MWD	0–15	Outwash S/glaciofluvial Si, L, S	Glaciofluvial	Mixed forest
CUNARD	2	MD	WD	0–18	Till L/ls bedrock	Bedrock	NHW
CURRAN	103	VD	SPD	0–5	Loess/alluvium Si	Alluvium	Mixed hardwood forest
CUSHING	7	VD	WD	20–35	Till L calc	Till	Mixed forest
CUTAWAY	0	VD	MWD	0–12	Outwash S/till L calc	Till	Mixed forest
CUTTRE	370	VD	SPD	0–3	Till C	Till	Boreal
DAIRYLAND	13	VD	MWD	0–20	Alluvium S-sk/till L	Till	Mixed forest
DAISYBAY	1	VD	VPD	0–1	Organic/till C	Till	Wet conifers
DAKOTA	148	VD	WD	0–18	Alluvium L/outwash S	Outwash	Prairie
DANCY	96	VD	PD	0–2	Alluvium L, S/till L	Till	Wet hardwoods
DARROCH	8	VD	SPD	0–3	Glaciolacustrine Si, L	Glaciolacustrine	Prairie
DAWSIL	170	VD	VPD	0–1	Organic/alluvium S	Alluvium	Wet conifers
DAWSON	603	VD	VPD	0–2	Organic/outwash S	Outwash	Wet conifers

(continued)

(continued)

Soil series	Area (km^2)	Soil depth class	Drainage class	Slope (%)	Parent material (surface)	Parent material (subsurface)	Vegetation
DECHAMPS	31	VD	SPD	0–2	Alluvium S	Alluvium	Mixed forest
DEERTON	7	MD	WD	0–35	Outwash S/ss bedrock	Bedrock	NHW
DEFORD	127	VD	VPD	0–2	Outwash S	Outwash	Wet mixed forest
DEL REY	2	VD	SPD	0–7	Glaciolacustrine SiCL	Glaciolacustrine	Mixed hardwood forest
DELLS	28	MD	SPD	0–3	Loess/alluvium S	Alluvium	Mixed hardwood forest
DELTON	83	VD	WD	0–15	Outwash S/glaciolacustrine L/C	Glaciolacustrine	Oak-hickory
DENOMIE	31	VD	WD	0–60	Till Si, L	Till	Mixed forest
DENROCK	1	D	SPD	0–2	Loess/lacustrine C/stratif alluvium	Alluvium	Prairie
DERINDA	5	MD	MWD	2–60	Loess/residuum calc shale	Bedrock	Oak-hickory
DICKINSON	96	VD	WD	0–30	Alluvium L	Alluvium	Prairie
DICKMAN	15	VD	SED	0–18	Mantle L/outwash G, S	Outwash	Prairie
DISHNO	9	D	MWD	0–35	Mantle L/till/meta bedrock	Gneiss bedrock	Boreal
DOBIE	64	MD	WD	2–30	Loess/residuum glauc ss	Bedrock	NHW
DODGE	495	VD	WD	0–20	Loess/till SL	Till	Mixed hardwood forest
DODGEVILLE	585	S	WD	0–30	Loess/paleosol ls/dolo ls bedrock	Bedrock	Oak-savanna
DODY	3	VD	VPD	0–2	Outwash S/glaciolacustrine C/ outwash S	Outwash	Marsh grasses, sedges
DOLPH	88	D	SPD	0–3	Loess/residuum C	Residuum	Wet hardwoods
DORA	5	VD	VPD	0–1	Organic/glaciolacustrine SiC	Glaciolacustrine	Wet mixed forest
DORCHESTER	0	VD	WD	0–5	Alluvium stratif calc	Alluvium	Prairie
DORERTON	394	VD	WD	12–65	Loess/frag dolostone	Bedrock	Mixed hardwood forest
DORITTY	31	VD	MWD	2–12	Loess/alluvium stratif	Alluvium	NHW
DOWNS	407	VD	WD	0–35	Loess	Loess	Oak-savanna
DRAMMEN	163	VD	SED	1–20	Alluvium S	Alluvium	Mixed forest
DRESDEN	246	VD	WD	0–30	Alluvium Si/outwash S	Alluvium	Oak-savanna
DRUMMER	39	VD	PD	0–2	Loess/outwash stratif	Outwash	Marsh grasses, sedges
DRYLANDING	68	S	SED	0–12	Alluvium L/mudst bedrock	Bedrock	Mixed forest
DUBUQUE	2835	MD	WD	2–60	Loess/residuum ls/ls bedrock	Bedrock	Oak-hickory
DUEL	3	MD	WD	1–12	Till S/ls bedrock	Bedrock	NHW
DUELM	4	VD	WD	0–3	Outwash S	Outwash	Oak-savanna
DUNBARTON	246	S	WD	2–60	Loess/residuum C/dolostone bedrock	Bedrock	Oak-hickory
DUNNBOT	47	VD	MWD	0–3	Alluvium L, S	Alluvium	NHW
DUNNVILLE	51	VD	WD	0–12	Alluvium L/S	Alluvium	Oak-savanna
DURAND	69	VD	WD	1–20	Loess/paleosol from till L	Till	Prairie
EAUCLAIRE	20	VD	MWD	1–6	Alluvium S/till L	Till	Mixed hardwood forest
EDMUND	258	S	WD	2–35	Mantle L/residuum C/dolo bedrock	Bedrock	Prairie

(continued)

(continued)

Soil series	Area (km^2)	Soil depth class	Drainage class	Slope (%)	Parent material (surface)	Parent material (subsurface)	Vegetation
EDWARDS	6	VD	VPD	0–2	Organic/marl	Marl	Wet mixed forest
ELBAVILLE	180	VD	WD	12–45	Loess/residuum L-sk	Residuum	Mixed oaks
ELBURN	260	VD	SPD	0–5	Outwash stratif	Outwash	Prairie
ELDERON	44	VD	SED	2–45	Till KS	Till	Mixed forest
ELEROY	7	D	MWD	2–30	Loess/residuum calc shale	Bedrock	Oak-hickory
ELEVA	541	MD	WD	1–60	Mantle L/ss bedrock	Bedrock	Mixed forest
ELEVASIL	257	MD	WD	2–60	Colluvium L/residuum S/ss bedrock	Bedrock	Mixed forest
ELKMOUND	322	S	WD	0–60	Mantle L/residuum ss	Residuum	Mixed oaks
ELLA	84	VD	MWD	0–12	Alluvium Si	Alluvium	Oak-savanna
ELLIOTT	147	VD	SPD	0–7	Loess/till SiCL	Till	Prairie
ELLWOOD	21	VD	MWD	1–15	Till L calc	Till	Mixed forest
ELM LAKE	274	MD	PD	0–2	Alluvium S/residuum L	Residuum	Wet hardwoods
ELVERS	16	VD	VPD	0–3	Alluvium Si/organic	Organic	Wet hardwoods
EMMERT	80	VD	ED	1–70	Outwash S	Outwash	Mixed forest
EMMET	537	VD	MWD	0–50	Till SL	Till	NHW-H
ENSLEY	95	VD	PD	0–2	Till L	Till	Wet mixed forest
ETTRICK	318	VD	PD	0–2	Alluvium Si	Alluvium	Wet hardwoods
FABIUS	36	VD	SPD	0–6	Till L, S	Till	Wet hardwoods
FAIRCHILD	237	MD	SPD	0–3	Alluvium S/residuum L	Residuum	Mixed forest
FAIRPORT	18	MD	MWD	0–25	Till Si/L/ls bedrock	Bedrock	NHW
FALLCREEK	117	VD	SPD	0–6	Till L	Till	NHW
FARRINGTON	17	VD	SPD	0–3	Outwash S	Outwash	Prairie
FAYETTE	2234	VD	WD	0–60	Loess	Loess	Oak-hickory
FENANDER	9	VD	VPD	0–2	Glaciolacustrine stratif L, S	Glaciolacustrine	Wet hardwoods
FENCE	79	VD	MWD	0–18	Glaciolacustrine stratif Si	Glaciolacustrine	Mixed forest
FENWOOD	328	D	WD	2–30	Loess/till L/ign-meta bedrock	Bedrock	NHW
FESTINA	14	VD	WD	0–12	Alluvium Si/stratif	Alluvium	Prairie
FINCHFORD	121	VD	ED	0–14	Outwash S	Outwash	Prairie
FISK	19	VD	SPD	0–3	Outwash S/outwash stratif	Outwash	Mixed forest
FIVEPOINTS	183	MD	WD	2–30	Loess/residuum C/dolostone bedrock	Bedrock	Oak-hickory
FLAGG	44	VD	WD	0–20	Loess/paleosol Illinoian till	Till	Mixed hardwood forest
FLAMBEAU	217	VD	WD	1–20	Till L	Till	NHW
FLINK	48	VD	SPD	0–4	Outwash S/glaciofluvial Si, L, S	Glaciofluvial	Mixed forest
FLOYD	19	VD	SPD	0–5	Mantle L/till L	Till	Prairie
FORADA	5	VD	PD	0–2	Mantle L/outwash G, S	Outwash	Prairie
FORDUM	595	VD	PD	0–2	Alluvium L, S	Alluvium	Wet hardwoods
FORKHORN	60	VD	WD	0–20	Alluvium L/outwash S	Alluvium	Oak-savanna
FOX	739	VD	WD	0–35	Loess/alluvium L/outwash S	Outwash	Mixed hardwood forest
FRECHETTE	73	VD	WD	2–35	Till L calc	Till	NHW
FREEON	2144	VD	MWD	0–20	Loess/till SL	Till	NHW
FREMSTADT	63	VD	WD	1–30	Till S	Till	NHW
FREYA	81	VD	SPD	0–3	Glaciolacustrine S/C	Glaciolacustrine	Mixed forest
FRIENDSHIP	530	VD	MWD	0–4	Outwash S	Outwash	Oak-pine

(continued)

(continued)

Soil series	Area (km^2)	Soil depth class	Drainage class	Slope (%)	Parent material (surface)	Parent material (subsurface)	Vegetation
FRIESLAND	37	VD	MWD	0–7	Eolian L/till Si	Till	Prairie
FROGCREEK	106	VD	MWD	2–15	Loess/alluvium L/till S, L	Till	NHW
GAASTRA	8	VD	SPD	0–4	Glaciofluvial Si, L	Glaciofluvial	Mixed forest
GALE	554	MD	WD	0–60	Loess/residuum S/ss bedrock	Bedrock	Oak-hickory
GAPHILL	67	VD	WD	30–60	Colluvium L/S/residuum S/ss bedrock	Bedrock	Mixed forest
GARDENVALE	10	D	WD	1–6	Eolian Si, L/residuum S	Residuum	NHW
GARNE	6	MD	SED	2–12	Outwash S/residuum L/glauc ss bedrock	Bedrock	Prairie
GARWIN	0	D	PD	0–2	Loess	Loess	Prairie
GASTROW	13	VD	SPD	0–3	Glaciolacustrine stratif Si, SL	Glaciolacustrine	Mixed forest
GAY	18	D	PD	0–3	Till L	Till	Boreal
GICHIGAMI	53	VD	MWD	0–6	Till Si, L	Till	Mixed forest
GIESE	28	VD	VPD	0–1	Mantle L/till SL	Till	Wet hardwoods
GILFORD	53	VD	PD	0–2	Outwash L/S	Outwash	Marsh grasses, sedges
GILLINGHAM	2	D	WD	6–20	Eolian S/alluvium L/glauc ss bedrock	Bedrock	Mixed hardwood forest
GLENDENNING	19	VD	SPD	0–4	Till SL	Till	NHW
GLENDORA	41	VD	VPD	0–2	Alluvium S	Alluvium	Wet hardwoods
GLENFLORA	17	VD	VPD	0–2	Loess/outwash S	Outwash	Mixed forest
GLIDDEN	64	VD	WD	0–30	Alluvium Si, L/outwash S	Alluvium	NHW-H
GOGEBIC	920	VD	MWD	1–55	Eolian L/till L, S	Till	NHW
GOODMAN	262	VD	WD	1–45	Loess/till L	Till	NHW
GOODWIT	142	VD	MWD	0–15	Loess/till L	Till	NHW
GOSIL	80	VD	ED	1–12	Eolian S/residuum S	Residuum	Oak-pine
GOTHAM	403	VD	SED	0–35	Glaciolacustrine S	Glaciolacustrine	Prairie
GRANBY	231	VD	PD	0–3	Outwash S	Outwash	Wet mixed forest
GRAYCALM	380	VD	SED	0–70	Outwash S	Outwash	Oak-pine
GRAYLING	587	VD	ED	0–45	Outwash S	Outwash	Oak-pine
GRAYS	54	VD	MWD	0–12	Loess/glaciolacust stratif calc	Glaciolacustrine	Oak-savanna
GREENRIDGE	36	VD	WD	4–20	Loess/residuum L/ss bedrock	Bedrock	Mixed hardwood forest
GREENWOOD	315	VD	VPD	0–2	Organic	Organic	Wet conifers
GRELLTON	73	VD	WD	0–20	Outwash L/till L	Till	Mixed hardwood forest
GRETTUM	171	VD	WD	0–30	Glaciolacustrine S	Glaciolacustrine	Mixed forest
GRISWOLD	190	VD	WD	0–20	Till SL calc	Till	Prairie
GUENTHER	19	VD	MWD	2–6	Alluvium S/till L	Till	Mixed forest
GULL POINT	1	VD	PD	0–2	Alluvium L/till L	Till	Wet mixed forest
HALDER	3	D	SPD	0–3	Mantle L/outwash G, S	Outwash	Mixed hardwood forest
HATLEY	90	VD	SPD	0–6	Loess/till SL	Till	Mixed forest
HAUGEN	322	VD	MWD	2–15	Till SL	Till	NHW
HAUSTRUP	13	S	SED	1–25	Loess/basalt bedrock	Bedrock	Mixed forest
HAYFIELD	30	VD	SPD	0–4	Mantle L/outwash G, S	Outwash	Oak-savanna

(continued)

(continued)

Soil series	Area (km^2)	Soil depth class	Drainage class	Slope (%)	Parent material (surface)	Parent material (subsurface)	Vegetation
HAYRIVER	208	MD	WD	2–50	Residuum L/S	Residuum	NHW
HEBRON	77	VD	MWD	0–6	Outwash L/glaciolacustrine C, Si	Outwash	Mixed oaks
HENNEPIN	4	VD	WD	10–70	Till L calc	Till	Mixed hardwood forest
HERBSTER	137	VD	SPD	0–3	Glaciolacustrine C	Glaciolacustrine	Mixed forest
HERSEY	57	VD	MWD	2–18	Loess/till L	Till	NHW
HESCH	20	MD	WD	0–45	Residuum L/S	Residuum	Oak-hickory
HIBBING	0	VD	MWD	3–45	Loess/till C	Till	Mixed forest
HILES	116	MD	MWD	1–12	Loess/residuum L	Residuum	NHW
HIXTON	566	MD	WD	2–60	Loess/residuum S/ss bedrock	Bedrock	Oak-hickory
HOCHHEIM	979	S	WD	0–30	Loess/till L	Till	NHW
HOOP	6	VD	SPD	0–3	Alluvium L/S	Alluvium	Prairie
HOOPESTON	48	VD	SPD	0–5	Outwash L, S	Outwash	Prairie
HORTONVILLE	1146	VD	WD	1–35	Till L calc	Till	NHW
HOUGHTON	1482	VD	VPD	0–2	Organic	Organic	Marsh grasses, sedges
HUBBARD	27	VD	ED	0–35	Outwash S	Outwash	Prairie
HUMBIRD	242	MD	MWD	1–20	Alluvium L/residuum C/ss bedrock	Bedrock	Mixed forest
HUNTSVILLE	153	VD	WD	0–6	Alluvium Si	Alluvium	Prairie
IMPACT	132	VD	ED	0–6	Alluvium S	Alluvium	Oak-pine
INDUS	2	D	PD	0–2	Glaciolacustrine C calc	Glaciolacustrine	Mixed forest
INGALLS	8	VD	SPD	0–6	Outwash S/glaciolacustrine stratif	Glaciolacustrine	Mixed forest
IONIA	33	VD	MWD	0–6	Mantle L/outwash G, S	Outwash	Mixed hardwood forest
IOSCO	185	VD	SPD	0–6	Glaciolacustrine S/L	Glaciolacustrine	NHW
IRONRUN	175	VD	SPD	0–3	Alluvium S	Alluvium	Mixed forest
ISHPEMING	87	MD	SED	0–70	Till S/ign bedrock	Bedrock	NHW
JACKSON	86	VD	MWD	0–9	Alluvium Si/S	Alluvium	Mixed hardwood forest
JASPER	8	VD	WD	0–18	Outwash L/stratif L	Outwash	Prairie
JEWETT	127	MD	WD	0–20	Eolian L/till L	Till	NHW
JOY	28	VD	SPD	0–5	Loess	Loess	Prairie
JUDA	9	VD	MWD	2–12	Loess/till SiCL	Till	Oak-savanna
JUDSON	84	VD	WD	0–12	Colluvium Si	Colluvium	Prairie
JUNEAU	48	VD	MWD	0–6	Alluvium Si/paleosol in till Si	Till	Oak-hickory
KALMARVILLE	74	VD	PD	0–1	Alluvium	Alluvium	Wet hardwoods
KANE	39	VD	SPD	0–3	Outwash Si/L/stratif G, S	Outwash	Prairie
KARLIN	480	VD	SED	0–75	Outwash S	Outwash	Mixed forest
KARLSBORG	57	VD	MWD	1–30	Glaciolacustrine S/C	Glaciolacustrine	Oak-pine
KATO	12	VD	VPD	0–2	Mantle Si/outwash S, G	Outwash	Prairie
KAUKAUNA	14	VD	MWD	0–4	Glaciolacustrine C/Si, vfs	Glaciolacustrine	Mixed hardwood forest
KEGONSA	72	VD	WD	0–6	Loess/outwash S	Outwash	Oak-savanna
KELLOGG	100	VD	WD	0–12	Glaciolacustrine S/C	Glaciolacustrine	Mixed forest
KELTNER	7	D	MWD	0–30	Loess/residuum shale	Bedrock	Prairie

(continued)

(continued)

Soil series	Area (km^2)	Soil depth class	Drainage class	Slope (%)	Parent material (surface)	Parent material (subsurface)	Vegetation
KENDALL	5	VD	SPD	0–5	Loess/outwash stratif L	Outwash	Oak-hickory
KENNAN	1139	D	WD	1–35	Till SL, LS	Till	Mixed forest
KEOWNS	0	VD	PD	0–3	GLACIOLACUSTRINE L, S	Glaciolacustrine	Wet hardwoods
KERT	308	MD	SPD	0–6	Loess/residuum L/ss bedrock	Bedrock	Mixed forest
KESHENA	17	VD	MWD	0–6	Till L calc	Till	NHW
KEVILAR	164	VD	MWD	0–12	Alluvium L/S/stratif L, S	Alluvium	Mixed forest
KEWAUNEE	2002	VD	WD	0–45	Loess/till C	Till	NHW
KEWEENAW	1278	VD	WD	0–70	Till S	Till	NHW-H
KEYESVILLE	6	MD	SED	20–65	Alluvium L/residuum C/ss bedrock	Bedrock	Oak-hickory
KIBBIE	67	VD	SPD	0–6	Glaciofluvial stratif L	Glaciofluvial	Mixed hardwood forest
KICKAPOO	71	D	MWD	0–6	Alluvium L	Alluvium	Mixed hardwood forest
KIDDER	644	VD	WD	0–35	Loess/till L	Till	Oak-hickory
KINGSVILLE	70	VD	VPD	0–2	Glaciolacustrine S	Glaciolacustrine	Wet hardwoods
KINROSS	102	VD	PD	0–3	Outwash S	Outwash	Wet mixed forest
KIVA	41	VD	WD	0–45	Glaciolacustrine S	Glaciolacustrine	Mixed forest
KNOWLES	76	MD	WD	0–30	Loess/till L	Till	Oak-hickory
KOLBERG	72	MD	WD	0–12	Till C/L/dolo bedrock	Bedrock	Mixed forest
KOMRO	14	VD	MWD	0–3	Outwash S	Outwash	Prairie
KOROBAGO	39	VD	SPD	0–3	Glaciolacustrine L	Glaciolacustrine	Mixed hardwood forest
KOST	2	VD	ED	0–18	Outwash S	Outwash	Oak-savanna
KRANSKI	103	VD	SED	2–30	Till S	Till	Oak-pine
LA FARGE	595	MD	WD	2–35	Loess/residuum/ss bedrock	Bedrock	Mixed hardwood forest
LABLATZ	1	VD	SPD	0–4	Till L calc	Till	NHW-H
LAMARTINE	264	VD	SPD	0–6	Loess/till L	Till	Mixed hardwood forest
LAMBEAU	8	VD	WD	6–20	Loess/alluvium L/residuum S/ss bedrock	Bedrock	Mixed hardwood forest
LAMONT	3	VD	WD	0–45	Eolian L, S	Eolian	Oak-hickory
LANGLADE	5	VD	WD	0–6	Alluvium Si/L/outwash S	Outwash	NHW
LAONA	32	MD	WD	5–35	Till L	Till	NHW
LAPEER	248	VD	WD	0–60	Till SL	Till	Oak-hickory
LAPOIN	7	MD	MWD	0–6	Alluvium L/till C/L	Till	Mixed forest
LARA	2	VD	MWD	0–6	Eolian S/glaciolacustrine C	Glaciolacustrine	Mixed forest
LAWLER	20	VD	SPD	0–6	Alluvium L/outwash S, G	Outwash	Prairie
LAWSON	74	VD	SPD	0–5	Alluvium S	Alluvium	Wet hardwoods
LENROOT	161	VD	MWD	0–4	Outwash S	Outwash	Oak-pine
LEOLA	47	VD	SPD	0–3	Outwash S	Outwash	Oak-pine
LERCH	48	VD	VPD	0–2	Glaciolacustrine C	Glaciolacustrine	Wet mixed forest
LEROY	98	VD	WD	2–30	Loess/till L calc	Till	Oak-hickory
LILAH	8	VD	ED	0–35	Outwash L/S	Outwash	Prairie

(continued)

(continued)

Soil series	Area (km^2)	Soil depth class	Drainage class	Slope (%)	Parent material (surface)	Parent material (subsurface)	Vegetation
LINDQUIST	61	VD	SED	0–50	Glaciofluvial S	Glaciofluvial	Mixed forest
LINDSTROM	50	D	WD	1–30	Loess	Loess	Oak-savanna
LINO	14	VD	SPD	0–3	Outwash S	Outwash	Oak-savanna
LOBO	6	VD	VPD	0–1	Organic	Organic	Wet conifers
LOCKE	8	VD	SPD	0–6	Till SL	Till	Mixed hardwood forest
LOMIRA	275	VD	WD	0–20	Loess/till L calc	Till	Oak-hickory
LONGRIE	150	MD	WD	0–35	Till L/ls bedrock	Bedrock	NHW
LORENZO	60	VD	WD	0–45	Outwash L/calc S, G	Outwash	Prairie
LOWS	81	VD	PD	0–2	Alluvium L/S	Alluvium	Mixed forest
LOXLEY	1333	VD	VPD	0–2	Organic	Organic	Wet conifers
LOYAL	1157	VD	MWD	1–12	Loess/till L	Till	NHW
LUDINGTON	170	MD	MWD	1–20	Alluvium S/residuum L/ss, sh bedrock	Bedrock	Oak-pine
LUNDEEN	1	MD	WD	2–25	Loess/basalt bedrock	Bedrock	Mixed forest
LUPTON	1578	VD	VPD	0–2	Organic	Organic	Wet mixed forest
LUTZKE	45	VD	SED	2–20	Mantle L/outwash G, S	Outwash	Mixed oaks
MAGNOR	2852	VD	SPD	0–6	Loess/till SL	Till	NHW
MAGROC	21	D	SPD	0–4	Loess/till L/meta bedrock	Bedrock	NHW
MAHALASVILLE	78	VD	PD	0–2	Loess/outwash L, S	Outwash	Wet hardwoods
MAHTOMEDI	474	VD	ED	0–45	Outwash S	Outwash	Oak-pine
MAINCREEK	27	VD	SPD	0–3	Loess/L diamicton/outwash stratif S, G	Outwash	Mixed forest
MAJIK	14	VD	SPD	0–3	Alluvium S	Alluvium	Mixed forest
MAKWA	74	VD	VPD	0–12	Alluvium L–sk/ glaciolacustrine L	Glaciolacustrine	Mixed forest
MANAWA	956	VD	SPD	0–6	Loess/till C	Till	Mixed forest
MANCELONA	146	VD	SED	0–70	Outwash S, G	Outwash	NHW
MANISTEE	56	VD	WD	0–50	Glaciolacustrine S/C	Glaciolacustrine	NHW
MANITOWISH	127	VD	MWD	0–18	Alluvium L/outwash S	Outwash	Mixed forest
MANN	6	D	VPD	0–2	Loess/till L	Till	Wet hardwoods
MAPLEHURST	67	VD	SPD	0–3	Loess/outwash stratif S	Outwash	Mixed forest
MARATHON	91	D	MWD	2–15	Loess/till LS	Till	NHW
MARCELLON	19	VD	SPD	0–3	Till SL calc	Till	Prairie
MARKESAN	34	VD	WD	2–20	Till L calc	Till	Oak-savanna
MARKEY	1449	VD	VPD	0–2	Organic/outwash S	Outwash	Wet mixed forest
MARKHAM	132	VD	MWD	0–20	Loess/till SiCL	Till	Oak-savanna
MARSHAN	124	VD	VPD	0–5	Mantle L/outwash G, S	Outwash	Prairie
MARSHFIELD	418	VD	PD	0–2	Loess/till L	Till	Wet Mixed forest
MARTINTON	69	VD	SPD	0–6	Glaciolacustrine stratif SiL	Glaciolacustrine	Prairie
MATHERTON	15	VD	SPD	0–6	Glaciofluvial L/outwash S, G	Outwash	Wet hardwoods
MAUMEE	5	VD	PD	0–2	Outwash S	Outwash	Marsh grasses, sedges
MAYVILLE	163	VD	MWD	0–15	Loess/till SL	Till	Oak-hickory
MCHENRY	576	VD	WD	0–30	Loess/till L	Till	Mixed hardwood forest
MEADLAND	255	VD	SPD	0–6	Till L	Till	Mixed forest

(continued)

(continued)

Soil series	Area (km^2)	Soil depth class	Drainage class	Slope (%)	Parent material (surface)	Parent material (subsurface)	Vegetation
MECAN	93	VD	WD	2–30	Till LS	Till	Oak-hickory
MECOSTA	2	VD	SED	0–35	Glaciolacustrine S	Glaciolacustrine	Mixed hardwood forest
MEDARY	10	VD	MWD	0–45	Loess/glaciolacust C	Glaciolacustrine	NHW
MEEHAN	780	VD	SPD	0–3	Alluvium S	Alluvium	Mixed forest
MEENON	22	VD	SPD	0–3	Outwash S/glaciolacustrine C	Glaciolacustrine	Mixed forest
MENAHGA	2090	VD	ED	0–55	Outwash S	Outwash	Oak-pine
MENASHA	49	VD	PD	0–2	Glaciolacustrine C	Glaciolacustrine	Wet hardwoods
MENDOTA	131	VD	WD	0–12	Loess/till L calc	Till	Prairie
MENOMIN	2	VD	MWD	0–3	Alluvium L/outwash S	Alluvium	Oak-savanna
MENOMINEE	346	VD	WD	6–70	Glaciofluvial S/till L	Till	NHW
MEQUITHY	33	MD	WD	0–15	Loess/till L/ign bedrock	Bedrock	NHW
MEQUON	92	VD	SPD	1–3	Loess/till SiCL	Till	Mixed hardwood forest
MERIDIAN	213	VD	WD	0–20	Alluvium L/outwash S	Outwash	Oak-savanna
MERIMOD	38	VD	MWD	0–6	Alluvium Si/L/S	Alluvium	Oak-savanna
MERIT	57	VD	WD	0–6	Alluvium Si/L/S	Alluvium	Oak-savanna
MERRILLAN	313	MD	SPD	0–6	Alluvium L/residuum C/ss bedrock	Bedrock	Mixed forest
METEA	50	VD	WD	0–25	Eolian S/till L	Till	Mixed hardwood forest
METONGA	63	MD	WD	1–60	Mantle L/till L/granite bedrock	Bedrock	NHW
MIAMI	480	VD	MWD	0–60	Loess/till L	Till	Mixed hardwood forest
MICHIGAMME	108	MD	WD	1–75	Mantle L/till L/meta bedrock	Bedrock	NHW-H
MICKLE	23	VD	WD	2–12	Alluvium Si	Alluvium	Mixed hardwood forest
MIFFLIN	13	D	WD	2–30	Residuum L/dolo bedrock	Bedrock	Mixed hardwood forest
MILACA	12	VD	MWD	2–45	Till L	Till	Mixed forest
MILFORD	45	VD	VPD	0–2	Glaciolacustrine SiCL	Glaciolacustrine	Marsh grasses, sedges
MILITARY	28	MD	WD	2–30	Till SL/ss bedrock	Bedrock	Mixed hardwood forest
MILLADORE	137	VD	PD	0–6	Loess/till L	Till	Mixed forest
MILLINGTON	6	VD	PD	0–2	Alluvium SL calc	Alluvium	Marsh grasses, sedges
MILTON	2	MD	WD	0–25	Loess/residuum C/ls bedrock	Bedrock	Oak-hickory
MINDORO	6	VD	MWD	0–3	Alluvium S	Alluvium	Prairie
MINOCQUA	562	VD	VPD	0–2	Alluvium Si, L/outwash S	Outwash	Wet mixed forest
MISKOAKI	15	VD	WD	2–70	Till C	Till	Boreal
MOBERG	26	D	SED	2–15	Mantle L/residuum G/granite bedrock	Bedrock	Mixed forest
MONICO	90	D	SPD	0–6	Mantle Si, L/till SL, LS	Till	Mixed forest
MONTELLO	17	VD	MWD	0–6	Glaciolacustrine SiCL	Glaciolacustrine	Prairie
MONTGOMERY	60	VD	VPD	0–1	Glaciolacustrine SICL	Glaciolacustrine	Wet hardwoods

(continued)

(continued)

Soil series	Area (km^2)	Soil depth class	Drainage class	Slope (%)	Parent material (surface)	Parent material (subsurface)	Vegetation
MOODIG	40	VD	SPD	0–4	Till SL	Till	Mixed forest
MOPPET	109	VD	MWD	0–3	Alluvium, L/S	Alluvium	Mixed forest
MOQUAH	91	VD	MWD	0–3	Alluvium L	Alluvium	Mixed forest
MORA	7	VD	SPD	0–6	Till L	Till	Mixed forest
MORGANLAKE	46	VD	MWD	0–35	Outwash S/till L	Till	NHW
MORLEY	416	VD	MWD	1–18	Loess/till SiCL, CL	Till	Mixed hardwood forest
MOROCCO	67	VD	SPD	0–3	Outwash S	Outwash	Mixed forest
MOSEL	75	VD	SPD	0–3	Outwash L/glaciolacustrine C, Si	Outwash	Oak-savanna
MOSHAWQUIT	3	VD	WD	2–15	Outwash S/till L c/outwash S	Outwash	Mixed forest
MOSINEE	172	D	WD	2–20	Mantle L/residuum GSL/ign-meta bedrock	Bedrock	Mixed forest
MOUNDVILLE	53	VD	WD	0–6	Outwash S	Outwash	Oak-savanna
MT. CARROLL	49	VD	WD	0–45	Loess	Loess	Oak-savanna
MUDLAKE	108	VD	SPD	1–6	Loess/till LS	Till	Mixed forest
MUNDELEIN	79	VD	SPD	0–5	Loess/outwash L stratif calc	Outwash	Prairie
MUNUSCONG	15	VD	PD	0–2	Glaciofluvial L/C calc	Glaciofluvial	Wet mixed forest
MUSCATINE	63	VD	SPD	0–5	Loess	Loess	Prairie
MUSCODA	2	D	WD	2–20	Outwash S/alluvium L/ residuum L/ss bedrock	Bedrock	Oak-savanna
MUSKEGO	7	VD	VPD	0–2	Organic/coprogeneous	Coprogeneous	Marsh grasses, sedges
MUSSEY	26	VD	VPD	0–2	Outwash L	Outwash	Mixed hardwood forest
MYLREA	98	D	SPD	0–6	Loess/till FSL/residuum S	Residuum	NHW
MYRTLE	17	VD	WD	2–20	LOESS/paleosol Illinoian till	Till	Oak-savanna
NADEAU	44	D	WD	0–45	Mantle L/outwash G, S	Outwash	Mixed forest
NAHMA	3	MD	PD	0–2	Till L/ls bedrock	Bedrock	Wet mixed forest
NAMUR	97	VS	ED	0–20	Loess/dolo bedrock	Bedrock	Mixed forest
NAVAN	81	VD	PD	0–3	Glaciolacustrine L/C, Si	Glaciolacustrine	Marsh grasses, sedges
NEBAGO	44	VD	SPD	0–4	Glaciolacustrine S/C	Glaciolacustrine	Mixed forest
NECONISH	6	VD	MWD	0–3	Outwash S	Outwash	Mixed forest
NEDA	14	VD	MWD	2–30	Loess/till L	Till	Mixed hardwood forest
NEENAH	43	VD	SPD	0–3	Glaciolacustrine C	Glaciolacustrine	Mixed forest
NENNO	44	MD	SPD	0–6	Loess/till L	Till	NHW
NEOPIT	26	VD	WD	0–6	Till L	Till	NHW-H
NESTER	0	VD	MWD	0–12	Till L	Till	NHW
NEWGLARUS	430	S	WD	2–45	Loess/residuum C/L/dolo bedrock	Bedrock	Mixed hardwood forest
NEWLANG	25	VD	PD	0–2	Alluvium S	Alluvium	Mixed forest
NEWOOD	847	VD	MWD	1–30	Till L	Till	NHW-H
NEWOT	294	VD	WD	10–45	Till L	Till	NHW-H
NEWSON	840	VD	VPD	0–2	Glaciolacustrine S	Glaciolacustrine	Mixed forest
NICHOLS	78	VD	MWD	0–25	Glaciolacustrine L stratif	Glaciolacustrine	NHW

(continued)

(continued)

Soil series	Area (km^2)	Soil depth class	Drainage class	Slope (%)	Parent material (surface)	Parent material (subsurface)	Vegetation
NICKIN	60	VD	WD	1–20	Mantle L/till L/residuum S/ss bedrock	Bedrock	Oak-savanna
NOKASIPPI	5	VD	VPD	0–1	Eolian S/till L	Till	Marsh grasses, sedges
NORDEN	1077	D	WD	2–65	Loess/residuum L/ss bedrock	Bedrock	Oak-hickory
NORGO	19	S	WD	2–20	Mantle L/ss bedrock	Bedrock	Oak-hickory
NORTHBEND	44	VD	SPD	0–3	Alluvium Si, L/S	Alluvium	Wet hardwoods
NORTHFIELD	149	S	WD	0–60	Mantle L/ss bedrock	Bedrock	Oak-hickory
NORTHMOUND	5	MD	SED	6–50	Loess/residuum L/ss bedrock	Bedrock	NHW
NOSEUM	7	VD	MWD	0–12	Glaciofluvial L/outwash S	Outwash	NHW
NUXMARUHANIXETE	3	VD	WD	1–12	Alluvium Si/L/outwash S	Outwash	Oak-savanna
NYMORE	33	VD	ED	0–25	Outwash S	Outwash	Oak-savanna
OAKVILLE	80	VD	ED	0–60	Eolian S	Eolian	Mixed forest
OCKLEY	19	VD	WD	0–30	Loess/outwash S, G	Outwash	Mixed forest
OCONTO	1	VD	WD	0–20	Alluvium L/outwash S	Outwash	Mixed hardwood forest
ODANAH	114	VD	WD	0–60	Till C	Till	Mixed forest
OESTERLE	397	VD	SPD	0–3	Alluvium L/outwash S	Outwash	Mixed forest
OGLE	47	VD	WD	2–18	Loess/paleosol Illinoian till	Till	Prairie
OKEE	111	VD	SED	1–25	Eolian S/outwash S	Outwash	Mixed oaks
OMEGA	88	VD	SED	0–25	Outwash S	Outwash	Mixed pines
OMENA	115	VD	WD	0–50	Till SL	Till	NHW
OMRO	43	VD	MWD	1–12	Till C/L	Till	NHW
ONAWAY	739	VD	MWD	0–60	Till L	Till	NHW
ORION	359	VD	SPD	0–5	Alluvium stratif	Alluvium	Wet hardwoods
ORONTO	24	VD	SPD	0–3	Till Si, L	Till	Mixed forest
OSHKOSH	174	VD	WD	0–30	Loess/glaciolacustrine C	Glaciolacustrine	Mixed hardwood forest
OSHTEMO	176	VD	WD	0–55	Outwash stratif L, S	Outwash	Oak-hickory
OSSIAN	146	VD	PD	0–4	Alluvium Si	Alluvium	Prairie
OSSMER	301	VD	SPD	0–3	Loess/alluvium L/outwash S	Outwash	Mixed forest
Other	2237						
OTTER	121	VD	PD	0–5	Alluvium Si	Alluvium	Marsh grasses, sedges
OTTERHOLT	66	VD	WD	1–20	Loess/till L	Till	NHW
OZAUKEE	255	MD	WD	0–35	Loess/till L	Till	Mixed hardwood forest
PADUS	2373	VD	WD	0–45	Alluvium L/outwash S	Outwash	NHW-H
PADWET	39	VD	MWD	0–6	Alluvium L/outwash S	Outwash	NHW-H
PADWOOD	55	VD	MWD	0–15	Alluvium L/outwash S	Outwash	NHW-H
PALMS	510	VD	VPD	0–6	Organic/L deposits	Alluvium	Marsh grasses, sedges
PALSGROVE	659	D	WD	2–30	Loess/residuum C/dolostone bedrock	Bedrock	Oak-hickory
PARDEEVILLE	8	VD	WD	2–30	Till SL calc	Till	Oak-savanna
PARKFALLS	49	VD	SPD	0–4	Alluvium L/till L	Till	NHW-H
PARTRIDGE	23	MD	SPD	0–3	Outwash S/ss bedrock	Bedrock	Oak-pine
PEARL	49	VD	MWD	0–6	Outwash L, S	Outwash	Oak-savanna
PECATONICA	92	VD	WD	0–30	Loess/paleosol till L	Till	Mixed hardwood forest

(continued)

(continued)

Soil series	Area (km^2)	Soil depth class	Drainage class	Slope (%)	Parent material (surface)	Parent material (subsurface)	Vegetation
PECORE	4	VD	WD	2–35	Till L/outwash S	Outwash	NHW
PEEBLES	10	VD	MWD	0–6	Till C	Till	Prairie
PELISSIER	41	VD	ED	1–50	Outwash S, G	Outwash	Mixed forest
PELKIE	37	VD	MWD	0–4	Alluvium S	Alluvium	NHW
PELLA	786	VD	PD	0–3	Mantle L/glaciolacustrine stratif L	Glaciolacustrine	Marsh grasses, sedges
PENCE	2343	VD	SED	0–50	Mantle L/outwash G, S	Outwash	NHW-H
PEPIN	191	VD	WD	2–30	Loess/residuum C/dolostone bedrock	Bedrock	NHW
PEQUAMING	34	D	SPD	0–4	Till S	Till	Mixed forest
PERCHLAKE	9	VD	SPD	0–3	Glaciolacustrine S	Glaciolacustrine	Mixed hardwood forest
PERIDA	24	VD	MWD	0–30	Outwash S/glaciolacustrine C/ outwash S	Outwash	Oak-pine
PEROTE	23	VD	WD	2–35	Till L/outwash S calc	Outwash	Mixed hardwood forest
PESABIC	283	VD	SPD	0–4	Till L	Till	NHW-H
PESHEKEE	17	S	WD	3–75	Mantle L/ign-meta bedrock	Bedrock	Mixed forest
PESHTIGO	2	VD	SPD	0–4	Till L calc	Till	NHW-H
PICKFORD	103	VD	PD	0–2	Glaciolacustrine C	Glaciolacustrine	Boreal
PILLOT	100	VD	WD	0–18	Loess/outwash L, S	Outwash	Prairie
PINCONNING	9	VD	PD	0–2	Glaciolacustrine S/C	Glaciolacustrine	Wet mixed forest
PLAINBO	188	MD	ED	1–45	Outwash S/ss bedrock	Bedrock	Oak-pine
PLAINFIELD	2432	VD	SED	0–70	Glaciolacustrine S	Glaciolacustrine	Oak-pine
PLANO	1122	VD	WD	0–12	Loess/till SL	Till	Prairie
PLEINE	100	VD	PD	0–2	Till L	Till	Boreal
PLOVER	58	VD	SPD	0–3	Glaciolacustrine stratif L	Glaciolacustrine	Mixed hardwood forest
PLUMCREEK	23	VD	WD	12–45	Alluvium Si, L/S	Alluvium	NHW
POINT	74	VD	SPD	1–6	Mantle S/till L	Till	Mixed forest
POMROY	27	VD	MWD	2–30	Outwash S/till L	Till	Mixed oaks
PONYCREEK	136	VD	PD	0–2	Alluvium S	Alluvium	Wet mixed forest
PORT BYRON	62	VD	WD	0–30	Loess	Loess	Prairie
PORTWING	228	VD	MWD	2–6	Glaciolacustrine C/S	Glaciolacustrine	Boreal
POSKIN	108	VD	SPD	0–3	Loess/outwash S	Outwash	Mixed forest
POY	123	VD	PD	0–2	Glaciolacustrine C/S	Glaciolacustrine	Marsh grasses, sedges
POYGAN	456	VD	PD	0–3	Till C	Till	Wet hardwoods
PRISSEL	70	VD	MWD	0–60	Alluvium S/stratif L, S	Alluvium	Oak-pine
PUCHYAN	21	VD	MWD	2–12	Eolian S/till L	Till	Oak-hickory
QUARDERER	14	VD	MWD	0–3	Alluvium SiL with buried soil	Alluvium	Prairie
RABE	11	VD	WD	2–35	Outwash S calc/till L	Till	NHW
RADFORD	80	VD	SPD	0–5	Alluvium SiL with buried soil	Alluvium	Prairie
RASSET	71	VD	WD	0–40	Mantle L/outwash G, S	Outwash	Prairie
REDRIM	2	S	ED	0–6	Residuum S/ss bedrock	Bedrock	Mixed forest
REEDSBURG	34	D	SPD	0–12	Loess/residuum C	Residuum	Oak-savanna
RENOVA	31	VD	WD	2–35	Mantle Si, L/till L	Till	Mixed hardwood forest

(continued)

(continued)

Soil series	Area (km^2)	Soil depth class	Drainage class	Slope (%)	Parent material (surface)	Parent material (subsurface)	Vegetation
RIB	110	VD	PD	0–2	Loess/outwash S	Outwash	Wet hardwoods
RIBHILL	43	MD	SED	2–40	Till Si/qtzite bedrock	Bedrock	NHW
RIBRIVER	9	VD	MWD	0–3	Loess/outwash strat S	Outwash	NHW
RICHFORD	581	VD	WD	0–20	Outwash L, S	Outwash	Oak-savanna
RICHWOOD	53	VD	WD	0–12	Loess/outwash stratif FS, MS	Outwash	Prairie
RIETBROCK	305	D	SPD	1–12	Loess/till L/meta bedrock	Bedrock	Mixed forest
RIFLE	136	VD	VPD	0–2	Organic	Organic	Wet conifers
RIMER	5	VD	SPD	0–6	Glaciolacustrine S/till C	Till	Mixed hardwood forest
RINGWOOD	181	VD	WD	0–12	Loess/till L	Till	Prairie
RIPON	8	MD	WD	0–12	Loess/till CL/dolo bedrock	Bedrock	Prairie
RITCHEY	47	S	WD	0–60	Till L/dolo bedrock	Bedrock	Oak-hickory
ROBAGO	30	VD	SPD	0–3	Glaciolacustrine stratif FSL	Glaciolacustrine	Mixed forest
ROBY	20	VD	SPD	0–7	Outwash stratif L, S	Outwash	Mixed hardwood forest
ROCKBLUFF	49	D	ED	30–60	Colluvium S/residuum S/ss bedrock	Bedrock	Oak-pine
ROCKBRIDGE	8	VD	WD	6–20	Loess/residuum L/ss bedrock	Bedrock	Mixed oaks
ROCKDAM	123	VD	MWD	0–3	Alluvium S	Alluvium	Mixed forest
ROCKERS	44	VD	SPD	0–6	Mantle S/till L	Till	Mixed forest
ROCKLAND	8	VD	WD	18–70	Colluvium L	Colluvium	Mixed forest
ROCKMARSH	4	VD	SPD	0–20	Alluvium L-sk, S-sk/till L-sk	Till	Mixed hardwood forest
ROCKTON	101	MD	WD	0–25	Mantle L/paleosol/ls bedrock	Bedrock	Prairie
RODMAN	153	VD	ED	2–70	Outwash S, G	Outwash	Prairie
RONDEAU	15	VD	VPD	0–1	Organic/marl	Marl	Marsh grasses, sedges
ROOT	1	D	VPD	1–4	Alluvium L/L-sk	Alluvium	Mixed hardwood forest
ROSCOMMON	379	VD	VPD	0–2	Glaciolacustrine S	Glaciolacustrine	Mixed forest
ROSHOLT	1501	VD	WD	0–45	Alluvium L/outwash stratif S	Outwash	NHW
ROTAMER	129	VD	WD	2–45	Till SL calc	Till	Oak-savanna
ROUSSEAU	291	VD	WD	0–70	Outwash S	Outwash	Mixed forest
ROWLEY	26	VD	SPD	0–4	Alluvium Si/S	Alluvium	Prairie
ROZELLVILLE	200	VD	WD	1–12	Loess/residuum L/ign-meta bedrock	Bedrock	NHW-H
ROZETTA	60	VD	WD	0–25	Loess	Loess	Oak-hickory
RUBICON	742	VD	ED	0–70	Outwash S	Outwash	Mixed pines
RUSE	6	S	VPD	0–2	Till L/ls bedrock	Bedrock	Mixed forest
RUSKTOWN	17	VD	MWD	0–3	Alluvium L/outwash S	Outwash	Oak-savanna
SABLE	94	VD	PD	0–2	Loess	Loess	Marsh grasses, sedges
SALTER	32	VD	WD	0–20	Glaciolacustrine stratif L, S	Glaciolacustrine	Mixed hardwood forest
SANBORG	228	VD	MWD	0–6	Till C	Till	Boreal
SANTIAGO	663	D	WD	1–45	Loess/till SL	Till	NHW
SARGEANT	4	VD	SPD	0–6	Loess/till SCL	Till	Mixed hardwood forest

(continued)

(continued)

Soil series	Area (km^2)	Soil depth class	Drainage class	Slope (%)	Parent material (surface)	Parent material (subsurface)	Vegetation
SARONA	3668	VD	WD	1–35	Till SL	Till	NHW-H
SARWET	72	VD	MWD	0–6	Till SL	Till	NHW-H
SATTRE	204	VD	WD	0–12	Alluvium L/outwash, S, G	Outwash	Oak-savanna
SAYBROOK	33	VD	MWD	0–20	Loess/till L	Till	Prairie
SAYLESVILLE	49	VD	WD	0–40	Glaciolacustrine Si, L	Glaciolacustrine	Mixed hardwood forest
SAYNER	1069	VD	ED	0–60	Outwash S, G	Outwash	Mixed forest
SCHAAT CREEK	1	VD	PD	0–2	Alluvium C	Alluvium	Wet mixed forest
SCHAPVILLE	1	MD	MWD	2–35	Loess/residuum C/ls bedrock	Bedrock	Prairie
SCHWEITZER	9	MD	WD	6–70	Eolian Si/till KL, GL, S	Till	NHW
SCOBA	19	VD	MWD	0–6	Alluvium L/outwash S	Outwash	NHW
SCONSIN	240	MD	MWD	0–6	Loess/L diamicton/outwash stratif S, G	Outwash	NHW
SCOTAH	47	VD	MWD	0–3	Alluvium S	Alluvium	Oak-pine
SCOTT LAKE	201	VD	MWD	0–6	Alluvium L/outwash S, G	Outwash	NHW
SEATON	1042	VD	WD	0–60	Loess	Loess	Oak-hickory
SEBBO	19	VD	MWD	1–6	Alluvium L, Si	Alluvium	Mixed oaks
SEBEWA	286	VD	VPD	0–3	Outwash L/S, G	Outwash	Wet hardwoods
SECHLER	4	VD	SPD	0–3	Alluvium L/S	Alluvium	Prairie
SEDGWICK	97	VD	SPD	0–15	Alluvium L/till C	Till	Mixed forest
SEELYEVILLE	1595	VD	VPD	0–15	Organic	Organic	Marsh grasses, sedges
SELKIRK	0	VD	SPD	0–6	Till C	Till	NHW
SEWARD	7	VD	MWD	0–6	Glaciolacustrine S/till C	Till	Mixed hardwood forest
SHAG	3	VD	PD	0–2	Glaciolacustrine Si	Glaciolacustrine	Mixed forest
SHAWANO	337	VD	ED	0–35	Outwash S	Outwash	Mixed forest
SHERRY	243	VD	VPD	0–3	Loess/till L/ign-meta bedrock	Bedrock	Wet mixed forest
SHIFFER	62	D	SPD	0–3	Glaciolacustrine L/S	Glaciolacustrine	Mixed hardwood forest
SHIOCTON	231	VD	SPD	0–3	Glaciolacustrine L	Glaciolacustrine	Prairie
SHULLSBURG	2	MD	SPD	1–25	Loess/shale	Bedrock	Oak-savanna
SILVERHILL	11	D	WD	1–6	Colluvium L/residuum S/ss bedrock	Bedrock	Mixed forest
SIMESCREEK	9	VD	ED	0–3	Alluvium S	Alluvium	Oak-pine
SIOUXCREEK	2	MD	SPD	0–3	Alluvium Si, L/residuum S/ss bedrock	Bedrock	Mixed hardwood forest
SIREN	9	VD	SPD	0–3	Alluvium L/glaciolacustrine C	Glaciolacustrine	NHW
SISSABAGAMA	6	VD	MWD	0–6	Outwash S/stratif S, L	Outwash	Mixed forest
SISSON	166	VD	WD	0–50	Glaciolacustrine L, Si	Glaciolacustrine	Oak-hickory
SKOG	2	VD	MWD	0–3	Alluvium L/S-sk	Alluvium	NHW
SKYBERG	6	VD	PD	0–2	Mantle Si, L/till L	Till	Prairie
SLIMLAKE	11	VD	MWD	0–3	Alluvium L/outwash stratif S	Outwash	Mixed forest
SMESTAD	16	VD	SPD	0–3	Glaciolacustrine S/L/C	Glaciolacustrine	Mixed hardwood forest
SODERBECK	0	D	SPD	0–2	Alluvium L-sk/S-sk/residuum S/ss bedrock	Bedrock	Mixed forest

(continued)

(continued)

Soil series	Area (km²)	Soil depth class	Drainage class	Slope (%)	Parent material (surface)	Parent material (subsurface)	Vegetation
SOGN	284	VS	SED	0–20	Residuum LS	Residuum	Prairie
SOLONA	562	VD	SPD	0–3	Till L	Till	NHW
SOONER	7	VD	SPD	0–3	Alluvium Si/L/S	Alluvium	Mixed forest
SOPERTON	44	MD	WD	15–35	Loess/till L, S	Till	NHW
SPARTA	222	VD	ED	0–40	Outwash S	Outwash	Prairie
SPEAR	14	VD	SPD	0–3	Glaciofluvial stratif SiL	Glaciofluvial	Marsh grasses, sedges
SPENCER	364	VD	MWD	0–12	Loess/till L	Till	NHW
SPIDERLAKE	45	VD	MWD	0–3	Alluvium Si, L/outwash S	Alluvium	NHW-H
SPINKS	8	VD	WD	0–70	Outwash S	Outwash	Oak-hickory
SPOONERHILL	46	VD	MWD	2–6	Alluvium L/S/till S	Till	Mixed forest
SPRINGSTEAD	142	VD	MWD	1–6	Outwash S/till S	Till	Boreal
ST. CHARLES	909	VD	WD	0–30	Loess/till SL	Till	Oak-hickory
STAMBAUGH	315	VD	WD	0–25	Loess/outwash S, G	Outwash	NHW-H
STANBERRY	171	VD	MWD	1–30	Alluvium L/till L, S	Till	NHW-H
STENGEL	6	VD	SPD	0–3	Outwash S/glaciolacustrine C/ outwash S	Outwash	Mixed hardwood forest
STINNETT	52	VD	SPD	0–4	Loess/alluvium L/till S, L	Till	Mixed forest
STRONGHURST	4	VD	SPD	0–6	Loess	Loess	Mixed hardwood forest
STURGEON	37	VD	SPD	0–1	Alluvium Si/S	Alluvium	Mixed forest
SULTZ	227	VD	WD	0–65	Outwash S/stratif L deposits	Outwash	Mixed forest
SUMMERVILLE	192	S	WD	0–45	Till L/ls bedrock	Bedrock	Mixed forest
SUNIA	3	VD	MWD	0–3	Glaciofluvial L/S	Glaciofluvial	NHW
SUPERIOR	71	VD	MWD	2–50	Glaciolacustrine L/C	Glaciolacustrine	NHW
SYLVESTER	12	MD	WD	2–12	Loess/residuum S/ss Bedrock	Bedrock	Prairie
SYMCO	345	VD	SPD	0–3	Till L calc	Till	NHW
SYMERTON	7	VD	MWD	0–10	Loess/glaciolacustrine L/Si	Glaciolacustrine	Prairie
TACOOSH	2	D	VPD	0–2	Organic/L material	Alluvium	Wet mixed forest
TAMA	621	VD	WD	0–20	Loess	Loess	Prairie
TARR	727	VD	SED	0–60	Pedisediment S/residuum S	Residuum	Oak-pine
TAWAS	109	VD	VPD	0–2	Organic/outwash S	Outwash	Wet mixed forest
TAYLOR	1	VD	MWD	0–45	Glaciolacustrine C calc	Glaciolacustrine	Mixed forest
TEDROW	87	VD	SPD	0–6	Glaciolacustrine S	Glaciolacustrine	Mixed hardwood forest
TELL	45	VD	WD	0–40	Loess/alluvium S	Alluvium	Mixed oaks
THACKERY	5	VD	MWD	0–6	Mantle L/outwash G, S	Outwash	Mixed hardwood forest
THERESA	589	MD	WD	0–30	Loess/till L	Till	Mixed hardwood forest
TILLEDA	235	VD	WD	1–35	Till L	Till	Mixed forest
TIMULA	9	VD	WD	2–60	Loess	Loess	Oak-hickory
TINT	28	VD	MWD	0–6	Alluvium S	Alluvium	Oak-pine
TINTSON	9	VD	MWD	0–6	Alluvium S/L	Alluvium	Oak-pine
TIPLER	175	VD	WD	0–3	Alluvium L/outwash S	Outwash	NHW-H
TODDVILLE	94	VD	MWD	0–6	Loess/alluvium S stratif	Alluvium	Prairie
TONKEY	9	VD	VPD	0–2	Glaciofluvial L, S stratif	Glaciofluvial	Mixed forest

(continued)

(continued)

Soil series	Area (km^2)	Soil depth class	Drainage class	Slope (%)	Parent material (surface)	Parent material (subsurface)	Vegetation
TOTAGATIC	88	VD	VPD	0–2	Alluvium S	Alluvium	Wet hardwoods
TOURTILLOTTE	3	VD	MWD	0–15	Alluvium S/glaciolacustrine Si stratif	Glaciolacustrine	Mixed forest
TRADELAKE	4	VD	MWD	1–35	Alluvium L/glaciolacustrine C/ outwash S	Outwash	Mixed hardwood forest
TREMPE	17	D	ED	0–12	Alluvium S	Alluvium	Prairie
TREMPEALEAU	3	MD	WD	0–6	Alluvium L/S	Alluvium	Prairie
TROXEL	105	VD	WD	0–4	Colluvium Si/till L	Till	Prairie
TULA	179	VD	SPD	0–4	Eolian L/till L	Till	Boreal
TUSCOLA	11	VD	MWD	0–12	Glaciolacustrine L, Si	Glaciolacustrine	Mixed hardwood forest
TUSTIN	37	VD	WD	0–35	Outwash S/L/till C	Till	Mixed forest
TWINMOUND	77	MD	ED	6–50	Residuum S/ss bedrock	Bedrock	Mixed oaks
URNE	712	MD	WD	2–65	Residuum L/glauc ss bedrock	Bedrock	Oak-hickory
VALTON	606	D	WD	1–45	Loess/residuum C/ls bedrock	Bedrock	Oak-hickory
VANCECREEK	12	VD	PD	0–2	Alluvium S	Alluvium	Marsh grasses, sedges
VANZILE	192	VD	MWD	0–6	Loess/outwash S	Outwash	NHW-H
VARNA	95	VD	MWD	1–18	Loess/till SiCL, CL	Till	Prairie
VEEDUM	254	MD	PD	0–2	Alluvium Si/residuum/ss bedrock	Bedrock	Mixed hardwood forest
VESPER	265	D	PD	0–2	Loess/residuum/sh, ss bedrock	Bedrock	Mixed hardwood forest
VILAS	1162	VD	ED	0–55	Outwash S	Outwash	Mixed forest
VIRGIL	19	VD	SPD	0–6	Loess/outwash L/till SL	Till	Oak-savanna
VLASATY	214	VD	MWD	1–12	Loess/till L	Till	Oak-hickory
WABENO	265	MD	MWD	1–18	Loess/till L, S	Till	NHW
WACOUSTA	154	VD	VPD	0–2	Glaciolacustrine Si	Glaciolacustrine	Marsh grasses, sedges
WAINOLA	279	VD	SPD	0–4	Glaciofluvial S	Glaciofluvial	Mixed hardwood forest
WAKEFIELD	54	S	MWD	1–18	Eolian L/till L	Till	NHW
WAKELEY	30	VD	VPD	0–2	Glaciolacustrine S/C	Glaciolacustrine	Wet conifers
WALLKILL	30	VD	VPD	0–3	Alluvium/organic	Organic	Wet hardwoods
WARMAN	2	VD	VPD	0–2	Outwash L/S, G	Outwash	Marsh grasses, sedges
WARSAW	226	VD	WD	0–15	Mantle L/outwash G, S	Outwash	Prairie
WASEPI	61	VD	SPD	0–6	Glaciofluvial L, S/S, G	Glaciofluvial	Wet hardwoods
WASHTENAW	8	D	PD	0–2	Alluvium Si/till L	Till	Wet hardwoods
WATSEKA	13	VD	SPD	0–4	Outwash S	Outwash	Prairie
WATTON	0	D	WD	0–40	Till L	Till	NHW-H
WAUCONDA	3	VD	SPD	0–5	Loess/outwash L calc	Outwash	Oak-savanna
WAUKEGAN	4	VD	WD	0–12	Loess/outwash S, G	Outwash	Prairie
WAUPACA	38	D	PD	0–2	Glaciolacustrine Si	Glaciolacustrine	Wet hardwoods
WAUSEON	31	VD	VPD	0–2	Glaciolacustrine L, S/till CL	Till	Wet hardwoods
WAUTOMA	50	VD	PD	0–2	Outwash S/glaciolacustrine L/C	Glaciolacustrine	Wet hardwoods
WAYKA	3	MD	SPD	0–4	Till L	Till	Mixed forest
WAYMOR	233	VD	WD	0–35	Loess/till L calc	Till	Mixed hardwood forest

(continued)

(continued)

Soil series	Area (km^2)	Soil depth class	Drainage class	Slope (%)	Parent material (surface)	Parent material (subsurface)	Vegetation
WEGA	15	D	SPD	0–3	Glaciolacustrine	Glaciolacustrine	Mixed hardwood forest
WESTVILLE	82	VD	MWD	2–30	Loess/till SL	Till	Oak-hickory
WHALAN	290	MD	WD	0–25	Till L/residuum C/ls bedrock	Bedrock	Mixed hardwood forest
WHEATLEY	7	VD	VPD	0–3	Glaciofluvial S, G	Glaciofluvial	Wet conifers
WHISKLAKE	54	VD	SPD	0–3	Alluvium Si/L/outwash S	Outwash	Mixed forest
WHITEHALL	14	VD	MWD	0–3	Alluvium Si/S	Alluvium	Prairie
WICKWARE	37	VD	WD	6–30	Loess/alluvium Si	Alluvium	NHW
WILDALE	104	VD	WD	2–45	Loess/residuum C/ls bedrock	Bedrock	Mixed hardwood forest
WILDWOOD	10	VD	VPD	0–1	Till C	Till	Marsh grasses, sedges
WILL	3	MD	PD	0–3	Outwash L/S, G	Outwash	Marsh grasses, sedges
WILLETTE	128	VD	VPD	0–2	Organic/C deposits	Alluvium	Wet hardwoods
WINDWARD	9	VD	SED	2–12	Eolian S	Eolian	Mixed hardwood forest
WINNEBAGO	25	VD	MWD	2–30	Loess/paleosol till SL Illinoian	Till	Prairie
WINNECONNE	115	VD	MWD	0–12	Glaciolacustrine C	Glaciolacustrine	Mixed hardwood forest
WINNESHIEK	9	MD	WD	0–35	Till CL/ls bedrock	Bedrock	Prairie
WINTERFIELD	65	VD	SPD	0–4	Alluvium S	Alluvium	Wet hardwoods
WITHEE	1186	VD	SPD	0–6	Loess/till L	Till	NHW
WORCESTER	192	VD	SPD	0–3	Alluvium L/outwash S	Outwash	Mixed forest
WORMET	65	VD	SPD	0–3	Alluvium L/outwash S	Outwash	Mixed forest
WORTHEN	83	VD	WD	0–12	Alluvium Si	Alluvium	Prairie
WORWOOD	14	VD	SPD	0–3	Alluvium L/outwash S	Outwash	Mixed forest
WOZNY	43	VD	VPD	0–2	Loess/alluvium L/till S, L	Till	Wet conifers
WURTSMITH	102	VD	MWD	0–12	Outwash S	Outwash	Oak-pine
WYEVILLE	76	VD	SPD	0–3	Eolian S/glaciolacustrine C	Glaciolacustrine	Mixed forest
WYKOFF	34	D	WD	2–20	mantle L, Si/till GSL	Till	Mixed hardwood forest
WYOCENA	363	VD	WD	2–45	Till S	Till	Mixed forest
YAHARA	94	VD	SPD	0–4	Glaciolacustrine Si, L, S	Glaciolacustrine	Oak-savanna
ZEBA	6	MD	SPD	0–3	Till L/ss bedrock	Bedrock	Mixed forest
ZITTAU	71	VD	SPD	0–3	Glaciolacustrine C/S	Glaciolacustrine	Mixed hardwood forest
ZURICH	85	VD	MWD	0–35	Loess/outwash L calc	Outwash	Oak-hickory

Appendix C
Thickness of Diagnostic Horizons of Wisconsin Soil Series

J.G. Bockheim and A.E. Hartemink, *The Soils of Wisconsin*,
World Soils Book Series, DOI 10.1007/978-3-319-52144-2

Series	Subgroup	Area (km^2)	Ochric	Mollic	Histic	Umbric	Albic	Argillic (cm)	Cambic	Spodic	Glossic	Fragipan	Calcic
ABBAYE	Alfic Oxyaquic Haplorthods	11	5				23	18		30			
ABSCO	Typic Udifluvents	236	10										
ABSCOTA	Oxyaquic Udipsamments	2	13										
ACKMORE	Mollic Fluvaquents	64	15										
ADDER	Terric Haplosaprists	23			56								
ADOLPH	Typic Epiaquolls	84		33					48				
ADRIAN	Terric Haplosaprists	507			86								
AFTAD	Oxyaquic Glossudalfs	53	33				15	76			23		
ALANGO	Chromic Vertic Epiaqualfs	0	5				20	46					
ALBAN	Typic Glossudalfs	66	81				31	31			61		
ALCONA	Alfic Haplorthods	47	8				5	13		8	94		
ALDO	Typic Udipsamments	25	23										
ALGANSEE	Aquic Udipsamments	166	20										
ALLENDALE	Alfic Epiaquods	153	25				17	13		16			
ALMENA	Aquic Glossudalfs	435	33				10	74			30		
ALPENA	Entic Hapludolls	34		18									
ALSTAD	Aquic Glossudalfs	2	46				15	78			23		
ALTDORF	Aeric Glossaqualfs	48	43				23	94			28		
AMASA	Typic Haplorthods	9	7.5				7.5			62			
AMERY	Haplic Glossudalfs	991	86				30	94			48		
AMNICON	Oxyaquic Vertic Glossudalfs	374	25				8	84			28		
ANGELICA	Aeric Endoaquepts	119	10						28				
ANIGON	Haplic Glossudalfs	259	36				10	51			15		
ANNALAKE	Alfic Oxyaquic Haplorthods	152	15				7	20		28	36		
ANNRIVER	Mollic Endoaqualfs	101	36					43					
ANTIGO	Haplic Glossudalfs	1242	30				25	54			18		
ANTON	Oxyaquic Vertic Glossudalfs	14	10					102			13		
ARBUTUS	Entic Haplorthods	1	13				8			30			
ARENZVILLE	Typic Udifluvents	464	25										
ARGONNE	Alfic Oxyaquic Fragiorthods	23	8				8	63		25	61	63	
ARLAND	Haplic Glossudalfs	291	38				18	53			33		
ARNHEIM	Typic Fluvaquents	203	13										
ASHDALE	Typic Argiudolls	210		38				92					
ASHIPPUN	Aquollic Hapludalfs	12	28					38					
ASHKUM	Typic Endoaquolls	223		30					107				
ASHWABAY	Alfic Oxyaquic Haplorthods	94	13				2.5	89		69			
ATTERBERRY	Udollic Endoaqualfs	5	33				10	79					

(continued)

(continued)

Series	Subgroup	Area (km^2)	Ochric	Mollic	Histic	Umbric	Albic	Argillic (cm)	Cambic	Spodic	Glossic	Fragipan	Calcic
AU GRES	Typic Endoaquods	384	13				5			45			
AUBURNDALE	Mollic Epiaqualfs	208	36				18	99					
AUGWOOD	Typic Epiaquods	13	8				6			45			
AUSABLE	Histic Humaquepts	13			25								
AZTALAN	Aquic Argiudolls	94		31				78					
BACH	Mollic Endoaquepts	26	20						71				
BADRIVER	Aeric Glossaqualfs	181	25					110			53		
BALMORAL	Oxyaquic Argiudolls	1		41				117					
BANAT	Aeric Endoaqualfs	1	15					41					
BARABOO	Oxyaquic Hapludalfs	113	13					73					
BARREMILLS	Pachic Argiudolls	43		72				95					
BARRONETT	Mollic Epiaqualfs	98	41				18	45					
BARRY	Typic Argiaquolls	32		28				63					
BASCO	Mollic Hapludalfs	53	28					56					
BATAVIA	Mollic Hapludalfs	11	43				7	84					
BEARPEN	Aquic Argiudolls	43		46				81					
BEARTREE	Lithic Endoaquolls	2		39									
BEAVERBAY	Alfic Oxyaquic Haplorthods	46	13				7.5	71		23	18		
BEAVERCREEK	Typic Udifluvents	26	13										
BEECHER	Udollic Epiaqualfs	7	23					48					
BELLECHESTER	Entic Hapludolls	3		41					66				
BELLEVILLE	Typic Endoaquolls	0		28									
BELLEVUE	Fluventic Hapludolls	336		46					45				
BERGLAND	Aeric Vertic Epiaqualfs	8	20					58					
BERTRAND	Typic Hapludalfs	7	25					114					
BESEMAN	Terric Haplosaprists	472			91								
BIGISLAND	Typic Hapludalfs	5	8					31					
BILLETT	Mollic Hapludalfs	55	33					86					
BILLYBOY	Oxyaquic Glossudalfs	97	51				41	38			38		
BILMOD	Mollic Hapludalfs	32	23					38					
BILSON	Mollic Hapludalfs	133	20					36					
BJORKLAND	Typic Epiaqualfs	9	64				43	33					
BLACKRIVER	Haplic Glossudalfs	15	52				23	86			41		
BLOUNT	Aeric Epiaqualfs	7	18					58					
BLUFFTON	Typic Endoaquolls	13		48					7.5				
BOAZ	Aeric Endoaquepts	52	23						66				
BOGUSCREEK	Mollic Udifluvents	1		127					18				

(continued)

(continued)

Series	Subgroup	Area (km^2)	Ochric	Mollic	Histic	Umbric	Albic	Argillic (cm)	Cambic	Spodic	Glossic	Fragipan	Calcic
BONDUEL	Aquollic Hapludalfs	71	36					28					
BOONE	Typic Quartzipsamments	617	20						33				
BOOTS	Typic Haplohemists	35			**152**								
BOPLAIN	Typic Udipsamments	15	23										
BOREA	Aeric Glossaqualfs	11	5					112			28		
BORTH	Mollic Oxyaquic Hapludalfs	47	20					56					
BOWSTRING	Fluvaquentic Haplosaprists	266			97								
BOYER	Typic Hapludalfs	398	46					40					
BRAHAM	Arenic Hapludalfs	0	71					52					
BRANDER	Oxyaquic Glossudalfs	18	43				18	46			30		
BRANSTAD	Oxyaquic Glossudalfs	4	36				13	122			79		
BREMS	Aquic Udipsamments	339	23										
BRENNYVILLE	Aquic Glossudalfs	1	20					69			7.5		
BREVORT	Mollic Endoaquents	79	20										
BRICE	Lamellic Hapludalfs	2	23					30					
BRIGGSVILLE	Typic Hapludalfs	102	18					81					
BRILL	Haplic Glossudalfs	102	48				30	48			20		
BRIMLEY	Argic Endoaquods	3	15				10	13		18			
BRINKMAN	Mollic Paleudalfs	44	23					180					
BRODALE	Entic Hapludolls	7		30									
BROOKSTON	Typic Argiaquolls	130		41				99					
BROWNCHURCH	Mollic Hapludalfs	58	15					142					
BROWNSTONE	Typic Haplorthods	6	25				25			28			
BRUCE	Mollic Endoaquepts	4	18						38				
BURKHARDT	Typic Hapludolls	182		25					18				
BUSHVILLE	Aquic Arenic Hapludalfs	1	61					15					
CABLE	Typic Epiaquepts	340	20				13		46				
CADIZ	Oxyaquic Hapludalfs	38	23					74					
CALAMINE	Typic Argiaquolls	9		52				52					
CAMPIA	Haplic Glossudalfs	57	30				10	58			30		
CAPITOLA	Aeric Epiaqualfs	707	18					28					
CARBONDALE	Hemic Haplosaprists	369			130								
CARYVILLE	Fluventic Hapludolls	12		40									
CASCO	Inceptic Hapludalfs	699	20					23					
CATHRO	Terric Haplosaprists	1823			58								
CEBANA	Mollic Glossaqualfs	567	69				48	56			36		
CERESCO	Fluvaquentic Hapludolls	36		30					46				

(continued)

(continued)

Series	Subgroup	Area (km^2)	Ochric	Mollic	Histic	Umbric	Albic	Argillic (cm)	Cambic	Spodic	Glossic	Fragipan	Calcic
CHABENEAU	Oxyaquic Haplorthods	11	7.5				7.5			43			
CHAMPION	Oxyaquic Fragiorthods	14	10				10			42			
CHANNAHON	Lithic Argiudolls	49		28				18					
CHANNING	Typic Endoaquods	9	10				2.5			43			
CHARLEVOIX	Argic Endoaquods	93	18				15	30		13			
CHASEBURG	Typic Udifluvents	277	10										
CHELMO	Umbric Epiaqualfs	3				23		38					
CHELSEA	Lamellic Udipsamments	74	91				28	15					
CHEQUAMEGON	Alfic Oxyaquic Haplorthods	341	18				13	46		25	33		
CHETEK	Inceptic Hapludalfs	750	25					25					
CHINWHISKER	Lamellic Oxyaquic Haplorthods	29	8				3	10		45			
CHIPPENY	Lithic Haplosaprists	2			51								
CHURCHTOWN	Mollic Hapludalfs	455	23					129					
CITYPOINT	Typic Haplosaprists	98			72								
CLEMENS	Aquic Dystric Eutrudepts	3	13						99				
CLYDE	Typic Endoaquolls	21		58					**142**				
COFFEEN	Fluvaquentic Hapludolls	37		30					62				
COFFTON	Fluvaquentic Hapludolls	10		28					68				
COLAND	Cumuliic Endoaquolls	1		102					30				
COLOMA	Lamellic Udipsamments	303	10					13	89				
COLWOOD	Typic Endoaquolls	168		30					61				
COMSTOCK	Aquic Glossudalfs	143	38				18	48			15		
CONOVER	Udollic Endoaqualfs	1	23					41					
CORMANT	Mollic Psammaquents	194	15										
CORNUCOPIA	Haplic Glossudalfs	103	7.5					89			18		
COSAD	Aquic Udorthents	12	23										
COUNCIL	Typic Hapludalfs	312	20					132					
CRESS	Humic Dystrudepts	195	7.5						30				
CREX	Oxyaquic Udipsamments	33	15										
CROMWELL	Typic Dystrudepts	238	7.7				2.5		30				
CROSSETT	Aquic Glossudalfs	4	23				25	155			25		
CROSWELL	Oxyaquic Haplorthods	518	13				8			28			
CROSWOOD	Oxyaquic Haplorthods	27	15				5			20			
CRYSTAL LAKE	Oxyaquic Glossudalfs	145	30				10	52			20		
CUBLAKE	Oxyaquic Haplorthods	118	10				2.5			48			
CUNARD	Typic Hapludalfs	2	10				18	33					
CURRAN	Udollic Endoaqualfs	103	38					97					

(continued)

(continued)

Series	Subgroup	Area (km²)	Ochric	Mollic	Histic	Umbric	Albic	Argillic (cm)	Cambic	Spodic	Glossic	Fragipan	Calcic
CUSHING	Haplic Glossudalfs	7	38				25	127			47		
CUTAWAY	Arenic Oxyaquic Hapludalfs	0	35				28	20					
CUTTRE	Aeric Glossaqualfs	370	7.5					89			23		
DAIRYLAND	Mollic Oxyaquic Hapludalfs	13	23					140					
DAISYBAY	Terric Haplohemists	1			89								
DAKOTA	Typic Argiudolls	148		36				43					
DANCY	Typic Glossaqualfs	96	25				15	23			28		
DARROCH	Aquic Argiudolls	8		38				36					
DAWSIL	Terric Haplosaprists	170			102								
DAWSON	Terric Haplosaprists	603			97								
DECHAMPS	Aquic Udifluvents	31	7.5										
DEERTON	Typic Haplorthods	7	23				20			42			
DEFORD	Typic Psammaquents	127	13										
DEL REY	Aeric Epiaqualfs	2	23				13	61					
DELLS	Aquollic Hapludalfs	28	42					43					
DELTON	Arenic Hapludalfs	83	42					18					
DENOMIE	Haplic Glossudalfs	31	18				7.5	185			15		
DENROCK	Aquic Argiudolls	1		33				94					
DERINDA	Oxyaquic Hapludalfs	5	30					34					
DICKINSON	Typic Hapludolls	96		46					45				
DICKMAN	Typic Hapludolls	15		30					18				
DISHNO	Oxyaquic Haplorthods	9	20				15			30			
DOBIE	Haplic Glossudalfs	64	20					74			28		
DODGE	Typic Hapludalfs	495	20					79					
DODGEVILLE	Typic Argiudolls	585		33				48					
DODY	Arenic Albaqualfs	3	52				15	36					
DOLPH	Aeric Glossaqualfs	88	20				5	43					
DORA	Terric Haplosaprists	5			82								
DORCHESTER	Typic Udifluvents	0	20										
DORERTON	Typic Hapludalfs	394	25					38					
DORITTY	Haplic Glossudalfs	31	30				7.5	66			15		
DOWNS	Mollic Hapludalfs	407	43				10	56					
DRAMMEN	Lamellic Hapludalfs	163	23					53					
DRESDEN	Mollic Hapludalfs	246	25					51					
DRUMMER	Typic Endoaquolls	39		48					71				
DRYLANDING	Lithic Hapludolls	68		18					20				
DUBUQUE	Typic Hapludalfs	2835	41					28					

(continued)

(continued)

Series	Subgroup	Area (km²)	Ochric	Mollic	Histic	Umbric	Albic	Argillic (cm)	Cambic	Spodic	Glossic	Fragipan	Calcic
DUEL	Entic Haplorthods	3	23				15			29			
DUELM	Oxyaquic Hapludolls	4		41									
DUNBARTON	Lithic Hapludalfs	246	18					23					
DUNNBOT	Mollic Udifluvents	47	23						23				
DUNNVILLE	Typic Hapludolls	51		30					30				
DURAND	Typic Argiudolls	69		33				135					
EAUCLAIRE	Alfic Oxyaquic Haplorthods	20	7.5				5	81		20			
EDMUND	Lithic Argiudolls	258		25				20					
EDWARDS	Limnic Haplosaprists	6			61								
ELBAVILLE	Glossic Hapludalfs	180	42					52					
ELBURN	Aquic Argiudolls	260		41				84					
ELDERON	Typic Dystrudepts	44	18						20				
ELEROY	Oxyaquic Hapludalfs	7	38				13	66					
ELEVA	Typic Hapludalfs	541	30					46					
ELEVASIL	Ultic Hapludalfs	257	5					62					
ELKMOUND	Typic Dystrudepts	322	15						15				
ELLA	Mollic Hapludalfs	84	20					163					
ELLIOTT	Aquic Argiudolls	147		28				76					
ELLWOOD	Oxyaquic Glossudalfs	21	20					183			18		
ELM LAKE	Humaqueptic Epiaquents	274	10										
ELVERS	Thapto-Histic Fluvaquents	16	18										
EMMERT	Typic Udorthents	80	30						64				
EMMET	Inceptic Hapludalfs	537	20				25	20	31				
ENSLEY	Aeric Endoaquepts	95	20						54				
ETTRICK	Fluvaquentic Endoaquolls	318		41					48				
FABIUS	Aquic Argiudolls	36		18				21					
FAIRCHILD	Ultic Epiaquods	237	28				20	15		18			
FAIRPORT	Glossic Hapludalfs	18	18				5	41			7.5		
FALLCREEK	Aquic Glossudalfs	117	30				15	86			20		
FARRINGTON	Aquic Hapludolls	17		46									
FAYETTE	Typic Hapludalfs	2234	36				20	50					
FENANDER	Udollic Epiaqualfs	9	23				15	30					
FENCE	Alfic Oxyaquic Haplorthods	79	15				10	43		20	56		
FENWOOD	Haplic Glossudalfs	328	20				25	64					
FESTINA	Mollic Hapludalfs	14	51					114					
FINCHFORD	Entic Hapludolls	121		46									
FISK	Aquic Dystric Eutrudepts	19	23						38				

(continued)

(continued)

Series	Subgroup	Area (km^2)	Ochric	Mollic	Histic	Umbric	Albic	Argillic (cm)	Cambic	Spodic	Glossic	Fragipan	Calcic
FIVEPOINTS	Typic Hapludalfs	183	18					71					
FLAGG	Typic Hapludalfs	44	41					183					
FLAMBEAU	Oxyaquic Glossudalfs	217	41					86			38		
FLINK	Typic Epiaquods	48	7.5				7.5			52			
FLOYD	Aquic Pachic Hapludolls	19		60					140				
FORADA	Typic Endoaquolls	5		40					30				
FORDUM	Mollic Fluvaquents	595	15										
FORKHORN	Mollic Hapludalfs	60	20					52					
FOX	Typic Hapludalfs	739	25					59					
FRECHETTE	Typic Glossudalfs	73	10					114			84		
FREEON	Oxyaquic Glossudalfs	2144	48				30	86			81		
FREMSTADT	Arenic Hapludalfs	63	8					30					
FREYA	Aquic Argiudolls	81		28				102					
FRIENDSHIP	Typic Udipsamments	530	20										
FRIESLAND	Typic Argiudolls	37		48				66					
FROGCREEK	Oxyaquic Glossudalfs	106	33				23	84			15		
GAASTRA	Argic Endoaquods	8	7.5				23	38		20			
GALE	Typic Hapludalfs	554	20					59					
GAPHILL	Typic Hapludalfs	67	23					54					
GARDENVALE	Mollic Hapludalfs	10	20					56					
GARNE	Typic Hapludolls	6		58					18				
GARWIN	Typic Endoaquolls	0		46					61				
GASTROW	Argic Endoaquods	13	15				7.5	41		10	53		
GAY	Aeric Endoaquepts	18	20				7.5		28				
GICHIGAMI	Oxyaquic Glossudalfs	53	18				13	124			15		
GIESE	Typic Humaquepts	28	25						64				
GILFORD	Typic Endoaquolls	53		36					47				
GILLINGHAM	Arenic Hapludalfs	2	20					53	53				
GLENDENNING	Aquic Glossudalfs	19	38				28	52			28		
GLENDORA	Mollic Psammaquents	41	20										
GLENFLORA	Mollic Glossaqualfs	17	53				38	76			25		
GLIDDEN	Alfic Haplorthods	64	20				13	23		25	23		
GOGEBIC	Alfic Oxyaquic Fragiorthods	920	7.5				7.5	30		38	18	48	
GOODMAN	Alfic Haplorthods	262	10				5	81		18			
GOODWIT	Alfic Oxyaquic Haplorthods	142	13				5	79		36	30		
GOSIL	Typic Quartzipsamments	80	23										
GOTHAM	Psammentic Hapludalfs	403	20					53					

(continued)

(continued)

Series	Subgroup	Area (km^2)	Ochric	Mollic	Histic	Umbric	Albic	Argillic (cm)	Cambic	Spodic	Glossic	Fragipan	Calcic
GRANBY	Typic Endoaquolls	231		25									
GRAYCALM	Lamellic Udipsamments	380	8					15					
GRAYLING	Typic Udipsamments	587	8						38				
GRAYS	Mollic Oxyaquic Hapludalfs	54	20					79					
GREENRIDGE	Typic Hapludalfs	36	23					152					
GREENWOOD	Typic Haplohemists	315			**152**								
GRELLTON	Typic Hapludalfs	73	13					61					
GRETTUM	Lamellic Udipsamments	171	80										
GRISWOLD	Typic Argiudolls	190		25				36					
GUENTHER	Alfic Oxyaquic Haplorthods	19	23					18		41	13		
GULL POINT	Typic Argiaquolls	1		82				18					
HALDER	Aquic Glossudalfs	3	15				10	53			13		
HATLEY	Aquic Glossudalfs	90	36					46			30		
HAUGEN	Oxyaquic Paleudalfs	322	83				15	129			76		
HAUSTRUP	Humic Lithic Dystrudepts	13				40							
HAYFIELD	Aquollic Hapludalfs	30	33					41					
HAYRIVER	Typic Hapludalfs	208	33				13	43					
HEBRON	Oxyaquic Hapludalfs	77	28				10	36					
HENNEPIN	Typic Eutrudepts	4	10						30				
HERBSTER	Aeric Glossaqualfs	137	25				13	114			7.5		
HERSEY	Mollic Hapludalfs	57	20					**272**					
HESCH	Typic Argiudolls	20		30				33					
HIBBING	Oxyaquic Glossudalfs	0	30					62			10		
HILES	Oxyaquic Glossudalfs	116	23					51			25		
HIXTON	Typic Hapludalfs	566	20					58					
HOCHHEIM	Typic Argiudolls	979		18				23					
HOOP	Aquic Argiudolls	6		28				30					
HOOPESTON	Aquic Hapludolls	48		36					61				
HORTONVILLE	Haplic Glossudalfs	1146	23					48			8		
HOUGHTON	Typic Haplosaprists	1482			**203**								
HUBBARD	Entic Hapludolls	27		50									
HUMBIRD	Oxyaquic Ultic Haplorthods	242	20					15		13			
HUNTSVILLE	Cumulic Hapludolls	153		69									
IMPACT	Humic Psammentic Dystrudepts	132				38							
INDUS	Vertic Epiaqualfs	2	13				7.5	36					
INGALLS	Typic Endoaquods	8	7.5				20			7.5			
IONIA	Oxyaquic Hapludalfs	33	20					40					

(continued)

(continued)

Series	Subgroup	Area (km^2)	Ochric	Mollic	Histic	Umbric	Albic	Argillic (cm)	Cambic	Spodic	Glossic	Fragipan	Calcic
IOSCO	Argic Endoaquods	185	18				5	16		48			
IRONRUN	Typic Endoaquods	175	25				20			20			
ISHPEMING	Entic Haplorthods	87	15				10			18			
JACKSON	Typic Hapludalfs	86	23					104					
JASPER	Typic Argiudolls	8		48				76					
JEWETT	Typic Hapludalfs	127	23				5	66					
JOY	Aquic Hapludolls	28		43					81				
JUDA	Mollic Oxyaquic Hapludalfs	9	33					58					
JUDSON	Cumulic Hapludolls	84		89					43				
JUNEAU	Typic Udifluvents	48	15										
KALMARVILLE	Mollic Fluvaquents	74	109										
KANE	Aquic Argiudolls	39		30				44					
KARLIN	Entic Haplorthods	480	7.5				7.5			57			
KARLSBORG	Arenic Oxyaquic Hapludalfs	57	23					52					
KATO	Typic Endoaquolls	12		53					25				
KAUKAUNA	Mollic Oxyaquic Hapludalfs	14	31					91					
KEGONSA	Mollic Hapludalfs	72	31					53					
KELLOGG	Alfic Oxyaquic Haplorthods	100	15				11	36		21	8		
KELTNER	Oxyaquic Argiudolls	7		33				71					
KENDALL	Aeric Endoaqualfs	5	36					94					
KENNAN	Haplic Glossudalfs	1139	41					81					
KEOWNS	Mollic Endoaquepts	0	23						40				
KERT	Aquic Glossudalfs	308	23					71			36		
KESHENA	Oxyaquic Glossudalfs	17	7.5					132			94		
KEVILAR	Mollic Hapludalfs	164	23					127					
KEWAUNEE	Typic Hapludalfs	2002	25					19					
KEWEENAW	Alfic Haplorthods	1278	25				25	56		71	56		
KEYESVILLE	Typic Dystrudepts	6	8						51				
KIBBIE	Aquollic Hapludalfs	67	28				10	58					
KICKAPOO	Typic Udifluvents	71	13										
KIDDER	Typic Hapludalfs	644	28						48				
KINGSVILLE	Mollic Psammaquents	70	20										
KINROSS	Typic Endoaquods	102	25				10			36			
KIVA	Entic Haplorthods	41	10				5			36			
KNOWLES	Typic Hapludalfs	76	31					68					
KOLBERG	Haplic Glossudalfs	72	20					54			13		
KOMRO	Entic Hapludolls	14		36									

(continued)

(continued)

Series	Subgroup	Area (km^2)	Ochric	Mollic	Histic	Umbric	Albic	Argillic (cm)	Cambic	Spodic	Glossic	Fragipan	Calcic
KOROBAGO	Aquic Eutrudepts	39	23						68				
KOST	Entic Hapludolls	2		43									
KRANSKI	Typic Hapludalfs	103	41					51					
LA FARGE	Typic Hapludalfs	595	20					58					
LABLATZ	Alfic Epiaquods	1	18				7.5	48		23	36		
LAMARTINE	Aquollic Hapludalfs	264	28					71					
LAMBEAU	Typic Hapludalfs	8	23					117					
LAMONT	Typic Hapludalfs	3	41				21	41					
LANGLADE	Haplic Glossudalfs	5	30				5	99					
LAONA	Alfic Fragiorthods	32	2.5				2.5	89		46	36	124	
LAPEER	Typic Hapludalfs	248	38					48					
LAPOIN	Alfic Oxyaquic Haplorthods	7	7.5				7.5	69		7.5	31		
LARA	Oxyaquic Argiudolls	2		25				122					
LAWLER	Aquic Hapludolls	20		48					49				
LAWSON	Aquic Cumulic Hapludolls	74		76									
LENROOT	Oxyaquic Udipsamments	161	10										
LEOLA	Aquic Arenic Hapludalfs	47	20					62					
LERCH	Vertic Epiaquepts	48	10						135				
LEROY	Typic Hapludalfs	98	23					33					
LILAH	Psammentic Hapludalfs	8	15				76						
LINDQUIST	Lamellic Haplorthods	61	13				8			20			
LINDSTROM	Cumulic Hapludolls	50		99					57				
LINO	Aquic Udipsamments	14	18										
LOBO	Hemic Sphagnofibrists	6			**152**								
LOCKE	Aquollic Hapludalfs	8	23					71					
LOMIRA	Typic Hapludalfs	275	36					50					
LONGRIE	Typic Haplorthods	150	10				13			46			
LORENZO	Typic Argiudolls	60		23				23					
LOWS	Mollic Endoaquepts	81	15						15				
LOXLEY	Typic Haplosaprists	1333			**152**								
LOYAL	Oxyaquic Glossudalfs	1157	52				28	64			25		
LUDINGTON	Oxyaquic Ultic Haplorthods	170	25				18	15		13			
LUNDEEN	Humic Dystrudepts	1				41			43				
LUPTON	Typic Haplosaprists	1578			**165**								
LUTZKE	Typic Hapludalfs	45	15					46					
MAGNOR	Aquic Glossudalfs	2852	41				23	107			25		
MAGROC	Aquic Glossudalfs	21	28				18	54			48		

(continued)

(continued)

Series	Subgroup	Area (km^2)	Ochric	Mollic	Histic	Umbric	Albic	Argillic (cm)	Cambic	Spodic	Glossic	Fragipan	Calcic
MAHALASVILLE	Typic Argiaquolls	78		30				69					
MAHTOMEDI	Typic Udipsamments	474	20										
MAINCREEK	Aquic Glossudalfs	27	46				25	25			30		
MAJIK	Aquic Quartzipsamments	14	18				7.5		18				40
MAKWA	Histic Humaquepts	74		20	20				27				
MANAWA	Aquollic Hapludalfs	956	31					45					
MANCELONA	Alfic Haplorthods	146	8				3	27		56			
MANISTEE	Alfic Haplorthods	56	20				8	20		43			
MANITOWISH	Oxyaquic Haplorthods	127	7.5				2.5			30			
MANN	Typic Epiaquolls	6		34					53				
MAPLEHURST	Aquic Glossudalfs	67	41				18	79			41		
MARATHON	Haplic Glossudalfs	91	38					52					
MARCELLON	Aquic Argiudolls	19		46				30					
MARKESAN	Typic Argiudolls	34		18				23					
MARKEY	Terric Haplosaprists	1449			81								
MARKHAM	Mollic Oxyaquic Hapludalfs	132	30					36					
MARSHAN	Typic Endoaquolls	124		46					30				
MARSHFIELD	Mollic Epiaqualfs	418	36				13	34					
MARTINTON	Aquic Argiudolls	69		43				64					
MATHERTON	Udollic Endoaqualfs	15	20				8	61					
MAUMEE	Typic Endoaquolls	5		58									
MAYVILLE	Oxyaquic Hapludalfs	163	31					50					
MCHENRY	Typic Hapludalfs	576	36					58					
MEADLAND	Aquic Glossudalfs	255	15					33			23		
MECAN	Typic Hapludalfs	93	31					88					
MECOSTA	Typic Udorthents	2	25										
MEDARY	Oxyaquic Hapludalfs	10	36					41					
MEEHAN	Aquic Udipsamments	780	10										
MEENON	Aquic Arenic Hapludalfs	22	23					13					
MENAHGA	Typic Udipsamments	2090	10										
MENASHA	Typic Epiaquolls	49		28					38				
MENDOTA	Typic Argiudolls	131		28				38					
MENOMIN	Mollic Hapludalfs	2	23					48					
MENOMINEE	Alfic Haplorthods	346	18				8	92		28	41		
MEQUITHY	Alfic Haplorthods	33	13				2.5	49		20	38		
MEQUON	Udollic Endoaqualfs	92	20					56					
MERIDIAN	Mollic Hapludalfs	213	23					58					

(continued)

(continued)

Series	Subgroup	Area (km^2)	Ochric	Mollic	Histic	Umbric	Albic	Argillic (cm)	Cambic	Spodic	Glossic	Fragipan	Calcic
MERIMOD	Mollic Hapludalfs	38	23					58					
MERIT	Mollic Hapludalfs	57	23					53					
MERRILLAN	Ultic Epiaquods	313	15				13	23		17			
METEA	Arenic Hapludalfs	50	71					41					
METONGA	Entic Haplorthods	63	10				2.5		7.5	53			
MIAMI	Oxyaquic Hapludalfs	480	20					59					
MICHIGAMME	Fragic Haplorthods	108	7.5				5			51		15	
MICKLE	Typic Argiudolls	23		30				135					
MIFFLIN	Typic Hapludalfs	13	20					48					
MILACA	Oxyaquic Glossudalfs	12	33				10	76			10		
MILFORD	Typic Endoaquolls	45		56					71				
MILITARY	Typic Hapludalfs	28	28					36					
MILLADORE	Aquic Glossudalfs	137	15				7.5	94			15		
MILLINGTON	Cumulic Endoaquolls	6		66					58				
MILTON	Typic Hapludalfs	2	20					54					
MINDORO	Humic Psammentic Dystrudepts	6				43							
MINOCQUA	Typic Endoaquepts	562	20				20		33				
MISKOAKI	Vertic Glossudalfs	15	10					109			15		
MOBERG	Typic Dystrudepts	26	10						41				
MONICO	Typic Endoaquods	90	18				7.5			36			
MONTELLO	Oxyaquic Argiudolls	17		31				33					
MONTGOMERY	Vertic Endoaquolls	60		38					59				
MOODIG	Alfic Epiaquods	40	13				5	102		44	79		
MOPPET	Oxyaquic Dystrudepts	109	25						74				
MOQUAH	Typic Udifluvents	91	13										
MORA	Aquic Glossudalfs	7	30				10	62			13		
MORGANLAKE	Alfic Oxyaquic Haplorthods	46	20				2.5	23		47			
MORLEY	Oxyaquic Hapludalfs	416	23					51					
MOROCCO	Aquic Udipsamments	67	36						33				
MOSEL	Aquollic Hapludalfs	75	56					35					
MOSHAWQUIT	Arenic Glossudalfs	3	7.5					46			56		
MOSINEE	Typic Dystrudepts	172	18						89				
MOUNDVILLE	Oxyaquic Hapludalfs	53	18					43					
MT. CARROLL	Mollic Hapludalfs	49	43				13	60					
MUDLAKE	Alfic Epiaquods	108	23				10	48		18	34		
MUNDELEIN	Aquic Argiudolls	79		43				36					
MUNUSCONG	Mollic Epiaquepts	15	20						56				

(continued)

(continued)

Series	Subgroup	Area (km^2)	Ochric	Mollic	Histic	Umbric	Albic	Argillic (cm)	Cambic	Spodic	Glossic	Fragipan	Calcic
MUSCATINE	Aquic Hapludolls	63		51					112				
MUSCODA	Lamellic Hapludalfs	2	25				18	46					
MUSKEGO	Limnic Haplosaprists	7			76								
MUSSEY	Typic Argiaquolls	26		23				23					
MYLREA	Aquic Dystrudepts	98	7.5				7.5		70				
MYRTLE	Mollic Hapludalfs	17	36				16	167					
NADEAU	Typic Hapludalfs	44	7.5					40					
NAHMA	Histic Humaquepts	3			23				10				
NAMUR	Lithic Hapludolls	97		15									
NAVAN	Typic Argiaquolls	81		31				43					
NEBAGO	Aquic Dystric Eutrudepts	44	23						36				
NECONISH	Oxyaquic Haplorthods	6	7.5				7.5			18			
NEDA	Mollic Oxyaquic Hapludalfs	14	28					36					
NEENAH	Aquollic Hapludalfs	43	18					38					
NENNO	Aquic Argiudolls	44		20				28					
NEOPIT	Oxyaquic Glossudalfs	26	7.5				7.5	119			107		
NESTER	Oxyaquic Glossudalfs	0	28					43			15		
NEWGLARUS	Typic Hapludalfs	430	23					81					
NEWLANG	Humaqueptic Psammaquents	25	7.5										
NEWOOD	Alfic Oxyaquic Haplorthods	847	13				2.5	43		20	51		
NEWOT	Alfic Haplorthods	294	13				7.5	76		28	46		
NEWSON	Humaqueptic Psammaquents	840	20										
NICHOLS	Oxyaquic Eutrudepts	78	20						46				
NICKIN	Typic Argiudolls	60		28				36					
NOKASIPPI	Typic Epiaquolls	5		35					33				
NORDEN	Typic Hapludalfs	1077	18					79					
NORGO	Lithic Hapludalfs	19	15					28					
NORTHBEND	Fluvaquentic Dystrudepts	44	18						69				
NORTHFIELD	Lithic Hapludalfs	149	15					25					
NORTHMOUND	Haplic Glossudalfs	5	10					48			33		
NOSEUM	Oxyaquic Haplorthods	7	6				6			28			
NUXMARUHANIXETE	Typic Argiudolls	3		28				97					
NYMORE	Typic Udipsamments	33	28										
OAKVILLE	Typic Udipsamments	80	8										
OCKLEY	Typic Hapludalfs	19	38					86					
OCONTO	Haplic Glossudalfs	1	23				5	43			15		
ODANAH	Haplic Glossudalfs	114	10					183			20		

(continued)

(continued)

Series	Subgroup	Area (km^2)	Ochric	Mollic	Histic	Umbric	Albic	Argillic (cm)	Cambic	Spodic	Glossic	Fragipan	Calcic
OESTERLE	Aquic Glossudalfs	397	28					52			10		
OGLE	Typic Argiudolls	47		28				155					
OKEE	Arenic Hapludalfs	111	61					25					
OMEGA	Typic Haplorthods	88	2.5				2.5			23			
OMENA	Haplic Glossudalfs	115	15				5	13			8		
OMRO	Typic Hapludalfs	43	20					51					
ONAWAY	Inceptic Hapludalfs	739	10				8	30	13				
ORION	Aquic Udifluvents	359	20										
ORONTO	Aeric Glossaqualfs	24	13					130			25		
OSHKOSH	Typic Hapludalfs	174	25					45					
OSHTEMO	Typic Hapludalfs	176	36					53					
OSSIAN	Typic Endoaquolls	146		58					145				
OSSMER	Aquic Glossudalfs	301	28				18	69			52		
OTTER	Cumulic Endoaquolls	121		69					35				
OTTERHOLT	Haplic Glossudalfs	66	43				20	97			43		
OZAUKEE	Oxyaquic Hapludalfs	255	25					33					
PADUS	Alfic Haplorthods	2373	7.5				2.5	30		40	48		
PADWET	Alfic Haplorthods	39	13				7.5	24		40	33		
PADWOOD	Alfic Oxyaquic Haplorthods	55	13				2.5	23		25	30		
PALMS	Terric Haplosaprists	510			89								
PALSGROVE	Typic Hapludalfs	659	28					112					
PARDEEVILLE	Mollic Hapludalfs	8	20					48					
PARKFALLS	Alfic Epiaquods	49	18				10	46		53	18		
PARTRIDGE	Aquic Udipsamments	23	28										
PEARL	Arenic Oxyaquic Hapludalfs	49	20					58					
PECATONICA	Typic Hapludalfs	92	48					99					
PECORE	Haplic Glossudalfs	4											
PEEBLES	Oxyaquic Argiudolls	10		28				23					
PELISSIER	Entic Haplorthods	41	10				10			38			
PELKIE	Oxyaquic Udipsamments	37	20										
PELLA	Typic Endoaquolls	786		43					54				
PENCE	Typic Haplorthods	2343	13				13			18			
PEPIN	Typic Hapludalfs	191	23					145					
PEQUAMING	Argic Endoaquods	34	15				10	20		25			
PERCHLAKE	Aquic Udipsamments	9	23										
PERIDA	Arenic Hapludalfs	24	23					74					
PEROTE	Haplic Glossudalfs	23	10					89			46		

(continued)

(continued)

Series	Subgroup	Area (km^2)	Ochric	Mollic	Histic	Umbric	Albic	Argillic (cm)	Cambic	Spodic	Glossic	Fragipan	Calcic
PESABIC	Alfic Epiaquods	283	13				2.5	76		20	51		
PESHEKEE	Lithic Haplorthods	17	7.5				5			33			
PESHTIGO	Aquic Glossudalfs	2	28				15	119			46		
PICKFORD	Aeric Epiaquepts	103	20				5		35				
PILLOT	Typic Argiudolls	100		38				43					
PINCONNING	Mollic Epiaquents	9	20										
PLAINBO	Typic Udipsamments	188	7.5										
PLAINFIELD	Typic Udipsamments	2432	18										
PLANO	Typic Argiudolls	1122		36				99					
PLEINE	Histic Humaquepts	100			23				62				
PLOVER	Aquic Glossudalfs	58	33					48			13		
PLUMCREEK	Typic Hapludalfs	23	18					74					
POINT	Aquic Glossudalfs	74	18				48	36			15		
POMROY	Arenic Oxyaquic Hapludalfs	27	56					23					
PONYCREEK	Humaqueptic Psammaquents	136	5										
PORT BYRON	Typic Hapludolls	62		33					81				
PORTWING	Oxyaquic Glossudalfs	228	10					107			13		
POSKIN	Aquic Glossudalfs	108	49				25	52			18		
POY	Typic Endoaquolls	123		31					55				
POYGAN	Typic Epiaquolls	456		33					36				
PRISSEL	Arenic Hapludalfs	70	23					20					
PUCHYAN	Arenic Oxyaquic Hapludalfs	21	23					48					
QUARDERER	Fluventic Hapludolls	14		33									
RABE	Arenic Glossudalfs	11	5					58			57		
RADFORD	Fluvaquentic Hapludolls	80											
RASSET	Typic Argiudolls	71		38				33					
REDRIM	Entic Lithic Haplorthods	2	20				18			18			
REEDSBURG	Aquic Paleudalfs	34	23					130					
RENOVA	Typic Hapludalfs	31	18				8	84					
RIB	Mollic Endoaqualfs	110	25				7.5	56					
RIBHILL	Haplic Glossudalfs	43	52				18	46			33		
RIBRIVER	Oxyaquic Glossudalfs	9	43				18	79			36		
RICHFORD	Arenic Hapludalfs	581	18					35					
RICHWOOD	Typic Argiudolls	53		46				71					
RIETBROCK	Aquic Glossudalfs	305	38					76			28		
RIFLE	Typic Haplohemists	136			**152**								
RIMER	Aquic Arenic Hapludalfs	5	58					44					

(continued)

(continued)

Series	Subgroup	Area (km^2)	Ochric	Mollic	Histic	Umbric	Albic	Argillic (cm)	Cambic	Spodic	Glossic	Fragipan	Calcic
RINGWOOD	Typic Argiudolls	181		30				61					
RIPON	Typic Argiudolls	8		28				58					
RITCHEY	Lithic Hapludalfs	47	20				10	21					
ROBAGO	Argic Endoaquods	30	18				13	28		25	46		
ROBY	Aquic Hapludalfs	20	38				10	43					
ROCKBLUFF	Typic Quartzipsamments	49	18										
ROCKBRIDGE	Typic Hapludalfs	8	15				5	69					
ROCKDAM	Entic Haplorthods	123	13				7.5			33			
ROCKERS	Argic Endoaquods	44	5					25		36	20		
ROCKLAND	Typic Eutrudepts	8	11						45				
ROCKMARSH	Aquollic Hapludalfs	4	20					58					
ROCKTON	Typic Argiudolls	101		38				41					
RODMAN	Typic Hapludolls	153		25					21				
RONDEAU	Limnic Haplosaprists	15			112								
ROOT	Mollic Fluvaquents	1											
ROSCOMMON	Mollic Psammaquents	379	23										
ROSHOLT	Haplic Glossudalfs	1501	33				13	54			18		
ROTAMER	Typic Argiudolls	129		20				28					
ROUSSEAU	Entic Haplorthods	291	20				18			23			
ROWLEY	Aquic Argiudolls	26		41				99					
ROZELLVILLE	Haplic Glossudalfs	200	23					66			13		
ROZETTA	Typic Hapludalfs	60	36				18	91					
RUBICON	Entic Haplorthods	742	15				13			30			
RUSE	Lithic Endoaquolls	6		18					20				
RUSKTOWN	Mollic Hapludalfs	17	23					41					
SABLE	Typic Endoaquolls	94		58					61				
SALTER	Typic Eutrudepts	32	23						30				
SANBORG	Oxyaquic Glossudalfs	228	13				13	102			30		
SANTIAGO	Haplic Glossudalfs	663	38				13	86			33		
SARGEANT	Aquic Glossudalfs	4	33					143			31		
SARONA	Alfic Haplorthods	3668	5				5	94		35	79		
SARWET	Alfic Oxyaquic Haplorthods	72	15				2.5	66		35	124		
SATTRE	Mollic Hapludalfs	204	43				13	46					
SAYBROOK	Oxyaquic Argiudolls	33		38				53					
SAYLESVILLE	Typic Hapludalfs	49	31				8	43					
SAYNER	Entic Haplorthods	1069	10				5			25			
SCHAAT CREEK	Aeric Endoaqualfs	1	13					84					

(continued)

(continued)

Series	Subgroup	Area (km^2)	Ochric	Mollic	Histic	Umbric	Albic	Argillic (cm)	Cambic	Spodic	Glossic	Fragipan	Calcic
SCHAPVILLE	Oxyaquic Argiudolls	1		25				51					
SCHWEITZER	Alfic Fragiorthods	9	13				10	86		41	102		
SCOBA	Haplic Glossudalfs	19	41					39			28		
SCONSIN	Oxyaquic Glossudalfs	240	69				2.5	18			41		
SCOTAH	Typic Udipsamments	47	10						46				
SCOTT LAKE	Oxyaquic Glossudalfs	201	30				7.5	41			23		
SEATON	Typic Hapludalfs	1042	23					74					
SEBBO	Mollic Hapludalfs	19	23					89					
SEBEWA	Typic Argiaquolls	286		28				55					
SECHLER	Aquic Humic Dystrudepts	4				31			25				
SEDGWICK	Alfic Epiaquods	97	20				7.5	163		20	7.5		
SEELYEVILLE	Typic Haplosaprists	1595			**203**								
SELKIRK	Aquic Glossudalfs	0	18				23	15					
SEWARD	Arenic Oxyaquic Hapludalfs	7	66					36					
SHAG	Typic Endoaquolls	3		28					35				
SHAWANO	Typic Udipsamments	337	28										
SHERRY	Udollic Endoaqualfs	243	20				13	52					
SHIFFER	Aquollic Hapludalfs	62	18					43					
SHIOCTON	Aquic Hapludolls	231		25					41				
SHULLSBURG	Aquic Argiudolls	2		43				62					
SILVERHILL	Ultic Hapludalfs	11	20					46					
SIMESCREEK	Typic Quartzipsamments	9	4										
SIOUXCREEK	Aquic Hapludults	2	36				15	46			30		
SIREN	Aquic Glossudalfs	9	30				10	208			15		
SISSABAGAMA	Oxyaquic Udipsamments	6	25										
SISSON	Typic Hapludalfs	166	23				13	40					
SKOG	Oxyaquic Udorthents	2	28										
SKYBERG	Mollic Epiaqualfs	6	36				18	71					
SLIMLAKE	Oxyaquic Dystrudepts	11	15						28				
SMESTAD	Aquic Argiudolls	16		25				163					
SODERBECK	Aquic Hapludolls	0		46					25				
SOGN	Lithic	284		23									
SOLONA	Aquic Argiudolls	562		23				46					
SOONER	Aquollic Hapludalfs	7	23					57					
SOPERTON	Alfic Fragiorthods	44	18				5	41		18	36	20	

(continued)

(continued)

Series	Subgroup	Area (km^2)	Ochric	Mollic	Histic	Umbric	Albic	Argillic (cm)	Cambic	Spodic	Glossic	Fragipan	Calcic
SPARTA	Entic Hapludolls	222		38									
SPEAR	Aquic Glossudalfs	14	5				10	64			64		
SPENCER	Oxyaquic Glossudalfs	364	56				7.5	66			46		
SPIDERLAKE	Alfic Oxyaquic Haplorthods	45	13				2.5	25		30	13		
SPINKS	Lamellic Hapludalfs	8	56					156					
SPOONERHILL	Oxyaquic Dystrudepts	46	7.5						23				
SPRINGSTEAD	Oxyaquic Haplorthods	142	7.5				2.5			58			
ST. CHARLES	Typic Hapludalfs	909	20					92					
STAMBAUGH	Alfic Haplorthods	315	10					43		35			
STANBERRY	Alfic Oxyaquic Haplorthods	171	7.5				5	20		40	33		
STENGEL	Aquic Arenic Hapludalfs	6	10					66					
STINNETT	Aquic Glossudalfs	52	46				13	84			56		
STRONGHURST	Aeric Endoaqualfs	4	38				10	81					
STURGEON	Aquic Udifluvents	37	20										
SULTZ	Entic Haplorthods	227	10				10			31			
SUMMERVILLE	Lithic Eutrudepts	192	5						36				
SUNIA	Oxyaquic Udipsamments	3	13						10				
SUPERIOR	Alfic Oxyaquic Haplorthods	71	15				7.5	30		20	13		
SYLVESTER	Typic Argiudolls	12		36				28					
SYMCO	Aquollic Hapludalfs	345	20					33					
SYMERTON	Oxyaquic Argiudolls	7		48				51					
TACOOSH	Terric Haplohemists	2			**102**								
TAMA	Typic Argiudolls	621		46				68					
TARR	Typic Quartzipsamments	727	10										
TAWAS	Terric Haplosaprists	109			79								
TAYLOR	Aquertic Glossudalfs	1	25				15	41			10		
TEDROW	Aquic Udipsamments	87	20						59				
TELL	Typic Hapludalfs	45	36					50					
THACKERY	Aquic Hapludalfs	5	30					72					
THERESA	Typic Hapludalfs	589	36					50					
TILLEDA	Haplic Glossudalfs	235	18					56			38		
TIMULA	Typic Eutrudepts	9	31						36				
TINT	Typic Quartzipsamments	28	23										
TINTSON	Oxyaquic Quartzipsamments	9	20										
TIPLER	Alfic Oxyaquic Haplorthods	175	13				5	35		35	18		

(continued)

(continued)

Series	Subgroup	Area (km^2)	Ochric	Mollic	Histic	Umbric	Albic	Argillic (cm)	Cambic	Spodic	Glossic	Fragipan	Calcic
TODDVILLE	Typic Argiudolls	94		49				78					
TONKEY	Mollic Endoaquepts	9	20						51				
TOTAGATIC	Typic Fluvaquents	88											
TOURTILLOTTE	Oxyaquic Udipsamments	3	7.5										
TRADELAKE	Oxyaquic Glossudalfs	4	33				7.5	99			20		
TREMPE	Entic Hapludolls	17		36									
TREMPEALEAU	Typic Argiudolls	3		28				33					
TROXEL	Pachic Argiudolls	105		84				102					
TULA	Argic Fragiaquods	179	20				7.5	64		52	86	86	
TUSCOLA	Aquic Hapludalfs	11	23					56					
TUSTIN	Arenic Hapludalfs	37	23					51					
TWINMOUND	Typic Quartzipsamments	77	5										
URNE	Dystric Eutrudepts	712	5						86				
VALTON	Mollic Paleudalfs	606	23					130					
VANCECREEK	Fluvaquentic Endoaquolls	12		41					84				
VANZILE	Alfic Haplorthods	192	10				7.5	41		23	41		
VARNA	Oxyaquic Argiudolls	95		30				76					
VEEDUM	Humic Epiaquepts	254	23						43				
VESPER	Humic Epiaquepts	265	20						46				
VILAS	Entic Haplorthods	1162	10				5			48			
VIRGIL	Udollic Endoaqualfs	19	33				15	117					
VLASATY	Aquic Glossudalfs	214	28					99			13		
WABENO	Alfic Oxyaquic Fragiorthods	265	5				5	109		15	23	79	
WACOUSTA	Typic Endoaquolls	154		43					21				
WAINOLA	Typic Endoaquods	279	30				25			16			
WAKEFIELD	Alfic Oxyaquic Fragiorthods	54	18					33		28	30	46	
WAKELEY	Aeric Epiaquents	30	48										
WALLKILL	Fluvaquentic Humaquepts	30		20					30				
WARMAN	Typic Endoaquolls	2		52					33				
WARSAW	Typic Argiudolls	226		36				46					
WASEPI	Aquollic Hapludalfs	61	43					33					
WASHTENAW	Aeric Fluvaquents	8	25										
WATSEKA	Aquic Hapludolls	13		25									
WATTON	Haplic Glossudalfs	0	2.5				13	94					
WAUCONDA	Udollic Endoaqualfs	3	36				13	40					

(continued)

(continued)

Series	Subgroup	Area (km^2)	Ochric	Mollic	Histic	Umbric	Albic	Argillic (cm)	Cambic	Spodic	Glossic	Fragipan	Calcic
WAUKEGAN	Typic Hapludolls	4		30					102				
WAUPACA	Mollic Fluvaquents	38	23										
WAUSEON	Typic Epiaquolls	31		33					58				
WAUTOMA	Mollic Epiaquents	50	20						28				
WAYKA	Typic Epiaquods	3	5				5			5	10		
WAYMOR	Haplic Glossudalfs	233	33					48			16		
WEGA	Aquic Udifluvents	15	23										
WESTVILLE	Typic Hapludalfs	82	20					107					
WHALAN	Typic Hapludalfs	290	23					30					
WHEATLEY	Mollic Psammaquents	7	18										
WHISKLAKE	Argic Endoaquods	54	15				7.5	48		20	33		
WHITEHALL	Typic Argiudolls	14		30				52					
WICKWARE	Haplic Glossudalfs	37	25					66			18		
WILDALE	Mollic Paleudalfs	104	23					130					
WILDWOOD	Histic Humaquepts	10			31				18				
WILL	Typic Endoaquolls	3		36					35				
WILLETTE	Terric Haplosaprists	128			81								
WINDWARD	Lamellic Quartzipsamments	9	23										
WINNEBAGO	Typic Argiudolls	25		38				74					
WINNECONNE	Mollic Oxyaquic Hapludalfs	115	18					38					
WINNESHIEK	Mollic Hapludalfs	9	41				10	20					
WINTERFIELD	Aquic Udipsamments	65	18										
WITHEE	Aquic Glossudalfs	1186	47				23	74			25		
WORCESTER	Argic Endoaquods	192	7.5				2.5	58		33	10		
WORMET	Typic Endoaquods	65	13				10			33			
WORTHEN	Cumulic Hapludolls	83		74					89				
WORWOOD	Alfic Epiaquods	14	10				2.5	25		18	20		
WOZNY	Typic Epiaqualfs	43	45				36	52					
WURTSMITH	Oxyaquic Udipsamments	102	10										
WYEVILLE	Aquic Arenic Hapludalfs	76	58						10				
WYKOFF	Typic Hapludalfs	34	20					28					
WYOCENA	Typic Hapludalfs	363	38					53					
YAHARA	Aquic Hapludolls	94		36					25				
ZEBA	Argic Endoaquods	6	13				7.5	30		20	23		
ZITTAU	Aquollic Hapludalfs	71	23					43					
ZURICH	Oxyaquic Hapludalfs	85	23				10	38					

(continued)

(continued)

Series	Subgroup	Area (km^2)	Ochric	Mollic	Histic	Umbric	Albic	Argillic (cm)	Cambic	Spodic	Glossic	Fragipan	Calcic
Other		2237											
		Avg.	24	39	95	36	13	62	48	31	33	62	40
		SD	14	17	50	7.6	9.5	37	28	14	22	33	
		Max	109	127	203	43	76	272	145	71	124	124	40
		Min	3	15	20	23	3	10	8	5	8	15	40
		Median	20	36	89	39	10	53	43	28	28	60	40
		Count	540	151	33	6	227	446	122	109	149	10	1

Appendix D
Area and Classification of Wisconsin Soil Series

J.G. Bockheim and A.E. Hartemink, *The Soils of Wisconsin*,
World Soils Book Series, DOI 10.1007/978-3-319-52144-2

Soil series	Area (km^2)	MLRAs	Soil regions	Order	Suborder	Great group	Subgroup	Particle-size class	Mineral class	CEC Activity class	Reaction class	Other	Soil-temperature class	Soil-moisture class
ABBAYE	11	93	I	Spodosols	Orthods	Haplorthods	Alfic Oxyaquic Haplorthods	Coarse-loamy	Mixed	Active			Frigid	Udic
ABSCO	236	105	Dr, J	Entisols	Fluvents	Udifluvents	Typic Udifluvents	Sandy	Siliceous				Mesic	Aquic
ABSCOTA	2	103, 105, 111C, 96, 97, 98	B, Dr, J	Entisols	Psamments	Udipsamments	Oxyaquic Udipsamments		Mixed				Mesic	Udic
ACKMORE	64	102B, 103, 107A, 107B, 108C, 108D, 109, 95B	B, J	Entisols	Aquents	Fluvaquents	Mollic Fluvaquents	Fine-silty	Mixed	Superactive	Nonacid		Mesic	Aquic
ADDER	23	105	Dr, J	Histosols	Saprists	Haplosaprists	Terric Haplosaprists	Sandy or sandy-skeletal	Siliceous		Euic		Mesic	Aquic
ADOLPH	84	90, 91, 93	F, J	Mollisols	Aquolls	Epiaquolls	Typic Epiaquolls	Coarse-loamy	Mixed	Superactive			Frigid	Aquic
ADRIAN	507	J	J	Histosols	Saprists	Haplosaprists	Terric Haplosaprists	Sandy or sandy-skeletal	Mixed		Euic		Mesic	Aquic
AFTAD	53	90A, 90B	F, G	Alfisols	Udalfs	Glossudalfs	Oxyaquic Glossudalfs	Coarse-loamy	Mixed	Superactive			Frigid	Udic
ALANGO	0	88, 91B	I, J	Alfisols	Aqualfs	Epiaqualfs	Chromic Vertic Epiaqualfs	Very fine	Smectitic				Frigid	Aquic
ALBAN	66	90A, 90B	C, E	Alfisols	Udalfs	Glossudalfs	Typic Glossudalfs	Coarse-loamy	Mixed	Superactive			Frigid	Udic
ALCONA	47	93, 94A, 94B	G, H	Spodosols	Orthods	Haplorthods	Alfic Haplorthods	Coarse-loamy	Mixed	Active			Frigid	Udic
ALDO	25	105	C	Entisols	Psamments	Udipsamments	Typic Udipsamments		Mixed				Mesic	Udic
ALGANSEE	166	103, 104, 105, 89, 91A, 96, 97, 98	C, J	Entisols	Psamments	Udipsamments	Aquic Udipsamments		Mixed				Mesic	Udic
ALLENDALE	153	90A, 92, 93A, 93B, 94A, 94B, 94C, 95A, 96, 98, 99	H, I, J	Spodosols	Aquods	Epiaquods	Alfic Epiaquods	Sandy over clayey	Mixed	Semiactive			Frigid	Aquic
ALMENA	435	90	F	Alfisols	Udalfs	Glossudalfs	Aquic Glossudalfs	Fine-silty	Mixed	Superactive			Frigid	Udic
ALPENA	34	93, 94A, 94B, 95A, 96	E	Mollisols	Udolls	Hapludolls	Entic Hapludolls	Sandy-skeletal	Mixed				Frigid	Udic
ALSTAD	2	90A, 90B, 91A, 91B, 93B, 94A	F	Alfisols	Udalfs	Glossudalfs	Aquic Glossudalfs	Fine-loamy	Mixed	Superactive			Frigid	Udic

(continued)

(continued)

Soil series	Area (km^2)	MLRAs	Soil regions	Order	Suborder	Great group	Subgroup	Particle-size class	Mineral class	CEC Activity class	Reaction class	Other	Soil-temperature class	Soil-moisture class
ALTDORF	48	90A, 90B	Fr, J	Alfisols	Aqualfs	Glossaqualfs	Aeric Glossaqualfs	Fine	Mixed	Active			Frigid	Aquic
AMASA	9		G	Spodosols	Orthods	Haplorthods	Typic Haplorthods	Coarse-loamy over sandy or sandy-skeletal	Mixed	Superactive			Frigid	Udic
AMERY	991	90A, 90B	Dr, F, G	Alfisols	Udalfs	Glossudalfs	Haplic Glossudalfs	Coarse-loamy	Mixed	Superactive			Frigid	Udic
AMNICON	374	92	I	Alfisols	Udalfs	Glossudalfs	Oxyaquic Vertic Glossudalfs	Very fine	Mixed	Active			Frigid	Udic
ANGELICA	119	90, 91, 93, 94A, 94B, 95A, 96	E, J	Inceptisols	Aquepts	Endoaquepts	Aeric Endoaquepts	Fine-loamy	Mixed	Active	Nonacid		Frigid	Aquic
ANIGON	259	90, 93	F, G	Alfisols	Udalfs	Glossudalfs	Haplic Glossudalfs	Fine-silty over sandy or sandy-skeletal	Mixed	Superactive			Frigid	Udic
ANNALAKE	152	90A, 92, 93B, 94A	G	Spodosols	Orthods	Haplorthods	Alfic Oxyaquic Haplorthods	Coarse-loamy	Mixed	Superactive			Frigid	Udic
ANNRIVER	101	90, 91	F, G, J	Alfisols	Aqualfs	Endoaqualfs	Mollic Endoaqualfs	Coarse-loamy over sandy or sandy-skeletal	Mixed	Superactive			Frigid	Aquic
ANTIGO	1242	90A, 90B, 91A, 91B, 93B	F, G	Alfisols	Udalfs	Glossudalfs	Haplic Glossudalfs	Coarse-loamy over sandy or sandy-skeletal	Mixed	Superactive			Frigid	Udic
ANTON	14	92	I	Alfisols	Udalfs	Glossudalfs	Oxyaquic Vertic Glossudalfs	Very fine	Mixed	Active			Frigid	Udic
ARBUTUS	1	89, 90B	Dr	Spodosols	Orthods	Haplorthods	Entic Haplorthods	Sandy	Siliceous				Frigid	Udic
ARENZVILLE	464	104, 105, 108, 115	A, J	Entisols	Fluvents	Udifluvents	Typic Udifluvents	Coarse-silty	Mixed	Superactive	Nonacid		Mesic	Udic
ARGONNE	23	90A	G	Spodosols	Orthods	Fragiorthods	Alfic Oxyaquic Fragiorthods	Coarse-loamy	Mixed	Superactive			Frigid	Udic
ARLAND	291	90	Dr, F	Alfisols	Udalfs	Glossudalfs	Haplic Glossudalfs	Coarse-loamy	Mixed	Superactive			Frigid	Udic
ARNHEIM	203	92, 93, 94A	I, J	Entisols	Aquents	Fluvaquents	Typic Fluvaquents	Coarse-loamy	Mixed	Superactive	Nonacid		Frigid	Aquic

(continued)

(continued)

Soil series	Area (km^2)	MLRAs	Soil regions	Order	Suborder	Great group	Subgroup	Particle-size class	Mineral class	CEC Activity class	Reaction class	Other	Soil-temperature class	Soil-moisture class
ASHDALE	210	104, 105, 108A, 108B, 115C, 95B	Am	Mollisols	Udolls	Argiudolls	Typic Argiudolls	Fine-silty	Mixed	Superactive			Mesic	Udic
ASHIPPUN	12	95B	B	Alfisols	Udalfs	Hapludalfs	Aquollic Hapludalfs	Fine-loamy	Mixed	Active			Mesic	Udic
ASHKUM	223	108A, 110, 95B, 97	Bm, J	Mollisols	Aquolls	Endoaquolls	Typic Endoaquolls	Fine	Mixed	Superactive			Mesic	Aquic
ASHWABAY	94	92	G, H	Spodosols	Orthods	Haplorthods	Alfic Oxyaquic Haplorthods	Sandy	Isotic				Frigid	Udic
ATTERBERRY	5	104, 105, 108A, 108B, 115B, 115C, 95A, 95B	B, J	Alfisols	Aqualfs	Endoaqualfs	Udollic Endoaqualfs	Fine-silty	Mixed	Superactive			Mesic	Aquic
AU GRES	384	142, 144A, 144B, 89, 90A, 90B, 91B, 92, 93B, 94A, 94B, 94C, 94D, 96, 98	E, H, J	Spodosols	Aquods	Endoaquods	Typic Endoaquods	Sandy	Mixed				Frigid	Aquic
AUBURNDALE	208	90A, 90B, 91B	F, J	Alfisols	Aqualfs	Epiaqualfs	Mollic Epiaqualfs	Fine-silty	Mixed	Superactive			Frigid	Aquic
AUGWOOD	13	94D	H, J	Spodosols	Aquods	Epiaquods	Typic Epiaquods	Sandy	Mixed				Frigid	Aquic
AUSABLE	13	90A, 94A, 96	H, J	Inceptisols	Aquepts	Humaquepts	Histic Humaquepts	Sandy	Mixed				Frigid	Aquic
AZTALAN	94	110, 95B	Bm, J	Mollisols	Udolls	Argiudolls	Aquic Argiudolls	Fine-loamy	Mixed	Superactive			Mesic	Udic
BACH	26	95A, 96, 98, 99	E, J	Inceptisols	Aquepts	Endoaquepts	Mollic Endoaquepts	Coarse-silty	Mixed	Semiactive	Calcareous		Mesic	Aquic
BADRIVER	181	92	I, J	Alfisols	Aqualfs	Glossaqualfs	Aeric Glossaqualfs	Fine	Mixed	Active			Frigid	Aquic
BALMORAL	1	105	Am	Mollisols	Udolls	Argiudolls	Oxyaquic Argiudolls	Fine-loamy	Mixed	Superactive			Mesic	Udic
BANAT	1	93, 94A, 94B, 95A	G, J	Alfisols	Aqualfs	Endoaqualfs	Aeric Endoaqualfs	Loamy-skeletal	Mixed	Active			Frigid	Aquic
BARABOO	113	105, 95B	A, B	Alfisols	Udalfs	Hapludalfs	Oxyaquic Hapludalfs	Fine-silty	Mixed	Superactive			Mesic	Udic
BARREMILLS	43	105	Am	Mollisols	Udolls	Argiudolls	Pachic Argiudolls	Fine-silty	Mixed	Superactive			Mesic	Udic
BARRONETT	98	90	F, G, J	Alfisols	Aqualfs	Epiaqualfs	Mollic Epiaqualfs	Fine-silty	Mixed	Superactive			Frigid	Aquic
BARRY	32	103, 111B, 111C, 95B, 97, 98, 99	Bm, J	Mollisols	Aquolls	Argiaquolls	Typic Argiaquolls	Fine-loamy	Mixed	Superactive			Mesic	Aquic
BASCO	53	105	Bm	Alfisols	Udalfs	Hapludalfs	Mollic Hapludalfs	Fine	Mixed	Superactive			Mesic	Udic

(continued)

(continued)

Soil series	Area (km^2)	MLRAs	Soil regions	Order	Suborder	Great group	Subgroup	Particle-size class	Mineral class	CEC Activity class	Reaction class	Other	Soil-temperature class	Soil-moisture class
BATAVIA	11	105, 108A, 108B, 110, 115C, 95B	B	Alfisols	Udalfs	Hapludalfs	Mollic Hapludalfs	Fine-silty	Mixed	Superactive			Mesic	Udic
BEARPEN	43	105	Dr	Mollisols	Udolls	Argiudolls	Aquic Argiudolls	Fine-silty	Mixed	Superactive			Mesic	Udic
BEARTREE	2	91B	G, J	Mollisols	Aquolls	Endoaquolls	Lithic Endoaquolls	Loamy-skeletal	Mixed	Superactive			Frigid	Aquic
BEAVERBAY	46		G	Spodosols	Orthods	Haplorthods	Alfic Oxyaquic Haplorthods	Coarse-loamy	Mixed	Superactive			Frigid	Udic
BEAVERCREEK	26	105	A, J	Entisols	Fluvents	Udifluvents	Typic Udifluvents	Loamy-skeletal	Mixed	Active	Nonacid		Mesic	Udic
BEECHER	7	108A, 110, 111C, 95B, 97	B, J	Alfisols	Aqualfs	Epiaqualfs	Udollic Epiaqualfs	Fine	Illitic				Mesic	Aquic
BELLECHESTER	3	104, 105	Am	Mollisols	Udolls	Hapludolls	Entic Hapludolls	Sandy	Mixed				Mesic	Udic
BELLEVILLE	0	103, 89, 97, 98, 99	Cm, J	Mollisols	Aquolls	Endoaquolls	Typic Endoaquolls	Sandy over loamy	Mixed	Active			Mesic	Aquic
BELLEVUE	336	95A, 95B	I	Mollisols	Udolls	Hapludolls	Fluventic Hapludolls	Fine-loamy	Mixed	Superactive			Mesic	Udic
BERGLAND	8	92, 93, 94A, 94B	I, J	Alfisols	Aqualfs	Epiaqualfs	Aeric Vertic Epiaqualfs	Very fine	Mixed	Semiactive			Frigid	Aquic
BERTRAND	7	105, 108, 109	A	Alfisols	Udalfs	Hapludalfs	Typic Hapludalfs	Fine-silty	Mixed	Superactive			Mesic	Udic
BESEMAN	472		J	Histosols	Saprists	Haplosaprists	Terric Haplosaprists	Loamy	Mixed		Dysic		Frigid	Aquic
BIGISLAND	5	90, 91B	H	Alfisols	Udalfs	Hapludalfs	Typic Hapludalfs	Sandy-skeletal	Isotic				Frigid	Udic
BILLETT	55	103, 104, 105, 108, 111, 115, 90, 91, 95B	C, Dr	Alfisols	Udalfs	Hapludalfs	Mollic Hapludalfs	Coarse-loamy	Mixed	Superactive			Mesic	Udic
BILLYBOY	97	90A, 90B	G	Alfisols	Udalfs	Glossudalfs	Oxyaquic Glossudalfs	Coarse-loamy over sandy or sandy-skeletal	Mixed	Superactive			Frigid	Udic
BILMOD	32	105, 91	Dr	Alfisols	Udalfs	Hapludalfs	Mollic Hapludalfs	Coarse-loamy	Siliceous	Superactive			Mesic	Udic
BILSON	133	105	Dr	Alfisols	Udalfs	Hapludalfs	Mollic Hapludalfs	Coarse-loamy	Siliceous	Superactive			Mesic	Udic
BJORKLAND	9	91	H, J	Alfisols	Aqualfs	Epiaqualfs	Typic Epiaqualfs	Sandy over clayey	Mixed OVER smectitic				Frigid	Aquic

(continued)

(continued)

Soil series	Area (km^2)	MLRAs	Soil regions	Order	Suborder	Great group	Subgroup	Particle-size class	Mineral class	CEC Activity class	Reaction class	Other	Soil-temperature class	Soil-moisture class
BLACKRIVER	15	90	F	Alfisols	Udalfs	Glossudalfs	Haplic Glossudalfs	Fine-silty	Mixed	Superactive			Frigid	Udic
BLOUNT	7	108A, 110, 111B, 95B, 97, 98, 99	B, J	Alfisols	Aqualfs	Epiaqualfs	Aeric Epiaqualfs	Fine	Illitic				Mesic	Aquic
BLUFFTON	13	57, 90, 91	F, J	Mollisols	Aquolls	Endoaquolls	Typic Endoaquolls	Fine-loamy	Mixed	Superactive			Frigid	Aquic
BOAZ	52	105	A, J	Inceptisols	Aquepts	Endoaquepts	Aeric Endoaquepts	Fine-silty	Mixed	Superactive	Nonacid		Mesic	Aquic
BOGUSCREEK	1	104, 105, 91	G, J	Entisols	Fluvents	Udifluvents	Mollic Udifluvents	Coarse-silty	Mixed	Superactive	Nonacid		Mesic	Udic
BONDUEL	71	94A, 94B, 95A	Er	Alfisols	Udalfs	Hapludalfs	Aquollic Hapludalfs	Fine-loamy	Mixed	Active			Frigid	Udic
BOONE	617	103, 104, 105, 107, 108, 90, 91	Dr	Entisols	Psamments	Quartzipsamments	Typic Quartzipsamments						Mesic	Udic
BOOTS	35	103, 104, 105, 141, 142, 91, 95B, 98	B, J	Histosols	Hemists	Haplohemists	Typic Haplohemists				Euic		Mesic	Aquic
BOPLAIN	15	105, 91	Dr	Entisols	Psamments	Udipsamments	Typic Udipsamments		Mixed			Uncoated	Mesic	Udic
BOREA	11	92	I, J	Alfisols	Aqualfs	Glossaqualfs	Aeric Glossaqualfs	Very fine	Mixed	Active			Frigid	Aquic
BORTH	47	95A	E, J	Alfisols	Udalfs	Hapludalfs	Mollic Oxyaquic Hapludalfs	Clayey over sandy or sandy-skeletal	Mixed	Active			Mesic	Udic
BOWSTRING	266	57, 88, 90, 94A	G, H, J	Histosols	Saprists	Haplosaprists	Fluvaquentic Haplosaprists				Euic		Frigid	Aquic
BOYER	398	110, 111B, 94A, 95A, 95B, 96, 97, 98, 99	B	Alfisols	Udalfs	Hapludalfs	Typic Hapludalfs	Coarse-loamy	Mixed	Semiactive			Mesic	Udic
BRAHAM	0	57, 88, 90, 91	I	Alfisols	Udalfs	Hapludalfs	Arenic Hapludalfs	Loamy	Mixed	Superactive			Frigid	Udic
BRANDER	18	90	F	Alfisols	Udalfs	Glossudalfs	Oxyaquic Glossudalfs	Fine-silty over sandy or sandy-skeletal	Mixed	Superactive			Frigid	Udic
BRANSTAD	4	103, 90	G	Alfisols	Udalfs	Glossudalfs	Oxyaquic Glossudalfs	Fine-loamy	Mixed	Superactive			Frigid	Udic

(continued)

(continued)

Soil series	Area (km^2)	MLRAs	Soil regions	Order	Suborder	Great group	Subgroup	Particle-size class	Mineral class	CEC Activity class	Reaction class	Other	Soil-temperature class	Soil-moisture class
BREMS	339	105, 110, 111B, 111C, 95A, 95B, 97, 98	B, C	Entisols	Psamments	Udipsamments	Aquic Udipsamments		Mixed				Mesic	Udic
BRENNYVILLE	1	90, 91	G	Alfisols	Udalfs	Glossudalfs	Aquic Glossudalfs	Coarse-loamy	Mixed	Superactive			Frigid	Udic
BREVORT	79	90A, 93B, 94A, 94B, 94C, 95A, 96, 98	E, J	Entisols	Aquents	Endoaquents	Mollic Endoaquents	Sandy over loamy	Mixed	Active	Nonacid		Frigid	Aquic
BRICE	2	105	B, Dr	Alfisols	Udalfs	Hapludalfs	Lamellic Hapludalfs	Coarse-loamy	Mixed	Superactive			Mesic	Udic
BRIGGSVILLE	102	89, 95A, 95B	I	Alfisols	Udalfs	Hapludalfs	Typic Hapludalfs	Fine	Mixed	Superactive			Mesic	Udic
BRILL	102	90	F	Alfisols	Udalfs	Glossudalfs	Haplic Glossudalfs	Fine-silty over sandy or sandy-skeletal	Mixed	Superactive			Frigid	Udic
BRIMLEY	3	93, 94A, 94B, 96	G, J	Spodosols	Aquods	Endoaquods	Argic Endoaquods	Fine-loamy	Mixed	Superactive			Frigid	Aquic
BRINKMAN	44	105	A, Dr	Alfisols	Udalfs	Paleudalfs	Mollic Paleudalfs	Fine-silty	Mixed	Superactive			Mesic	Udic
BRODALE	7	104, 105, 90, 91	A, Dr	Mollisols	Udolls	Hapludolls	Entic Hapludolls	Loamy-skeletal	Carbonatic				Mesic	Udic
BROOKSTON	130	110, 111A, 111C, 95B, 97, 98, 99	B, J	Mollisols	Aquolls	Argiaquolls	Typic Argiaquolls	Fine-loamy	Mixed	Superactive			Mesic	Aquic
BROWNCHURCH	58		A, Dr	Alfisols	Udalfs	Hapludalfs	Mollic Hapludalfs	Coarse-loamy	Mixed	Superactive			Mesic	Udic
BROWNSTONE	6	92	G, H	Spodosols	Orthods	Haplorthods	Typic Haplorthods	Sandy-skeletal	Mixed				Frigid	Udic
BRUCE	4	90, 93, 94A, 94B	G, E, J	Inceptisols	Aquepts	Endoaquepts	Mollic Endoaquepts	Fine-loamy	Mixed	Superactive	Nonacid		Frigid	Aquic
BURKHARDT	182	104, 105, 108, 115C, 90	C	Mollisols	Udolls	Hapludolls	Typic Hapludolls	Sandy	Mixed				Mesic	Udic
BUSHVILLE	1	88, 90, 91	G	Alfisols	Udalfs	Hapludalfs	Aquic Arenic Hapludalfs	Loamy	Mixed	Superactive			Frigid	Udic
CABLE	340	90, 93	F, G, J	Inceptisols	Aquepts	Epiaquepts	Typic Epiaquepts	Coarse-loamy	Isotic		Nonacid		Frigid	Aquic
CADIZ	38	110, 95B	A	Alfisols	Udalfs	Hapludalfs	Oxyaquic Hapludalfs	Fine-silty	Mixed	Superactive			Mesic	Udic
CALAMINE	9	103, 104, 105, 95B	Am, J	Mollisols	Aquolls	Argiaquolls	Typic Argiaquolls	Fine	Mixed	Superactive			Mesic	Aquic

(continued)

(continued)

Soil series	Area (km^2)	MLRAs	Soil regions	Order	Suborder	Great group	Subgroup	Particle-size class	Mineral class	CEC Activity class	Reaction class	Other	Soil-temperature class	Soil-moisture class
CAMPIA	57	90	F	Alfisols	Udalfs	Glossudalfs	Haplic Glossudalfs	Fine-silty	Mixed	Superactive			Frigid	Udic
CAPITOLA	707	90, 93	F, G, J	Alfisols	Aqualfs	Epiaqualfs	Aeric Epiaqualfs	Coarse-loamy	Mixed	Superactive			Frigid	Aquic
CARBONDALE	369	140, 142, 143, 147, 90, 93, 94A, 94B, 95A, 96	J	Histosols	Saprists	Haplosaprists	Hemic Haplosaprists				Euic		Frigid	Aquic
CARYVILLE	12	105, 90	Cm, J	Mollisols	Udolls	Hapludolls	Fluventic Hapludolls	Sandy	Mixed				Frigid	Udic
CASCO	699	108A, 108B, 110, 111A, 111B, 111C, 111D, 115C, 124, 95A, 95B, 97	B, I	Alfisols	Udalfs	Hapludalfs	Inceptic Hapludalfs	Fine-loamy over sandy or sandy-skeletal	Mixed	Superactive			Mesic	Udic
CATHRO	1823	J	A, J	Histosols	Saprists	Haplosaprists	Terric Haplosaprists	Loamy	Mixed		Euic		Frigid	Aquic
CEBANA	567	90	F, J	Alfisols	Aqualfs	Glossaqualfs	Mollic Glossaqualfs	Coarse-loamy	Mixed	Superactive			Frigid	Aquic
CERESCO	36	105, 111D, 115C, 97, 98, 99	Dr	Mollisols	Udolls	Hapludolls	Fluvaquentic Hapludolls	Coarse-loamy	Mixed	Superactive			Mesic	Udic
CHABENEAU	11		G	Spodosols	Orthods	Haplorthods	Oxyaquic Haplorthods	Coarse-loamy over sandy or sandy-skeletal	Mixed	Superactive			Frigid	Udic
CHAMPION	14	93	G	Spodosols	Orthods	Fragiorthods	Oxyaquic Fragiorthods	Coarse-loamy	Mixed	Superactive			Frigid	Udic
CHANNAHON	49	104, 105, 108A, 110, 111B, 95A, 95B, 99	Bm	Mollisols	Udolls	Argiudolls	Lithic Argiudolls	Loamy	Mixed	Superactive			Mesic	Udic
CHANNING	9		G	Spodosols	Aquods	Endoaquods	Typic Endoaquods	Coarse-loamy over sandy or sandy-skeletal	Mixed	Superactive			Frigid	Aquic

(continued)

(continued)

Soil series	Area (km^2)	MLRAs	Soil regions	Order	Suborder	Great group	Subgroup	Particle-size class	Mineral class	CEC Activity class	Reaction class	Other	Soil-temperature class	Soil-moisture class
CHARLEVOIX	93	93B, 94A, 94B, 94C, 95A, 96	G, J	Spodosols	Aquods	Endoaquods	Argic Endoaquods	Coarse-loamy	Mixed	Semiactive			Frigid	Aquic
CHASEBURG	277	105	A, J	Entisols	Fluvents	Udifluvents	Typic Udifluvents	Coarse-silty	Mixed	Superactive	Nonacid		Mesic	Udic
CHELMO	3	91	E, G, J	Alfisols	Aqualfs	Epiaqualfs	Umbric Epiaqualfs	Clayey over sandy or sandy-skeletal	Smectitic OVER mixed				Frigid	Aquic
CHELSEA	74	103, 104, 105, 108C, 109, 110, 111B, 111C, 115A, 115C, 94A, 95B, 97, 98, 99	C	Entisols	Psamments	Udipsamments	Lamellic Udipsamments		Mixed				Mesic	Udic
CHEQUAMEGON	341	90	G	Spodosols	Orthods	Haplorthods	Alfic Oxyaquic Haplorthods	Coarse-loamy	Mixed	Superactive			Frigid	Udic
CHETEK	750	90, 91	Dr, G, H	Alfisols	Udalfs	Hapludalfs	Inceptic Hapludalfs	Coarse-loamy	Mixed	Superactive			Frigid	Udic
CHINWHISKER	29	90A, 94A, 96	G	Spodosols	Orthods	Haplorthods	Lamellic Oxyaquic Haplorthods	Sandy	Mixed				Frigid	Udic
CHIPPENY	2	101, 93, 94A, 94B, 96	H, J	Histosols	Saprists	Haplosaprists	Lithic Haplosaprists				Euic		Frigid	Aquic
CHURCHTOWN	455	105	A	Alfisols	Udalfs	Hapludalfs	Mollic Hapludalfs	Fine-silty	Mixed	Superactive			Mesic	Udic
CITYPOINT	98	90, 91	I, J	Histosols	Saprists	Haplosaprists	Typic Haplosaprists				Dysic		Frigid	Aquic
CLEMENS	3	90, 91	G	Inceptisols	Udepts	Eutrudepts	Aquic Dystric Eutrudepts	Loamy-skeletal	Mixed	Superactive			Frigid	Udic
CLYDE	21	104, 108B, 108C, 90B	E, G, J	Mollisols	Aquolls	Endoaquolls	Typic Endoaquolls	Fine-loamy	Mixed	Superactive			Mesic	Aquic
COFFEEN	37	105, 108A, 108B, 113, 114B, 115B, 115C	Am	Mollisols	Udolls	Hapludolls	Fluvaquentic Hapludolls	Coarse-silty	Mixed	Superactive			Mesic	Udic
COFFTON	10	105	Dr	Mollisols	Udolls	Hapludolls	Fluvaquentic Hapludolls	Coarse-silty	Mixed	Superactive			Mesic	Udic
COLAND	1		Am	Mollisols	Aquolls	Endoaquolls	Cumuliic Endoaquolls	Fine-loamy	Mixed	Superactive			Mesic	Aquic

(continued)

(continued)

Soil series	Area (km^2)	MLRAs	Soil regions	Order	Suborder	Great group	Subgroup	Particle-size class	Mineral class	CEC Activity class	Reaction class	Other	Soil-temperature class	Soil-moisture class
COLOMA	303	104, 105, 108A, 108B, 111B, 111C, 111D, 115C, 89, 95A, 95B, 96, 97, 98	C	Entisols	Psamments	Udipsamments	Lamellic Udipsamments		Mixed				Mesic	Udic
COLWOOD	168	105, 111B, 95B, 97, 98, 99	Am, Bm, J	Mollisols	Aquolls	Endoaquolls	Typic Endoaquolls	Fine-loamy	Mixed	Active			Mesic	Aquic
COMSTOCK	143	90	F, G	Alfisols	Udalfs	Glossudalfs	Aquic Glossudalfs	Fine-silty	Mixed	Superactive			Frigid	Udic
CONOVER	1	110, 111B, 111C, 97, 98, 99	B, J	Alfisols	Aqualfs	Endoaqualfs	Udollic Endoaqualfs	Fine-loamy	Mixed	Active			Mesic	Aquic
CORMANT	194	56, 88, 90, 94A, 95A	H, J	Entisols	Aquents	Psammaquents	Mollic Psammaquents		Mixed				Frigid	Aquic
CORNUCOPIA	103		I	Alfisols	Udalfs	Glossudalfs	Haplic Glossudalfs	Fine	Mixed	Active			Frigid	Udic
COSAD	12	101, 144A, 95A, 98	I	Entisols	Orthents	Udorthents	Aquic Udorthents	Sandy over clayey	Mixed	Superactive	Nonacid		Mesic	Udic
COUNCIL	312	105	Dr	Alfisols	Udalfs	Hapludalfs	Typic Hapludalfs	Coarse-loamy	Mixed	Superactive			Mesic	Udic
CRESS	195	90, 92	H	Inceptisols	Udepts	Dystrudepts	Humic Dystrudepts	Sandy	Mixed				Frigid	Udic
CREX	33	90, 94A	H	Entisols	Psamments	Udipsamments	Oxyaquic Udipsamments		Mixed				Frigid	Udic
CROMWELL	238	88, 90, 92, 93	H	Inceptisols	Udepts	Dystrudepts	Typic Dystrudepts	Sandy	Isotic				Frigid	Udic
CROSSETT	4	94A	G	Alfisols	Udalfs	Glossudalfs	Aquic Glossudalfs	Fine-loamy	Mixed	Active			Frigid	Udic
CROSWELL	518	90A, 90B, 91B, 92, 93B, 94A, 94B, 94C, 96, 98	H	Spodosols	Orthods	Haplorthods	Oxyaquic Haplorthods	Sandy	Isotic				Frigid	Udic
CROSWOOD	27	93, 94A	H	Spodosols	Orthods	Haplorthods	Oxyaquic Haplorthods	Sandy	Mixed				Frigid	Udic
CRYSTAL LAKE	145	90	F, G	Alfisols	Udalfs	Glossudalfs	Oxyaquic Glossudalfs	Fine-silty	Mixed	Superactive			Frigid	Udic
CUBLAKE	118	93, 94A	G, H	Spodosols	Orthods	Haplorthods	Oxyaquic Haplorthods	Sandy	Mixed				Frigid	Udic
CUNARD	2	93B, 94B, 94C, 95A	E, G	Alfisols	Udalfs	Hapludalfs	Typic Hapludalfs	Coarse-loamy	Mixed	Active			Frigid	Udic
CURRAN	103	105, 91	A, J	Alfisols	Aqualfs	Endoaqualfs	Udollic Endoaqualfs	Fine-silty	Mixed	Superactive			Mesic	Aquic

(continued)

(continued)

Soil series	Area (km^2)	MLRAs	Soil regions	Order	Suborder	Great group	Subgroup	Particle-size class	Mineral class	CEC Activity class	Reaction class	Other	Soil-temperature class	Soil-moisture class
CUSHING	7	103, 90	F	Alfisols	Udalfs	Glossudalfs	Haplic Glossudalfs	Fine-loamy	Mixed	Superactive			Frigid	Udic
CUTAWAY	0	57, 88, 90, 93	G	Alfisols	Udalfs	Hapludalfs	Arenic Oxyaquic Hapludalfs	Loamy	Mixed	Superactive			Frigid	Udic
CUTTRE	370		I, J	Alfisols	Aqualfs	Glossaqualfs	Aeric Glossaqualfs	Very fine	Mixed	Active			Frigid	Aquic
DAIRYLAND	13	90, 91	H	Alfisols	Udalfs	Hapludalfs	Mollic Oxyaquic Hapludalfs	Sandy-skeletal	Mixed				Frigid	Udic
DAISYBAY	1	88, 92, 93	I, J	Histosols	Hemists	Haplohemists	Terric Haplohemists	Clayey	Smectitic		Euic		Frigid	Aquic
DAKOTA	148	103, 104, 105, 108, 115, 90, 91, 95B	Am, Cm	Mollisols	Udolls	Argiudolls	Typic Argiudolls	Fine-loamy over sandy or sandy-skeletal	Mixed	Superactive			Mesic	Udic
DANCY	96	90	C, Fr, J	Alfisols	Aqualfs	Glossaqualfs	Typic Glossaqualfs	Fine-loamy	Mixed	Superactive			Frigid	Aquic
DARROCH	8	110, 111C, 111D, 95B, 97, 99	Bm	Mollisols	Udolls	Argiudolls	Aquic Argiudolls	Fine-loamy	Mixed	Superactive			Mesic	Udic
DAWSIL	170	91	C, J	Histosols	Saprists	Haplosaprists	Terric Haplosaprists	Sandy or sandy-skeletal	Siliceous		Dysic		Frigid	Aquic
DAWSON	603		J	Histosols	Saprists	Haplosaprists	Terric Haplosaprists	Sandy or sandy-skeletal	Mixed		Dysic		Frigid	Aquic
DECHAMPS	31		F, G, H, J	Entisols	Fluvents	Udifluvents	Aquic Udifluvents	Sandy	Mixed				Frigid	Udic
DEERTON	7	92, 93, 94B	H	Spodosols	Orthods	Haplorthods	Typic Haplorthods	Sandy	Mixed				Frigid	Udic
DEFORD	127	142, 93, 94A, 94B, 95A, 96	H, J	Entisols	Aquents	Psammaquents	Typic Psammaquents		Mixed				Frigid	Aquic
DEL REY	2	108A, 110, 111A, 111B, 111C, 94A, 95B, 97, 98, 99	B, J	Alfisols	Aqualfs	Epiaqualfs	Aeric Epiaqualfs	Fine	Illitic				Mesic	Aquic
DELLS	28	104, 105	A	Alfisols	Udalfs	Hapludalfs	Aquollic Hapludalfs	Fine-silty over sandy or sandy-skeletal	Mixed	Superactive			Mesic	Udic

(continued)

(continued)

Soil series	Area (km^2)	MLRAs	Soil regions	Order	Suborder	Great group	Subgroup	Particle-size class	Mineral class	CEC Activity class	Reaction class	Other	Soil-temperature class	Soil-moisture class
DELTON	83	91	C, E	Alfisols	Udalfs	Hapludalfs	Arenic Hapludalfs	Loamy	Mixed	Active			Mesic	Udic
DENOMIE	31	92	G	Alfisols	Udalfs	Glossudalfs	Haplic Glossudalfs	Fine-silty	Mixed	Active			Frigid	Udic
DENROCK	1	105, 108B, 109	Am	Mollisols	Udolls	Argiudolls	Aquic Argiudolls	Fine	Mixed	Superactive			Mesic	Udic
DERINDA	5	104, 105, 108B, 115C, 90B, 95B	A	Alfisols	Udalfs	Hapludalfs	Oxyaquic Hapludalfs	Fine	Mixed	Active			Mesic	Udic
DICKINSON	96	102A, 103, 104, 105, 106, 107A, 107B, 108A, 108B, 108C, 108D, 113, 115A, 115C, 95B	Am	Mollisols	Udolls	Hapludolls	Typic Hapludolls	Coarse-loamy	Mixed	Superactive			Mesic	Udic
DICKMAN	15	102B, 103, 104, 107, 95B	Bm	Mollisols	Udolls	Hapludolls	Typic Hapludolls	Sandy	Mixed				Mesic	Udic
DISHNO	9		G	Spodosols	Orthods	Haplorthods	Oxyaquic Haplorthods	Coarse-loamy over sandy or sandy-skeletal	Mixed	Superactive			Frigid	Udic
DOBIE	64	90	Dr	Alfisols	Udalfs	Glossudalfs	Haplic Glossudalfs	Fine-silty	Mixed	Superactive			Frigid	Udic
DODGE	495	108A, 108B, 110, 115C, 95B	B	Alfisols	Udalfs	Hapludalfs	Typic Hapludalfs	Fine-silty	Mixed	Superactive			Mesic	Udic
DODGEVILLE	585	104, 105, 108, 95B	Am	Mollisols	Udolls	Argiudolls	Typic Argiudolls	Fine-silty over clayey	Mixed	Superactive			Mesic	Udic
DODY	3	90, 94A	I	Alfisols	Aqualfs	Albaqualfs	Arenic Albaqualfs	Clayey	Smectitic				Frigid	Aquic
DOLPH	88	90	Fr, J	Alfisols	Aqualfs	Glossaqualfs	Aeric Glossaqualfs	Fine	Mixed	Superactive			Frigid	Aquic
DORA	5	56, 57, 88, 93	I, J	Histosols	Saprists	Haplosaprists	Terric Haplosaprists	Clayey	Smectitic		Euic		Frigid	Aquic
DORCHESTER	0	103, 105, 108, 95B	B, J	Entisols	Fluvents	Udifluvents	Typic Udifluvents	Fine-silty	Mixed	Superactive	Calcareous		Mesic	Udic
DORERTON	394	104, 105, 90	A	Alfisols	Udalfs	Hapludalfs	Typic Hapludalfs	Loamy-skeletal	Mixed	Active			Mesic	Udic

(continued)

(continued)

Soil series	Area (km^2)	MLRAs	Soil regions	Order	Suborder	Great group	Subgroup	Particle-size class	Mineral class	CEC Activity class	Reaction class	Other	Soil-temperature class	Soil-moisture class
DORITTY	31	90	Dr	Alfisols	Udalfs	Glossudalfs	Haplic Glossudalfs	Fine-silty	Mixed	Superactive			Frigid	Udic
DOWNS	407	104, 105, 108, 109, 115, 95B	A	Alfisols	Udalfs	Hapludalfs	Mollic Hapludalfs	Fine-silty	Mixed	Superactive			Mesic	Udic
DRAMMEN	163	105, 91	Dr, H	Alfisols	Udalfs	Hapludalfs	Lamellic Hapludalfs	Sandy	Mixed				Mesic	Udic
DRESDEN	246	108A, 110, 95A, 95B	Bm	Alfisols	Udalfs	Hapludalfs	Mollic Hapludalfs	Fine-loamy over sandy or sandy-skeletal	Mixed	Active			Mesic	Udic
DRUMMER	39	108A, 108B, 110, 111A, 111D, 95B, 97	Bm, J	Mollisols	Aquolls	Endoaquolls	Typic Endoaquolls	Fine-silty	Mixed	Superactive			Mesic	Aquic
DRYLANDING	68		G	Mollisols	Udolls	Hapludolls	Lithic Hapludolls	Loamy-skeletal	Mixed	Superactive			Frigid	Udic
DUBUQUE	2835	104, 105, 108, 95B	A, B	Alfisols	Udalfs	Hapludalfs	Typic Hapludalfs	Fine-silty	Mixed	Superactive			Mesic	Udic
DUEL	3	93B, 94A, 94B, 95A	Er, J	Spodosols	Orthods	Haplorthods	Entic Haplorthods	Sandy	Mixed				Frigid	Udic
DUELM	4	103, 90, 91, 91B	C, H	Mollisols	Udolls	Hapludolls	Oxyaquic Hapludolls	Sandy	Mixed				Frigid	Udic
DUNBARTON	246	105, 108, 95B	A	Alfisols	Udalfs	Hapludalfs	Lithic Hapludalfs	Clayey	Smectitic				Mesic	Udic
DUNNBOT	47	105, 91	Dr, J	Entisols	Fluvents	Udifluvents	Mollic Udifluvents	Coarse-loamy	Mixed	Superactive	Nonacid		Mesic	Udic
DUNNVILLE	51	90, 91	Dr, G	Mollisols	Udolls	Hapludolls	Typic Hapludolls	Coarse-loamy	Mixed	Superactive			Frigid	Udic
DURAND	69	105, 108B, 95B	Bm	Mollisols	Udolls	Argiudolls	Typic Argiudolls	Fine-loamy	Mixed	Superactive			Mesic	Udic
EAUCLAIRE	20	90	Dr	Spodosols	Orthods	Haplorthods	Alfic Oxyaquic Haplorthods	Sandy	Mixed				Frigid	Udic
EDMUND	258	105, 95B	Am	Mollisols	Udolls	Argiudolls	Lithic Argiudolls	Clayey	Smectitic				Mesic	Udic
EDWARDS	6	101, 111B, 111C, 94A, 95A, 95B, 96, 97, 98, 99	J	Histosols	Saprists	Haplosaprists	Limnic Haplosaprists		Marly		Euic		Mesic	Aquic
ELBAVILLE	180		A	Alfisols	Udalfs	Hapludalfs	Glossic Hapludalfs	Fine-loamy		Superactive			Mesic	Udic
ELBURN	260	108A, 108B, 110, 95B	Bm	Mollisols	Udolls	Argiudolls	Aquic Argiudolls	Fine-silty	Mixed	Superactive			Mesic	Udic

(continued)

(continued)

Soil series	Area (km^2)	MLRAs	Soil regions	Order	Suborder	Great group	Subgroup	Particle-size class	Mineral class	CEC Activity class	Reaction class	Other	Soil-temperature class	Soil-moisture class
ELDERON	44	90, 93	E, G	Inceptisols	Udepts	Dystrudepts	Typic Dystrudepts	Sandy-skeletal	Mixed				Frigid	Udic
ELEROY	7	105, 108B, 95B	A	Alfisols	Udalfs	Hapludalfs	Oxyaquic Hapludalfs	Fine-silty	Mixed	Superactive			Mesic	Udic
ELEVA	541	104, 105, 108, 111, 90, 95A, 95B, 98	B, Dr	Alfisols	Udalfs	Hapludalfs	Typic Hapludalfs	Coarse-loamy	Mixed	Active			Mesic	Udic
ELEVASIL	257	105	Dr	Alfisols	Udalfs	Hapludalfs	Ultic Hapludalfs	Coarse-loamy	Siliceous	Active			Mesic	Udic
ELKMOUND	322	105, 90, 91	Dr	Inceptisols	Udepts	Dystrudepts	Typic Dystrudepts	Loamy	Mixed	Superactive			Mesic	Udic
ELLA	84	104, 105, 108C	F, Dr	Alfisols	Udalfs	Hapludalfs	Mollic Hapludalfs	Fine-silty	Mixed	Superactive			Mesic	Udic
ELLIOTT	147	108A, 110, 111B, 111C, 95B, 97	Bm	Mollisols	Udolls	Argiudolls	Aquic Argiudolls	Fine	Illitic				Mesic	Udic
ELLWOOD	21	94A	G	Alfisols	Udalfs	Glossudalfs	Oxyaquic Glossudalfs	Fine-loamy	Mixed	Active			Frigid	Udic
ELM LAKE	274	90, 91	Dr, J	Entisols	Aquents	Epiaquents	Humaqueptic Epiaquents	Sandy over loamy	Siliceous	Semiactive	Acid		Frigid	Aquic
ELVERS	16	105, 108B, 144A, 144B, 95B, 97	B, J	Entisols	Aquents	Fluvaquents	Thapto-Histic Fluvaquents	Coarse-silty	Mixed	Superactive	Nonacid		Mesic	Aquic
EMMERT	80	57, 90, 91, 93	G	Entisols	Orthents	Udorthents	Typic Udorthents	Sandy-skeletal	Mixed				Frigid	Udic
EMMET	537	93B, 94A, 94B, 94C, 95A, 96, 98	Er, J	Alfisols	Udalfs	Hapludalfs	Inceptic Hapludalfs	Coarse-loamy	Mixed	Active			Frigid	Udic
ENSLEY	95	141, 142, 93B, 94A, 94B, 94C, 95A, 96, 98	G, J	Inceptisols	Aquepts	Endoaquepts	Aeric Endoaquepts	Coarse-loamy	Mixed	Active	Nonacid		Frigid	Aquic
ETTRICK	318	105, 91	Am, J	Mollisols	Aquolls	Endoaquolls	Fluvaquentic Endoaquolls	Fine-silty	Mixed	Superactive			Mesic	Aquic
FABIUS	36	95A, 95B, 98, 99	Bm	Mollisols	Udolls	Argiudolls	Aquic Argiudolls	Fine-loamy over sandy or sandy-skeletal	Mixed	Semiactive			Mesic	Udic
FAIRCHILD	237	90, 91	Dr, J	Spodosols	Aquods	Epiaquods	Ultic Epiaquods	Sandy over loamy	Siliceous	Semiactive			Frigid	Aquic

(continued)

(continued)

Soil series	Area (km^2)	MLRAs	Soil regions	Order	Suborder	Great group	Subgroup	Particle-size class	Mineral class	CEC Activity class	Reaction class	Other	Soil-temperature class	Soil-moisture class
FAIRPORT	18	90, 93, 94A, 95A	E, G	Alfisols	Udalfs	Hapludalfs	Glossic Hapludalfs	Fine-loamy	Mixed	Active			Mesic	Udic
FALLCREEK	117	90	Dr, F	Alfisols	Udalfs	Glossudalfs	Aquic Glossudalfs	Coarse-loamy	Mixed	Superactive			Frigid	Udic
FARRINGTON	17	105	Cm, Dr	Mollisols	Udolls	Hapludolls	Aquic Hapludolls	Sandy	Mixed				Mesic	Udic
FAYETTE	2234	104, 105, 108, 109, 114, 115, 95B	A, Dr	Alfisols	Udalfs	Hapludalfs	Typic Hapludalfs	Fine-silty	Mixed	Superactive			Mesic	Udic
FENANDER	9		F, G, J	Alfisols	Aqualfs	Epiaqualfs	Udollic Epiaqualfs	Coarse-loamy	Mixed	Superactive			Frigid	Aquic
FENCE	79	92, 93, 94A, 95A	G, H	Spodosols	Orthods	Haplorthods	Alfic Oxyaquic Haplorthods	Coarse-silty	Mixed	Superactive			Frigid	Udic
FENWOOD	328	90, 91	Fr	Alfisols	Udalfs	Glossudalfs	Haplic Glossudalfs	Fine-loamy	Mixed	Superactive			Frigid	Udic
FESTINA	14	105, 108, 109	Am, J	Alfisols	Udalfs	Hapludalfs	Mollic Hapludalfs	Fine-silty	Mixed	Superactive			Mesic	Udic
FINCHFORD	121	104, 105, 108C, 115C	Dr	Mollisols	Udolls	Hapludolls	Entic Hapludolls	Sandy	Mixed				Mesic	Udic
FISK	19	89, 95A, 95B	C, E	Inceptisols	Udepts	Eutrudepts	Aquic Dystric Eutrudepts	Sandy over loamy	Mixed	Superactive			Mesic	Udic
FIVEPOINTS	183		A	Alfisols	Udalfs	Hapludalfs	Typic Hapludalfs	Clayey over loamy-skeletal	Mixed	Superactive			Mesic	Udic
FLAGG	44	105, 108B, 95B	B	Alfisols	Udalfs	Hapludalfs	Typic Hapludalfs	Fine-silty	Mixed	Superactive			Mesic	Udic
FLAMBEAU	217	90	F	Alfisols	Udalfs	Glossudalfs	Oxyaquic Glossudalfs	Fine-loamy	Mixed	Superactive			Frigid	Udic
FLINK	48	92, 93, 94A	H, I, J	Spodosols	Aquods	Epiaquods	Typic Epiaquods	Sandy	Mixed				Frigid	Aquic
FLOYD	19	104, 105, 90B	F	Mollisols	Udolls	Hapludolls	Aquic Pachic Hapludolls	Fine-loamy	Mixed	Superactive			Mesic	Udic
FORADA	5	102A, 56, 57, 88, 90, 91, 94A	G, J	Mollisols	Aquolls	Endoaquolls	Typic Endoaquolls	Coarse-loamy	Mixed	Superactive			Frigid	Aquic
FORDUM	595	102A, 57, 88, 90, 91, 93, 94A, 95A	J	Entisols	Aquents	Fluvaquents	Mollic Fluvaquents	Coarse-loamy	Mixed	Superactive	Nonacid		Frigid	Aquic
FORKHORN	60	105	Dr	Alfisols	Udalfs	Hapludalfs	Mollic Hapludalfs	Coarse-loamy	Mixed	Active			Mesic	Udic

(continued)

(continued)

Soil series	Area (km^2)	MLRAs	Soil regions	Order	Suborder	Great group	Subgroup	Particle-size class	Mineral class	CEC Activity class	Reaction class	Other	Soil-temperature class	Soil-moisture class
FOX	739	108A, 108B, 110, 111A, 111B, 111C, 111D, 114A, 114B, 95B, 97, 98, 99	B, I	Alfisols	Udalfs	Hapludalfs	Typic Hapludalfs	Fine-loamy over sandy or sandy-skeletal	Mixed	Superactive			Mesic	Udic
FRECHETTE	73	90, 93, 94A	G	Alfisols	Udalfs	Glossudalfs	Typic Glossudalfs	Coarse-loamy	Mixed	Active			Frigid	Udic
FREEON	2144	90, 91	F, G	Alfisols	Udalfs	Glossudalfs	Oxyaquic Glossudalfs	Coarse-loamy	Mixed	Superactive			Frigid	Udic
FREMSTADT	63	90	H	Alfisols	Udalfs	Hapludalfs	Arenic Hapludalfs	Sandy	Mixed				Frigid	Udic
FREYA	81	91	H	Mollisols	Udolls	Argiudolls	Aquic Argiudolls	Sandy over clayey	Mixed OVER smectitic				Frigid	Udic
FRIENDSHIP	530	57, 88, 90, 91, 94A	C, H	Entisols	Psamments	Udipsamments	Typic Udipsamments		Mixed				Frigid	Udic
FRIESLAND	37	108A, 95B	Bm	Mollisols	Udolls	Argiudolls	Typic Argiudolls	Fine-loamy	Mixed	Superactive			Mesic	Udic
FROGCREEK	106	90	G	Alfisols	Udalfs	Glossudalfs	Oxyaquic Glossudalfs	Coarse-loamy	Mixed	Superactive			Frigid	Udic
GAASTRA	8	90, 93, 94A	G, H, J	Spodosols	Aquods	Endoaquods	Argic Endoaquods	Coarse-loamy	Mixed	Active			Frigid	Aquic
GALE	554	104, 105, 108, 90, 91	Dr	Alfisols	Udalfs	Hapludalfs	Typic Hapludalfs	Fine-silty over sandy or sandy-skeletal	Mixed	Superactive			Mesic	Udic
GAPHILL	67	105	A, Dr	Alfisols	Udalfs	Hapludalfs	Typic Hapludalfs	Coarse-loamy	Siliceous	Active			Mesic	Udic
GARDENVALE	10	105	Dr	Alfisols	Udalfs	Hapludalfs	Mollic Hapludalfs	Fine-loamy over sandy or sandy-skeletal	Mixed	Superactive			Mesic	Udic
GARNE	6	105	Cm	Mollisols	Udolls	Hapludolls	Typic Hapludolls	Sandy over loamy	Mixed	Active			Mesic	Udic
GARWIN	0	104, 105, 108C	Am, J	Mollisols	Aquolls	Endoaquolls	Typic Endoaquolls	Fine-silty	Mixed	Superactive			Mesic	Aquic

(continued)

(continued)

Soil series	Area (km^2)	MLRAs	Soil regions	Order	Suborder	Great group	Subgroup	Particle-size class	Mineral class	CEC Activity class	Reaction class	Other	Soil-temperature class	Soil-moisture class
GASTROW	13	93	G, J	Spodosols	Aquods	Endoaquods	Argic Endoaquods	Coarse-loamy	Mixed	Active			Frigid	Aquic
GAY	18		G	Inceptisols	Aquepts	Endoaquepts	Aeric Endoaquepts	Coarse-loamy	Mixed	Active	Nonacid		Frigid	Aquic
GICHIGAMI	53	92	G	Alfisols	Udalfs	Glossudalfs	Oxyaquic Glossudalfs	Fine-silty	Mixed	Superactive			Frigid	Udic
GIESE	28	88, 90, 93	G, J	Inceptisols	Aquepts	Humaquepts	Typic Humaquepts	Coarse-loamy	Isotic		Nonacid		Frigid	Aquic
GILFORD	53	108A, 108B, 110, 111B, 111C, 115C, 95B, 97, 98, 99	Bm, J	Mollisols	Aquolls	Endoaquolls	Typic Endoaquolls	Coarse-loamy	Mixed	Superactive			Mesic	Aquic
GILLINGHAM	2		A	Alfisols	Udalfs	Hapludalfs	Arenic Hapludalfs	Loamy	Mixed	Superactive			Mesic	Udic
GLENDENNING	19	90	G	Alfisols	Udalfs	Glossudalfs	Aquic Glossudalfs	Coarse-loamy	Mixed	Superactive			Frigid	Udic
GLENDORA	41	105, 111B, 89, 96, 97, 98, 99	J	Entisols	Aquents	Psammaquents	Mollic Psammaquents		Mixed				Mesic	Aquic
GLENFLORA	17		F, J	Alfisols	Aqualfs	Glossaqualfs	Mollic Glossaqualfs	Fine-silty	Mixed	Superactive			Frigid	Aquic
GLIDDEN	64		G	Spodosols	Orthods	Haplorthods	Alfic Haplorthods	Coarse-loamy	Mixed	Superactive			Frigid	Udic
GOGEBIC	920	92, 93	G, I	Spodosols	Orthods	Fragiorthods	Alfic Oxyaquic Fragiorthods	Coarse-loamy	Isotic				Frigid	Udic
GOODMAN	262	90, 93, 94A	F, G	Spodosols	Orthods	Haplorthods	Alfic Haplorthods	Coarse-loamy	Mixed	Superactive			Frigid	Udic
GOODWIT	142	90, 93, 94A	G	Spodosols	Orthods	Haplorthods	Alfic Oxyaquic Haplorthods	Coarse-loamy	Mixed	Superactive			Frigid	Udic
GOSIL	80	105	Dr	Entisols	Psamments	Quartzipsamments	Typic Quartzipsamments						Mesic	Udic
GOTHAM	403	103, 104, 105, 90, 91	Cm	Alfisols	Udalfs	Hapludalfs	Psammentic Hapludalfs		Mixed				Mesic	Udic
GRANBY	231	101, 103, 105, 110, 111B, 111C, 141, 142, 144A, 91B, 95A, 95B, 97, 98, 99	Cm, E, J	Mollisols	Aquolls	Endoaquolls	Typic Endoaquolls	Sandy	Mixed				Mesic	Aquic
GRAYCALM	380	57, 88, 90A, 90B, 93A, 94A, 94C, 96, 98	H	Entisols	Psamments	Udipsamments	Lamellic Udipsamments		Isotic			Coated	Frigid	Udic

(continued)

(continued)

Soil series	Area (km²)	MLRAs	Soil regions	Order	Suborder	Great group	Subgroup	Particle-size class	Mineral class	CEC Activity class	Reaction class	Other	Soil-temperature class	Soil-moisture class
GRAYLING	587	93A, 93B, 94A, 94B, 94C, 95A, 98	H	Entisols	Psamments	Udipsamments	Typic Udipsamments		Isotic				Frigid	Udic
GRAYS	54	103, 110, 90B, 91A, 95A, 95B	B	Alfisols	Udalfs	Hapludalfs	Mollic Oxyaquic Hapludalfs	Fine-silty	Mixed	Superactive			Mesic	Udic
GREENRIDGE	36	105	A, Dr	Alfisols	Udalfs	Hapludalfs	Typic Hapludalfs	Fine-silty	Mixed	Superactive			Mesic	Udic
GREENWOOD	315	143, 144A, 144B, 57, 88, 90, 91, 92, 93, 94A, 94B, 96	J	Histosols	Hemists	Haplohemists	Typic Haplohemists				Dysic		Frigid	Aquic
GRELLTON	73	95B	B	Alfisols	Udalfs	Hapludalfs	Typic Hapludalfs	Fine-loamy	Mixed	Superactive			Mesic	Udic
GRETTUM	171	90, 94A	H	Entisols	Psamments	Udipsamments	Lamellic Udipsamments		Mixed				Frigid	Udic
GRISWOLD	190	105, 108A, 108B, 110, 111B, 111C, 95A, 95B, 98	Bm	Mollisols	Udolls	Argiudolls	Typic Argiudolls	Fine-loamy	Mixed	Superactive			Mesic	Udic
GUENTHER	19	90	C, Dr	Spodosols	Orthods	Haplorthods	Alfic Oxyaquic Haplorthods	Sandy over loamy	Mixed	Superactive			Frigid	Udic
GULL POINT	1	92, 93B	G, J	Mollisols	Aquolls	Argiaquolls	Typic Argiaquolls	Fine-loamy	Mixed	Superactive			Frigid	Aquic
HALDER	3	90, 91	G	Alfisols	Udalfs	Glossudalfs	Aquic Glossudalfs	Fine-loamy over sandy or sandy-skeletal	Mixed	Superactive			Frigid	Udic
HATLEY	90	90, 93	F	Alfisols	Udalfs	Glossudalfs	Aquic Glossudalfs	Coarse-loamy	Mixed	Superactive			Frigid	Udic
HAUGEN	322	90	G	Alfisols	Udalfs	Paleudalfs	Oxyaquic Paleudalfs	Coarse-loamy	Mixed	Superactive			Frigid	Udic
HAUSTRUP	13		G	Inceptisols	Udepts	Dystrudepts	Humic Lithic Dystrudepts	Loamy	Isotic				Frigid	Udic
HAYFIELD	30	103, 104, 105, 95B	B	Alfisols	Udalfs	Hapludalfs	Aquollic Hapludalfs	Fine-loamy over sandy or sandy-skeletal	Mixed	Superactive			Mesic	Udic

(continued)

(continued)

Soil series	Area (km^2)	MLRAs	Soil regions	Order	Suborder	Great group	Subgroup	Particle-size class	Mineral class	CEC Activity class	Reaction class	Other	Soil-temperature class	Soil-moisture class
HAYRIVER	208	90	Dr	Alfisols	Udalfs	Hapludalfs	Typic Hapludalfs	Coarse-loamy	Mixed	Active			Frigid	Udic
HEBRON	77	110, 95B	B, J	Alfisols	Udalfs	Hapludalfs	Oxyaquic Hapludalfs	Fine-loamy	Mixed	Superactive		Shallow	Mesic	Udic
HENNEPIN	4	108A, 108B, 110, 111A, 111B, 111C, 111D, 111E, 114B, 115C, 95B	B	Inceptisols	Udepts	Eutrudepts	Typic Eutrudepts	Fine-loamy	Mixed	Active			Mesic	Udic
HERBSTER	137		I, J	Alfisols	Aqualfs	Glossaqualfs	Aeric Glossaqualfs	Fine	Mixed	Active			Frigid	Aquic
HERSEY	57	104, 105	Dr	Alfisols	Udalfs	Hapludalfs	Mollic Hapludalfs	Fine-silty	Mixed	Superactive			Mesic	Udic
HESCH	20	103, 105, 108A, 110, 90	Dr	Mollisols	Udolls	Argiudolls	Typic Argiudolls	Coarse-loamy	Mixed	Active			Mesic	Udic
HIBBING	0	90, 91, 92, 93, 94A, 95A	I	Alfisols	Udalfs	Glossudalfs	Oxyaquic Glossudalfs	Fine	Mixed	Active			Frigid	Udic
HILES	116	90	Dr, F	Alfisols	Udalfs	Glossudalfs	Oxyaquic Glossudalfs	Fine-loamy	Mixed	Superactive			Frigid	Udic
HIXTON	566	104, 105, 108, 90, 98	A, Dr	Alfisols	Udalfs	Hapludalfs	Typic Hapludalfs	Fine-loamy over sandy or sandy-skeletal	Mixed	Superactive			Mesic	Udic
HOCHHEIM	979	95A, 95B	Bm	Mollisols	Udolls	Argiudolls	Typic Argiudolls	Fine-loamy	Mixed	Active			Mesic	Udic
HOOP	6	105	Dr	Mollisols	Udolls	Argiudolls	Aquic Argiudolls	Coarse-loamy	Siliceous	Active			Mesic	Udic
HOOPESTON	48	103, 104, 105, 108A, 108B, 108C, 109, 115C, 95B, 97	Dr	Mollisols	Udolls	Hapludolls	Aquic Hapludolls	Coarse-loamy	Mixed	Superactive			Mesic	Udic
HORTONVILLE	1146	95A	E, I	Alfisols	Udalfs	Glossudalfs	Haplic Glossudalfs	Fine-loamy	Mixed	Active			Mesic	Udic
HOUGHTON	1482		J	Histosols	Saprists	Haplosaprists	Typic Haplosaprists				Euic		Mesic	Aquic
HUBBARD	27	102A, 57, 90A, 90B, 91A, 91B	Cm	Mollisols	Udolls	Hapludolls	Entic Hapludolls	Sandy	Mixed				Frigid	Udic

(continued)

(continued)

Soil series	Area (km^2)	MLRAs	Soil regions	Order	Suborder	Great group	Subgroup	Particle-size class	Mineral class	CEC Activity class	Reaction class	Other	Soil-temperature class	Soil-moisture class
HUMBIRD	242	90, 91	Dr	Spodosols	Orthods	Haplorthods	Oxyaquic Ultic Haplorthods	Coarse-loamy over clayey	Mixed	Semiactive			Frigid	Udic
HUNTSVILLE	153	103, 104, 105, 108A, 108B, 108C, 108D, 113, 114B, 115A, 115B, 115C, 95B	J	Mollisols	Udolls	Hapludolls	Cumulic Hapludolls	Fine-silty	Mixed	Superactive			Mesic	Udic
IMPACT	132	105, 91	Dr	Inceptisols	Udepts	Dystrudepts	Humic Psammentic Dystrudepts		Siliceous				Mesic	Udic
INDUS	2	57, 88	I, J	Alfisols	Aqualfs	Epiaqualfs	Vertic Epiaqualfs	Very fine	Smectitic				Frigid	Aquic
INGALLS	8	90, 93, 94A, 94B	G, J	Spodosols	Aquods	Endoaquods	Typic Endoaquods	Sandy over loamy	Mixed	Active			Frigid	Aquic
IONIA	33	111B, 95B, 98	B	Alfisols	Udalfs	Hapludalfs	Oxyaquic Hapludalfs	Fine-loamy over sandy or sandy-skeletal	Mixed	Semiactive			Mesic	Udic
IOSCO	185	90A, 93A, 93B, 94A, 94B, 94C, 95A, 96, 98	G, H, J	Spodosols	Aquods	Endoaquods	Argic Endoaquods	Sandy over loamy	Mixed	Active			Frigid	Aquic
IRONRUN	175	90, 91	Dr, J	Spodosols	Aquods	Endoaquods	Typic Endoaquods	Sandy	Siliceous				Frigid	Aquic
ISHPEMING	87	92, 93, 94A, 94B	H	Spodosols	Orthods	Haplorthods	Entic Haplorthods	Sandy	Mixed				Frigid	Udic
JACKSON	86	104, 105, 108, 91	A, Dr	Alfisols	Udalfs	Hapludalfs	Typic Hapludalfs	Fine-silty	Mixed	Superactive			Mesic	Udic
JASPER	8	108A, 108B, 110, 111D, 115C, 95B	Bm	Mollisols	Udolls	Argiudolls	Typic Argiudolls	Fine-loamy	Mixed	Superactive			Mesic	Udic
JEWETT	127	90, 91	F	Alfisols	Udalfs	Hapludalfs	Typic Hapludalfs	Fine-loamy	Mixed	Superactive			Frigid	Udic
JOY	28	104, 105, 107B, 108B, 115C, 95B	Bm	Mollisols	Udolls	Hapludolls	Aquic Hapludolls	Fine-silty	Mixed	Superactive			Mesic	Udic
JUDA	9	95B	B	Alfisols	Udalfs	Hapludalfs	Mollic Oxyaquic Hapludalfs	Fine-silty	Mixed	Superactive			Mesic	Udic

(continued)

(continued)

Soil series	Area (km^2)	MLRAs	Soil regions	Order	Suborder	Great group	Subgroup	Particle-size class	Mineral class	CEC Activity class	Reaction class	Other	Soil-temperature class	Soil-moisture class
JUDSON	84	102C, 103, 104, 105, 106, 107A, 107B, 108C, 108D	A, J	Mollisols	Udolls	Hapludolls	Cumulic Hapludolls	Fine-silty	Mixed	Superactive			Mesic	Udic
JUNEAU	48	95B	B, J	Entisols	Fluvents	Udifluvents	Typic Udifluvents	Coarse-silty	Mixed	Superactive	Nonacid		Mesic	Udic
KALMARVILLE	74	103, 104, 105, 90, 91	Dr, J	Entisols	Aquents	Fluvaquents	Mollic Fluvaquents	Coarse-loamy	Mixed	Superactive	Nonacid		Mesic	Aquic
KANE	39	108A, 108B, 110, 111A, 111B, 95B, 97	Bm	Mollisols	Udolls	Argiudolls	Aquic Argiudolls	Fine-loamy over sandy or sandy-skeletal	Mixed	Superactive			Mesic	Udic
KARLIN	480	92, 93, 94A, 94B	G, H	Spodosols	Orthods	Haplorthods	Entic Haplorthods	Sandy	Mixed				Frigid	Udic
KARLSBORG	57	91	I	Alfisols	Udalfs	Hapludalfs	Arenic Oxyaquic Hapludalfs	Clayey	Smectitic				Frigid	Udic
KATO	12	103, 104, 105, 107	Bm, J	Mollisols	Aquolls	Endoaquolls	Typic Endoaquolls	Fine-silty over sandy or sandy-skeletal	Mixed	Superactive			Mesic	Aquic
KAUKAUNA	14	95A	E, I	Alfisols	Udalfs	Hapludalfs	Mollic Oxyaquic Hapludalfs	Clayey over loamy	Mixed	Active			Mesic	Udic
KEGONSA	72	104, 105, 95B	B	Alfisols	Udalfs	Hapludalfs	Mollic Hapludalfs	Fine-silty over sandy or sandy-skeletal	Mixed	Superactive			Mesic	Udic
KELLOGG	100	92, 94A, 96	I	Spodosols	Orthods	Haplorthods	Alfic Oxyaquic Haplorthods	Sandy over clayey	Mixed	Active			Frigid	Udic
KELTNER	7	105, 108B, 95B	Am	Mollisols	Udolls	Argiudolls	Oxyaquic Argiudolls	Fine-silty	Mixed	Superactive			Mesic	Udic
KENDALL	5	108A, 108B, 110, 111D, 114B, 115B, 115C, 95B	B, J	Alfisols	Aqualfs	Endoaqualfs	Aeric Endoaqualfs	Fine-silty	Mixed	Superactive			Mesic	Aquic

(continued)

(continued)

Soil series	Area (km²)	MLRAs	Soil regions	Order	Suborder	Great group	Subgroup	Particle-size class	Mineral class	CEC Activity class	Reaction class	Other	Soil-temperature class	Soil-moisture class
KENNAN	1139	90, 93, 95A	F, G	Alfisols	Udalfs	Glossudalfs	Haplic Glossudalfs	Coarse-loamy	Mixed	Superactive			Frigid	Udic
KEOWNS	0	89, 95A, 95B, 98	E, J	Inceptisols	Aquepts	Endoaquepts	Mollic Endoaquepts	Coarse-loamy	Mixed	Superactive	Nonacid		Mesic	Aquic
KERT	308	90	Dr	Alfisols	Udalfs	Glossudalfs	Aquic Glossudalfs	Fine-loamy	Mixed	Superactive			Frigid	Udic
KESHENA	17	95A	E, G	Alfisols	Udalfs	Glossudalfs	Oxyaquic Glossudalfs	Fine-loamy	Mixed	Active			Frigid	Udic
KEVILAR	164	105, 91	Dr	Alfisols	Udalfs	Hapludalfs	Mollic Hapludalfs	Coarse-loamy	Mixed	Active			Mesic	Udic
KEWAUNEE	2002	110, 89, 95A, 95B	E, I	Alfisols	Udalfs	Hapludalfs	Typic Hapludalfs	Fine	Mixed	Active			Mesic	Udic
KEWEENAW	1278	90, 92, 93	H	Spodosols	Orthods	Haplorthods	Alfic Haplorthods	Sandy	Mixed				Frigid	Udic
KEYESVILLE	6		B, Dr	Inceptisols	Udepts	Dystrudepts	Typic Dystrudepts	Loamy-skeletal	Mixed	Active			Mesic	Udic
KIBBIE	67	111B, 95A, 95B, 96, 97, 98, 99	E	Alfisols	Udalfs	Hapludalfs	Aquollic Hapludalfs	Fine-loamy	Mixed	Active			Mesic	Udic
KICKAPOO	71	105, 113, 115, 116A, 116B	D, J	Entisols	Fluvents	Udifluvents	Typic Udifluvents	Coarse-loamy	Mixed	Superactive	Nonacid		Mesic	Udic
KIDDER	644	108B, 110, 95A, 95B, 98	B	Alfisols	Udalfs	Hapludalfs	Typic Hapludalfs	Fine-loamy	Mixed	Active			Mesic	Udic
KINGSVILLE	70	139, 95A, 97, 98, 99	C, J	Entisols	Aquents	Psammaquents	Mollic Psammaquents		Mixed				Mesic	Aquic
KINROSS	102	90, 93, 94A, 94B	G, H, J	Spodosols	Aquods	Endoaquods	Typic Endoaquods	Sandy	Mixed				Frigid	Aquic
KIVA	41	93B, 94B, 94C, 95A, 96	E	Spodosols	Orthods	Haplorthods	Entic Haplorthods	Sandy	Mixed				Frigid	Udic
KNOWLES	76	95A, 95B	B	Alfisols	Udalfs	Hapludalfs	Typic Hapludalfs	Fine-silty	Mixed	Superactive			Mesic	Udic
KOLBERG	72	95A	Er, I	Alfisols	Udalfs	Glossudalfs	Haplic Glossudalfs	Fine	Mixed	Active			Frigid	Udic
KOMRO	14	105	Dr	Mollisols	Udolls	Hapludolls	Entic Hapludolls	Sandy	Mixed				Mesic	Udic
KOROBAGO	39	89, 95A	I	Inceptisols	Udepts	Eutrudepts	Aquic Eutrudepts	Coarse-loamy over clayey	Mixed	Active			Mesic	Udic
KOST	2	90, 91	H	Mollisols	Udolls	Hapludolls	Entic Hapludolls	Sandy	Mixed				Frigid	Udic
KRANSKI	103	91	C	Alfisols	Udalfs	Hapludalfs	Typic Hapludalfs	Sandy	Mixed				Mesic	Udic
LA FARGE	595	105	A, Dr	Alfisols	Udalfs	Hapludalfs	Typic Hapludalfs	Fine-silty	Mixed	Superactive			Mesic	Udic

(continued)

(continued)

Soil series	Area (km^2)	MLRAs	Soil regions	Order	Suborder	Great group	Subgroup	Particle-size class	Mineral class	CEC Activity class	Reaction class	Other	Soil-temperature class	Soil-moisture class
LABLATZ	1		G, J	Spodosols	Aquods	Epiaquods	Alfic Epiaquods	Coarse-loamy	Mixed	Active			Frigid	Aquic
LAMARTINE	264	95A, 95B	B, I	Alfisols	Udalfs	Hapludalfs	Aquollic Hapludalfs	Fine-silty	Mixed	Superactive			Mesic	Udic
LAMBEAU	8	105	A	Alfisols	Udalfs	Hapludalfs	Typic Hapludalfs	Fine-silty	Mixed	Superactive			Mesic	Udic
LAMONT	3	103, 104, 105, 108B, 108C, 114B, 115A, 115B, 115C, 95B	A	Alfisols	Udalfs	Hapludalfs	Typic Hapludalfs	Coarse-loamy	Mixed	Superactive			Mesic	Udic
LANGLADE	5	90	G	Alfisols	Udalfs	Glossudalfs	Haplic Glossudalfs	Coarse-loamy	Mixed	Superactive			Frigid	Udic
LAONA	32	93	G	Spodosols	Orthods	Fragiorthods	Alfic Fragiorthods	Coarse-loamy	Mixed	Superactive			Frigid	Udic
LAPEER	248	89, 95B, 98	B, I	Alfisols	Udalfs	Hapludalfs	Typic Hapludalfs	Coarse-loamy	Mixed	Semiactive			Mesic	Udic
LAPOIN	7		I	Spodosols	Orthods	Haplorthods	Alfic Oxyaquic Haplorthods	Fine	Mixed	Active			Frigid	Udic
LARA	2	91	H	Mollisols	Udolls	Argiudolls	Oxyaquic Argiudolls	Sandy over clayey	Mixed OVER smectitic				Frigid	Udic
LAWLER	20	104, 105, 108B, 108C	A, G	Mollisols	Udolls	Hapludolls	Aquic Hapludolls	Fine-loamy over sandy or sandy-skeletal	Mixed	Superactive			Mesic	Udic
LAWSON	74	104, 105, 108, 109, 110, 111, 114, 115, 91, 95B	Am	Mollisols	Udolls	Hapludolls	Aquic Cumulic Hapludolls	Fine-silty	Mixed	Superactive			Mesic	Udic
LENROOT	161	91	H	Entisols	Psamments	Udipsamments	Oxyaquic Udipsamments		Mixed				Frigid	Udic
LEOLA	47	91	C, G	Alfisols	Udalfs	Hapludalfs	Aquic Arenic Hapludalfs	Loamy	Mixed	Active			Mesic	Udic
LERCH	48		I, J	Inceptisols	Aquepts	Epiaquepts	Vertic Epiaquepts	Very fine	Mixed	Active	Nonacid		Frigid	Aquic
LEROY	98	95B	B	Alfisols	Udalfs	Hapludalfs	Typic Hapludalfs	Fine-loamy	Mixed	Active			Mesic	Udic

(continued)

(continued)

Soil series	Area (km^2)	MLRAs	Soil regions	Order	Suborder	Great group	Subgroup	Particle-size class	Mineral class	CEC Activity class	Reaction class	Other	Soil-temperature class	Soil-moisture class
LILAH	8		Dr	Alfisols	Udalfs	Hapludalfs	Psammentic Hapludalfs	Sandy	Mixed				Mesic	Udic
LINDQUIST	61	90A, 94A, 96	G	Spodosols	Orthods	Haplorthods	Lamellic Haplorthods	Sandy	Mixed				Frigid	Udic
LINDSTROM	50	104, 105, 90, 91	Am	Mollisols	Udolls	Hapludolls	Cumulic Hapludolls	Fine-silty	Mixed	Superactive			Mesic	Udic
LINO	14	90, 91	H	Entisols	Psamments	Udipsamments	Aquic Udipsamments		Mixed				Frigid	Udic
LOBO	6	57, 88, 90, 93, 95A	E, I, J	Histosols	Fibrists	Sphagnofibrists	Hemic Sphagnofibrists				Dysic		Frigid	Aquic
LOCKE	8	111B, 95B, 98	A	Alfisols	Udalfs	Hapludalfs	Aquollic Hapludalfs	Fine-loamy	Mixed	Semiactive			Mesic	Udic
LOMIRA	275	95A, 95B	B	Alfisols	Udalfs	Hapludalfs	Typic Hapludalfs	Fine-silty	Mixed	Superactive			Mesic	Udic
LONGRIE	150	93B, 94B, 94C, 95A	Er, I	Spodosols	Orthods	Haplorthods	Typic Haplorthods	Coarse-loamy	Mixed	Superactive			Frigid	Udic
LORENZO	60	108A, 108B, 111A, 91A, 95A, 95B, 97	Bm	Mollisols	Udolls	Argiudolls	Typic Argiudolls	Fine-loamy over sandy or sandy-skeletal	Mixed	Active			Mesic	Udic
LOWS	81	105, 90, 91	J	Inceptisols	Aquepts	Endoaquepts	Mollic Endoaquepts	Fine-loamy over sandy or sandy-skeletal	Mixed	Superactive	Nonacid		Frigid	Aquic
LOXLEY	1333		J	Histosols	Saprists	Haplosaprists	Typic Haplosaprists				Dysic		Frigid	Aquic
LOYAL	1157	90	F	Alfisols	Udalfs	Glossudalfs	Oxyaquic Glossudalfs	Fine-loamy	Mixed	Superactive			Frigid	Udic
LUDINGTON	170	90, 91	Dr	Spodosols	Orthods	Haplorthods	Oxyaquic Ultic Haplorthods	Sandy over loamy	Siliceous	Semiactive			Frigid	Udic
LUNDEEN	1		F	Inceptisols	Udepts	Dystrudepts	Humic Dystrudepts	Coarse-silty	Isotic				Frigid	Udic
LUPTON	1578	J	J	Histosols	Saprists	Haplosaprists	Typic Haplosaprists				Euic		Frigid	Aquic
LUTZKE	45	95A, 95B	B, I	Alfisols	Udalfs	Hapludalfs	Typic Hapludalfs	Loamy-skeletal	Mixed	Active			Mesic	Udic
MAGNOR	2852	90	F, G	Alfisols	Udalfs	Glossudalfs	Aquic Glossudalfs	Coarse-loamy	Mixed	Superactive			Frigid	Udic

(continued)

(continued)

Soil series	Area (km^2)	MLRAs	Soil regions	Order	Suborder	Great group	Subgroup	Particle-size class	Mineral class	CEC Activity class	Reaction class	Other	Soil-temperature class	Soil-moisture class
MAGROC	21	90	G	Alfisols	Udalfs	Glossudalfs	Aquic Glossudalfs	Coarse-loamy	Mixed	Superactive			Frigid	Udic
MAHALASVILLE	78	111A, 111B, 111C, 111D, 121, 95B	Bm, J	Mollisols	Aquolls	Argiaquolls	Typic Argiaquolls	Fine-silty	Mixed	Superactive			Mesic	Aquic
MAHTOMEDI	474	57, 88, 90, 91	H	Entisols	Psamments	Udipsamments	Typic Udipsamments		Mixed				Frigid	Udic
MAINCREEK	27	90	F	Alfisols	Udalfs	Glossudalfs	Aquic Glossudalfs	Coarse-loamy	Mixed	Superactive			Frigid	Udic
MAJIK	14	105	Dr	Entisols	Psamments	Quartzipsamments	Aquic Quartzipsamments						Mesic	Udic
MAKWA	74	91	J	Inceptisols	Aquepts	Humaquepts	Histic Humaquepts	Loamy-skeletal	Isotic		Nonacid	Coated	Frigid	Aquic
MANAWA	956	89, 95A, 95B	I	Alfisols	Udalfs	Hapludalfs	Aquollic Hapludalfs	Fine	Mixed	Active			Mesic	Udic
MANCELONA	146	93B, 94A, 94B, 94C, 95A, 96, 98	H	Spodosols	Orthods	Haplorthods	Alfic Haplorthods	Sandy	Mixed				Frigid	Udic
MANISTEE	56	92, 93B, 94A, 94B, 94C, 95A, 96, 98	I	Spodosols	Orthods	Haplorthods	Alfic Haplorthods	Sandy over clayey	Mixed	Active			Frigid	Udic
MANITOWISH	127	93, 94A	G, H	Spodosols	Orthods	Haplorthods	Oxyaquic Haplorthods	Sandy	Mixed				Frigid	Udic
MANN	6	90	F, J	Mollisols	Aquolls	Epiaquolls	Typic Epiaquolls	Fine-loamy	Mixed	Superactive			Frigid	Aquic
MAPLEHURST	67	90	F	Alfisols	Udalfs	Glossudalfs	Aquic Glossudalfs	Fine-silty	Mixed	Superactive			Frigid	Udic
MARATHON	91	90	Fr	Alfisols	Udalfs	Glossudalfs	Haplic Glossudalfs	Coarse-loamy	Mixed	Superactive			Frigid	Udic
MARCELLON	19	103, 91, 95B	Bm	Mollisols	Udolls	Argiudolls	Aquic Argiudolls	Fine-loamy	Mixed	Superactive			Mesic	Udic
MARKESAN	34	95B	Bm	Mollisols	Udolls	Argiudolls	Typic Argiudolls	Fine-loamy	Mixed	Active			Mesic	Udic
MARKEY	1449		J	Histosols	Saprists	Haplosaprists	Terric Haplosaprists	Sandy or sandy-skeletal	Mixed		Euic		Frigid	Aquic
MARKHAM	132	110, 111C, 111D, 95B, 97	B	Alfisols	Udalfs	Hapludalfs	Mollic Oxyaquic Hapludalfs	Fine	Illitic				Mesic	Udic
MARSHAN	124	103, 104, 105, 108, 115, 91, 95B	J	Mollisols	Aquolls	Endoaquolls	Typic Endoaquolls	Fine-loamy over sandy or sandy-skeletal	Mixed	Superactive			Mesic	Aquic

(continued)

(continued)

Soil series	Area (km^2)	MLRAs	Soil regions	Order	Suborder	Great group	Subgroup	Particle-size class	Mineral class	CEC Activity class	Reaction class	Other	Soil-temperature class	Soil-moisture class
MARSHFIELD	418	90	F, J	Alfisols	Aqualfs	Epiaqualfs	Mollic Epiaqualfs	Fine-loamy	Mixed	Superactive			Frigid	Aquic
MARTINTON	69	110, 95B, 97	Bm	Mollisols	Udolls	Argiudolls	Aquic Argiudolls	Fine	Illitic				Mesic	Udic
MATHERTON	15	110, 111B, 111C, 95A, 95B, 97, 98	B, I, J	Alfisols	Aqualfs	Endoaqualfs	Udollic Endoaqualfs	Fine-loamy over sandy or sandy-skeletal	Mixed	Superactive			Mesic	Aquic
MAUMEE	5	110, 111B, 111C, 94A, 95B, 97, 98	Bm, J	Mollisols	Aquolls	Endoaquolls	Typic Endoaquolls	Sandy	Mixed				Mesic	Aquic
MAYVILLE	163	95B	B, E	Alfisols	Udalfs	Hapludalfs	Oxyaquic Hapludalfs	Fine-silty	Mixed	Superactive			Mesic	Udic
MCHENRY	576	95B	B	Alfisols	Udalfs	Hapludalfs	Typic Hapludalfs	Fine-loamy	Mixed	Superactive			Mesic	Udic
MEADLAND	255	90	Fr	Alfisols	Udalfs	Glossudalfs	Aquic Glossudalfs	Fine-loamy	Mixed	Superactive			Frigid	Udic
MECAN	93	89, 95B	C	Alfisols	Udalfs	Hapludalfs	Typic Hapludalfs	Coarse-loamy	Mixed	Active			Mesic	Udic
MECOSTA	2	95A, 98	C	Entisols	Orthents	Udorthents	Typic Udorthents	Sandy-skeletal	Mixed				Mesic	Udic
MEDARY	10	105	A	Alfisols	Udalfs	Hapludalfs	Oxyaquic Hapludalfs	Fine	Mixed	Superactive			Mesic	Udic
MEEHAN	780	57, 88, 90, 91, 94A, 95A	C	Entisols	Psamments	Udipsamments	Aquic Udipsamments		Mixed				Frigid	Udic
MEENON	22	91	I	Alfisols	Udalfs	Hapludalfs	Aquic Arenic Hapludalfs	Clayey	Smectitic				Frigid	Udic
MENAHGA	2090	57, 88, 90, 91, 94A, 94B, 95A	H	Entisols	Psamments	Udipsamments	Typic Udipsamments		Mixed				Frigid	Udic
MENASHA	49	89, 95A, 95B	J	Mollisols	Aquolls	Epiaquolls	Typic Epiaquolls	Very fine	Mixed	Active			Mesic	Aquic
MENDOTA	131	95B	Bm	Mollisols	Udolls	Argiudolls	Typic Argiudolls	Fine-silty	Mixed	Superactive			Mesic	Udic
MENOMIN	2	105, 90	Dr	Alfisols	Udalfs	Hapludalfs	Mollic Hapludalfs	Fine-loamy over sandy or sandy-skeletal	Mixed	Superactive			Mesic	Udic

(continued)

(continued)

Soil series	Area (km^2)	MLRAs	Soil regions	Order	Suborder	Great group	Subgroup	Particle-size class	Mineral class	CEC Activity class	Reaction class	Other	Soil-temperature class	Soil-moisture class
MENOMINEE	346	90A, 93B, 94A, 94B, 94C, 95A, 96, 98	E, G, H	Spodosols	Orthods	Haplorthods	Alfic Haplorthods	Sandy over loamy	Mixed	Active			Frigid	Udic
MEQUITHY	33	90	G	Spodosols	Orthods	Haplorthods	Alfic Haplorthods	Coarse-loamy	Mixed	Superactive			Frigid	Udic
MEQUON	92	110, 95B	B, J	Alfisols	Aqualfs	Endoaqualfs	Udollic Endoaqualfs	Fine	Mixed	Superactive			Mesic	Aquic
MERIDIAN	213	104, 105, 89, 90, 91, 95B	A, C	Alfisols	Udalfs	Hapludalfs	Mollic Hapludalfs	Fine-loamy over sandy or sandy-skeletal	Mixed	Superactive			Mesic	Udic
MERIMOD	38	105	Dr	Alfisols	Udalfs	Hapludalfs	Mollic Hapludalfs	Fine-loamy over sandy or sandy-skeletal	Mixed	Superactive			Mesic	Udic
MERIT	57	105	Dr	Alfisols	Udalfs	Hapludalfs	Mollic Hapludalfs	Fine-loamy over sandy or sandy-skeletal	Mixed	Superactive			Mesic	Udic
MERRILLAN	313	90, 91	Dr, J	Spodosols	Aquods	Epiaquods	Ultic Epiaquods	Coarse-loamy over clayey	Mixed	Semiactive			Frigid	Aquic
METEA	50	108A, 110, 111B, 111C, 95B, 97, 98, 99	B	Alfisols	Udalfs	Hapludalfs	Arenic Hapludalfs	Loamy	Mixed	Active			Mesic	Udic
METONGA	63	93, 94A	G	Spodosols	Orthods	Haplorthods	Entic Haplorthods	Coarse-loamy	Mixed	Superactive			Frigid	Udic
MIAMI	480	102B, 108A, 110, 111A, 111D, 114A, 115C, 95B, 97, 98	B	Alfisols	Udalfs	Hapludalfs	Oxyaquic Hapludalfs	Fine-loamy	Mixed	Active			Mesic	Udic
MICHIGAMME	108	92, 93, 94A	G	Spodosols	Orthods	Haplorthods	Fragic Haplorthods	Coarse-loamy	Mixed	Superactive			Frigid	Udic
MICKLE	23		Am	Mollisols	Udolls	Argiudolls	Typic Argiudolls	Fine-silty	Mixed	Superactive			Mesic	Udic

(continued)

(continued)

Soil series	Area (km^2)	MLRAs	Soil regions	Order	Suborder	Great group	Subgroup	Particle-size class	Mineral class	CEC Activity class	Reaction class	Other	Soil-temperature class	Soil-moisture class
MIFFLIN	13	105	A	Alfisols	Udalfs	Hapludalfs	Typic Hapludalfs	Fine-loamy	Mixed	Superactive			Mesic	Udic
MILACA	12	90, 91	D, F, G	Alfisols	Udalfs	Glossudalfs	Oxyaquic Glossudalfs	Coarse-loamy	Mixed	Superactive			Frigid	Udic
MILFORD	45	108A, 110, 111A, 111B, 111D, 95B, 97, 98, 99	Bm, J	Mollisols	Aquolls	Endoaquolls	Typic Endoaquolls	Fine	Mixed	Superactive			Mesic	Aquic
MILITARY	28	105, 95B	B	Alfisols	Udalfs	Hapludalfs	Typic Hapludalfs	Fine-loamy	Mixed	Active			Mesic	Udic
MILLADORE	137	90	F	Alfisols	Udalfs	Glossudalfs	Aquic Glossudalfs	Fine-loamy	Mixed	Superactive			Frigid	Udic
MILLINGTON	6	102A, 102B, 103, 108A, 108B, 110, 95B, 97	Bm, J	Mollisols	Aquolls	Endoaquolls	Cumulic Endoaquolls	Fine-loamy	Mixed	Superactive	Calcareous		Mesic	Aquic
MILTON	2	111A, 111B, 111D, 111E, 114A, 95B, 99	B	Alfisols	Udalfs	Hapludalfs	Typic Hapludalfs	Fine	Mixed	Active			Mesic	Udic
MINDORO	6	105	Dr	Inceptisols	Udepts	Dystrudepts	Humic Psammentic Dystrudepts		Siliceous				Mesic	Udic
MINOCQUA	562	90, 93, 94A	J	Inceptisols	Aquepts	Endoaquepts	Typic Endoaquepts	Coarse-loamy over sandy or sandy-skeletal	Mixed	Superactive	Nonacid		Frigid	Aquic
MISKOAKI	15	92	I	Alfisols	Udalfs	Glossudalfs	Vertic Glossudalfs	Very fine	Mixed	Active			Frigid	Udic
MOBERG	26	90	Fr, H	Inceptisols	Udepts	Dystrudepts	Typic Dystrudepts	Sandy-skeletal	Mixed				Frigid	Udic
MONICO	90	90, 93	G, J	Spodosols	Aquods	Endoaquods	Typic Endoaquods	Coarse-loamy	Mixed	Superactive			Frigid	Aquic
MONTELLO	17	89, 95B	Bm, J	Mollisols	Udolls	Argiudolls	Oxyaquic Argiudolls	Fine	Mixed	Superactive			Mesic	Udic
MONTGOMERY	60	110, 111A, 111B, 111C, 113, 114A, 114B, 115A, 120A, 122, 139, 95B, 99	Bm, J	Mollisols	Aquolls	Endoaquolls	Vertic Endoaquolls	Fine	Mixed	Active			Mesic	Aquic
MOODIG	40	90, 93	G, J	Spodosols	Aquods	Epiaquods	Alfic Epiaquods	Coarse-loamy	Mixed	Superactive			Frigid	Aquic

(continued)

(continued)

Soil series	Area (km^2)	MLRAs	Soil regions	Order	Suborder	Great group	Subgroup	Particle-size class	Mineral class	CEC Activity class	Reaction class	Other	Soil-temperature class	Soil-moisture class
MOPPET	109	90	F	Inceptisols	Udepts	Dystrudepts	Oxyaquic Dystrudepts	Coarse-loamy	Mixed	Superactive			Frigid	Udic
MOQUAH	91	90, 92, 93, 95A	G, J	Entisols	Fluvents	Udifluvents	Typic Udifluvents	Coarse-loamy	Mixed	Superactive	Nonacid		Frigid	Udic
MORA	7	90, 91	G	Alfisols	Udalfs	Glossudalfs	Aquic Glossudalfs	Coarse-loamy	Mixed	Superactive			Frigid	Udic
MORGANLAKE	46	93, 94A	H	Spodosols	Orthods	Haplorthods	Alfic Oxyaquic Haplorthods	Sandy over loamy	Mixed	Active			Frigid	Udic
MORLEY	416	108A, 110, 111A, 111B, 111C, 111D, 111E, 115C, 95A, 95B, 97, 98, 99	B	Alfisols	Udalfs	Hapludalfs	Oxyaquic Hapludalfs	Fine	Illitic				Mesic	Udic
MOROCCO	67	105, 110, 111B, 111C, 95A, 97, 98	C, J	Entisols	Psamments	Udipsamments	Aquic Udipsamments		Mixed				Mesic	Udic
MOSEL	75	89, 95A, 95B	B, I	Alfisols	Udalfs	Hapludalfs	Aquollic Hapludalfs	Fine-loamy	Mixed	Superactive			Mesic	Udic
MOSHAWQUIT	3	93, 94A	G	Alfisols	Udalfs	Glossudalfs	Arenic Glossudalfs	Loamy	Mixed	Active			Frigid	Udic
MOSINEE	172	90, 91	Fr	Inceptisols	Udepts	Dystrudepts	Typic Dystrudepts	Loamy-skeletal	Mixed	Superactive			Frigid	Udic
MOUNDVILLE	53	105, 89, 90A, 95B	B	Alfisols	Udalfs	Hapludalfs	Oxyaquic Hapludalfs	Sandy	Mixed				Mesic	Udic
MT. CARROLL	49	104, 105, 108B, 115C, 95B	A, B, Dr	Alfisols	Udalfs	Hapludalfs	Mollic Hapludalfs	Fine-silty	Mixed	Superactive			Mesic	Udic
MUDLAKE	108	93	G, J	Spodosols	Aquods	Epiaquods	Alfic Epiaquods	Coarse-loamy	Mixed	Superactive			Frigid	Aquic
MUNDELEIN	79	110, 95A, 95B, 97	Bm, I	Mollisols	Udolls	Argiudolls	Aquic Argiudolls	Fine-silty	Mixed	Superactive			Mesic	Udic
MUNUSCONG	15	142, 92, 93B, 94A, 96, 98	G, J	Inceptisols	Aquepts	Epiaquepts	Mollic Epiaquepts	Coarse-loamy over clayey	Mixed	Active	Nonacid		Frigid	Aquic
MUSCATINE	63	104, 105, 108C, 115C, 95B	Am	Mollisols	Udolls	Hapludolls	Aquic Hapludolls	Fine-silty	Mixed	Superactive			Mesic	Udic
MUSCODA	2		Dr	Alfisols	Udalfs	Hapludalfs	Lamellic Hapludalfs	Coarse-loamy	Mixed	Superactive			Mesic	Udic
MUSKEGO	7	102B, 103, 104, 105, 108A, 108B, 110, 111B, 111C, 111E, 114A, 144A, 95B, 97, 98	B, J	Histosols	Saprists	Haplosaprists	Limnic Haplosaprists		Coprogenous		Euic		Mesic	Aquic

(continued)

(continued)

Soil series	Area (km^2)	MLRAs	Soil regions	Order	Suborder	Great group	Subgroup	Particle-size class	Mineral class	CEC Activity class	Reaction class	Other	Soil-temperature class	Soil-moisture class
MUSSEY	26	95A, 95B, 98	E, I, J	Mollisols	Aquolls	Argiaquolls	Typic Argiaquolls	Fine-loamy over sandy or sandy-skeletal	Mixed	Semiactive			Mesic	Aquic
MYLREA	98	90	Fr	Inceptisols	Udepts	Dystrudepts	Aquic Dystrudepts	Coarse-loamy	Mixed	Superactive			Frigid	Udic
MYRTLE	17	105, 108A, 108B, 95B	A	Alfisols	Udalfs	Hapludalfs	Mollic Hapludalfs	Fine-silty	Mixed	Superactive			Mesic	Udic
NADEAU	44	93, 94A, 95A, 96	G, H	Alfisols	Udalfs	Hapludalfs	Typic Hapludalfs	Coarse-loamy	Mixed	Active			Frigid	Udic
NAHMA	3	93B, 94B, 95A	G, J	Inceptisols	Aquepts	Humaquepts	Histic Humaquepts	Coarse-loamy	Mixed	Active	Nonacid		Frigid	Aquic
NAMUR	97	94A, 95A	Er	Mollisols	Udolls	Hapludolls	Lithic Hapludolls	Loamy	Mixed	Active			Frigid	Udic
NAVAN	81	110, 95A, 95B	B, I, J	Mollisols	Aquolls	Argiaquolls	Typic Argiaquolls	Fine-loamy	Mixed	Superactive			Mesic	Aquic
NEBAGO	44	95A	E, G, I	Inceptisols	Udepts	Eutrudepts	Aquic Dystric Eutrudepts	Sandy over clayey	Mixed	Active			Mesic	Udic
NECONISH	6	94A	H	Spodosols	Orthods	Haplorthods	Oxyaquic Haplorthods	Sandy	Isotic				Frigid	Udic
NEDA	14	95B	B	Alfisols	Udalfs	Hapludalfs	Mollic Oxyaquic Hapludalfs	Fine-loamy	Mixed	Active			Mesic	Udic
NEENAH	43	95A, 95B	I	Alfisols	Udalfs	Hapludalfs	Aquollic Hapludalfs	Very fine	Mixed	Active			Mesic	Udic
NENNO	44	95B	Bm	Mollisols	Udolls	Argiudolls	Aquic Argiudolls	Fine-loamy	Mixed	Active			Mesic	Udic
NEOPIT	26	90	G	Alfisols	Udalfs	Glossudalfs	Oxyaquic Glossudalfs	Coarse-loamy	Mixed	Superactive			Frigid	Udic
NESTER	0	94A, 94B, 95A, 96, 98	E	Alfisols	Udalfs	Glossudalfs	Oxyaquic Glossudalfs	Fine	Mixed	Semiactive			Frigid	Udic
NEWGLARUS	430	105, 111, 115	A	Alfisols	Udalfs	Hapludalfs	Typic Hapludalfs	Fine-silty over clayey	Mixed	Superactive			Mesic	Udic
NEWLANG	25	105	Dr, J	Entisols	Aquents	Psammaquents	Humaqueptic Psammaquents		Siliceous				Mesic	Aquic
NEWOOD	847	90	G	Spodosols	Orthods	Haplorthods	Alfic Oxyaquic Haplorthods	Coarse-loamy	Isotic				Frigid	Udic
NEWOT	294	90	F, G	Spodosols	Orthods	Haplorthods	Alfic Haplorthods	Coarse-loamy	Mixed	Superactive			Frigid	Udic
NEWSON	840	105, 88, 90, 91, 92, 93, 94A	J	Entisols	Aquents	Psammaquents	Humaqueptic Psammaquents		Mixed				Frigid	Aquic

(continued)

(continued)

Soil series	Area (km^2)	MLRAs	Soil regions	Order	Suborder	Great group	Subgroup	Particle-size class	Mineral class	CEC Activity class	Reaction class	Other	Soil-temperature class	Soil-moisture class
NICHOLS	78	95A	E, I	Inceptisols	Udepts	Eutrudepts	Oxyaquic Eutrudepts	Coarse-silty	Mixed	Superactive			Mesic	Udic
NICKIN	60	90	Am, Dr	Mollisols	Udolls	Argiudolls	Typic Argiudolls	Fine-loamy over sandy or sandy-skeletal	Mixed over siliceous	Active			Frigid	Udic
NOKASIPPI	5	90, 91	G, J	Mollisols	Aquolls	Epiaquolls	Typic Epiaquolls	Coarse-loamy	Mixed	Superactive			Frigid	Aquic
NORDEN	1077	105	Dr	Alfisols	Udalfs	Hapludalfs	Typic Hapludalfs	Fine-loamy	Mixed	Superactive			Mesic	Udic
NORGO	19	90	Dr	Alfisols	Udalfs	Hapludalfs	Lithic Hapludalfs	Loamy	Mixed	Active			Frigid	Udic
NORTHBEND	44	105, 91	Dr	Inceptisols	Udepts	Dystrudepts	Fluvaquentic Dystrudepts	Coarse-loamy over sandy or sandy-skeletal	Mixed	Active			Mesic	Udic
NORTHFIELD	149	105, 108A	Dr	Alfisols	Udalfs	Hapludalfs	Lithic Hapludalfs	Loamy	Mixed	Active			Mesic	Udic
NORTHMOUND	5	90	Dr	Alfisols	Udalfs	Glossudalfs	Haplic Glossudalfs	Loamy-skeletal	Mixed	Superactive			Frigid	Udic
NOSEUM	7	93, 94A	H	Spodosols	Orthods	Haplorthods	Oxyaquic Haplorthods	Sandy	Isotic				Frigid	Udic
NUXMARUHANIXETE	3	105	Am	Mollisols	Udolls	Argiudolls	Typic Argiudolls	Fine-loamy	Mixed	Superactive			Mesic	Udic
NYMORE	33	57, 88, 90, 91, 94A	C, H	Entisols	Psamments	Udipsamments	Typic Udipsamments		Mixed				Frigid	Udic
OAKVILLE	80	101, 108B, 110, 111C, 115B, 115C, 139, 142, 143, 144A, 95A, 95B, 97, 98, 99	C, E, H, I	Entisols	Psamments	Udipsamments	Typic Udipsamments		Mixed				Mesic	Udic
OCKLEY	19	108A, 111A, 111B, 111D, 111E, 114A, 95B, 97, 98	A	Alfisols	Udalfs	Hapludalfs	Typic Hapludalfs	Fine-loamy	Mixed	Active			Mesic	Udic
OCONTO	1	95A	G	Alfisols	Udalfs	Glossudalfs	Haplic Glossudalfs	Coarse-loamy	Mixed	Active			Frigid	Udic
ODANAH	114	92	I	Alfisols	Udalfs	Glossudalfs	Haplic Glossudalfs	Fine	Mixed	Active			Frigid	Udic

(continued)

(continued)

Soil series	Area (km^2)	MLRAs	Soil regions	Order	Suborder	Great group	Subgroup	Particle-size class	Mineral class	CEC Activity class	Reaction class	Other	Soil-temperature class	Soil-moisture class
OESTERLE	397	90, 91	F, G	Alfisols	Udalfs	Glossudalfs	Aquic Glossudalfs	Coarse-loamy	Mixed	Superactive			Frigid	Udic
OGLE	47	108A, 108B, 115C	Bm	Mollisols	Udolls	Argiudolls	Typic Argiudolls	Fine-silty	Mixed	Superactive			Mesic	Udic
OKEE	111	111B, 89, 95B, 97, 98	C	Alfisols	Udalfs	Hapludalfs	Arenic Hapludalfs	Loamy	Mixed	Active			Mesic	Udic
OMEGA	88	90, 91, 92, 93, 94A	E, G, H	Spodosols	Orthods	Haplorthods	Typic Haplorthods	Sandy	Mixed			Uncoated	Frigid	Udic
OMENA	115		E	Alfisols	Udalfs	Glossudalfs	Haplic Glossudalfs	Coarse-loamy	Mixed	Active			Frigid	Udic
OMRO	43	95A, 95B	E, I	Alfisols	Udalfs	Hapludalfs	Typic Hapludalfs	Clayey over loamy	Mixed	Active			Mesic	Udic
ONAWAY	739	93B, 94A, 94B, 94C, 95A, 96	B, E, I	Alfisols	Udalfs	Hapludalfs	Inceptic Hapludalfs	Fine-loamy	Mixed	Active			Frigid	Udic
ORION	359	104, 105, 108, 113, 114B, 115, 89, 90B, 95B	J	Entisols	Fluvents	Udifluvents	Aquic Udifluvents	Coarse-silty	Mixed	Superactive	Nonacid		Mesic	Udic
ORONTO	24	92	F, G, J	Alfisols	Aqualfs	Glossaqualfs	Aeric Glossaqualfs	Fine-silty	Mixed	Active			Frigid	Aquic
OSHKOSH	174	95A, 95B	E, I	Alfisols	Udalfs	Hapludalfs	Typic Hapludalfs	Very fine	Mixed	Active			Mesic	Udic
OSHTEMO	176	111B, 111C, 111D, 124, 139, 95B, 97, 98, 99	C	Alfisols	Udalfs	Hapludalfs	Typic Hapludalfs	Coarse-loamy	Mixed	Active			Mesic	Udic
OSSIAN	146	104, 105, 108, 95B	Bm, I, J	Mollisols	Aquolls	Endoaquolls	Typic Endoaquolls	Fine-silty	Mixed	Superactive			Mesic	Aquic
OSSMER	301	90, 91	G	Alfisols	Udalfs	Glossudalfs	Aquic Glossudalfs	Coarse-loamy over sandy or sandy-skeletal	Mixed	Superactive			Frigid	Udic
OTTER	121	103, 104, 105, 107B, 108A, 108B, 115B, 115C, 95B	J	Mollisols	Aquolls	Endoaquolls	Cumulic Endoaquolls	Fine-silty	Mixed	Superactive			Mesic	Aquic
OTTERHOLT	66	90, 91	F	Alfisols	Udalfs	Glossudalfs	Haplic Glossudalfs	Fine-silty	Mixed	Superactive			Frigid	Udic

(continued)

(continued)

Soil series	Area (km^2)	MLRAs	Soil regions	Order	Suborder	Great group	Subgroup	Particle-size class	Mineral class	CEC Activity class	Reaction class	Other	Soil-temperature class	Soil-moisture class
OZAUKEE	255	108B, 97	B	Alfisols	Udalfs	Hapludalfs	Oxyaquic Hapludalfs	Fine	Illitic				Mesic	Udic
PADUS	2373	90, 93, 94A	F, G	Spodosols	Orthods	Haplorthods	Alfic Haplorthods	Coarse-loamy	Mixed	Superactive			Frigid	Udic
PADWET	39	90, 93, 94A	G	Spodosols	Orthods	Haplorthods	Alfic Haplorthods	Coarse-loamy	Mixed	Superactive			Frigid	Udic
PADWOOD	55	93, 94A	G	Spodosols	Orthods	Haplorthods	Alfic Oxyaquic Haplorthods	Coarse-loamy	Mixed	Superactive			Frigid	Udic
PALMS	510	J	J	Histosols	Saprists	Haplosaprists	Terric Haplosaprists	Loamy	Mixed		Euic		Mesic	Aquic
PALSGROVE	659	105, 108B, 95B	A	Alfisols	Udalfs	Hapludalfs	Typic Hapludalfs	Fine-silty	Mixed	Superactive			Mesic	Udic
PARDEEVILLE	8	89	B	Alfisols	Udalfs	Hapludalfs	Mollic Hapludalfs	Coarse-loamy	Mixed	Active			Mesic	Udic
PARKFALLS	49		G, J	Spodosols	Aquods	Epiaquods	Alfic Epiaquods	Coarse-loamy	Mixed	Superactive			Frigid	Aquic
PARTRIDGE	23	105, 91	C	Entisols	Psamments	Udipsamments	Aquic Udipsamments		Mixed				Mesic	Udic
PEARL	49	91	C, G	Alfisols	Udalfs	Hapludalfs	Arenic Oxyaquic Hapludalfs	Loamy	Mixed	Superactive			Mesic	Udic
PECATONICA	92	105, 108A, 108B, 95B	B	Alfisols	Udalfs	Hapludalfs	Typic Hapludalfs	Fine-loamy	Mixed	Superactive			Mesic	Udic
PECORE	4	93	G	Alfisols	Udalfs	Glossudalfs	Haplic Glossudalfs	Fine-loamy	Mixed	Active			Frigid	Udic
PEEBLES	10	95B	Bm	Mollisols	Udolls	Argiudolls	Oxyaquic Argiudolls	Very fine	Mixed	Active			Mesic	Udic
PELISSIER	41	90, 93, 94A	G	Spodosols	Orthods	Haplorthods	Entic Haplorthods	Sandy-skeletal	Mixed				Frigid	Udic
PELKIE	37	90, 92, 93, 95A	H, J	Entisols	Psamments	Udipsamments	Oxyaquic Udipsamments		Mixed				Frigid	Udic
PELLA	786	108A, 108B, 110, 111A, 111B, 111C, 111D, 115C, 95A, 95B, 97, 98, 99	Bm, J	Mollisols	Aquolls	Endoaquolls	Typic Endoaquolls	Fine-silty	Mixed	Superactive			Mesic	Aquic
PENCE	2343	90, 93, 94A	F, G, H	Spodosols	Orthods	Haplorthods	Typic Haplorthods	Sandy	Isotic				Frigid	Udic
PEPIN	191	105	A, Dr	Alfisols	Udalfs	Hapludalfs	Typic Hapludalfs	Fine-silty	Mixed	Superactive			Mesic	Udic
PEQUAMING	34	93, 94A, 94B	H, J	Spodosols	Aquods	Endoaquods	Argic Endoaquods	Sandy	Mixed				Frigid	Aquic
PERCHLAKE	9	91	H	Entisols	Psamments	Udipsamments	Aquic Udipsamments		Mixed				Frigid	Udic
PERIDA	24	91	H, I	Alfisols	Udalfs	Hapludalfs	Arenic Hapludalfs	Clayey	Smectitic				Frigid	Udic
PEROTE	23	93, 94A	G	Alfisols	Udalfs	Glossudalfs	Haplic Glossudalfs	Coarse-loamy	Mixed	Active			Frigid	Udic

(continued)

(continued)

Soil series	Area (km^2)	MLRAs	Soil regions	Order	Suborder	Great group	Subgroup	Particle-size class	Mineral class	CEC Activity class	Reaction class	Other	Soil-temperature class	Soil-moisture class
PESABIC	283	90	G, J	Spodosols	Aquods	Epiaquods	Alfic Epiaquods	Coarse-loamy	Mixed	Superactive			Frigid	Aquic
PESHEKEE	17	93	G	Spodosols	Orthods	Haplorthods	Lithic Haplorthods	Loamy	Mixed	Semiactive			Frigid	Udic
PESHTIGO	2	95A	G	Alfisols	Udalfs	Glossudalfs	Aquic Glossudalfs	Fine-loamy	Mixed	Active			Frigid	Udic
PICKFORD	103	92, 93, 94A, 94B	I, J	Inceptisols	Aquepts	Epiaquepts	Aeric Epiaquepts	Fine	Mixed	Active	Nonacid		Frigid	Aquic
PILLOT	100	105, 108	Am, F	Mollisols	Udolls	Argiudolls	Typic Argiudolls	Fine-silty over sandy or sandy-skeletal	Mixed	Superactive			Mesic	Udic
PINCONNING	9	92, 93B, 94A, 94B, 94C, 95A, 96, 98	E, J	Entisols	Aquents	Epiaquents	Mollic Epiaquents	Sandy over clayey	Mixed	Semiactive	Nonacid		Frigid	Aquic
PLAINBO	188	105, 90, 91	C, Dr, H	Entisols	Psamments	Udipsamments	Typic Udipsamments		Mixed				Frigid	Udic
PLAINFIELD	2432	101, 103, 104, 105, 107B, 108B, 109, 110, 111, 115A, 115C, 124, 141, 142, 143, 144A, 91A, 95A, 95B, 97, 98	C	Entisols	Psamments	Udipsamments	Typic Udipsamments		Mixed				Mesic	Udic
PLANO	1122	108A, 108B, 110, 115A, 115C, 95A, 95B	Bm	Mollisols	Udolls	Argiudolls	Typic Argiudolls	Fine-silty	Mixed	Superactive			Mesic	Udic
PLEINE	100	93	G, J	Inceptisols	Aquepts	Humaquepts	Histic Humaquepts	Coarse-loamy	Mixed	Superactive	Nonacid		Frigid	Aquic
PLOVER	58	90, 93, 94A	F, G	Alfisols	Udalfs	Glossudalfs	Aquic Glossudalfs	Coarse-loamy	Mixed	Superactive			Frigid	Udic
PLUMCREEK	23	105	Dr	Alfisols	Udalfs	Hapludalfs	Typic Hapludalfs	Fine-loamy	Mixed	Superactive			Mesic	Udic
POINT	74	90	Fr	Alfisols	Udalfs	Glossudalfs	Aquic Glossudalfs	Fine-loamy	Mixed	Superactive			Frigid	Udic
POMROY	27	90, 91, 93	G	Alfisols	Udalfs	Hapludalfs	Arenic Oxyaquic Hapludalfs	Loamy	Mixed	Superactive			Frigid	Udic
PONYCREEK	136	90, 91	C, Dr, J	Entisols	Aquents	Psammaquents	Humaqueptic Psammaquents		Siliceous				Frigid	Aquic

(continued)

(continued)

Soil series	Area (km^2)	MLRAs	Soil regions	Order	Suborder	Great group	Subgroup	Particle-size class	Mineral class	CEC Activity class	Reaction class	Other	Soil-temperature class	Soil-moisture class
PORT BYRON	62	103, 104, 105, 108B, 108C, 115C, 95B	Am	Mollisols	Udolls	Hapludolls	Typic Hapludolls	Fine-silty	Mixed	Superactive			Mesic	Udic
PORTWING	228		I	Alfisols	Udalfs	Glossudalfs	Oxyaquic Glossudalfs	Fine	Mixed	Active			Frigid	Udic
POSKIN	108	90	F	Alfisols	Udalfs	Glossudalfs	Aquic Glossudalfs	Fine-silty over sandy or sandy-skeletal	Mixed	Superactive			Frigid	Udic
POY	123	95A, 95B, 97, 98	C, I, J	Mollisols	Aquolls	Endoaquolls	Typic Endoaquolls	Clayey over sandy or sandy-skeletal	Mixed	Active			Mesic	Aquic
POYGAN	456	89, 95A, 95B	E, I, J	Mollisols	Aquolls	Epiaquolls	Typic Epiaquolls	Fine	Mixed	Active			Mesic	Aquic
PRISSEL	70	105, 91	B, Dr	Alfisols	Udalfs	Hapludalfs	Arenic Hapludalfs	Loamy	Mixed	Active			Mesic	Udic
PUCHYAN	21	95A, 95B	B	Alfisols	Udalfs	Hapludalfs	Arenic Oxyaquic Hapludalfs	Loamy	Mixed	Superactive			Mesic	Udic
QUARDERER	14	90	Cm	Mollisols	Udolls	Hapludolls	Fluventic Hapludolls	Coarse-silty	Mixed	Superactive			Frigid	Udic
RABE	11	93, 94A	G	Alfisols	Udalfs	Glossudalfs	Arenic Glossudalfs	Loamy	Mixed	Active			Frigid	Udic
RADFORD	80	104, 105, 107A, 108A, 108B, 108C, 109, 113, 115C, 95B	Bm	Mollisols	Udolls	Hapludolls	Fluvaquentic Hapludolls	Fine-silty	Mixed	Superactive			Mesic	Udic
RASSET	71	103, 104, 105	Am, Dr	Mollisols	Udolls	Argiudolls	Typic Argiudolls	Coarse-loamy	Mixed	Superactive			Mesic	Udic
REDRIM	2	92	G, H	Spodosols	Orthods	Haplorthods	Entic Lithic Haplorthods	Sandy-skeletal	Mixed				Frigid	Udic
REEDSBURG	34	105	A, Dr	Alfisols	Udalfs	Paleudalfs	Aquic Paleudalfs	Fine-silty	Mixed	Superactive			Mesic	Udic
RENOVA	31	104, 105, 90B	G	Alfisols	Udalfs	Hapludalfs	Typic Hapludalfs	Fine-loamy	Mixed	Superactive			Mesic	Udic
RIB	110	90	J	Alfisols	Aqualfs	Endoaqualfs	Mollic Endoaqualfs	Fine-silty over sandy or sandy-skeletal	Mixed	Superactive			Frigid	Aquic

(continued)

(continued)

Soil series	Area (km^2)	MLRAs	Soil regions	Order	Suborder	Great group	Subgroup	Particle-size class	Mineral class	CEC Activity class	Reaction class	Other	Soil-temperature class	Soil-moisture class
RIBHILL	43	90	Fr, G	Alfisols	Udalfs	Glossudalfs	Haplic Glossudalfs	Loamy-skeletal	Mixed	Superactive			Frigid	Udic
RIBRIVER	9	90	F	Alfisols	Udalfs	Glossudalfs	Oxyaquic Glossudalfs	Fine-silty	Mixed	Superactive			Frigid	Udic
RICHFORD	581	90, 91	Cm	Alfisols	Udalfs	Hapludalfs	Arenic Hapludalfs	Loamy	Mixed	Superactive			Mesic	Udic
RICHWOOD	53	104, 105, 108, 109, 90	Am	Mollisols	Udolls	Argiudolls	Typic Argiudolls	Fine-silty	Mixed	Superactive			Mesic	Udic
RIETBROCK	305	90	Fr	Alfisols	Udalfs	Glossudalfs	Aquic Glossudalfs	Fine-loamy	Mixed	Superactive			Frigid	Udic
RIFLE	136	102A, 141, 143, 144B, 55B, 56, 57, 88, 90, 91, 93, 94A, 94B, 96, 98	J	Histosols	Hemists	Haplohemists	Typic Haplohemists				Euic		Frigid	Aquic
RIMER	5	105, 111B, 139, 89, 95A, 95B, 97, 98, 99	A	Alfisols	Udalfs	Hapludalfs	Aquic Arenic Hapludalfs	Loamy	Mixed	Active			Mesic	Udic
RINGWOOD	181	95B	Bm	Mollisols	Udolls	Argiudolls	Typic Argiudolls	Fine-loamy	Mixed	Superactive			Mesic	Udic
RIPON	8	104, 105, 108A, 108B, 110, 90B, 95B	Bm	Mollisols	Udolls	Argiudolls	Typic Argiudolls	Fine-silty	Mixed	Superactive			Mesic	Udic
RITCHEY	47	108A, 110, 111A, 111B, 90B, 95A, 95B, 99	B, E, F	Alfisols	Udalfs	Hapludalfs	Lithic Hapludalfs	Loamy	Mixed	Superactive			Mesic	Udic
ROBAGO	30	92, 93, 94A	G, J	Spodosols	Aquods	Endoaquods	Argic Endoaquods	Coarse-loamy	Mixed	Superactive			Frigid	Aquic
ROBY	20	110, 114B, 115A, 115B	C, Dr	Alfisols	Udalfs	Hapludalfs	Aquic Hapludalfs	Coarse-loamy	Mixed	Superactive			Mesic	Udic
ROCKBLUFF	49	105	Dr	Entisols	Psamments	Quartzipsamments	Typic Quartzipsamments						Mesic	Udic
ROCKBRIDGE	8	105	A	Alfisols	Udalfs	Hapludalfs	Typic Hapludalfs	Fine-loamy	Mixed	Superactive			Mesic	Udic
ROCKDAM	123	90, 91	Dr	Spodosols	Orthods	Haplorthods	Entic Haplorthods	Sandy	Siliceous				Frigid	Udic
ROCKERS	44	90	G, J	Spodosols	Aquods	Endoaquods	Argic Endoaquods	Coarse-loamy	Mixed	Active			Frigid	Aquic
ROCKLAND	8	92	G	Inceptisols	Udepts	Eutrudepts	Typic Eutrudepts	Fine-loamy	Mixed	Active			Frigid	Udic
ROCKMARSH	4	90, 91	G	Alfisols	Udalfs	Hapludalfs	Aquollic Hapludalfs	Loamy-skeletal	Mixed	Superactive			Frigid	Udic

(continued)

(continued)

Soil series	Area (km^2)	MLRAs	Soil regions	Order	Suborder	Great group	Subgroup	Particle-size class	Mineral class	CEC Activity class	Reaction class	Other	Soil-temperature class	Soil-moisture class
ROCKTON	101	103, 104, 105, 108, 110, 90B, 91, 95B, 98	Bm	Mollisols	Udolls	Argiudolls	Typic Argiudolls	Fine-loamy	Mixed	Superactive			Mesic	Udic
RODMAN	153	108B, 110, 111A, 111D, 114A, 95A, 95B, 97, 98	Bm	Mollisols	Udolls	Hapludolls	Typic Hapludolls	Sandy-skeletal	Mixed				Mesic	Udic
RONDEAU	15	56, 57, 88, 90, 91, 94A, 95A, 96	A	Histosols	Saprists	Haplosaprists	Limnic Haplosaprists		Marly		Euic		Frigid	Aquic
ROOT	1		A	Entisols	Aquents	Fluvaquents	Mollic Fluvaquents	Coarse-loamy	Mixed	Active	Nonacid		Mesic	Aquic
ROSCOMMON	379	57, 90, 91, 93, 94A, 94B, 95A, 96	J	Entisols	Aquents	Psammaquents	Mollic Psammaquents		Mixed				Frigid	Aquic
ROSHOLT	1501	90	G, F	Alfisols	Udalfs	Glossudalfs	Haplic Glossudalfs	Coarse-loamy	Mixed	Superactive			Frigid	Udic
ROTAMER	129	95B	Bm	Mollisols	Udolls	Argiudolls	Typic Argiudolls	Fine-loamy	Mixed	Active			Mesic	Udic
ROUSSEAU	291	93B, 94A, 94B, 94C, 95A, 98	E, H	Spodosols	Orthods	Haplorthods	Entic Haplorthods	Sandy	Mixed				Frigid	Udic
ROWLEY	26	105, 108	Am	Mollisols	Udolls	Argiudolls	Aquic Argiudolls	Fine-silty	Mixed	Superactive			Mesic	Udic
ROZELLVILLE	200	90	Fr	Alfisols	Udalfs	Glossudalfs	Haplic Glossudalfs	Fine-loamy	Mixed	Superactive			Frigid	Udic
ROZETTA	60	105, 108A, 108B, 108C, 115B, 115C, 89, 95B	A	Alfisols	Udalfs	Hapludalfs	Typic Hapludalfs	Fine-silty	Mixed	Superactive			Mesic	Udic
RUBICON	742	90A, 92, 93B, 94A, 94B, 94C, 96, 98	G, H	Spodosols	Orthods	Haplorthods	Entic Haplorthods	Sandy	Mixed				Frigid	Udic
RUSE	6	142, 93, 94A, 94B, 95A, 96	E, J	Mollisols	Aquolls	Endoaquolls	Lithic Endoaquolls	Loamy	Mixed	Active			Frigid	Aquic
RUSKTOWN	17	105	Dr	Alfisols	Udalfs	Hapludalfs	Mollic Hapludalfs	Coarse-loamy	Mixed	Active			Mesic	Udic
SABLE	94	104, 105, 108A, 108B, 110, 111D, 115C, 95B	J	Mollisols	Aquolls	Endoaquolls	Typic Endoaquolls	Fine-silty	Mixed	Superactive			Mesic	Aquic
SALTER	32	95A, 95B	B	Inceptisols	Udepts	Eutrudepts	Typic Eutrudepts	Coarse-loamy	Mixed	Superactive			Mesic	Udic

(continued)

(continued)

Soil series	Area (km^2)	MLRAs	Soil regions	Order	Suborder	Great group	Subgroup	Particle-size class	Mineral class	CEC Activity class	Reaction class	Other	Soil-temperature class	Soil-moisture class
SANBORG	228	92	I	Alfisols	Udalfs	Glossudalfs	Oxyaquic Glossudalfs	Fine	Mixed	Active			Frigid	Udic
SANTIAGO	663	90, 91, 93	F, G	Alfisols	Udalfs	Glossudalfs	Haplic Glossudalfs	Coarse-loamy	Mixed	Superactive			Frigid	Udic
SARGEANT	4	104, 90B	F	Alfisols	Udalfs	Glossudalfs	Aquic Glossudalfs	Fine-loamy	Mixed	Superactive			Mesic	Udic
SARONA	3668	90, 93, 94A	G, H	Spodosols	Orthods	Haplorthods	Alfic Haplorthods	Coarse-loamy	Mixed	Superactive			Frigid	Udic
SARWET	72	90, 93, 94A	G	Spodosols	Orthods	Haplorthods	Alfic Oxyaquic Haplorthods	Coarse-loamy	Mixed	Superactive			Frigid	Udic
SATTRE	204	103, 104, 105, 108C, 90A, 90B	F, G	Alfisols	Udalfs	Hapludalfs	Mollic Hapludalfs	Fine-loamy over sandy or sandy-skeletal	Mixed	Superactive			Mesic	Udic
SAYBROOK	33	108A, 110, 95B	Bm	Mollisols	Udolls	Argiudolls	Oxyaquic Argiudolls	Fine-silty	Mixed	Superactive			Mesic	Udic
SAYLESVILLE	49	110, 95B, 97, 98	B	Alfisols	Udalfs	Hapludalfs	Typic Hapludalfs	Fine	Illitic				Mesic	Udic
SAYNER	1069	90, 93, 94A	G, H	Spodosols	Orthods	Haplorthods	Entic Haplorthods	Sandy	Mixed				Frigid	Udic
SCHAAT CREEK	1	92, 93B	I, J	Alfisols	Aqualfs	Endoaqualfs	Aeric Endoaqualfs	Fine	Mixed	Active			Frigid	Aquic
SCHAPVILLE	1	104, 105, 95B	Am	Mollisols	Udolls	Argiudolls	Oxyaquic Argiudolls	Fine	Mixed	Active			Mesic	Udic
SCHWEITZER	9	93	G	Spodosols	Orthods	Fragiorthods	Alfic Fragiorthods	Coarse-loamy	Mixed	Superactive			Frigid	Udic
SCOBA	19	90, 93	G	Alfisols	Udalfs	Glossudalfs	Haplic Glossudalfs	Coarse-loamy	Mixed	Superactive			Frigid	Udic
SCONSIN	240	90, 93	G	Alfisols	Udalfs	Glossudalfs	Oxyaquic Glossudalfs	Coarse-loamy	Mixed	Superactive			Frigid	Udic
SCOTAH	47	104, 105	Dr	Entisols	Psamments	Udipsamments	Typic Udipsamments		Mixed			Coated	Mesic	Udic
SCOTT LAKE	201	90, 93	G	Alfisols	Udalfs	Glossudalfs	Oxyaquic Glossudalfs	Coarse-loamy	Mixed	Superactive			Frigid	Udic
SEATON	1042	104, 105, 108C, 115C, 90B	A, Dr	Alfisols	Udalfs	Hapludalfs	Typic Hapludalfs	Fine-silty	Mixed	Superactive			Mesic	Udic
SEBBO	19	105	Dr	Alfisols	Udalfs	Hapludalfs	Mollic Hapludalfs	Fine-loamy	Mixed	Superactive			Mesic	Udic
SEBEWA	286	110, 111B, 111C, 95A, 95B, 97, 98	Bm, J	Mollisols	Aquolls	Argiaquolls	Typic Argiaquolls	Fine-loamy over sandy or sandy-skeletal	Mixed	Superactive			Mesic	Aquic

(continued)

(continued)

Soil series	Area (km^2)	MLRAs	Soil regions	Order	Suborder	Great group	Subgroup	Particle-size class	Mineral class	CEC Activity class	Reaction class	Other	Soil-temperature class	Soil-moisture class
SECHLER	4	105	Cm	Inceptisols	Udepts	Dystrudepts	Aquic Humic Dystrudepts	Coarse-loamy	Siliceous	Superactive			Mesic	Udic
SEDGWICK	97	92	G, I, J	Spodosols	Aquods	Epiaquods	Alfic Epiaquods	Coarse-loamy over clayey	Mixed	Active			Frigid	Aquic
SEELYEVILLE	1595	J		Histosols	Saprists	Haplosaprists	Typic Haplosaprists				Euic		Frigid	Aquic
SELKIRK	0	94A, 94B, 98	E	Alfisols	Udalfs	Glossudalfs	Aquic Glossudalfs	Fine	Mixed	Semiactive			Frigid	Udic
SEWARD	7	111B, 139, 95B, 97, 98, 99	B	Alfisols	Udalfs	Hapludalfs	Arenic Oxyaquic Hapludalfs	Loamy	Mixed	Active			Mesic	Udic
SHAG	3	93	F, G, J	Mollisols	Aquolls	Endoaquolls	Typic Endoaquolls	Coarse-silty	Mixed	Active			Frigid	Aquic
SHAWANO	337	90, 91, 95A	E, J	Entisols	Psamments	Udipsamments	Typic Udipsamments		Mixed				Frigid	Udic
SHERRY	243	90	Fr, J	Alfisols	Aqualfs	Endoaqualfs	Udollic Endoaqualfs	Fine-loamy	Mixed	Superactive			Frigid	Aquic
SHIFFER	62	105	Dr	Alfisols	Udalfs	Hapludalfs	Aquollic Hapludalfs	Fine-loamy over sandy or sandy-skeletal	Mixed	Superactive			Mesic	Udic
SHIOCTON	231	95A	E, I	Mollisols	Udolls	Hapludolls	Aquic Hapludolls	Coarse-silty	Mixed	Superactive			Frigid	Udic
SHULLSBURG	2		Am	Mollisols	Udolls	Argiudolls	Aquic Argiudolls	Fine	Mixed	Superactive			Mesic	Udic
SILVERHILL	11	105	Dr	Alfisols	Udalfs	Hapludalfs	Ultic Hapludalfs	Coarse-loamy	Siliceous	Active			Mesic	Udic
SIMESCREEK	9	90	Dr	Entisols	Psamments	Quartzipsamments	Typic Quartzipsamments						Frigid	Udic
SIOUXCREEK	2	90	Dr	Ultisols	Udults	Hapludults	Aquic Hapludults	Coarse-loamy over sandy or sandy-skeletal	Mixed	Active			Frigid	Udic
SIREN	9	90, 91	G	Alfisols	Udalfs	Glossudalfs	Aquic Glossudalfs	Fine-loamy over clayey	Mixed OVER smectitic	Superactive			Frigid	Udic
SISSABAGAMA	6	91	H	Entisols	Psamments	Udipsamments	Oxyaquic Udipsamments		Mixed				Frigid	Udic

(continued)

(continued)

Soil series	Area (km^2)	MLRAs	Soil regions	Order	Suborder	Great group	Subgroup	Particle-size class	Mineral class	CEC Activity class	Reaction class	Other	Soil-temperature class	Soil-moisture class
SISSON	166	105, 110, 111B, 95A, 95B, 96, 98, 99	B, E, I	Alfisols	Udalfs	Hapludalfs	Typic Hapludalfs	Fine-loamy	Mixed	Semiactive			Mesic	Udic
SKOG	2		H	Entisols	Orthents	Udorthents	Oxyaquic Udorthents	Sandy-skeletal	Mixed				Frigid	Udic
SKYBERG	6	104, 90B	G, J	Alfisols	Aqualfs	Epiaqualfs	Mollic Epiaqualfs	Fine-loamy	Mixed	Superactive			Mesic	Aquic
SLIMLAKE	11	90, 91	H	Inceptisols	Udepts	Dystrudepts	Oxyaquic Dystrudepts	Sandy	Mixed				Frigid	Udic
SMESTAD	16	91	G	Mollisols	Udolls	Argiudolls	Aquic Argiudolls	Coarse-loamy over clayey	Mixed over smectitic	Superactive			Frigid	Udic
SODERBECK	0		G	Mollisols	Udolls	Hapludolls	Aquic Hapludolls	Loamy-skeletal	Mixed	Active			Frigid	Udic
SOGN	284		Am	Mollisols	Ustolls	Haplustolls	Lithic	Loamy	Mixed	Superactive			Mesic	Udic
SOLONA	562	93, 94B, 95A	E	Mollisols	Udolls	Argiudolls	Aquic Argiudolls	Coarse-loamy	Mixed	Superactive			Frigid	Udic
SOONER	7	105	Dr	Alfisols	Udalfs	Hapludalfs	Aquollic Hapludalfs	Fine-loamy over sandy or sandy-skeletal	Mixed	Superactive			Mesic	Udic
SOPERTON	44	93	G	Spodosols	Orthods	Fragiorthods	Alfic Fragiorthods	Coarse-loamy	Mixed	Active			Frigid	Udic
SPARTA	222	103, 104, 105, 108, 109, 110, 115, 90, 91, 95B, 98	Cm	Mollisols	Udolls	Hapludolls	Entic Hapludolls	Sandy	Mixed				Mesic	Udic
SPEAR	14	93	G	Alfisols	Udalfs	Glossudalfs	Aquic Glossudalfs	Coarse-silty	Mixed	Superactive			Frigid	Udic
SPENCER	364	90, 91	F	Alfisols	Udalfs	Glossudalfs	Oxyaquic Glossudalfs	Fine-silty	Mixed	Superactive			Frigid	Udic
SPIDERLAKE	45		G	Spodosols	Orthods	Haplorthods	Alfic Oxyaquic Haplorthods	Coarse-loamy	Mixed	Superactive			Frigid	Udic
SPINKS	8	105, 111B, 95B, 96, 97, 98, 99	B	Alfisols	Udalfs	Hapludalfs	Lamellic Hapludalfs	Sandy	Mixed				Mesic	Udic
SPOONERHILL	46	91, 93	H	Inceptisols	Udepts	Dystrudepts	Oxyaquic Dystrudepts	Sandy	Mixed				Frigid	Udic
SPRINGSTEAD	142		G, H	Spodosols	Orthods	Haplorthods	Oxyaquic Haplorthods	Sandy	Isotic				Frigid	Udic

(continued)

(continued)

Soil series	Area (km^2)	MLRAs	Soil regions	Order	Suborder	Great group	Subgroup	Particle-size class	Mineral class	CEC Activity class	Reaction class	Other	Soil-temperature class	Soil-moisture class
ST. CHARLES	909	108A, 108B, 110, 111D, 115C, 95B	A, B	Alfisols	Udalfs	Hapludalfs	Typic Hapludalfs	Fine-silty	Mixed	Superactive			Mesic	Udic
STAMBAUGH	315	92, 93	G	Spodosols	Orthods	Haplorthods	Alfic Haplorthods	Coarse-silty over sandy or sandy-skeletal	Mixed	Superactive			Frigid	Udic
STANBERRY	171	90	G	Spodosols	Orthods	Haplorthods	Alfic Oxyaquic Haplorthods	Coarse-loamy	Mixed	Superactive			Frigid	Udic
STENGEL	6	91	I	Alfisols	Udalfs	Hapludalfs	Aquic Arenic Hapludalfs	Clayey	Smectitic				Frigid	Udic
STINNETT	52		G	Alfisols	Udalfs	Glossudalfs	Aquic Glossudalfs	Coarse-loamy	Mixed	Superactive			Frigid	Udic
STRONGHURST	4	104, 105, 108A, 108B, 108C, 115C, 95B	A, J	Alfisols	Aqualfs	Endoaqualfs	Aeric Endoaqualfs	Fine-silty	Mixed	Superactive			Mesic	Aquic
STURGEON	37	90, 91, 92, 93	J	Entisols	Fluvents	Udifluvents	Aquic Udifluvents	Coarse-silty over sandy or sandy-skeletal	Mixed	Superactive	Nonacid		Frigid	Udic
SULTZ	227	90, 92, 93, 94A, 94B	H	Spodosols	Orthods	Haplorthods	Entic Haplorthods	Sandy	Mixed				Frigid	Udic
SUMMERVILLE	192	142, 93B, 94A, 94B, 94C, 95A, 96	Er, I	Inceptisols	Udepts	Eutrudepts	Lithic Eutrudepts	Loamy	Mixed	Active			Frigid	Udic
SUNIA	3	90, 93	H	Entisols	Psamments	Udipsamments	Oxyaquic Udipsamments		Mixed				Frigid	Udic
SUPERIOR	71	92, 93, 94B	I	Spodosols	Orthods	Haplorthods	Alfic Oxyaquic Haplorthods	Coarse-loamy over clayey	Mixed	Active			Frigid	Udic
SYLVESTER	12	105	Am	Mollisols	Udolls	Argiudolls	Typic Argiudolls	Fine-silty over sandy or sandy-skeletal	Mixed	Superactive			Mesic	Udic

(continued)

(continued)

Soil series	Area (km^2)	MLRAs	Soil regions	Order	Suborder	Great group	Subgroup	Particle-size class	Mineral class	CEC Activity class	Reaction class	Other	Soil-temperature class	Soil-moisture class
SYMCO	345	95A, 95B	E, G, I	Alfisols	Udalfs	Hapludalfs	Aquollic Hapludalfs	Fine-loamy	Mixed	Active			Mesic	Udic
SYMERTON	7	108A, 110, 111D, 95B, 97	Bm	Mollisols	Udolls	Argiudolls	Oxyaquic Argiudolls	Fine-loamy	Mixed	Superactive			Mesic	Udic
TACOOSH	2	56, 57, 88, 90, 93	J	Histosols	Hemists	Haplohemists	Terric Haplohemists	Loamy	Mixed		Euic		Frigid	Aquic
TAMA	621	104, 105, 108B, 108C, 95B	Am	Mollisols	Udolls	Argiudolls	Typic Argiudolls	Fine-silty	Mixed	Superactive			Mesic	Udic
TARR	727	105, 91	C, Dr	Entisols	Psamments	Quartzipsamments	Typic Quartzipsamments					Uncoated	Mesic	Udic
TAWAS	109	57, 88, 90, 93, 94A, 94B, 96	E, J	Histosols	Saprists	Haplosaprists	Terric Haplosaprists	Sandy or sandy-skeletal	Mixed		Euic		Frigid	Aquic
TAYLOR	1	57, 88, 90	I	Alfisols	Udalfs	Glossudalfs	Aquertic Glossudalfs	Very fine	Smectitic				Frigid	Udic
TEDROW	87	111B, 95A, 95B, 97, 98, 99	B	Entisols	Psamments	Udipsamments	Aquic Udipsamments		Mixed				Mesic	Udic
TELL	45	104, 105, 108A, 108B, 108C, 115C, 89, 95B	A	Alfisols	Udalfs	Hapludalfs	Typic Hapludalfs	Fine-silty over sandy or sandy-skeletal	Mixed	Superactive			Mesic	Udic
THACKERY	5	111A, 111B, 111C, 111D, 111E, 114A, 95B	A	Alfisols	Udalfs	Hapludalfs	Aquic Hapludalfs	Fine-loamy	Mixed	Active			Mesic	Udic
THERESA	589	95B	B, I	Alfisols	Udalfs	Hapludalfs	Typic Hapludalfs	Fine-loamy	Mixed	Superactive			Mesic	Udic
TILLEDA	235	90, 94A, 95A	E	Alfisols	Udalfs	Glossudalfs	Haplic Glossudalfs	Fine-loamy	Mixed	Active			Frigid	Udic
TIMULA	9	104, 105, 107B, 108B, 115C, 90B	A	Inceptisols	Udepts	Eutrudepts	Typic Eutrudepts	Coarse-silty	Mixed	Superactive			Mesic	Udic
TINT	28	105	Dr	Entisols	Psamments	Quartzipsamments	Typic Quartzipsamments					Uncoated	Mesic	Udic
TINTSON	9	105	Dr	Entisols	Psamments	Quartzipsamments	Oxyaquic Quartzipsamments					Uncoated	Mesic	Udic
TIPLER	175	90, 93, 94A	G	Spodosols	Orthods	Haplorthods	Alfic Oxyaquic Haplorthods	Coarse-loamy	Mixed	Superactive			Frigid	Udic

(continued)

(continued)

Soil series	Area (km^2)	MLRAs	Soil regions	Order	Suborder	Great group	Subgroup	Particle-size class	Mineral class	CEC Activity class	Reaction class	Other	Soil-temperature class	Soil-moisture class
TODDVILLE	94	105	Am	Mollisols	Udolls	Argiudolls	Typic Argiudolls	Fine-silty	Mixed	Superactive			Mesic	Udic
TONKEY	9	90A, 91B, 93B, 94A, 94B, 96, 98, 99	G, J	Inceptisols	Aquepts	Endoaquepts	Mollic Endoaquepts	Coarse-loamy	Mixed	Semiactive	Nonacid		Frigid	Aquic
TOTAGATIC	88	91, 93	H, J	Entisols	Aquents	Fluvaquents	Typic Fluvaquents	Sandy	Mixed				Frigid	Aquic
TOURTILLOTTE	3	90, 93, 94A	H	Entisols	Psamments	Udipsamments	Oxyaquic Udipsamments		Mixed				Frigid	Udic
TRADELAKE	4	90	G, H	Alfisols	Udalfs	Glossudalfs	Oxyaquic Glossudalfs	Coarse-loamy over clayey	Mixed OVER smectitic	Superactive			Frigid	Udic
TREMPE	17	105	Cm	Mollisols	Udolls	Hapludolls	Entic Hapludolls	Sandy	Mixed				Mesic	Udic
TREMPEALEAU	3	105, 108	Cm	Mollisols	Udolls	Argiudolls	Typic Argiudolls	Fine-loamy over sandy or sandy-skeletal	Mixed	Superactive			Mesic	Udic
TROXEL	105	110, 95B, 98	Bm	Mollisols	Udolls	Argiudolls	Pachic Argiudolls	Fine-silty	Mixed	Superactive			Mesic	Udic
TULA	179	93	G, J	Spodosols	Aquods	Fragiaquods	Argic Fragiaquods	Coarse-loamy	Mixed	Superactive			Frigid	Aquic
TUSCOLA	11	110, 111B, 111D, 111E, 95B, 97, 98, 99	B	Alfisols	Udalfs	Hapludalfs	Aquic Hapludalfs	Fine-loamy	Mixed	Active			Mesic	Udic
TUSTIN	37	95A, 95B, 97, 98	E, I, J	Alfisols	Udalfs	Hapludalfs	Arenic Hapludalfs	Clayey	Mixed	Active			Mesic	Udic
TWINMOUND	77	90	Dr	Entisols	Psamments	Quartzipsamments	Typic Quartzipsamments						Frigid	Udic
URNE	712	105, 95B	Dr	Inceptisols	Udepts	Eutrudepts	Dystric Eutrudepts	Coarse-loamy	Mixed	Active			Mesic	Udic
VALTON	606	105	A	Alfisols	Udalfs	Paleudalfs	Mollic Paleudalfs	Fine-silty	Mixed	Superactive			Mesic	Udic
VANCECREEK	12	90	Dr, J	Mollisols	Aquolls	Endoaquolls	Fluvaquentic Endoaquolls	Fine-silty	Mixed	Superactive			Frigid	Aquic
VANZILE	192	93	G	Spodosols	Orthods	Haplorthods	Alfic Haplorthods	Coarse-silty over sandy or sandy-skeletal	Mixed	Superactive			Frigid	Udic

(continued)

(continued)

Soil series	Area (km^2)	MLRAs	Soil regions	Order	Suborder	Great group	Subgroup	Particle-size class	Mineral class	CEC Activity class	Reaction class	Other	Soil-temperature class	Soil-moisture class
VARNA	95	108A, 108B, 110, 111C, 111D, 115C, 95B, 97	Bm	Mollisols	Udolls	Argiudolls	Oxyaquic Argiudolls	Fine	Illitic				Mesic	Udic
VEEDUM	254	90	Dr, J	Inceptisols	Aquepts	Epiaquepts	Humic Epiaquepts	Fine-loamy	Mixed	Superactive	Acid		Frigid	Aquic
VESPER	265	90	Dr, J	Inceptisols	Aquepts	Epiaquepts	Humic Epiaquepts	Fine-loamy over sandy or sandy-skeletal	Mixed	Superactive	Acid		Frigid	Aquic
VILAS	1162	90, 92, 93, 94A	G, H	Spodosols	Orthods	Haplorthods	Entic Haplorthods	Sandy	Isotic				Frigid	Udic
VIRGIL	19	108A, 108B, 115C, 95A, 95B	B, J	Alfisols	Aqualfs	Endoaqualfs	Udollic Endoaqualfs	Fine-silty	Mixed	Superactive			Mesic	Aquic
VLASATY	214	104, 90B	A	Alfisols	Udalfs	Glossudalfs	Aquic Glossudalfs	Fine-loamy	Mixed	Superactive			Mesic	Udic
WABENO	265	93, 94A	G	Spodosols	Orthods	Fragiorthods	Alfic Oxyaquic Fragiorthods	Coarse-loamy	Mixed	Superactive			Frigid	Udic
WACOUSTA	154	103, 95B	Bm, J	Mollisols	Aquolls	Endoaquolls	Typic Endoaquolls	Fine-silty	Mixed	Superactive			Mesic	Aquic
WAINOLA	279	142, 90A, 93B, 94A, 94B, 94C, 95A, 96, 99	E, H, J	Spodosols	Aquods	Endoaquods	Typic Endoaquods	Sandy	Mixed				Frigid	Aquic
WAKEFIELD	54	93	G	Spodosols	Orthods	Fragiorthods	Alfic Oxyaquic Fragiorthods	Coarse-loamy	Isotic				Frigid	Udic
WAKELEY	30	94A, 94B, 96	H, I, J	Entisols	Aquents	Epiaquents	Aeric Epiaquents	Sandy over clayey	Mixed	Semiactive	Nonacid		Frigid	Aquic
WALLKILL	30	101, 105, 110, 111, 114, 115, 139, 140, 144A, 91, 95A, 95B, 97, 98	J	Inceptisols	Aquepts	Humaquepts	Fluvaquentic Humaquepts	Fine-loamy	Mixed	Superactive	Nonacid		Mesic	Aquic
WARMAN	2	57, 88, 90	G, H, J	Mollisols	Aquolls	Endoaquolls	Typic Endoaquolls	Coarse-loamy over sandy or sandy-skeletal	Mixed	Superactive			Frigid	Aquic

(continued)

(continued)

Soil series	Area (km^2)	MLRAs	Soil regions	Order	Suborder	Great group	Subgroup	Particle-size class	Mineral class	CEC Activity class	Reaction class	Other	Soil-temperature class	Soil-moisture class
WARSAW	226	104, 105, 108A, 108B, 110, 111A, 111B, 111C, 111D, 115A, 115C, 95B, 98	Bm	Mollisols	Udolls	Argiudolls	Typic Argiudolls	Fine-loamy over sandy or sandy-skeletal	Mixed	Superactive			Mesic	Udic
WASEPI	61	110, 95B, 97, 98, 99	B, I	Alfisols	Udalfs	Hapludalfs	Aquollic Hapludalfs	Coarse-loamy	Mixed	Semiactive			Mesic	Udic
WASHTENAW	8	110, 111A, 111B, 111C, 111D, 114B, 115A, 95B, 97, 98, 99	J	Entisols	Aquents	Fluvaquents	Aeric Fluvaquents	Fine-loamy	Mixed	Active	Nonacid		Mesic	Aquic
WATSEKA	13	103, 104, 105, 108B, 108C, 110, 111C, 115C, 95B, 97, 98	Bm	Mollisols	Udolls	Hapludolls	Aquic Hapludolls	Sandy	Mixed				Mesic	Udic
WATTON	0	92, 93, 94A	G	Alfisols	Udalfs	Glossudalfs	Haplic Glossudalfs	Fine-loamy	Mixed	Semiactive			Frigid	Udic
WAUCONDA	3	105, 110, 95B, 97	B, J	Alfisols	Aqualfs	Endoaqualfs	Udollic Endoaqualfs	Fine-silty	Mixed	Superactive			Mesic	Aquic
WAUKEGAN	4	103, 104, 105, 108B, 108C, 90B, 91A	A, F	Mollisols	Udolls	Hapludolls	Typic Hapludolls	Fine-silty over sandy or sandy-skeletal	Mixed	Superactive			Mesic	Udic
WAUPACA	38	88, 90, 95A	E, J	Entisols	Aquents	Fluvaquents	Mollic Fluvaquents	Coarse-silty	Mixed	Superactive	Nonacid		Frigid	Aquic
WAUSEON	31	111B, 95A, 95B, 97, 98, 99	E, J	Mollisols	Aquolls	Epiaquolls	Typic Epiaquolls	Coarse-loamy over clayey	Mixed OVER illitic	Superactive			Mesic	Aquic
WAUTOMA	50	91	C, J	Entisols	Aquents	Epiaquents	Mollic Epiaquents	Sandy over clayey	Mixed	Active	Nonacid		Mesic	Aquic
WAYKA	3	93, 94A	G, J	Spodosols	Aquods	Epiaquods	Typic Epiaquods	Coarse-loamy	Isotic				Frigid	Aquic
WAYMOR	233	95A, 95B	E, I	Alfisols	Udalfs	Glossudalfs	Haplic Glossudalfs	Fine-loamy	Mixed	Active			Mesic	Udic
WEGA	15	88, 90, 91, 95A	E, J	Entisols	Fluvents	Udifluvents	Aquic Udifluvents	Coarse-silty	Mixed	Superactive	Nonacid		Frigid	Udic
WESTVILLE	82	108B, 115C, 95B	B	Alfisols	Udalfs	Hapludalfs	Typic Hapludalfs	Fine-loamy	Mixed	Superactive			Mesic	Udic

(continued)

(continued)

Soil series	Area (km^2)	MLRAs	Soil regions	Order	Suborder	Great group	Subgroup	Particle-size class	Mineral class	CEC Activity class	Reaction class	Other	Soil-temperature class	Soil-moisture class
WHALAN	290	104, 105, 108, 110, 90, 95A, 95B	B, E, G	Alfisols	Udalfs	Hapludalfs	Typic Hapludalfs	Fine-loamy	Mixed	Superactive			Mesic	Udic
WHEATLEY	7	90A, 93B, 94A, 94B, 94C, 95A, 96, 98	J	Entisols	Aquents	Psammaquents	Mollic Psammaquents		Mixed				Frigid	Aquic
WHISKLAKE	54	93	G, J	Spodosols	Aquods	Endoaquods	Argic Endoaquods	Coarse-loamy over sandy or sandy-skeletal	Mixed	Superactive			Frigid	Aquic
WHITEHALL	14	105	Dr	Mollisols	Udolls	Argiudolls	Typic Argiudolls	Fine-silty over sandy or sandy-skeletal	Mixed	Superactive			Mesic	Udic
WICKWARE	37	90	Dr	Alfisols	Udalfs	Glossudalfs	Haplic Glossudalfs	Fine-silty	Mixed	Superactive			Frigid	Udic
WILDALE	104	105	Dr	Alfisols	Udalfs	Paleudalfs	Mollic Paleudalfs	Fine	Mixed	Active			Mesic	Udic
WILDWOOD	10	57, 88, 93	I, J	Inceptisols	Aquepts	Humaquepts	Histic Humaquepts	Very fine	Smectitic		Nonacid		Frigid	Aquic
WILL	3	108A, 108B, 110, 115C, 95A, 95B, 97	Bm, J	Mollisols	Aquolls	Endoaquolls	Typic Endoaquolls	Fine-loamy over sandy or sandy-skeletal	Mixed	Superactive			Mesic	Aquic
WILLETTE	128	101, 111B, 139, 142, 95A, 95B, 98	J	Histosols	Saprists	Haplosaprists	Terric Haplosaprists	Clayey	Illitic		Euic		Mesic	Aquic
WINDWARD	9		Dr	Entisols	Psamments	Quartzipsamments	Lamellic Quartzipsamments						Mesic	Udic
WINNEBAGO	25	105, 95B	Bm	Mollisols	Udolls	Argiudolls	Typic Argiudolls	Fine-loamy	Mixed	Superactive			Mesic	Udic
WINNECONNE	115	95A	E, I	Alfisols	Udalfs	Hapludalfs	Mollic Oxyaquic Hapludalfs	Very fine	Mixed	Active		Coated	Mesic	Udic
WINNESHIEK	9	104, 105, 95B, 98	B	Alfisols	Udalfs	Hapludalfs	Mollic Hapludalfs	Fine-loamy	Mixed	Superactive			Mesic	Udic
WINTERFIELD	65	90, 91, 92, 93, 94A, 94B, 96	Dr, H	Entisols	Psamments	Udipsamments	Aquic Udipsamments		Mixed				Frigid	Udic

(continued)

(continued)

Soil series	Area (km^2)	MLRAs	Soil regions	Order	Suborder	Great group	Subgroup	Particle-size class	Mineral class	CEC Activity class	Reaction class	Other	Soil-temperature class	Soil-moisture class
WITHEE	1186	90	F	Alfisols	Udalfs	Glossudalfs	Aquic Glossudalfs	Fine-loamy	Mixed	Superactive			Frigid	Udic
WORCESTER	192	93, 94A	G, H, J	Spodosols	Aquods	Endoaquods	Argic Endoaquods	Coarse-loamy	Mixed	Superactive			Frigid	Aquic
WORMET	65	93, 94A	G, H, J	Spodosols	Aquods	Endoaquods	Typic Endoaquods	Sandy	Mixed				Frigid	Aquic
WORTHEN	83	104, 105, 108A, 108B, 114B, 115B, 115C, 95B	J	Mollisols	Udolls	Hapludolls	Cumulic Hapludolls	Fine-silty	Mixed	Superactive			Mesic	Udic
WORWOOD	14	90, 93, 94A	G, J	Spodosols	Aquods	Epiaquods	Alfic Epiaquods	Coarse-loamy	Mixed	Superactive			Frigid	Aquic
WOZNY	43		G, J	Alfisols	Aqualfs	Epiaqualfs	Typic Epiaqualfs	Coarse-loamy	Mixed	Superactive			Frigid	Aquic
WURTSMITH	102	57, 90A, 91B, 94A	H	Entisols	Psamments	Udipsamments	Oxyaquic Udipsamments		Mixed			Uncoated	Frigid	Udic
WYEVILLE	76	91, 99	C, E	Alfisols	Udalfs	Hapludalfs	Aquic Arenic Hapludalfs	Clayey	Mixed	Active			Mesic	Udic
WYKOFF	34	104, 105, 90	G	Alfisols	Udalfs	Hapludalfs	Typic Hapludalfs	Fine-loamy	Mixed	Superactive			Mesic	Udic
WYOCENA	363	89, 95A, 95B	C, G	Alfisols	Udalfs	Hapludalfs	Typic Hapludalfs	Coarse-loamy	Mixed	Semiactive			Mesic	Udic
YAHARA	94	110, 89, 95A, 95B	Bm, I	Mollisols	Udolls	Hapludolls	Aquic Hapludolls	Coarse-loamy	Mixed	Superactive			Mesic	Udic
ZEBA	6	93	G, J	Spodosols	Aquods	Endoaquods	Argic Endoaquods	Coarse-loamy	Mixed	Active			Frigid	Aquic
ZITTAU	71	95A, 95B	E, G, I	Alfisols	Udalfs	Hapludalfs	Aquollic Hapludalfs	Clayey over sandy or sandy-skeletal	Mixed	Active			Mesic	Udic
ZURICH	85	110, 139, 95A, 95B, 97, 99	B, E, I	Alfisols	Udalfs	Hapludalfs	Oxyaquic Hapludalfs	Fine-silty	Mixed	Superactive			Mesic	Udic
Other	2237													
Avg.	137077													

Appendix E
Some Miscellaneous Land Types and Soil Taxa Identified in Wisconsin

Land types	Soil taxa
Alluvial land	Aquents
Beaches	Fluvaquents
Borrow pit	Haplosaprists and Psammaquents
Clayey land	Histosols, ponded
Coal pit (or pile)	Humaquepts–Fluvaquepts
Cut and fill	Psammaquents
Dams	Saprists and Aquents
Dumps	Udifluvents
Dune land	Udorthents
Gravel pit	
Gullied land	
Landfill	
Limestone quarry	
Loamy land	
Made land	
Marsh	
Mine pits and dumps	
Miscellaneous water	
Peat and muck	
Pits	
Quarry	
Riverwash	
Rock land (or outcrop)	
Rough, broken land	
Rubble land	
Sand pit	
Sandy alluvial land	
Spoil area	
Terrace escarpments	
Urban land	
Water	

J.G. Bockheim and A.E. Hartemink, *The Soils of Wisconsin*,
World Soils Book Series, DOI 10.1007/978-3-319-52144-2

Appendix F
Benchmark Soil Series in Wisconsin

J.G. Bockheim and A.E. Hartemink, *The Soils of Wisconsin*,
World Soils Book Series, DOI 10.1007/978-3-319-52144-2

Series name	State TL	Other states using	MLRAs using	Order	Suborder	Great group	Subgroup	Particle-size class	Mineral class	CEC class	Reaction class	Other family Class	Soil-temperature class
AHMEEK	MN	WI	90, 93	Inceptisols	Udepts	Eutrudepts	Dystric Eutrudepts	Coarse-loamy	Isotic				Frigid
ALMENA	WI	MN	90	Alfisols	Udalfs	Glossudalfs	Aquic Glossudalfs	Fine-silty	Mixed	Superactive			Frigid
AMERY	WI		90A, 90B	Alfisols	Udalfs	Glossudalfs	Haplic Glossudalfs	Coarse-loamy	Mixed	Superactive			Frigid
ANTIGO	WI	MN	90A, 90B, 91A, 91B, 93B	Alfisols	Udalfs	Glossudalfs	Haplic Glossudalfs	Coarse-loamy over sandy or sandy-skeletal	Mixed	Superactive			Frigid
ARENZVILLE	WI	IA, IL, MN	104, 105, 108, 115	Entisols	Fluvents	Udifluvents	Typic Udifluvents	Coarse-silty	Mixed	Superactive	Nonacid		Mesic
ASHKUM	IL	IN, MI, WI	108A, 110, 95B, 97	Mollisols	Aquolls	Endoaquolls	Typic Endoaquolls	Fine	Mixed	Superactive			Mesic
BERTRAND	WI	IA, IL, MN	105, 108, 109	Alfisols	Udalfs	Hapludalfs	Typic Hapludalfs	Fine-silty	Mixed	Superactive			Mesic
BILLETT	IL	IA, IN, MN, WI	103, 104, 105, 108, 111, 115, 90, 91, 95B	Alfisols	Udalfs	Hapludalfs	Mollic Hapludalfs	Coarse-loamy	Mixed	Superactive			Mesic
BLOUNT	OH	IL, IN, MI, WI	108A, 110, 111B, 95B, 97, 98, 99	Alfisols	Aqualfs	Epiaqualfs	Aeric Epiaqualfs	Fine	Illitic				Mesic
BOONE	WI	IA, IL, MN, NE	103, 104, 105, 107, 108, 90, 91	Entisols	Psamments	Quartzipsamments	Typic Quartzipsamments					Uncoated	Mesic
BOYER	MI	IL, IN, OH, WI	110, 111B, 94A, 95A, 95B, 96, 97, 98, 99	Alfisols	Udalfs	Hapludalfs	Typic Hapludalfs	Coarse-loamy	Mixed	Semiactive			Mesic
BROOKSTON	IN	MI, OH, WI	110, 111A, 111C, 95B, 97, 98, 99	Mollisols	Aquolls	Argiaquolls	Typic Argiaquolls	Fine-loamy	Mixed	Superactive			Mesic
CARBONDALE	MI	MN, NY, VT, WI	140, 142, 143, 147, 90, 93, 94A, 94B, 95A, 96	Histosols	Saprists	Haplosaprists	Hemic Haplosaprists				Euic		Frigid
CHASEBURG	WI	IA, MN	105	Entisols	Fluvents	Udifluvents	Typic Udifluvents	Coarse-silty	Mixed	Superactive	Nonacid		Mesic

(continued)

(continued)

Series name	State TL	Other states using	MLRAs using	Order	Suborder	Great group	Subgroup	Particle-size class	Mineral class	CEC class	Reaction class	Other family Class	Soil-temperature class
CHETEK	WI	MN	90, 91	Alfisols	Udalfs	Hapludalfs	Inceptic Hapludalfs	Coarse-loamy	Mixed	Superactive			Frigid
CLYDE	IA	IL, MN, WI	104, 108B, 108C, 90B	Mollisols	Aquolls	Endoaquolls	Typic Endoaquolls	Fine-loamy	Mixed	Superactive			Mesic
COLOMA	WI	IA, IL, IN, MI, MN	104, 105, 108A, 108B, 111B, 111C, 111D, 115C, 89, 95A, 95B, 96, 97, 98	Entisols	Psamments	Udipsamments	Lamellic Udipsamments		Mixed				Mesic
CUSHING	WI	MN	103, 90	Alfisols	Udalfs	Glossudalfs	Haplic Glossudalfs	Fine-loamy	Mixed	Superactive			Frigid
DEL REY	IL	IN, MI, OH, WI	108A, 110, 111A, 111B, 111C, 94A, 95B, 97, 98, 99	Alfisols	Aqualfs	Epiaqualfs	Aeric Epiaqualfs	Fine	Illitic				Mesic
DODGEVILLE	WI	IA, IL, MN	104, 105, 108, 95B	Mollisols	Udolls	Argiudolls	Typic Argiudolls	Fine-silty over clayey	Mixed	Superactive			Mesic
DOWNS	IA	IL, MN, WI	104, 105, 108, 109, 115, 95B	Alfisols	Udalfs	Hapludalfs	Mollic Hapludalfs	Fine-silty	Mixed	Superactive			Mesic
DRUMMER	IL	IN, OH, WI	108A, 108B, 110, 111A, 111D, 95B, 97	Mollisols	Aquolls	Endoaquolls	Typic Endoaquolls	Fine-silty	Mixed	Superactive			Mesic
ELBURN	IL	WI	108A, 108B, 110, 95B	Mollisols	Udolls	Argiudolls	Aquic Argiudolls	Fine-silty	Mixed	Superactive			Mesic
ELLIOTT	IL	IN, OH, WI	108A, 110, 111B, 111C, 95B, 97	Mollisols	Udolls	Argiudolls	Aquic Argiudolls	Fine	Illitic				Mesic
EMMET	MI	WI	93B, 94A, 94B, 94C, 95A, 96, 98	Alfisols	Udalfs	Hapludalfs	Inceptic Hapludalfs	Coarse-loamy	Mixed	Active			Frigid
FAYETTE	IA	IL, MN, WI	104, 105, 108, 109, 114, 115, 95B	Alfisols	Udalfs	Hapludalfs	Typic Hapludalfs	Fine-silty	Mixed	Superactive			Mesic
FLOYD	IA	MN, WI	104, 105, 90B	Mollisols	Udolls	Hapludolls	Aquic Pachic Hapludolls	Fine-loamy	Mixed	Superactive			Mesic

(continued)

(continued)

Series name	State TL	Other states using	MLRAs using	Order	Suborder	Great group	Subgroup	Particle-size class	Mineral class	CEC class	Reaction class	Other family Class	Soil-temperature class
FOX	WI	IL, IN, MI, OH	108A, 108B, 110, 111A, 111B, 111C, 111D, 114A, 114B, 95B, 97, 98, 99	Alfisols	Udalfs	Hapludalfs	Typic Hapludalfs	Fine-loamy over sandy or sandy-skeletal	Mixed	Superactive			Mesic
GALE	WI	IA, IL, MN	104, 105, 108, 90, 91	Alfisols	Udalfs	Hapludalfs	Typic Hapludalfs	Fine-silty over sandy or sandy-skeletal	Mixed	Superactive			Mesic
GILFORD	IN	IA, IL, MI, MO, OH, WI	108A, 108B, 110, 111B, 111C, 115C, 95B, 97, 98, 99	Mollisols	Aquolls	Endoaquolls	Typic Endoaquolls	Coarse-loamy	Mixed	Superactive			Mesic
GRANBY	MI	IA, IL, IN, MN, NY, OH, WI	101, 103, 105, 110, 111B, 111C, 141, 142, 144A, 91B, 95A, 95B, 97, 98, 99	Mollisols	Aquolls	Endoaquolls	Typic Endoaquolls	Sandy	Mixed				Mesic
GRAYLING	MI	WI	93A, 93B, 94A, 94B, 94C, 95A, 98	Entisols	Psamments	Udipsamments	Typic Udipsamments		Isotic				Frigid
GRISWOLD	IL	IN, WI	105, 108A, 108B, 110, 111B, 111C, 95A, 95B, 98	Mollisols	Udolls	Argiudolls	Typic Argiudolls	Fine-loamy	Mixed	Superactive			Mesic
HIBBING	MN	WI	90, 91, 92, 93, 94A, 95A	Alfisols	Udalfs	Glossudalfs	Oxyaquic Glossudalfs	Fine	Mixed	Active			Frigid
HIXTON	WI	IA, MI, MN	104, 105, 108, 90, 98	Alfisols	Udalfs	Hapludalfs	Typic Hapludalfs	Fine-loamy over sandy or sandy-skeletal	Mixed	Superactive			Mesic

(continued)

(continued)

Series name	State TL	Other states using	MLRAs using	Order	Suborder	Great group	Subgroup	Particle-size class	Mineral class	CEC class	Reaction class	Other family Class	Soil-temperature class
HOCHHEIM	WI		95A, 95B	Mollisols	Udolls	Argiudolls	Typic Argiudolls	Fine-loamy	Mixed	Active			Mesic
HOOPESTON	IL	IA, MN, MO, WI	103, 104, 105, 108A, 108B, 108C, 109, 115C, 95B, 97	Mollisols	Udolls	Hapludolls	Aquic Hapludolls	Coarse-loamy	Mixed	Superactive			Mesic
HORTONVILLE	WI		95A	Alfisols	Udalfs	Glossudalfs	Haplic Glossudalfs	Fine-loamy	Mixed	Active			Mesic
HOUGHTON	MI	IA, IL, IN, MN, WI	103, 104, 105, 108A, 108B, 108C, 110, 111A, 111B, 111C, 111D, 115A, 89, 95A, 95B, 96, 97, 98, 99	Histosols	Saprists	Haplosaprists	Typic Haplosaprists				Euic		Mesic
HUBBARD	MN	WI	102A, 57, 90A, 90B, 91A, 91B	Mollisols	Udolls	Hapludolls	Entic Hapludolls	Sandy	Mixed				Frigid
INDUS	MN	WI	57, 88	Alfisols	Aqualfs	Epiaqualfs	Vertic Epiaqualfs	Very fine	Smectitic				Frigid
JUDSON	IA	KS, MN, MO, NE, WI	102C, 103, 104, 105, 106, 107A, 107B, 108C, 108D	Mollisols	Udolls	Hapludolls	Cumulic Hapludolls	Fine-silty	Mixed	Superactive			Mesic
KEWAUNEE	WI		110, 89, 95A, 95B	Alfisols	Udalfs	Hapludalfs	Typic Hapludalfs	Fine	Mixed	Active			Mesic
KIDDER	WI	IL, MI	108B, 110, 95A, 95B, 98	Alfisols	Udalfs	Hapludalfs	Typic Hapludalfs	Fine-loamy	Mixed	Active			Mesic
LAMARTINE	WI	IL	95A, 95B	Alfisols	Udalfs	Hapludalfs	Aquollic Hapludalfs	Fine-silty	Mixed	Superactive			Mesic
LOXLEY	MI	ME, MN, NY, WI	105, 143, 144B, 88, 90, 91, 92, 93, 94A, 94B	Histosols	Saprists	Haplosaprists	Typic Haplosaprists				Dysic		Frigid
MANAWA	WI		89, 95A, 95B	Alfisols	Udalfs	Hapludalfs	Aquollic Hapludalfs	Fine	Mixed	Active			Mesic
MAUMEE	IN	IL, MI, WI	110, 111B, 111C, 94A, 95B, 97, 98	Mollisols	Aquolls	Endoaquolls	Typic Endoaquolls	Sandy	Mixed				Mesic

(continued)

(continued)

Series name	State TL	Other states using	MLRAs using	Order	Suborder	Great group	Subgroup	Particle-size class	Mineral class	CEC class	Reaction class	Other family Class	Soil-temperature class
MERIDIAN	WI	MN	104, 105, 89, 90, 91, 95B	Alfisols	Udalfs	Hapludalfs	Mollic Hapludalfs	Fine-loamy over sandy or sandy-skeletal	Mixed	Superactive			Mesic
MIAMI	IN	IL, MI, OH, WI	102B, 108A, 110, 111A, 111D, 114A, 115C, 95B, 97, 98	Alfisols	Udalfs	Hapludalfs	Oxyaquic Hapludalfs	Fine-loamy	Mixed	Active			Mesic
MILACA	MN	WI	90, 91	Alfisols	Udalfs	Glossudalfs	Oxyaquic Glossudalfs	Coarse-loamy	Mixed	Superactive			Frigid
MILFORD	IL	IN, OH, WI	108A, 110, 111A, 111B, 111D, 95B, 97, 98, 99	Mollisols	Aquolls	Endoaquolls	Typic Endoaquolls	Fine	Mixed	Superactive			Mesic
MORLEY	IN	IL, MI, OH, WI	108A, 110, 111A, 111B, 111C, 111D, 111E, 115C, 95A, 95B, 97, 98, 99	Alfisols	Udalfs	Hapludalfs	Oxyaquic Hapludalfs	Fine	Illitic				Mesic
MUSCATINE	IA	IL, MN, WI	104, 105, 108C, 115C, 95B	Mollisols	Udolls	Hapludolls	Aquic Hapludolls	Fine-silty	Mixed	Superactive			Mesic
NESTER	MI	WI	94A, 94B, 95A, 96, 98	Alfisols	Udalfs	Glossudalfs	Oxyaquic Glossudalfs	Fine	Mixed	Semiactive			Frigid
NEWSON	WI	MN	105, 88, 90, 91, 92, 93, 94A	Entisols	Aquents	Psammaquents	Humaqueptic Psammaquents		Mixed				Frigid
OAKVILLE	IN	IL, MI, NY, OH, WI	101, 108B, 110, 111C, 115B, 115C, 139, 142, 143, 144A, 95A, 95B, 97, 98, 99	Entisols	Psamments	Udipsamments	Typic Udipsamments		Mixed				Mesic
OCKLEY	IN	IL, MI, OH, WI	108A, 111A, 111B, 111D, 111E, 114A, 95B, 97, 98	Alfisols	Udalfs	Hapludalfs	Typic Hapludalfs	Fine-loamy	Mixed	Active			Mesic

(continued)

(continued)

Series name	State TL	Other states using	MLRAs using	Order	Suborder	Great group	Subgroup	Particle-size class	Mineral class	CEC class	Reaction class	Other family Class	Soil-temperature class
OKEE	WI	MI	111B, 89, 95B, 97, 98	Alfisols	Udalfs	Hapludalfs	Arenic Hapludalfs	Loamy	Mixed	Active			Mesic
ONAWAY	MI	WI	93B, 94A, 94B, 94C, 95A, 96	Alfisols	Udalfs	Hapludalfs	Inceptic Hapludalfs	Fine-loamy	Mixed	Active			Frigid
ONTONAGON	MI	MN, WI	90, 92, 93, 94A, 94B	Alfisols	Udalfs	Glossudalfs	Haplic Glossudalfs	Very fine	Mixed	Semiactive			Frigid
PALMS	MI	IA, IL, IN, MA, MN, NY, PA, VA, WI	101, 102A, 103, 104, 105, 108A, 108B, 110, 111A, 111B, 111C, 111D, 115C, 127, 140, 142, 144A, 144B, 145, 89, 95A, 95B, 97, 98, 99	Histosols	Saprists	Haplosaprists	Terric Haplosaprists	Loamy	Mixed		Euic		Mesic
PELLA	IL	IN, MI, WI	108A, 108B, 110, 111A, 111B, 111C, 111D, 115C, 95A, 95B, 97, 98, 99	Mollisols	Aquolls	Endoaquolls	Typic Endoaquolls	Fine-silty	Mixed	Superactive			Mesic
PICKFORD	MI	WI	92, 93, 94A, 94B	Inceptisols	Aquepts	Epiaquepts	Aeric Epiaquepts	Fine	Mixed	Active	Nonacid		Frigid
PLAINFIELD	WI	IL, IN, MI, MN, MO, NY, OH	101, 103, 104, 105, 107B, 108B, 109, 110, 111, 115A, 115C, 124, 141, 142, 143, 144A, 91A, 95A, 95B, 97, 98	Entisols	Psamments	Udipsamments	Typic Udipsamments		Mixed				Mesic
POYGAN	WI		89, 95A, 95B	Mollisols	Aquolls	Epiaquolls	Typic Epiaquolls	Fine	Mixed	Active			Mesic
RONNEBY	MN	WI	90, 91	Alfisols	Aqualfs	Glossaqualfs	Aeric Glossaqualfs	Coarse-loamy	Mixed	Superactive			Frigid
ROSCOMMON	MI	MN, WI	57, 90, 91, 93, 94A, 94B, 95A, 96	Entisols	Aquents	Psammaquents	Mollic Psammaquents		Mixed				Frigid
ROSHOLT	WI	MN	90	Alfisols	Udalfs	Glossudalfs	Haplic Glossudalfs	Coarse-loamy	Mixed	Superactive			Frigid

(continued)

(continued)

Series name	State TL	Other states using	MLRAs using	Order	Suborder	Great group	Subgroup	Particle-size class	Mineral class	CEC class	Reaction class	Other family Class	Soil-temperature class
ROZETTA	IL	IA, WI	105, 108A, 108B, 108C, 115B, 115C, 89, 95B	Alfisols	Udalfs	Hapludalfs	Typic Hapludalfs	Fine-silty	Mixed	Superactive			Mesic
RUBICON	MI	WI	90A, 92, 93B, 94A, 94B, 94C, 96, 98	Spodosols	Orthods	Haplorthods	Entic Haplorthods	Sandy	Mixed				Frigid
SABLE	IL	IA, IN, WI	104, 105, 108A, 108B, 110, 111D, 115C, 95B	Mollisols	Aquolls	Endoaquolls	Typic Endoaquolls	Fine-silty	Mixed	Superactive			Mesic
SANTIAGO	WI	MN	90, 91, 93	Alfisols	Udalfs	Glossudalfs	Haplic Glossudalfs	Coarse-loamy	Mixed	Superactive			Frigid
SAYBROOK	IL	WI	108A, 110, 95B	Mollisols	Udolls	Argiudolls	Oxyaquic Argiudolls	Fine-silty	Mixed	Superactive			Mesic
SEATON	IL	IA, MN, WI	104, 105, 108C, 115C, 90B	Alfisols	Udalfs	Hapludalfs	Typic Hapludalfs	Fine-silty	Mixed	Superactive			Mesic
SEBEWA	MI	IN, WI	110, 111B, 111C, 95A, 95B, 97, 98	Mollisols	Aquolls	Argiaquolls	Typic Argiaquolls	Fine-loamy over sandy or sandy-skeletal	Mixed	Superactive			Mesic
SEELYEVILLE	MN	ND, WI	55B, 56, 57, 88, 90, 91, 93, 94A, 95A	Histosols	Saprists	Haplosaprists	Typic Haplosaprists				Euic		Frigid
SPARTA	WI	IA, IL, IN, MI, MN, OH	103, 104, 105, 108, 109, 110, 115, 90, 91, 95B, 98	Mollisols	Udolls	Hapludolls	Entic Hapludolls	Sandy	Mixed				Mesic
SPINKS	MI	IN, MN, OH, WI	105, 111B, 95B, 96, 97, 98, 99	Alfisols	Udalfs	Hapludalfs	Lamellic Hapludalfs	Sandy	Mixed				Mesic
ST. CHARLES	IL	IN, WI	108A, 108B, 110, 111D, 115C, 95B	Alfisols	Udalfs	Hapludalfs	Typic Hapludalfs	Fine-silty	Mixed	Superactive			Mesic
TAMA	IA	IL, IN, MN, WI	104, 105, 108B, 108C, 95B	Mollisols	Udolls	Argiudolls	Typic Argiudolls	Fine-silty	Mixed	Superactive			Mesic

(continued)

(continued)

Series name	State TL	Other states using	MLRAs using	Order	Suborder	Great group	Subgroup	Particle-size class	Mineral class	CEC class	Reaction class	Other family Class	Soil-temperature class
TAYLOR	MN	WI	57, 88, 90	Alfisols	Udalfs	Glossudalfs	Aquertic Glossudalfs	Very fine	Smectitic				Frigid
THERESA	WI		95B	Alfisols	Udalfs	Hapludalfs	Typic Hapludalfs	Fine-loamy	Mixed	Superactive			Mesic
URNE	WI		105, 95B	Inceptisols	Udepts	Eutrudepts	Dystric Eutrudepts	Coarse-loamy	Mixed	Active			Mesic
VALTON	WI	IA, MN	105	Alfisols	Udalfs	Paleudalfs	Mollic Paleudalfs	Fine-silty	Mixed	Superactive			Mesic
VARNA	IL	IN, WI	108A, 108B, 110, 111C, 111D, 115C, 95B, 97	Mollisols	Udolls	Argiudolls	Oxyaquic Argiudolls	Fine	Illitic				Mesic
WAUKEGAN	MN	IA, IL, WI	103, 104, 105, 108B, 108C, 90B, 91A	Mollisols	Udolls	Hapludolls	Typic Hapludolls	Fine-silty over sandy or sandy-skeletal	Mixed	Superactive			Mesic
WITHEE	WI		90	Alfisols	Udalfs	Glossudalfs	Aquic Glossudalfs	Fine-loamy	Mixed	Superactive			Frigid
WYOCENA	WI		89, 95A, 95B	Alfisols	Udalfs	Hapludalfs	Typic Hapludalfs	Coarse-loamy	Mixed	Semiactive			Mesic

Appendix G
Endemic, Rare, and Endangered Soils of Wisconsin

Series Name	Endemic[a]	Rare[b]	Family
ADDER	Y	Y	SANDY OR SANDY-SKELETAL, SILICEOUS, EUIC, MESIC TERRIC HAPLOSAPRISTS
ALBAN	Y	Y	COARSE-LOAMY, MIXED, SUPERACTIVE, FRIGID TYPIC GLOSSUDALFS
ANGELICA	Y	N	FINE-LOAMY, MIXED, ACTIVE, NONACID, FRIGID AERIC ENDOAQUEPTS
ANNRIVER	Y	N	COARSE-LOAMY OVER SANDY OR SANDY-SKELETAL, MIXED, SUPERACTIVE, FRIGID MOLLIC ENDOAQUALFS
ANTIGO	Y	N	COARSE-LOAMY OVER SANDY OR SANDY-SKELETAL, MIXED, SUPERACTIVE, FRIGID HAPLIC GLOSSUDALFS
BANAT	Y	Y	LOAMY-SKELETAL, MIXED, ACTIVE, FRIGID AERIC ENDOAQUALFS
BEARTREE	Y	Y	LOAMY-SKELETAL, MIXED, SUPERACTIVE, FRIGID LITHIC ENDOAQUOLLS
BIGISLAND	Y	Y	SANDY-SKELETAL, ISOTIC, FRIGID TYPIC HAPLUDALFS
BILLYBOY	Y	Y	COARSE-LOAMY OVER SANDY OR SANDY-SKELETAL, MIXED, SUPERACTIVE, FRIGID OXYAQUIC GLOSSUDALFS
BJORKLAND	Y	Y	SANDY OVER CLAYEY, MIXED OVER SMECTITIC, FRIGID TYPIC EPIAQUALFS
BOAZ	Y	Y	FINE-SILTY, MIXED, SUPERACTIVE, NONACID, MESIC AERIC ENDOAQUEPTS
BOGUSCREEK	Y	Y	COARSE-SILTY, MIXED, SUPERACTIVE, NONACID, MESIC MOLLIC UDIFLUVENTS
BONDUEL	Y	Y	FINE-LOAMY, MIXED, ACTIVE, FRIGID AQUOLLIC HAPLUDALFS
BORTH	Y	Y	CLAYEY OVER SANDY OR SANDY-SKELETAL, MIXED, ACTIVE, MESIC MOLLIC OXYAQUIC HAPLUDALFS
BOWSTRING	Y	N	EUIC, FRIGID FLUVAQUENTIC HAPLOSAPRISTS
BRANDER	Y	Y	FINE-SILTY OVER SANDY OR SANDY-SKELETAL, MIXED, SUPERACTIVE, FRIGID OXYAQUIC GLOSSUDALFS
BRICE	Y	Y	COARSE-LOAMY, MIXED, SUPERACTIVE, MESIC LAMELLIC HAPLUDALFS
BRODALE	Y	Y	LOAMY-SKELETAL, CARBONATIC, MESIC ENTIC HAPLUDOLLS
CARYVILLE	Y	Y	SANDY, MIXED, FRIGID FLUVENTIC HAPLUDOLLS
CHAMPION	Y	N	COARSE-LOAMY, MIXED, SUPERACTIVE, FRIGID OXYAQUIC FRAGIORTHODS
CHANNING	Y	Y	COARSE-LOAMY OVER SANDY OR SANDY-SKELETAL, MIXED, SUPERACTIVE, FRIGID TYPIC ENDOAQUODS
CHELMO	Y	Y	CLAYEY OVER SANDY OR SANDY-SKELETAL, SMECTITIC OVER MIXED, FRIGID UMBRIC EPIAQUALFS
CHINWHISKER	Y	Y	SANDY, MIXED, FRIGID LAMELLIC OXYAQUIC HAPLORTHODS
CLEMENS	Y	Y	LOAMY-SKELETAL, MIXED, SUPERACTIVE, FRIGID AQUIC DYSTRIC EUTRUDEPTS
CROMWELL	Y	N	SANDY, ISOTIC, FRIGID TYPIC DYSTRUDEPTS
DAIRYLAND	Y	Y	SANDY-SKELETAL, MIXED, FRIGID MOLLIC OXYAQUIC HAPLUDALFS
DAISYBAY	Y	Y	CLAYEY, SMECTITIC, EUIC, FRIGID TERRIC HAPLOHEMISTS
DAWSIL	Y	N	SANDY OR SANDY-SKELETAL, SILICEOUS, DYSIC, FRIGID TERRIC HAPLOSAPRISTS
DECHAMPS	Y	Y	SANDY, MIXED, FRIGID AQUIC UDIFLUVENTS
DELLS	Y	Y	FINE-SILTY OVER SANDY OR SANDY-SKELETAL, MIXED, SUPERACTIVE, MESIC AQUOLLIC HAPLUDALFS

(continued)

J.G. Bockheim and A.E. Hartemink, *The Soils of Wisconsin*,
World Soils Book Series, DOI 10.1007/978-3-319-52144-2

(continued)

Series Name	Endemic[a]	Rare[b]	Family
DENOMIE	Y	Y	FINE-SILTY, MIXED, ACTIVE, FRIGID HAPLIC GLOSSUDALFS
DODGEVILLE	Y	N	FINE-SILTY OVER CLAYEY, MIXED, SUPERACTIVE, MESIC TYPIC ARGIUDOLLS
DODY	Y	Y	CLAYEY, SMECTITIC, FRIGID ARENIC ALBAQUALFS
DOLPH	Y	Y	FINE, MIXED, SUPERACTIVE, FRIGID AERIC GLOSSAQUALFS
DRESDEN	Y	N	FINE-LOAMY OVER SANDY OR SANDY-SKELETAL, MIXED, ACTIVE, MESIC MOLLIC HAPLUDALFS
DRYLANDING	Y	Y	LOAMY-SKELETAL, MIXED, SUPERACTIVE, FRIGID LITHIC HAPLUDOLLS
DUNNVILLE	Y	Y	COARSE-LOAMY, MIXED, SUPERACTIVE, FRIGID TYPIC HAPLUDOLLS
EAUCLAIRE	Y	Y	SANDY, MIXED, FRIGID ALFIC OXYAQUIC HAPLORTHODS
EDMUND	Y	N	CLAYEY, SMECTITIC, MESIC LITHIC ARGIUDOLLS
ELKMOUND	Y	N	LOAMY, MIXED, SUPERACTIVE, MESIC, SHALLOW TYPIC DYSTRUDEPTS
ELM LAKE	Y	N	SANDY OVER LOAMY, SILICEOUS, SEMIACTIVE, ACID, FRIGID HUMAQUEPTIC EPIAQUENTS
EMMET	Y	N	COARSE-LOAMY, MIXED, ACTIVE, FRIGID INCEPTIC HAPLUDALFS
FAIRCHILD	Y	N	SANDY OVER LOAMY, SILICEOUS, SEMIACTIVE, FRIGID ULTIC EPIAQUODS
FENCE	Y	Y	COARSE-SILTY, MIXED, SUPERACTIVE, FRIGID ALFIC OXYAQUIC HAPLORTHODS
FISK	Y	Y	SANDY OVER LOAMY, MIXED, SUPERACTIVE, MESIC AQUIC DYSTRIC EUTRUDEPTS
FIVEPOINTS	Y	N	CLAYEY OVER LOAMY-SKELETAL, MIXED, SUPERACTIVE, MESIC TYPIC HAPLUDALFS
FLOYD	Y	Y	FINE-LOAMY, MIXED, SUPERACTIVE, MESIC AQUIC PACHIC HAPLUDOLLS
FORDUM	Y	N	COARSE-LOAMY, MIXED, SUPERACTIVE, NONACID, FRIGID MOLLIC FLUVAQUENTS
FRECHETTE	Y	Y	COARSE-LOAMY, MIXED, ACTIVE, FRIGID TYPIC GLOSSUDALFS
FREMSTADT	Y	Y	SANDY, MIXED, FRIGID ARENIC HAPLUDALFS
FREYA	Y	Y	SANDY OVER CLAYEY, MIXED OVER SMECTITIC, FRIGID AQUIC ARGIUDOLLS
GAPHILL	Y	Y	COARSE-LOAMY, SILICEOUS, ACTIVE, MESIC TYPIC HAPLUDALFS
GARNE	Y	Y	SANDY OVER LOAMY, MIXED, ACTIVE, MESIC TYPIC HAPLUDOLLS
GLENFLORA	Y	Y	FINE-SILTY, MIXED, SUPERACTIVE, FRIGID MOLLIC GLOSSAQUALFS
GUENTHER	Y	Y	SANDY OVER LOAMY, MIXED, SUPERACTIVE, FRIGID ALFIC OXYAQUIC HAPLORTHODS
HALDER	Y	Y	FINE-LOAMY OVER SANDY OR SANDY-SKELETAL, MIXED, SUPERACTIVE, FRIGID AQUIC GLOSSUDALFS
HAUGEN	Y	N	COARSE-LOAMY, MIXED, SUPERACTIVE, FRIGID OXYAQUIC PALEUDALFS
HAUSTRUP	Y	Y	LOAMY, ISOTIC, FRIGID HUMIC LITHIC DYSTRUDEPTS
HOOP	Y	Y	COARSE-LOAMY, SILICEOUS, ACTIVE, MESIC AQUIC ARGIUDOLLS
HUMBIRD	Y	N	COARSE-LOAMY OVER CLAYEY, MIXED, SEMIACTIVE, FRIGID OXYAQUIC ULTIC HAPLORTHODS
INDUS	Y	N	VERY FINE, SMECTITIC, FRIGID VERTIC EPIAQUALFS
IRONRUN	Y	N	SANDY, SILICEOUS, FRIGID TYPIC ENDOAQUODS
KARLSBORG	Y	Y	CLAYEY, SMECTITIC, FRIGID ARENIC OXYAQUIC HAPLUDALFS
KAUKAUNA	Y	Y	CLAYEY OVER LOAMY, MIXED, ACTIVE, MESIC MOLLIC OXYAQUIC HAPLUDALFS
KELLOGG	Y	N	SANDY OVER CLAYEY, MIXED, ACTIVE, FRIGID ALFIC OXYAQUIC HAPLORTHODS
KEYESVILLE	Y	Y	LOAMY-SKELETAL, MIXED, ACTIVE, MESIC TYPIC DYSTRUDEPTS
KOROBAGO	Y	Y	COARSE-LOAMY OVER CLAYEY, MIXED, ACTIVE, MESIC AQUIC EUTRUDEPTS
KRANSKI	Y	N	SANDY, MIXED, MESIC TYPIC HAPLUDALFS
LABLATZ	Y	Y	COARSE-LOAMY, MIXED, ACTIVE, FRIGID ALFIC EPIAQUODS
LAPOIN	Y	Y	FINE, MIXED, ACTIVE, FRIGID ALFIC OXYAQUIC HAPLORTHODS
LARA	Y	Y	SANDY OVER CLAYEY, MIXED OVER SMECTITIC, FRIGID OXYAQUIC ARGIUDOLLS
LERCH	Y	Y	VERY FINE, MIXED, ACTIVE, NONACID, FRIGID VERTIC EPIAQUEPTS
LOWS	Y	Y	FINE-LOAMY OVER SANDY OR SANDY-SKELETAL, MIXED, SUPERACTIVE, NONACID, FRIGID MOLLIC ENDOAQUEPTS

(continued)

(continued)

Series Name	Endemic[a]	Rare[b]	Family
LUDINGTON	Y	N	SANDY OVER LOAMY, SILICEOUS, SEMIACTIVE, FRIGID OXYAQUIC ULTIC HAPLORTHODS
LUNDEEN	Y	Y	COARSE-SILTY, ISOTIC, FRIGID HUMIC DYSTRUDEPTS
MAJIK	Y	Y	MESIC, COATED AQUIC QUARTZIPSAMMENTS
MAKWA	Y	Y	LOAMY-SKELETAL, ISOTIC, NONACID, FRIGID HISTIC HUMAQUEPTS
MANISTEE	Y	Y	SANDY OVER CLAYEY, MIXED, ACTIVE, FRIGID ALFIC HAPLORTHODS
MANN	Y	Y	FINE-LOAMY, MIXED, SUPERACTIVE, FRIGID TYPIC EPIAQUOLLS
MARKHAM	Y	N	FINE, ILLITIC, MESIC MOLLIC OXYAQUIC HAPLUDALFS
MARSHFIELD	Y	N	FINE-LOAMY, MIXED, SUPERACTIVE, FRIGID MOLLIC EPIAQUALFS
MATHERTON	Y	Y	FINE-LOAMY OVER SANDY OR SANDY-SKELETAL, MIXED, SUPERACTIVE, MESIC UDOLLIC ENDOAQUALFS
MEENON	Y	Y	CLAYEY, SMECTITIC, FRIGID AQUIC ARENIC HAPLUDALFS
MENASHA	Y	Y	VERY FINE, MIXED, ACTIVE, MESIC TYPIC EPIAQUOLLS
MEQUON	Y	Y	FINE, MIXED, SUPERACTIVE, MESIC UDOLLIC ENDOAQUALFS
MERRILLAN	Y	N	COARSE-LOAMY OVER CLAYEY, MIXED, SEMIACTIVE, FRIGID ULTIC EPIAQUODS
METONGA	Y	Y	COARSE-LOAMY, MIXED, SUPERACTIVE, FRIGID ENTIC HAPLORTHODS
MICHIGAMME	Y	N	COARSE-LOAMY, MIXED, SUPERACTIVE, FRIGID FRAGIC HAPLORTHODS
MINOCQUA	Y	N	COARSE-LOAMY OVER SANDY OR SANDY-SKELETAL, MIXED, SUPERACTIVE, NONACID, FRIGID TYPIC ENDOAQUEPTS
MISKOAKI	Y	Y	VERY FINE, MIXED, ACTIVE, FRIGID VERTIC GLOSSUDALFS
MONICO	Y	Y	COARSE-LOAMY, MIXED, SUPERACTIVE, FRIGID TYPIC ENDOAQUODS
MONTELLO	Y	Y	FINE, MIXED, SUPERACTIVE, MESIC OXYAQUIC ARGIUDOLLS
MOPPET	Y	N	COARSE-LOAMY, MIXED, SUPERACTIVE, FRIGID OXYAQUIC DYSTRUDEPTS
MOQUAH	Y	Y	COARSE-LOAMY, MIXED, SUPERACTIVE, NONACID, FRIGID TYPIC UDIFLUVENTS
MOSINEE	Y	N	LOAMY-SKELETAL, MIXED, SUPERACTIVE, FRIGID TYPIC DYSTRUDEPTS
MOUNDVILLE	Y	Y	SANDY, MIXED, MESIC OXYAQUIC HAPLUDALFS
MYLREA	Y	Y	COARSE-LOAMY, MIXED, SUPERACTIVE, FRIGID AQUIC DYSTRUDEPTS
NEBAGO	Y	Y	SANDY OVER CLAYEY, MIXED, ACTIVE, MESIC AQUIC DYSTRIC EUTRUDEPTS
NEENAH	Y	Y	VERY FINE, MIXED, ACTIVE, MESIC AQUOLLIC HAPLUDALFS
NEWLANG	Y	Y	SILICEOUS, MESIC HUMAQUEPTIC PSAMMAQUENTS
NEWSON	Y	N	MIXED, FRIGID HUMAQUEPTIC PSAMMAQUENTS
NICHOLS	Y	Y	COARSE-SILTY, MIXED, SUPERACTIVE, MESIC OXYAQUIC EUTRUDEPTS
NICKIN	Y	Y	FINE-LOAMY OVER SANDY OR SANDY-SKELETAL, MIXED OVER SILICEOUS, ACTIVE, FRIGID TYPIC ARGIUDOLLS
NORGO	Y	Y	LOAMY, MIXED, ACTIVE, FRIGID LITHIC HAPLUDALFS
NORTHBEND	Y	Y	COARSE-LOAMY OVER SANDY OR SANDY-SKELETAL, MIXED, ACTIVE, MESIC FLUVAQUENTIC DYSTRUDEPTS
OMRO	Y	Y	CLAYEY OVER LOAMY, MIXED, ACTIVE, MESIC TYPIC HAPLUDALFS
ONAWAY	Y	N	FINE-LOAMY, MIXED, ACTIVE, FRIGID INCEPTIC HAPLUDALFS
ORONTO	Y	Y	FINE-SILTY, MIXED, ACTIVE, FRIGID AERIC GLOSSAQUALFS
OSSMER	Y	N	COARSE-LOAMY OVER SANDY OR SANDY-SKELETAL, MIXED, SUPERACTIVE, FRIGID AQUIC GLOSSUDALFS
PEEBLES	Y	Y	VERY FINE, MIXED, ACTIVE, MESIC OXYAQUIC ARGIUDOLLS
PERIDA	Y	Y	CLAYEY, SMECTITIC, FRIGID ARENIC HAPLUDALFS
PICKFORD	Y	N	FINE, MIXED, ACTIVE, NONACID, FRIGID AERIC EPIAQUEPTS
PONYCREEK	Y	N	SILICEOUS, FRIGID HUMAQUEPTIC PSAMMAQUENTS
POSKIN	Y	N	FINE-SILTY OVER SANDY OR SANDY-SKELETAL, MIXED, SUPERACTIVE, FRIGID AQUIC GLOSSUDALFS
POYGAN	Y	N	FINE, MIXED, ACTIVE, MESIC TYPIC EPIAQUOLLS
QUARDERER	Y	Y	COARSE-SILTY, MIXED, SUPERACTIVE, FRIGID FLUVENTIC HAPLUDOLLS
REDRIM	Y	Y	SANDY-SKELETAL, MIXED, FRIGID ENTIC LITHIC HAPLORTHODS

(continued)

(continued)

Series Name	Endemic[a]	Rare[b]	Family
REEDSBURG	Y	Y	FINE-SILTY, MIXED, SUPERACTIVE, MESIC AQUIC PALEUDALFS
RIB	Y	N	FINE-SILTY OVER SANDY OR SANDY-SKELETAL, MIXED, SUPERACTIVE, FRIGID MOLLIC ENDOAQUALFS
ROCKMARSH	Y	Y	LOAMY-SKELETAL, MIXED, SUPERACTIVE, FRIGID AQUOLLIC HAPLUDALFS
SECHLER	Y	Y	COARSE-LOAMY, SILICEOUS, SUPERACTIVE, MESIC AQUIC HUMIC DYSTRUDEPTS
SEDGWICK	Y	Y	COARSE-LOAMY OVER CLAYEY, MIXED, ACTIVE, FRIGID ALFIC EPIAQUODS
SHAG	Y	Y	COARSE-SILTY, MIXED, ACTIVE, FRIGID TYPIC ENDOAQUOLLS
SHERRY	Y	N	FINE-LOAMY, MIXED, SUPERACTIVE, FRIGID UDOLLIC ENDOAQUALFS
SIOUXCREEK	Y	Y	COARSE-LOAMY OVER SANDY OR SANDY-SKELETAL, MIXED, ACTIVE, FRIGID AQUIC HAPLUDULTS
SIREN	Y	Y	FINE-LOAMY OVER CLAYEY, MIXED OVER SMECTITIC, SUPERACTIVE, FRIGID AQUIC GLOSSUDALFS
SKOG	Y	Y	SANDY-SKELETAL, MIXED, FRIGID OXYAQUIC UDORTHENTS
SMESTAD	Y	Y	COARSE-LOAMY OVER CLAYEY, MIXED OVER SMECTITIC, SUPERACTIVE, FRIGID AQUIC ARGIUDOLLS
SODERBECK	Y	Y	LOAMY-SKELETAL, MIXED, ACTIVE, FRIGID AQUIC HAPLUDOLLS
SPEAR	Y	Y	COARSE-SILTY, MIXED, SUPERACTIVE, FRIGID AQUIC GLOSSUDALFS
STAMBAUGH	Y	N	COARSE-SILTY OVER SANDY OR SANDY-SKELETAL, MIXED, SUPERACTIVE, FRIGID ALFIC HAPLORTHODS
STENGEL	Y	Y	CLAYEY, SMECTITIC, FRIGID AQUIC ARENIC HAPLUDALFS
STURGEON	Y	Y	COARSE-SILTY OVER SANDY OR SANDY-SKELETAL, MIXED, SUPERACTIVE, NONACID, FRIGID AQUIC UDIFLUVENTS
SUPERIOR	Y	Y	COARSE-LOAMY OVER CLAYEY, MIXED, ACTIVE, FRIGID ALFIC OXYAQUIC HAPLORTHODS
TAYLOR	Y	Y	VERY FINE, SMECTITIC, FRIGID AQUERTIC GLOSSUDALFS
TINTSON	Y	Y	MESIC, UNCOATED OXYAQUIC QUARTZIPSAMMENTS
TOTAGATIC	Y	Y	SANDY, MIXED, FRIGID TYPIC FLUVAQUENTS
TRADELAKE	Y	Y	COARSE-LOAMY OVER CLAYEY, MIXED OVER SMECTITIC, SUPERACTIVE, FRIGID OXYAQUIC GLOSSUDALFS
TULA	Y	N	COARSE-LOAMY, MIXED, SUPERACTIVE, FRIGID ARGIC FRAGIAQUODS
VARNA	Y	Y	FINE, ILLITIC, MESIC OXYAQUIC ARGIUDOLLS
VEEDUM	Y	N	FINE-LOAMY, MIXED, SUPERACTIVE, ACID, FRIGID HUMIC EPIAQUEPTS
VESPER	Y	N	FINE-LOAMY OVER SANDY OR SANDY-SKELETAL, MIXED, SUPERACTIVE, ACID, FRIGID HUMIC EPIAQUEPTS
WARMAN	Y	Y	COARSE-LOAMY OVER SANDY OR SANDY-SKELETAL, MIXED, SUPERACTIVE, FRIGID TYPIC ENDOAQUOLLS
WATTON	Y	Y	FINE-LOAMY, MIXED, SEMIACTIVE, FRIGID HAPLIC GLOSSUDALFS
WAUTOMA	Y	Y	SANDY OVER CLAYEY, MIXED, ACTIVE, NONACID, MESIC MOLLIC EPIAQUENTS
WAYKA	Y	Y	COARSE-LOAMY, ISOTIC, FRIGID TYPIC EPIAQUODS
WHISKLAKE	Y	Y	COARSE-LOAMY OVER SANDY OR SANDY-SKELETAL, MIXED, SUPERACTIVE, FRIGID ARGIC ENDOAQUODS
WILDALE	Y	N	FINE, MIXED, ACTIVE, MESIC MOLLIC PALEUDALFS
WINDWARD	Y	Y	MESIC, COATED LAMELLIC QUARTZIPSAMMENTS
WINNECONNE	Y	N	VERY FINE, MIXED, ACTIVE, MESIC MOLLIC OXYAQUIC HAPLUDALFS
WOZNY	Y	Y	COARSE-LOAMY, MIXED, SUPERACTIVE, FRIGID TYPIC EPIAQUALFS
WYEVILLE	Y	Y	CLAYEY, MIXED, ACTIVE, MESIC AQUIC ARENIC HAPLUDALFS
ZITTAU	Y	Y	CLAYEY OVER SANDY OR SANDY-SKELETAL, MIXED, ACTIVE, MESIC AQUOLLIC HAPLUDALFS

[a]Endemic = the only family of its kind in the USA

[b]Y = yes; N = no; Rare = <10,000 ha; endangered = rare and occurs only in Wisconsin

Appendix H
Inactive Soil Series in Wisconsin

J.G. Bockheim and A.E. Hartemink, *The Soils of Wisconsin*,
World Soils Book Series, DOI 10.1007/978-3-319-52144-2

Series name	State TL	Other States	MLRAs Using	Order	Suborder	Great group	Subgroup	Particle-size class	Mineral class	CEC Activity class	Reaction class	Soil-temperature class
AKAN	WI		105	Inceptisols	Aquepts	Endoaquepts	Typic Endoaquepts	Fine-silty	Mixed	Superactive	Nonacid	Mesic
ALTOONA	WI			Alfisols	Aqualfs	Glossaqualfs	Aeric Glossaqualfs	Fine-loamy over sandy or sandy-skeletal	Mixed			Frigid
BAYPORT	WI											
BIBON	WI			Spodosols	Orthods	Haplorthods	Typic Haplorthods	Sandy	Mixed			Frigid
BUTTERNUT	WI		90A	Spodosols	Orthods	Haplorthods	Alfic Oxyaquic Haplorthods	Coarse-loamy	Isotic			Frigid
DRYBURG	MI	WI	93	Spodosols	Orthods	Haplorthods	Alfic Haplorthods	Coarse-loamy over clayey	Mixed	Active		Frigid
EAUPLEINE	WI			Alfisols	Boralfs	Glossoboralfs	Typic Glossoboralfs	Coarse-loamy	Mixed			
FOXPAW	WI			Spodosols	Aquods	Endoaquods	Typic Endoaquods	Coarse-loamy	Isotic			Frigid
HINES	WI											
LA CROSSE	WI											
MELLEN	WI											
NORWALK	WI			Alfisols	Udalfs	Hapludalfs	Typic Hapludalfs	Fine-silty over clayey	Mixed			Mesic
OGDEN	WI	OH	111, 114	Histosols	Saprists	Medisaprists	Terric Medisaprists	Clayey	Illitic		Euic	Mesic
ONAMIA	WI			Alfisols	Boralfs	Glossoboralfs	Typic Glossoboralfs	Fine-loamy over sandy or sandy-skeletal	Mixed			
ORIENTA	WI			Spodosols	Orthods	Haplorthods	Aquic Haplorthods	Sandy	Mixed			Frigid
OSSEO	WI											
PEEKSVILLE	WI		90A	Spodosols	Aquods	Epiaquods	Alfic Epiaquods	Coarse-loamy	Isotic			Frigid
ROSEMOUNT	WI			Mollisols	Udolls	Argiudolls	Typic Argiudolls	Coarse-loamy	Mixed			Mesic
SHANAGOLDEN	WI		90A	Spodosols	Orthods	Haplorthods	Alfic Oxyaquic Haplorthods	Coarse-loamy	Isotic			Frigid
SUAMICO	WI			Histosols	Saprists	Borosaprists	Terric Borosaprists	Clayey	Mixed		Euic	
TAHQUAMENON	WI			Histosols	Hemists	Borohemists	Hydric Borohemists				Dysic	
TORCH	WI		90A	Spodosols	Aquods	Epiaquods	Alfic Epiaquods	Coarse-loamy	Isotic			Frigid

Appendix I
Characteristics and Yield Potential of Wisconsin Soil Series

J.G. Bockheim and A.E. Hartemink, *The Soils of Wisconsin*,
World Soils Book Series, DOI 10.1007/978-3-319-52144-2

Soil series[a]	Soil YP[b]	Soil Depth[c]	Drainage class[d]	Slope (%)	Soil area (km^2)	Soil region	Order	Suborder	Great group	Subgroup	Particle-size class	Mineral class	CEC Activity class	Reaction class	Soil-temperature class	Soil-moisture class
ABBAYE	M	MD	MWD	1–70	11	I	Spodosols	Orthods	Haplorthods	Alfic Oxyaquic Haplorthods	Coarse-loamy	Mixed	Active		Frigid	Udic
ABSCO	S	VD	MWD	0–3	236	Dr, J	Entisols	Fluvents	Udifluvents	Typic Udifluvents	Sandy	Siliceous			Mesic	Aquic
ABSCOTA	S	VD	MWD	0–6	2	B, Dr, J	Entisols	Psamments	Udipsamments	Oxyaquic Udipsamments		Mixed			Mesic	Udic
ACKMORE	H	VD	SPD	0–5	64	B, J	Entisols	Aquents	Fluvaquents	Mollic Fluvaquents	Fine-silty	Mixed	Superactive	Nonacid	Mesic	Aquic
ADDER	H	VD	VPD	0–1	23	Dr, J	Histosols	Saprists	Haplosaprists	Terric Haplosaprists	Sandy or sandy-skeletal	Siliceous		Euic	Mesic	Aquic
ADOLPH	M	VD	VPD	0–2	84	F, J	Mollisols	Aquolls	Epiaquolls	Typic Epiaquolls	Coarse-loamy	Mixed	Superactive		Frigid	Aquic
ADRIAN	H	VD	VPD	0–1	507	J	Histosols	Saprists	Haplosaprists	Terric Haplosaprists	Sandy or sandy-skeletal	Mixed		Euic	Mesic	Aquic
AFTAD	H	VD	MWD	0–12	53	F, G	Alfisols	Udalfs	Glossudalfs	Oxyaquic Glossudalfs	Coarse-loamy	Mixed	Superactive		Frigid	Udic
ALANGO	H	VD	SPD	1–4	0	I, J	Alfisols	Aqualfs	Epiaqualfs	Chromic Vertic Epiaqualfs	Very fine	Smectitic			Frigid	Ustic
ALBAN	H	VD	WD	0–30	66	C, E	Alfisols	Udalfs	Glossudalfs	Typic Glossudalfs	Coarse-loamy	Mixed	Superactive		Frigid	Udic
ALCONA	H	VD	WD	0–70	47	G, H	Spodosols	Orthods	Haplorthods	Alfic Haplorthods	Coarse-loamy	Mixed	Active		Frigid	Aquic
ALDO	S	VD	MWD	0–3	25	C	Entisols	Psamments	Udipsamments	Typic Udipsamments		Mixed			Mesic	Udic
ALGANSEE	S	VD	SPD	0–4	166	C, J	Entisols	Psamments	Udipsamments	Aquic Udipsamments		Mixed			Mesic	Udic
ALLENDALE	S	VD	SPD	0–12	153	H, I, J	Spodosols	Aquods	Epiaquods	Alfic Epiaquods	Sandy over clayey	Mixed	Semiactive		Frigid	Udic
ALMENA	H	VD	SPD	0–6	435	F	Alfisols	Udalfs	Glossudalfs	Aquic Glossudalfs	Fine-silty	Mixed	Superactive		Frigid	Aquic
ALPENA	S	VD	SED	0–50	34	E	Mollisols	Udolls	Hapludolls	Entic Hapludolls	Sandy-skeletal	Mixed			Frigid	Udic
ALSTAD	H	VD	SPD	0–4	2	F	Alfisols	Udalfs	Glossudalfs	Aquic Glossudalfs	Fine-loamy	Mixed	Superactive		Frigid	Udic
ALTDORF	M	VD	PD	0–2	48	Fr, J	Alfisols	Aqualfs	Glossaqualfs	Aeric Glossaqualfs	Fine	Mixed	Active		Frigid	Aquic
AMASA	M	VD	WD	0–70	9	G	Spodosols	Orthods	Haplorthods	Typic Haplorthods	Coarse-loamy over sandy or sandy-skeletal	Mixed	Superactive		Frigid	Udic
AMERY	M	VD	WD	1–45	991	Dr, F, G	Alfisols	Udalfs	Glossudalfs	Haplic Glossudalfs	Coarse-loamy	Mixed	Superactive		Frigid	Udic
AMNICON	H	VD	MWD	0–15	374	I	Alfisols	Udalfs	Glossudalfs	Oxyaquic Vertic Glossudalfs	Very fine	Mixed	Active		Frigid	Udic
ANGELICA	M	VD	VPD	0–2	119	E, J	Inceptisols	Aquepts	Endoaquepts	Aeric Endoaquepts	Fine-loamy	Mixed	Active	Nonacid	Frigid	Aquic

(continued)

(continued)

Soil series[a]	Soil YP[b]	Soil Depth[c]	Drainage class[d]	Slope (%)	Soil area (km^2)	Soil region	Order	Suborder	Great group	Subgroup	Particle-size class	Mineral class	CEC Activity class	Reaction class	Soil-temperature class	Soil-moisture class
ANIGON	H	VD	WD	0–25	259	F, G	Alfisols	Udalfs	Glossudalfs	Haplic Glossudalfs	Fine-silty over sandy or sandy-skeletal	Mixed	Superactive		Frigid	Udic
ANKENY	H	VD	WD	0–5	0	Bm	Mollisols	Udolls	Hapludolls	Cumulic Hapludolls	Coarse-loamy	Mixed	Superactive		Mesic	Udic
ANNALAKE	H	VD	MWD	0–18	152	G	Spodosols	Orthods	Haplorthods	Alfic Oxyaquic Haplorthods	Coarse-loamy	Mixed	Superactive		Frigid	Udic
ANNRIVER	M	VD	VPD	0–2	101	F, G, J	Alfisols	Aqualfs	Endoaqualfs	Mollic Endoaqualfs	Coarse-loamy over sandy or sandy-skeletal	Mixed	Superactive		Frigid	Udic
ANTIGO	H	VD	WD	0–30	1242	F, G	Alfisols	Udalfs	Glossudalfs	Haplic Glossudalfs	Coarse-loamy over sandy or sandy-skeletal	Mixed	Superactive		Frigid	Udic
ANTON	H	VD	MWD	0–15	14	I	Alfisols	Udalfs	Glossudalfs	Oxyaquic Vertic Glossudalfs	Very fine	Mixed	Active		Frigid	Aquic
ARBUTUS	S	MD	ED	2–6	1	Dr	Spodosols	Orthods	Haplorthods	Entic Haplorthods	Sandy	Siliceous			Frigid	Udic
ARENZVILLE	H	VD	MWD	0–5	464	A, J	Entisols	Fluvents	Udifluvents	Typic Udifluvents	Coarse-silty	Mixed	Superactive	Nonacid	Mesic	Udic
ARGONNE	H	MD	MWD	0–15	23	G	Spodosols	Orthods	Fragiorthods	Alfic Oxyaquic Fragiorthods	Coarse-loamy	Mixed	Superactive		Frigid	Udic
ARLAND	M	MD	WD	1–35	291	Dr, F	Alfisols	Udalfs	Glossudalfs	Haplic Glossudalfs	Coarse-loamy	Mixed	Superactive		Frigid	Udic
ARNHEIM	M	D	PD	0–1	203	I, J	Entisols	Aquents	Fluvaquents	Typic Fluvaquents	Coarse-loamy	Mixed	Superactive	Nonacid	Frigid	Aquic
ASHDALE	H	D	WD	0–20	210	Am	Mollisols	Udolls	Argiudolls	Typic Argiudolls	Fine-silty	Mixed	Superactive		Mesic	Udic
ASHIPPUN	H	VD	SPD	0–6	12	B	Alfisols	Udalfs	Hapludalfs	Aquollic Hapludalfs	Fine-loamy	Mixed	Active		Mesic	Udic
ASHKUM	M	VD	PD	0–3	223	Bm, J	Mollisols	Aquolls	Endoaquolls	Typic Endoaquolls	Fine	Mixed	Superactive		Mesic	Aquic
ASHWABAY	S	VD	MWD	0–45	94	G, H	Spodosols	Orthods	Haplorthods	Alfic Oxyaquic Haplorthods	Sandy	Isotic			Frigid	Udic
ATTERBERRY	H	VD	SPD	0–6	5	B, J	Alfisols	Aqualfs	Endoaqualfs	Udollic Endoaqualfs	Fine-silty	Mixed	Superactive		Mesic	Aquic
AU GRES	S	VD	SPD	0–6	384	E, H, J	Spodosols	Aquods	Endoaquods	Typic Endoaquods	Sandy	Mixed			Frigid	Aquic
AUBURNDALE	M	VD	PD	0–3	208	F, J	Alfisols	Aqualfs	Epiaqualfs	Mollic Epiaqualfs	Fine-silty	Mixed	Superactive		Frigid	Aquic
AUGWOOD	S	VD	SPD	0–3	13	H, J	Spodosols	Aquods	Epiaquods	Typic Epiaquods	Sandy	Mixed			Frigid	Udic
AUSABLE	M	VD	VPD	0–2	13	H, J	Inceptisols	Aquepts	Humaquepts	Histic Humaquepts	Sandy	Mixed			Frigid	Aquic
AZTALAN	H	VD	SPD	0–6	94	Bm, J	Mollisols	Udolls	Argiudolls	Aquic Argiudolls	Fine-loamy	Mixed	Superactive		Mesic	Udic

(continued)

(continued)

Soil series[a]	Soil YP[b]	Soil Depth[c]	Drainage class[d]	Slope (%)	Soil area (km^2)	Soil region	Order	Suborder	Great group	Subgroup	Particle-size class	Mineral class	CEC Activity class	Reaction class	Soil-temperature class	Soil-moisture class
BACH	M	VD	PD	0–2	26	E, J	Inceptisols	Aquepts	Endoaquepts	Mollic Endoaquepts	Coarse-silty	Mixed	Semiactive	Calcareous	Mesic	Aquic
BADRIVER	H	VD	SPD	0–3	181	I, J	Alfisols	Aqualfs	Glossaqualfs	Aeric Glossaqualfs	Fine	Mixed	Active		Frigid	Aquic
BALMORAL	H	VD	MWD	0–3	1	Am	Mollisols	Udolls	Argiudolls	Oxyaquic Argiudolls	Fine-loamy	Mixed	Superactive		Mesic	Udic
BANAT	M	D	SPD	0–4	1	G, J	Alfisols	Aqualfs	Endoaqualfs	Aeric Endoaqualfs	Loamy-skeletal	Mixed	Active		Frigid	Aquic
BARABOO	M	D	MWD	2–60	113	A, B	Alfisols	Udalfs	Hapludalfs	Oxyaquic Hapludalfs	Fine-silty	Mixed	Superactive		Mesic	Udic
BARREMILLS	H	VD	MWD	1–6	43	Am	Mollisols	Udolls	Argiudolls	Pachic Argiudolls	Fine-silty	Mixed	Superactive		Mesic	Udic
BARRONETT	M	VD	PD	0–2	98	F, G, J	Alfisols	Aqualfs	Epiaqualfs	Mollic Epiaqualfs	Fine-silty	Mixed	Superactive		Frigid	Aquic
BARRY	M	Vd	PD	0–3	32	Bm, J	Mollisols	Aquolls	Argiaquolls	Typic Argiaquolls	Fine-loamy	Mixed	Superactive		Mesic	Udic
BASCO	M	MD	WD	2–30	53	Bm	Alfisols	Udalfs	Hapludalfs	Mollic Hapludalfs	Fine	Mixed	Superactive		Mesic	Udic
BATAVIA	H	VD	WD	0–12	11	B	Alfisols	Udalfs	Hapludalfs	Mollic Hapludalfs	Fine-silty	Mixed	Superactive		Mesic	Udic
BEARPEN	H	VD	SPD	0–3	43	Dr	Mollisols	Udolls	Argiudolls	Aquic Argiudolls	Fine-silty	Mixed	Superactive		Mesic	Udic
BEARTREE	M	S	VPD	0–2	2	G, J	Mollisols	Aquolls	Endoaquolls	Lithic Endoaquolls	Loamy-skeletal	Mixed	Superactive		Frigid	Aquic
BEAVERBAY	H	VD	MWD	2–30	46	G	Spodosols	Orthods	Haplorthods	Alfic Oxyaquic Haplorthods	Coarse-loamy	Mixed	Superactive		Frigid	Aquic
BEAVERCREEK	H	VD	WD	1–15	26	A, J	Entisols	Fluvents	Udifluvents	Typic Udifluvents	Loamy-skeletal	Mixed	Active	Nonacid	Mesic	Udic
BEECHER	H	VD	SPD	0–6	7	B, J	Alfisols	Aqualfs	Epiaqualfs	Udollic Epiaqualfs	Fine	Illitic			Mesic	Udic
BELLECHESTER	S	VD	ED	12–90	3	Am	Mollisols	Udolls	Hapludolls	Entic Hapludolls	Sandy	Mixed			Mesic	Udic
BELLEVILLE	S	VD	VPD	0–2	0	Cm, J	Mollisols	Aquolls	Endoaquolls	Typic Endoaquolls	Sandy over loamy	Mixed	Active		Mesic	Aquic
BELLEVUE	H	VD	MWD	0–3	336	I	Mollisols	Udolls	Hapludolls	Fluventic Hapludolls	Fine-loamy	Mixed	Superactive		Mesic	Udic
BERGLAND	M	VD	PD	0–2	8	I, J	Alfisols	Aqualfs	Epiaqualfs	Aeric Vertic Epiaqualfs	Very fine	Mixed	Semiactive		Frigid	Aquic
BERTRAND	H	VD	WD	0–35	7	A	Alfisols	Udalfs	Hapludalfs	Typic Hapludalfs	Fine-silty	Mixed	Superactive		Mesic	Udic
BESEMAN	M	VD	VPD	0–1	472	J	Histosols	Saprists	Haplosaprists	Terric Haplosaprists	Loamy	Mixed		Dysic	Frigid	Aquic
BIGISLAND	S	VD	SED	15–45	5	H	Alfisols	Udalfs	Hapludalfs	Typic Hapludalfs	Sandy-skeletal	Isotic			Frigid	Aquic
BILLETT	H	VD	WD	0–20	55	C, Dr	Alfisols	Udalfs	Hapludalfs	Mollic Hapludalfs	Coarse-loamy	Mixed	Superactive		Mesic	Aquic
BILLYBOY	H	VD	WD	0–3	97	G	Alfisols	Udalfs	Glossudalfs	Oxyaquic Glossudalfs	Coarse-loamy over sandy or sandy-skeletal	Mixed	Superactive		Frigid	Udic

(continued)

(continued)

Soil series[a]	Soil YP[b]	Soil Depth[c]	Drainage class[d]	Slope (%)	Soil area (km^2)	Soil region	Order	Suborder	Great group	Subgroup	Particle-size class	Mineral class	CEC Activity class	Reaction class	Soil-temperature class	Soil-moisture class
BILMOD	H	VD	WD	0–3	32	Dr	Alfisols	Udalfs	Hapludalfs	Mollic Hapludalfs	Coarse-loamy	Siliceous	Superactive		Mesic	Udic
BILSON	H	VD	WD	0–20	133	Dr	Alfisols	Udalfs	Hapludalfs	Mollic Hapludalfs	Coarse-loamy	Siliceous	Superactive		Mesic	Udic
BJORKLAND	M	VD	VPD	0–2	9	H, J	Alfisols	Aqualfs	Epiaqualfs	Typic Epiaqualfs	Sandy over clayey	Mixed over smectitic			Frigid	Aquic
BLACKHAMMER	H	D	WD	2–20	19	A	Alfisols	Udalfs	Hapludalfs	Typic Hapludalfs	Fine-silty	Mixed	Superactive		Mesic	Udic
BLACKRIVER	H	VD	MWD	0–6	15	F	Alfisols	Udalfs	Glossudalfs	Haplic Glossudalfs	Fine-silty	Mixed	Superactive		Frigid	Udic
BLOUNT	H	VD	SPD	0–6	7	B, J	Alfisols	Aqualfs	Epiaqualfs	Aeric Epiaqualfs	Fine	Illitic			Mesic	Udic
BLUFFTON	M	VD	VPD	0–2	13	F, J	Mollisols	Aquolls	Endoaquolls	Alfisols+Z71:AE7	Fine-loamy	Mixed	Superactive		Frigid	Aquic
BOAZ	H	VD	SPD	0–3	52	A, J	Inceptisols	Aquepts	Endoaquepts	Aeric Endoaquepts	Fine-silty	Mixed	Superactive	Nonacid	Mesic	Aquic
BOGUSCREEK	H	VD	WD	0–3	1	G, J	Entisols	Fluvents	Udifluvents	Mollic Udifluvents	Coarse-silty	Mixed	Superactive	Nonacid	Mesic	Udic
BONDUEL	M	MD	SPD	0–3	71	Er	Alfisols	Udalfs	Hapludalfs	Aquollic Hapludalfs	Fine-loamy	Mixed	Active		Frigid	Udic
BOONE	S	MD	ED	1–90	617	Dr	Entisols	Psamments	Quartzipsamments	Typic Quartzipsamments					Mesic	Udic
BOOTS	H	VD	VPD	0–2	35	B, J	Histosols	Hemists	Haplohemists	Typic Haplohemists				Euic	Mesic	Aquic
BOPLAIN	S	MD	ED	0–60	15	Dr	Entisols	Psamments	Udipsamments	Typic Udipsamments		Mixed			Mesic	Udic
BOREA	H	VD	SPD	0–3	11	I, J	Alfisols	Aqualfs	Glossaqualfs	Aeric Glossaqualfs	Very fine	Mixed	Active		Frigid	Udic
BORTH	M	VD	MWD	0–6	47	E, J	Alfisols	Udalfs	Hapludalfs	Mollic Oxyaquic Hapludalfs	Clayey over sandy or sandy-skeletal	Mixed	Active		Mesic	Udic
BOWSTRING	M	VD	VPD	0–1	266	G, H, J	Histosols	Saprists	Haplosaprists	Fluvaquentic Haplosaprists				Euic	Frigid	Aquic
BOYER	S	VD	WD	0–50	398	B	Alfisols	Udalfs	Hapludalfs	Typic Hapludalfs	Coarse-loamy	Mixed	Semiactive		Mesic	Aquic
BRAHAM	S	VD	WD	6–30	0	I	Alfisols	Udalfs	Hapludalfs	Arenic Hapludalfs	Loamy	Mixed	Superactive		Frigid	Aquic
BRANDER	H	VD	MWD	0–3	18	F	Alfisols	Udalfs	Glossudalfs	Oxyaquic Glossudalfs	Fine-silty over sandy or sandy-skeletal	Mixed	Superactive		Frigid	Udic
BRANSTAD	H	VD	MWD	2–20	4	G	Alfisols	Udalfs	Glossudalfs	Oxyaquic Glossudalfs	Fine-loamy	Mixed	Superactive		Frigid	Udic
BREMS	S	VD	MWD	0–8	339	B, C	Entisols	Psamments	Udipsamments	Aquic Udipsamments		Mixed			Mesic	Udic
BRENNYVILLE	H	VD	SPD	0–6	1	G	Alfisols	Udalfs	Glossudalfs	Aquic Glossudalfs	Coarse-loamy	Mixed	Superactive		Frigid	Udic

(continued)

(continued)

Soil series[a]	Soil YP[b]	Soil Depth[c]	Drainage class[d]	Slope (%)	Soil area (km^2)	Soil region	Order	Suborder	Great group	Subgroup	Particle-size class	Mineral class	CEC Activity class	Reaction class	Soil-temperature class	Soil-moisture class
BREVORT	S	VD	VPD	0–2	79	E, J	Entisols	Aquents	Endoaquents	Mollic Endoaquents	Sandy over loamy	Mixed	Active	Nonacid	Frigid	Aquic
BRICE	S	VD	WD	2–6	2	B, Dr	Alfisols	Udalfs	Hapludalfs	Lamellic Hapludalfs	Coarse-loamy	Mixed	Superactive		Mesic	Udic
BRIGGSVILLE	H	VD	WD	0–20	102	I	Alfisols	Udalfs	Hapludalfs	Typic Hapludalfs	Fine	Mixed	Superactive		Mesic	Udic
BRILL	H	VD	MWD	0–6	102	F	Alfisols	Udalfs	Glossudalfs	Haplic Glossudalfs	Fine-silty over sandy or sandy-skeletal	Mixed	Superactive		Frigid	Udic
BRIMLEY	H	D	SPD	0–6	3	G, J	Spodosols	Aquods	Endoaquods	Argic Endoaquods	Fine-loamy	Mixed	Superactive		Frigid	Udic
BRINKMAN	H	VD	MWD	2–12	44	A, Dr	Alfisols	Udalfs	Paleudalfs	Mollic Paleudalfs	Fine-silty	Mixed	Superactive		Mesic	Udic
BRODALE	M	VD	ED	6–90	7	A, Dr	Mollisols	Udolls	Hapludolls	Entic Hapludolls	Loamy-skeletal	Carbonatic			Mesic	Udic
BROOKSTON	M	VD	PD	0–3	130	B, J	Mollisols	Aquolls	Argiaquolls	Typic Argiaquolls	Fine-loamy	Mixed	Superactive		Mesic	Udic
BROWNCHURCH	H	VD	WD	12–35	58	A, Dr	Alfisols	Udalfs	Hapludalfs	Mollic Hapludalfs	Coarse-loamy	Mixed	Superactive		Mesic	Udic
BROWNSTONE	S	MD	ED	0–15	6	G, H	Spodosols	Orthods	Haplorthods	Typic Haplorthods	Sandy-skeletal	Mixed			Frigid	Aquic
BRUCE	M	D	PD	0–2	4	G, E, J	Inceptisols	Aquepts	Endoaquepts	Mollic Endoaquepts	Fine-loamy	Mixed	Superactive	Nonacid	Frigid	Udic
BURKHARDT	S	VD	SED	0–30	182	C	Mollisols	Udolls	Hapludolls	Typic Hapludolls	Sandy	Mixed			Mesic	Udic
BUSHVILLE	S	VD	SPD	0–6	1	G	Alfisols	Udalfs	Hapludalfs	Aquic Arenic Hapludalfs	Loamy	Mixed	Superactive		Frigid	Udic
CABLE	M	VD	VPD	0–2	340	F, G, J	Inceptisols	Aquepts	Epiaquepts	Typic Epiaquepts	Coarse-loamy	Isotic		Nonacid	Frigid	Aquic
CADIZ	H	VD	MWD	2–12	38	A	Alfisols	Udalfs	Hapludalfs	Oxyaquic Hapludalfs	Fine-silty	Mixed	Superactive		Mesic	Aquic
CALAMINE	M	D	PD	0–12	9	Am, J	Mollisols	Aquolls	Argiaquolls	Typic Argiaquolls	Fine	Mixed	Superactive		Mesic	Aquic
CAMPIA	H	VD	WD	0–25	57	F	Alfisols	Udalfs	Glossudalfs	Haplic Glossudalfs	Fine-silty	Mixed	Superactive		Frigid	Udic
CAPITOLA	M	VD	PD	0–2	707	F, G, J	Alfisols	Aqualfs	Epiaqualfs	Aeric Epiaqualfs	Coarse-loamy	Mixed	Superactive		Frigid	Aquic
CARBONDALE	M	VD	VPD	0–2	369	J	Histosols	Saprists	Haplosaprists	Hemic Haplosaprists				Euic	Frigid	Aquic
CARYVILLE	H	D	WD	0–3	12	Cm, J	Mollisols	Udolls	Hapludolls	Fluventic Hapludolls	Sandy	Mixed			Frigid	Udic
CASCO	M	VD	SED	0–70	699	B, I	Alfisols	Udalfs	Hapludalfs	Inceptic Hapludalfs	Fine-loamy over sandy or sandy-skeletal	Mixed	Superactive		Mesic	Udic
CATHRO	M	VD	VPD	0–2	1823	A, J	Histosols	Saprists	Haplosaprists	Terric Haplosaprists	Loamy	Mixed		Euic	Frigid	Aquic

(continued)

(continued)

Soil series[a]	Soil YP[b]	Soil Depth[c]	Drainage class[d]	Slope (%)	Soil area (km^2)	Soil region	Order	Suborder	Great group	Subgroup	Particle-size class	Mineral class	CEC Activity class	Reaction class	Soil-temperature class	Soil-moisture class
CEBANA	M	VD	VPD	0–2	567	F, J	Alfisols	Aqualfs	Glossaqualfs	Mollic Glossaqualfs	Coarse-loamy	Mixed	Superactive		Frigid	Aquic
CERESCO	H	VD	SPD	0–3	36	Dr	Mollisols	Udolls	Hapludolls	Fluvaquentic Hapludolls	Coarse-loamy	Mixed	Superactive		Mesic	Udic
CHABENEAU	M	VD	MWD	0–18	11	G	Spodosols	Orthods	Haplorthods	Oxyaquic Haplorthods	Coarse-loamy over sandy or sandy-skeletal	Mixed	Superactive		Frigid	Udic
CHAMPION	M	VD	WD	0–70	14	G	Spodosols	Orthods	Fragiorthods	Oxyaquic Fragiorthods	Coarse-loamy	Mixed	Superactive		Frigid	Udic
CHANNAHON	M	S	WD	0–25	49	B	Mollisols	Udolls	Argiudolls	Lithic Argiudolls	Loamy	Mixed	Superactive		Mesic	Udic
CHANNING	M	VD	SPD	0–3	9	G	Spodosols	Aquods	Endoaquods	Typic Endoaquods	Coarse-loamy over sandy or sandy-skeletal	Mixed	Superactive		Frigid	Aquic
CHARLEVOIX	H	VD	SPD	0–6	93	G, J	Spodosols	Aquods	Endoaquods	Argic Endoaquods	Coarse-loamy	Mixed	Semiactive		Frigid	Udic
CHASEBURG	H	VD	WD	0–15	277	A, J	Entisols	Fluvents	Udifluvents	Typic Udifluvents	Coarse-silty	Mixed	Superactive	Nonacid	Mesic	Udic
CHELMO	M	VD	PD	0–2	3	E, G, J	Alfisols	Aqualfs	Epiaqualfs	Umbric Epiaqualfs	Clayey over sandy or sandy-skeletal	Smectitic over mixed			Frigid	Aquic
CHELSEA	S	VD	ED	0–45	74	C	Entisols	Psamments	Udipsamments	Lamellic Udipsamments		Mixed			Mesic	Udic
CHEQUAMEGON	H	VD	MWD	2–30	341	G	Spodosols	Orthods	Haplorthods	Alfic Oxyaquic Haplorthods	Coarse-loamy	Mixed	Superactive		Frigid	Aquic
CHETEK	M	VD	SED	0–45	750	Dr, G, H	Alfisols	Udalfs	Hapludalfs	Inceptic Hapludalfs	Coarse-loamy	Mixed	Superactive		Frigid	Aquic
CHINWHISKER	S	VD	MWD	0–4	29	G	Spodosols	Orthods	Haplorthods	Lamellic Oxyaquic Haplorthods	Sandy	Mixed			Frigid	Udic
CHIPPENY	M	MD	VPD	0–12	2	H, J	Histosols	Saprists	Haplosaprists	Lithic Haplosaprists				Euic	Frigid	Aquic
CHURCHTOWN	H	VD	WD	6–30	455	A	Alfisols	Udalfs	Hapludalfs	Mollic Hapludalfs	Fine-silty	Mixed	Superactive		Mesic	Udic
CITYPOINT	M	MD	VPD	0–1	98	I, J	Histosols	Saprists	Haplosaprists	Typic Haplosaprists				Dysic	Frigid	Aquic
CLEMENS	M	VD	SPD	0–2	3	G	Inceptisols	Udepts	Eutrudepts	Aquic Dystric Eutrudepts	Loamy-skeletal	Mixed	Superactive		Frigid	Udic
CLYDE	M	VD	VPD	0–4	21	E, G, J	Mollisols	Aquolls	Endoaquolls	Typic Endoaquolls	Fine-loamy	Mixed	Superactive		Mesic	Aquic
COFFEEN	H	VD	SPD	0–2	37	Am	Mollisols	Udolls	Hapludolls	Fluvaquentic Hapludolls	Coarse-silty	Mixed	Superactive		Mesic	Udic
COFFTON	H	VD	SPD	0–3	10	Dr	Mollisols	Udolls	Hapludolls	Fluvaquentic Hapludolls	Coarse-silty	Mixed	Superactive		Mesic	Udic

(continued)

(continued)

Soil series[a]	Soil YP[b]	Soil Depth[c]	Drainage class[d]	Slope (%)	Soil area (km^2)	Soil region	Order	Suborder	Great group	Subgroup	Particle-size class	Mineral class	CEC Activity class	Reaction class	Soil-temperature class	Soil-moisture class
COLAND	M	VD	PD	0–5	1	Am	Mollisols	Aquolls	Endoaquolls	Cumuliic Endoaquolls	Fine-loamy	Mixed	Superactive		Mesic	Aquic
COLOMA	S	VD	SED	0–70	303	C	Entisols	Psamments	Udipsamments	Lamellic Udipsamments		Mixed			Mesic	Udic
COLWOOD	M	VD	VPD	0–3	168	Am, Bm, J	Mollisols	Aquolls	Endoaquolls	Typic Endoaquolls	Fine-loamy	Mixed	Active		Mesic	Aquic
COMSTOCK	H	VD	SPD	0–3	143	F, G	Alfisols	Udalfs	Glossudalfs	Aquic Glossudalfs	Fine-silty	Mixed	Superactive		Frigid	Udic
CONOVER	H	VD	SPD	0–6	1	B, J	Alfisols	Aqualfs	Endoaqualfs	Udollic Endoaqualfs	Fine-loamy	Mixed	Active		Mesic	Udic
CORMANT	S	VD	VPD	0–3	194	H, J	Entisols	Aquents	Psammaquents	Mollic Psammaquents		Mixed			Frigid	Udic
CORNUCOPIA	H	VD	WD	2–45	103	I	Alfisols	Udalfs	Glossudalfs	Haplic Glossudalfs	Fine	Mixed	Active		Frigid	Udic
COSAD	S	VD	SPD	0–8	12	I	Entisols	Orthents	Udorthents	Aquic Udorthents	Sandy over clayey	Mixed	Superactive	Nonacid	Mesic	Udic
COUNCIL	H	VD	WD	2–50	312	Dr	Alfisols	Udalfs	Hapludalfs	Typic Hapludalfs	Coarse-loamy	Mixed	Superactive		Mesic	Aquic
CRESS	S	VD	SED	0–35	195	H	Inceptisols	Udepts	Dystrudepts	Humic Dystrudepts	Sandy	Mixed			Frigid	Aquic
CREX	S	VD	MWD	0–3	33	H	Entisols	Psamments	Udipsamments	Oxyaquic Udipsamments		Mixed			Frigid	Udic
CROMWELL	S	VD	SED	0–40	238	H	Inceptisols	Udepts	Dystrudepts	Typic Dystrudepts	Sandy	Isotic			Frigid	Udic
CROSSETT	H	VD	SPD	0–6	4	G	Alfisols	Udalfs	Glossudalfs	Aquic Glossudalfs	Fine-loamy	Mixed	Active		Frigid	Udic
CROSWELL	S	VD	MWD	0–12	518	H	Spodosols	Orthods	Haplorthods	Oxyaquic Haplorthods	Sandy	Isotic			Frigid	Udic
CROSWOOD	S	VD	MWD	0–6	27	H	Spodosols	Orthods	Haplorthods	Oxyaquic Haplorthods	Sandy	Mixed			Frigid	Udic
CRYSTAL LAKE	H	VD	MWD	0–45	145	F, G	Alfisols	Udalfs	Glossudalfs	Oxyaquic Glossudalfs	Fine-silty	Mixed	Superactive		Frigid	Udic
CUBLAKE	S	VD	MWD	0–15	118	G, H	Spodosols	Orthods	Haplorthods	Oxyaquic Haplorthods	Sandy	Mixed			Frigid	Udic
CUNARD	M	MD	WD	0–18	2	E, G	Alfisols	Udalfs	Hapludalfs	Typic Hapludalfs	Coarse-loamy	Mixed	Active		Frigid	Udic
CURRAN	H	VD	SPD	0–5	103	A, J	Alfisols	Aqualfs	Endoaqualfs	Udollic Endoaqualfs	Fine-silty	Mixed	Superactive		Mesic	Udic
CUSHING	H	VD	WD	20–35	7	F	Alfisols	Udalfs	Glossudalfs	Haplic Glossudalfs	Fine-loamy	Mixed	Superactive		Frigid	Udic
CUTAWAY	S	VD	MWD	0–12	0	G	Alfisols	Udalfs	Hapludalfs	Arenic Oxyaquic Hapludalfs	Loamy	Mixed	Superactive		Frigid	Udic
CUTTRE	H	VD	SPD	0–3	370	I, J	Alfisols	Aqualfs	Glossaqualfs	Aeric Glossaqualfs	Very fine	Mixed	Active		Frigid	Aquic
DAIRYLAND	S	VD	MWD	0–20	13	H	Alfisols	Udalfs	Hapludalfs	Mollic Oxyaquic Hapludalfs	Sandy-skeletal	Mixed			Frigid	Udic
DAISYBAY	M	VD	VPD	0–1	1	I, J	Histosols	Hemists	Haplohemists	Terric Haplohemists	Clayey	Smectitic		Euic	Frigid	Aquic

(continued)

(continued)

Soil series[a]	Soil YP[b]	Soil Depth[c]	Drainage class[d]	Slope (%)	Soil area (km^2)	Soil region	Order	Suborder	Great group	Subgroup	Particle-size class	Mineral class	CEC Activity class	Reaction class	Soil-temperature class	Soil-moisture class
DAKOTA	H	VD	WD	0–18	148	Am, Cm	Mollisols	Udolls	Argiudolls	Typic Argiudolls	Fine-loamy over sandy or sandy-skeletal	Mixed	Superactive		Mesic	Udic
DANCY	M	VD	PD	0–2	96	C, Fr, J	Alfisols	Aqualfs	Glossaqualfs	Typic Glossaqualfs	Fine-loamy	Mixed	Superactive		Frigid	Aquic
DARROCH	H	VD	SPD	0–3	8	Bm	Mollisols	Udolls	Argiudolls	Aquic Argiudolls	Fine-loamy	Mixed	Superactive		Mesic	Udic
DAWSIL	M	VD	VPD	0–1	170	C, J	Histosols	Saprists	Haplosaprists	Terric Haplosaprists	Sandy or sandy-skeletal	Siliceous		Dysic	Frigid	Aquic
DAWSON	M	VD	VPD	0–2	603	J	Histosols	Saprists	Haplosaprists	Terric Haplosaprists	Sandy or sandy-skeletal	Mixed		Dysic	Frigid	Aquic
DECHAMPS	S	VD	SPD	0–2	31	F, G, H, J	Entisols	Fluvents	Udifluvents	Aquic Udifluvents	Sandy	Mixed			Frigid	Udic
DEERTON	S	MD	WD	0–35	7	H	Spodosols	Orthods	Haplorthods	Typic Haplorthods	Sandy	Mixed			Frigid	Udic
DEFORD	S	VD	VPD	0–2	127	H, J	Entisols	Aquents	Psammaquents	Typic Psammaquents		Mixed			Frigid	Aquic
DEL REY	H	VD	SPD	0–7	2	B, J	Alfisols	Aqualfs	Epiaqualfs	Aeric Epiaqualfs	Fine	Illitic			Mesic	Aquic
DELLS	H	MD	SPD	0–3	28	A	Alfisols	Udalfs	Hapludalfs	Aquollic Hapludalfs	Fine-silty over sandy or sandy-skeletal	Mixed	Superactive		Mesic	Aquic
DELTON	S	VD	WD	0–15	83	C, E	Alfisols	Udalfs	Hapludalfs	Arenic Hapludalfs	Loamy	Mixed	Active		Mesic	Udic
DENOMIE	H	VD	WD	0–60	31	G	Alfisols	Udalfs	Glossudalfs	Haplic Glossudalfs	Fine-silty	Mixed	Active		Frigid	Udic
DENROCK	H	D	SPD	0–2	1	Am	Mollisols	Udolls	Argiudolls	Aquic Argiudolls	Fine	Mixed	Superactive		Mesic	Udic
DERINDA	M	MD	MWD	2–60	5	A	Alfisols	Udalfs	Hapludalfs	Oxyaquic Hapludalfs	Fine	Mixed	Active		Mesic	Udic
DICKINSON	M	VD	WD	0–30	96	Am	Mollisols	Udolls	Hapludolls	Typic Hapludolls	Coarse-loamy	Mixed	Superactive		Mesic	Udic
DICKMAN	S	VD	SED	0–18	15	Bm	Mollisols	Udolls	Hapludolls	Typic Hapludolls	Sandy	Mixed			Mesic	Udic
DILLON	S	VD	SPD	0–4	2	A	Inceptisols	Ustepts	Haplustepts	Oxyaquic Haplustepts	Fine-loamy over sandy or sandy-skeletal	Mixed	Superactive		Frigid	Ustic
DISHNO	M	D	MWD	0–35	9	G	Spodosols	Orthods	Haplorthods	Oxyaquic Haplorthods	Coarse-loamy over sandy or sandy-skeletal		Superactive		Frigid	Udic
DOBIE	H	MD	WD	2–30	64	Dr	Alfisols	Udalfs	Glossudalfs	Haplic Glossudalfs	Fine-silty	Mixed	Superactive		Frigid	Udic
DODGE	H	VD	WD	0–20	495	B	Alfisols	Udalfs	Hapludalfs	Typic Hapludalfs	Fine-silty	Mixed	Superactive		Mesic	Udic

(continued)

(continued)

Soil series[a]	Soil YP[b]	Soil Depth[c]	Drainage class[d]	Slope (%)	Soil area (km^2)	Soil region	Order	Suborder	Great group	Subgroup	Particle-size class	Mineral class	CEC Activity class	Reaction class	Soil-temperature class	Soil-moisture class
DODGEVILLE	M	S	WD	0–30	585	Am	Mollisols	Udolls	Argiudolls	Typic Argiudolls	Fine-silty over clayey	Mixed	Superactive		Mesic	Udic
DODY	S	VD	VPD	0–2	3	I	Alfisols	Aqualfs	Albaqualfs	Arenic Albaqualfs	Clayey	Smectitic			Frigid	Aquic
DOLPH	H	D	SPD	0–3	88	Fr, J	Alfisols	Aqualfs	Glossaqualfs	Aeric Glossaqualfs	Fine	Mixed	Superactive		Frigid	Aquic
DORA	M	VD	VPD	0–1	5	I, J	Histosols	Saprists	Haplosaprists	Terric Haplosaprists	Clayey	Smectitic		Euic	Frigid	Aquic
DORCHESTER	H	VD	WD	0–5	0	B, J	Entisols	Fluvents	Udifluvents	Typic Udifluvents	Fine-silty	Mixed	Superactive	Calcareous	Mesic	Udic
DORERTON	M	VD	WD	12–65	394	A	Alfisols	Udalfs	Hapludalfs	Typic Hapludalfs	Loamy-skeletal	Mixed	Active		Mesic	Udic
DORITTY	H	VD	MWD	2–12	31	Dr	Alfisols	Udalfs	Glossudalfs	Haplic Glossudalfs	Fine-silty	Mixed	Superactive		Frigid	Udic
DOWNS	H	VD	WD	0–35	407	A	Alfisols	Udalfs	Hapludalfs	Mollic Hapludalfs	Fine-silty	Mixed	Superactive		Mesic	Udic
DRAMMEN	S	VD	SED	1–20	163	Dr, H	Alfisols	Udalfs	Hapludalfs	Lamellic Hapludalfs	Sandy	Mixed			Mesic	Udic
DRESDEN	H	VD	WD	0–30	246	Bm	Alfisols	Udalfs	Hapludalfs	Mollic Hapludalfs	Fine-loamy over sandy or sandy-skeletal	Mixed	Active		Mesic	Udic
DRUMMER	M	VD	PD	0–2	39	Bm, J	Mollisols	Aquolls	Endoaquolls	Typic Endoaquolls	Fine-silty	Mixed	Superactive		Mesic	Aquic
DRYLANDING	M	S	SED	0–12	68	G	Mollisols	Udolls	Hapludolls	Lithic Hapludolls	Loamy-skeletal	Mixed	Superactive		Frigid	Udic
DUBUQUE	M	MD	WD	2–60	2835	A, B	Alfisols	Udalfs	Hapludalfs	Typic Hapludalfs	Fine-silty	Mixed	Superactive		Mesic	Udic
DUEL	S	MD	WD	1–12	3	Er, J	Spodosols	Orthods	Haplorthods	Entic Haplorthods	Sandy	Mixed			Frigid	Udic
DUELM	S	VD	WD	0–3	4	C, H	Mollisols	Udolls	Hapludolls	Oxyaquic Hapludolls	Sandy	Mixed			Frigid	Udic
DUNBARTON	M	S	WD	2–60	246	A	Alfisols	Udalfs	Hapludalfs	Lithic Hapludalfs	Clayey	Smectitic			Mesic	Udic
DUNNBOT	H	VD	MWD	0–3	47	Dr, J	Entisols	Fluvents	Udifluvents	Mollic Udifluvents	Coarse-loamy	Mixed	Superactive	Nonacid	Mesic	Udic
DUNNVILLE	H	VD	WD	0–12	51	Dr, G	Mollisols	Udolls	Hapludolls	Typic Hapludolls	Coarse-loamy	Mixed	Superactive		Frigid	Udic
DURAND	H	VD	WD	1–20	69	Bm	Mollisols	Udolls	Argiudolls	Typic Argiudolls	Fine-loamy	Mixed	Superactive		Mesic	Udic
EAUCLAIRE	S	VD	MWD	1–6	20	Dr	Spodosols	Orthods	Haplorthods	Alfic Oxyaquic Haplorthods	Sandy	Mixed			Frigid	Udic
EDMUND	M	S	WD	2–35	258	Am	Mollisols	Udolls	Argiudolls	Lithic Argiudolls	Clayey	Smectitic			Mesic	Udic
EDWARDS	H	VD	VPD	0–2	6	J	Histosols	Saprists	Haplosaprists	Limnic Haplosaprists		Marly		Euic	Mesic	Aquic
ELBAVILLE	M	VD	WD	12–45	180	A	Alfisols	Udalfs	Hapludalfs	Glossic Hapludalfs	Fine-loamy		Superactive		Mesic	Udic
ELBURN	H	VD	SPD	0–5	260	Bm	Mollisols	Udolls	Argiudolls	Aquic Argiudolls	Fine-silty	Mixed	Superactive		Mesic	Udic
ELDERON	S	VD	SED	2–45	44	E, G	Inceptisols	Udepts	Dystrudepts	Typic Dystrudepts	Sandy-skeletal	Mixed			Frigid	Udic

(continued)

(continued)

Soil series[a]	Soil YP[b]	Soil Depth[c]	Drainage class[d]	Slope (%)	Soil area (km^2)	Soil region	Order	Suborder	Great group	Subgroup	Particle-size class	Mineral class	CEC Activity class	Reaction class	Soil-temperature class	Soil-moisture class
ELEROY	H	D	MWD	2–30	7	A	Alfisols	Udalfs	Hapludalfs	Oxyaquic Hapludalfs	Fine-silty	Mixed	Superactive		Mesic	Udic
ELEVA	M	MD	WD	1–60	541	B, Dr	Alfisols	Udalfs	Hapludalfs	Typic Hapludalfs	Coarse-loamy	Mixed	Active		Mesic	Udic
ELEVASIL	M	MD	WD	2–60	257	Dr	Alfisols	Udalfs	Hapludalfs	Ultic Hapludalfs	Coarse-loamy	Siliceous	Active		Mesic	Udic
ELKMOUND	M	S	WD	0–60	322	Dr	Inceptisols	Udepts	Dystrudepts	Typic Dystrudepts	Loamy	Mixed	Superactive		Mesic	Udic
ELLA	H	VD	MWD	0–12	84	F, Dr	Alfisols	Udalfs	Hapludalfs	Mollic Hapludalfs	Fine-silty	Mixed	Superactive		Mesic	Udic
ELLIOTT	H	VD	SPD	0–7	147	Bm	Mollisols	Udolls	Argiudolls	Aquic Argiudolls	Fine	Illitic			Mesic	Udic
ELLWOOD	H	VD	MWD	1–15	21	G	Alfisols	Udalfs	Glossudalfs	Oxyaquic Glossudalfs	Fine-loamy	Mixed	Active		Frigid	Udic
ELM LAKE	S	MD	PD	0–2	274	Dr, J	Entisols	Aquents	Epiaquents	Humaqueptic Epiaquents	Sandy over loamy	Siliceous	Semiactive	Acid	Frigid	Aquic
ELVERS	M	VD	VPD	0–3	16	B, J	Entisols	Aquents	Fluvaquents	Thapto-Histic Fluvaquents	Coarse-silty	Mixed	Superactive	Nonacid	Mesic	Aquic
EMMERT	S	VD	ED	1–70	80	G	Entisols	Orthents	Udorthents	Typic Udorthents	Sandy-skeletal	Mixed			Frigid	Udic
EMMET	H	VD	MWD	0–50	537	Er, J	Alfisols	Udalfs	Hapludalfs	Inceptic Hapludalfs	Coarse-loamy	Mixed	Active		Frigid	Udic
ENSLEY	M	VD	PD	0–2	95	G, J	Inceptisols	Aquepts	Endoaquepts	Aeric Endoaquepts	Coarse-loamy	Mixed	Active	Nonacid	Frigid	Aquic
ETTRICK	M	VD	PD	0–2	318	Am, J	Mollisols	Aquolls	Endoaquolls	Fluvaquentic Endoaquolls	Fine-silty	Mixed	Superactive		Mesic	Aquic
FABIUS	M	VD	SPD	0–6	36	Bm	Mollisols	Udolls	Argiudolls	Aquic Argiudolls	Fine-loamy over sandy or sandy-skeletal	Mixed	Semiactive		Mesic	Udic
FAIRCHILD	S	MD	SPD	0–3	237	Dr, J	Spodosols	Aquods	Epiaquods	Ultic Epiaquods	Sandy over loamy	Siliceous	Semiactive		Frigid	Udic
FAIRPORT	M	MD	MWD	0–25	18	E, G	Alfisols	Udalfs	Hapludalfs	Glossic Hapludalfs	Fine-loamy	Mixed	Active		Mesic	Udic
FALLCREEK	H	VD	SPD	0–6	117	Dr, F	Alfisols	Udalfs	Glossudalfs	Aquic Glossudalfs	Coarse-loamy	Mixed	Superactive		Frigid	Udic
FARRINGTON	S	VD	SPD	0–3	17	Cm, Dr	Mollisols	Udolls	Hapludolls	Aquic Hapludolls	Sandy	Mixed			Mesic	Udic
FAYETTE	H	VD	WD	0–60	2234	A, Dr	Alfisols	Udalfs	Hapludalfs	Typic Hapludalfs	Fine-silty	Mixed	Superactive		Mesic	Udic
FENANDER	M	VD	VPD	0–2	9	F, G, J	Alfisols	Aqualfs	Epiaqualfs	Udollic Epiaqualfs	Coarse-loamy	Mixed	Superactive		Frigid	Aquic
FENCE	H	VD	MWD	0–18	79	G, H	Spodosols	Orthods	Haplorthods	Alfic Oxyaquic Haplorthods	Coarse-silty	Mixed	Superactive		Frigid	Aquic
FENWOOD	H	D	WD	2–30	328	Fr	Alfisols	Udalfs	Glossudalfs	Haplic Glossudalfs	Fine-loamy	Mixed	Superactive		Frigid	Udic

(continued)

(continued)

Soil series[a]	Soil YP[b]	Soil Depth[c]	Drainage class[d]	Slope (%)	Soil area (km^2)	Soil region	Order	Suborder	Great group	Subgroup	Particle-size class	Mineral class	CEC Activity class	Reaction class	Soil-temperature class	Soil-moisture class
FESTINA	H	VD	WD	0–12	14	Am, J	Alfisols	Udalfs	Hapludalfs	Mollic Hapludalfs	Fine-silty	Mixed	Superactive		Mesic	Udic
FINCHFORD	S	VD	ED	0–14	121	Dr	Mollisols	Udolls	Hapludolls	Entic Hapludolls	Sandy	Mixed			Mesic	Udic
FISK	S	VD	SPD	0–3	19	C, E	Inceptisols	Udepts	Eutrudepts	Aquic Dystric Eutrudepts	Sandy over loamy	Mixed	Superactive		Mesic	Udic
FIVEPOINTS	M	MD	WD	2–30	183	A	Alfisols	Udalfs	Hapludalfs	Typic Hapludalfs	Clayey over loamy-skeletal	Mixed	Superactive		Mesic	Udic
FLAGG	H	VD	WD	0–20	44	B	Alfisols	Udalfs	Hapludalfs	Typic Hapludalfs	Fine-silty	Mixed	Superactive		Mesic	Udic
FLAMBEAU	H	VD	WD	1–20	217	F	Alfisols	Udalfs	Glossudalfs	Oxyaquic Glossudalfs	Fine-loamy	Mixed	Superactive		Frigid	Udic
FLINK	S	VD	SPD	0–4	48	H, I, J	Spodosols	Aquods	Epiaquods	Typic Epiaquods	Sandy	Mixed			Frigid	Udic
FLOYD	H	VD	SPD	0–5	19	F	Mollisols	Udolls	Hapludolls	Aquic Pachic Hapludolls	Fine-loamy	Mixed	Superactive		Mesic	Udic
FORADA	M	VD	PD	0–2	5	G, J	Mollisols	Aquolls	Endoaquolls	Typic Endoaquolls	Coarse-loamy	Mixed	Superactive		Frigid	Aquic
FORDUM	M	VD	PD	0–2	595	J	Entisols	Aquents	Fluvaquents	Mollic Fluvaquents	Coarse-loamy	Mixed	Superactive	Nonacid	Frigid	Aquic
FORKHORN	M	VD	WD	0–20	60	Dr	Alfisols	Udalfs	Hapludalfs	Mollic Hapludalfs	Coarse-loamy	Mixed	Active		Mesic	Udic
FOX	H	VD	WD	0–35	739	B, I	Alfisols	Udalfs	Hapludalfs	Typic Hapludalfs	Fine-loamy over sandy or sandy-skeletal	Mixed	Superactive		Mesic	Udic
FRECHETTE	H	VD	WD	2–35	73	G	Alfisols	Udalfs	Glossudalfs	Typic Glossudalfs	Coarse-loamy	Mixed	Active		Frigid	Udic
FREEON	H	VD	MWD	0–20	2144	F, G	Alfisols	Udalfs	Glossudalfs	Oxyaquic Glossudalfs	Coarse-loamy	Mixed	Superactive		Frigid	Udic
FREMSTADT	S	VD	WD	1–30	63	H	Alfisols	Udalfs	Hapludalfs	Arenic Hapludalfs	Sandy	Mixed			Frigid	Udic
FREYA	S	VD	SPD	0–3	81	H	Mollisols	Udolls	Argiudolls	Aquic Argiudolls	Sandy over clayey	Mixed over smectitic			Frigid	Aquic
FRIENDSHIP	S	VD	MWD	0–4	530	C, H	Entisols	Psamments	Udipsamments	Typic Udipsamments		Mixed			Frigid	Udic
FRIESLAND	H	VD	MWD	0–7	37	Bm	Mollisols	Udolls	Argiudolls	Typic Argiudolls	Fine-loamy	Mixed	Superactive		Mesic	Udic
FROGCREEK	H	VD	MWD	2–15	106	G	Alfisols	Udalfs	Glossudalfs	Oxyaquic Glossudalfs	Coarse-loamy	Mixed	Superactive		Frigid	Udic
GAASTRA	H	VD	SPD	0–4	8	G, H, J	Spodosols	Aquods	Endoaquods	Argic Endoaquods	Coarse-loamy	Mixed	Active		Frigid	Udic
GALE	H	MD	WD	0–60	554	Dr	Alfisols	Udalfs	Hapludalfs	Typic Hapludalfs	Fine-silty over sandy or sandy-skeletal	Mixed	Superactive		Mesic	Udic

(continued)

(continued)

Soil series[a]	Soil YP[b]	Soil Depth[c]	Drainage class[d]	Slope (%)	Soil area (km^2)	Soil region	Order	Suborder	Great group	Subgroup	Particle-size class	Mineral class	CEC Activity class	Reaction class	Soil-temperature class	Soil-moisture class
GAPHILL	M	VD	WD	30–60	67	A, Dr	Alfisols	Udalfs	Hapludalfs	Typic Hapludalfs	Coarse-loamy	Siliceous	Active		Mesic	Aquic
GARDENVALE	H	D	WD	1–6	10	Dr	Alfisols	Udalfs	Hapludalfs	Mollic Hapludalfs	Fine-loamy over sandy or sandy-skeletal	Mixed	Superactive		Mesic	Udic
GARNE	S	MD	SED	2–12	6	Cm	Mollisols	Udolls	Hapludolls	Typic Hapludolls	Sandy over loamy	Mixed	Active		Mesic	Udic
GARWIN	M	D	PD	0–2	0	Am, J	Mollisols	Aquolls	Endoaquolls	Typic Endoaquolls	Fine-silty	Mixed	Superactive		Mesic	Aquic
GASTROW	H	VD	SPD	0–3	13	G, J	Spodosols	Aquods	Endoaquods	Argic Endoaquods	Coarse-loamy	Mixed	Active		Frigid	Udic
GAY	M	D	PD	0–3	18	G	Inceptisols	Aquepts	Endoaquepts	Aeric Endoaquepts	Coarse-loamy	Mixed	Active	Nonacid	Frigid	Aquic
GICHIGAMI	H	VD	MWD	0–6	53	G	Alfisols	Udalfs	Glossudalfs	Oxyaquic Glossudalfs	Fine-silty	Mixed	Superactive		Frigid	Aquic
GIESE	M	VD	VPD	0–1	28	G, J	Inceptisols	Aquepts	Humaquepts	Typic Humaquepts	Coarse-loamy	Isotic		Nonacid	Frigid	Aquic
GILFORD	M	VD	PD	0–2	53	Bm, J	Mollisols	Aquolls	Endoaquolls	Typic Endoaquolls	Coarse-loamy	Mixed	Superactive		Mesic	Aquic
GILLINGHAM	S	D	WD	6–20	2	A	Alfisols	Udalfs	Hapludalfs	Arenic Hapludalfs	Loamy	Mixed	Superactive		Mesic	Udic
GLENDENNING	H	VD	SPD	0–4	19	G	Alfisols	Udalfs	Glossudalfs	Aquic Glossudalfs	Coarse-loamy	Mixed	Superactive		Frigid	Udic
GLENDORA	S	VD	VPD	0–2	41	J	Entisols	Aquents	Psammaquents	Mollic Psammaquents		Mixed			Mesic	Aquic
GLENFLORA	M	VD	VPD	0–2	17	F, J	Alfisols	Aqualfs	Glossaqualfs	Mollic Glossaqualfs	Fine-silty	Mixed	Superactive		Frigid	Udic
GLIDDEN	H	VD	WD	0–30	64	G	Spodosols	Orthods	Haplorthods	Alfic Haplorthods	Coarse-loamy	Mixed	Superactive		Frigid	Udic
GOGEBIC	M	VD	MWD	1–55	920	G, I	Spodosols	Orthods	Fragiorthods	Alfic Oxyaquic Fragiorthods	Coarse-loamy	Isotic			Frigid	Udic
GOODMAN	H	VD	WD	1–45	262	F, G	Spodosols	Orthods	Haplorthods	Alfic Haplorthods	Coarse-loamy	Mixed	Superactive		Frigid	Udic
GOODWIT	H	VD	MWD	0–15	142	G	Spodosols	Orthods	Haplorthods	Alfic Oxyaquic Haplorthods	Coarse-loamy	Mixed	Superactive		Frigid	Udic
GOSIL	S	VD	ED	1–12	80	Dr	Entisols	Psamments	Quartzipsamments	Typic Quartzipsamments					Mesic	Udic
GOTHAM	S	VD	SED	0–35	403	Cm	Alfisols	Udalfs	Hapludalfs	Psammentic Hapludalfs		Mixed			Mesic	Udic
GRANBY	S	VD	PD	0–3	231	Cm, E, J	Mollisols	Aquolls	Endoaquolls	Typic Endoaquolls	Sandy	Mixed			Mesic	Aquic
GRAYCALM	S	VD	SED	0–70	380	H	Entisols	Psamments	Udipsamments	Lamellic Udipsamments		Isotic			Frigid	Udic
GRAYLING	S	VD	ED	0–45	587	H	Entisols	Psamments	Udipsamments	Typic Udipsamments		Isotic			Frigid	Udic

(continued)

(continued)

Soil series[a]	Soil YP[b]	Soil Depth[c]	Drainage class[d]	Slope (%)	Soil area (km^2)	Soil region	Order	Suborder	Great group	Subgroup	Particle-size class	Mineral class	CEC Activity class	Reaction class	Soil-temperature class	Soil-moisture class
GRAYS	H	VD	MWD	0–12	54	B	Alfisols	Udalfs	Hapludalfs	Mollic Oxyaquic Hapludalfs	Fine-silty	Mixed	Superactive		Mesic	Udic
GREENRIDGE	H	VD	WD	4–20	36	A, Dr	Alfisols	Udalfs	Hapludalfs	Typic Hapludalfs	Fine-silty	Mixed	Superactive		Mesic	Udic
GREENWOOD	M	VD	VPD	0–2	315	J	Histosols	Hemists	Haplohemists	Typic Haplohemists				Dysic	Frigid	Aquic
GRELLTON	H	VD	WD	0–20	73	B	Alfisols	Udalfs	Hapludalfs	Typic Hapludalfs	Fine-loamy	Mixed	Superactive		Mesic	Udic
GRETTUM	S	VD	WD	0–30	171	H	Entisols	Psamments	Udipsamments	Lamellic Udipsamments		Mixed			Frigid	Aquic
GRISWOLD	H	VD	WD	0–20	190	Bm	Mollisols	Udolls	Argiudolls	Typic Argiudolls	Fine-loamy	Mixed	Superactive		Mesic	Udic
GUENTHER	S	VD	MWD	2–6	19	C, Dr	Spodosols	Orthods	Haplorthods	Alfic Oxyaquic Haplorthods	Sandy over loamy	Mixed	Superactive		Frigid	Udic
GULL POINT	M	VD	PD	0–2	1	G, J	Mollisols	Aquolls	Argiaquolls	Typic Argiaquolls	Fine-loamy	Mixed	Superactive		Frigid	Aquic
HALDER	H	D	SPD	0–3	3	G	Alfisols	Udalfs	Glossudalfs	Aquic Glossudalfs	Fine-loamy over sandy or sandy-skeletal	Mixed	Superactive		Frigid	Udic
HATLEY	H	VD	SPD	0–6	90	F	Alfisols	Udalfs	Glossudalfs	Aquic Glossudalfs	Coarse-loamy	Mixed	Superactive		Frigid	Udic
HAUGEN	H	VD	MWD	2–15	322	G	Alfisols	Udalfs	Paleudalfs	Oxyaquic Paleudalfs	Coarse-loamy	Mixed	Superactive		Frigid	Udic
HAUSTRUP	M	S	SED	1–25	13	G	Inceptisols	Udepts	Dystrudepts	Humic Lithic Dystrudepts	Loamy	Isotic			Frigid	Udic
HAYFIELD	H	VD	SPD	0–4	30	B	Alfisols	Udalfs	Hapludalfs	Aquollic Hapludalfs	Fine-loamy over sandy or sandy-skeletal	Mixed	Superactive		Mesic	Udic
HAYRIVER	M	MD	WD	2–50	208	Dr	Alfisols	Udalfs	Hapludalfs	Typic Hapludalfs	Coarse-loamy	Mixed	Active		Frigid	Udic
HEBRON	H	VD	MWD	0–6	77	B, J	Alfisols	Udalfs	Hapludalfs	Oxyaquic Hapludalfs	Fine-loamy	Mixed	Superactive		Mesic	Udic
HENNEPIN	H	VD	WD	10–70	4	B	Inceptisols	Udepts	Eutrudepts	Typic Eutrudepts	Fine-loamy	Mixed	Active		Mesic	Udic
HERBSTER	H	VD	SPD	0–3	137	I, J	Alfisols	Aqualfs	Glossaqualfs	Aeric Glossaqualfs	Fine	Mixed	Active		Frigid	Udic
HERSEY	H	VD	MWD	2–18	57	Dr	Alfisols	Udalfs	Hapludalfs	Mollic Hapludalfs	Fine-silty	Mixed	Superactive		Mesic	Udic
HESCH	H	MD	WD	0–45	20	Dr	Mollisols	Udolls	Argiudolls	Typic Argiudolls	Coarse-loamy	Mixed	Active		Mesic	Udic
HIBBING	H	VD	MWD	3–45	0	I	Alfisols	Udalfs	Glossudalfs	Oxyaquic Glossudalfs	Fine	Mixed	Active		Frigid	Aquic

(continued)

(continued)

Soil series[a]	Soil YP[b]	Soil Depth[c]	Drainage class[d]	Slope (%)	Soil area (km^2)	Soil region	Order	Suborder	Great group	Subgroup	Particle-size class	Mineral class	CEC Activity class	Reaction class	Soil-temperature class	Soil-moisture class
HILES	H	MD	MWD	1–12	116	Dr, F	Alfisols	Udalfs	Glossudalfs	Oxyaquic Glossudalfs	Fine-loamy	Mixed	Superactive		Frigid	Udic
HIXTON	M	MD	WD	2–60	566	A, Dr	Alfisols	Udalfs	Hapludalfs	Typic Hapludalfs	Fine-loamy over sandy or sandy-skeletal	Mixed	Superactive		Mesic	Udic
HOCHHEIM	H	S	WD	0–30	979	Bm	Mollisols	Udolls	Argiudolls	Typic Argiudolls	Fine-loamy	Mixed	Active		Mesic	Udic
HOOP	M	VD	SPD	0–3	6	Dr	Mollisols	Udolls	Argiudolls	Aquic Argiudolls	Coarse-loamy	Siliceous	Active		Mesic	Udic
HOOPESTON	H	VD	SPD	0–5	48	Dr	Mollisols	Udolls	Hapludolls	Aquic Hapludolls	Coarse-loamy	Mixed	Superactive		Mesic	Udic
HORTONVILLE	H	VD	WD	1–35	1146	E, I	Alfisols	Udalfs	Glossudalfs	Haplic Glossudalfs	Fine-loamy	Mixed	Active		Mesic	Udic
HOUGHTON	H	VD	VPD	0–2	1482	J	Histosols	Saprists	Haplosaprists	Typic Haplosaprists				Euic	Mesic	Aquic
HUBBARD	S	VD	ED	0–35	27	Cm	Mollisols	Udolls	Hapludolls	Entic Hapludolls	Sandy	Mixed			Frigid	Udic
HUMBIRD	M	MD	MWD	1–20	242	Dr	Spodosols	Orthods	Haplorthods	Oxyaquic Ultic Haplorthods	Coarse-loamy over clayey	Mixed	Semiactive		Frigid	Udic
HUNTSVILLE	H	VD	WD	0–6	153	J	Mollisols	Udolls	Hapludolls	Cumulic Hapludolls	Fine-silty	Mixed	Superactive		Mesic	Udic
IMPACT	S	VD	ED	0–6	132	Dr	Inceptisols	Udepts	Dystrudepts	Humic Psammentic Dystrudepts		Siliceous			Mesic	Udic
INDUS	M	D	PD	0–2	2	I, J	Alfisols	Aqualfs	Epiaqualfs	Vertic Epiaqualfs	Very fine	Smectitic			Frigid	Udic
INGALLS	S	VD	SPD	0–6	8	G, J	Spodosols	Aquods	Endoaquods	Typic Endoaquods	Sandy over loamy	Mixed	Active		Frigid	Aquic
IONIA	H	VD	MWD	0–6	33	B	Alfisols	Udalfs	Hapludalfs	Oxyaquic Hapludalfs	Fine-loamy over sandy or sandy-skeletal	Mixed	Semiactive		Mesic	Udic
IOSCO	S	VD	SPD	0–6	185	G, H, J	Spodosols	Aquods	Endoaquods	Argic Endoaquods	Sandy over loamy	Mixed	Active		Frigid	Udic
IRONRUN	S	VD	SPD	0–3	175	Dr, J	Spodosols	Aquods	Endoaquods	Typic Endoaquods	Sandy	Siliceous			Frigid	Aquic
ISHPEMING	S	MD	SED	0–70	87	H	Spodosols	Orthods	Haplorthods	Entic Haplorthods	Sandy	Mixed			Frigid	Aquic
JACKSON	H	VD	MWD	0–9	86	A, Dr	Alfisols	Udalfs	Hapludalfs	Typic Hapludalfs	Fine-silty	Mixed	Superactive		Mesic	Udic
JASPER	H	VD	WD	0–18	8	Bm	Mollisols	Udolls	Argiudolls	Typic Argiudolls	Fine-loamy	Mixed	Superactive		Mesic	Udic
JEWETT	H	MD	WD	0–20	127	F	Alfisols	Udalfs	Hapludalfs	Typic Hapludalfs	Fine-loamy	Mixed	Superactive		Frigid	Udic

(continued)

(continued)

Soil series[a]	Soil YP[b]	Soil Depth[c]	Drainage class[d]	Slope (%)	Soil area (km^2)	Soil region	Order	Suborder	Great group	Subgroup	Particle-size class	Mineral class	CEC Activity class	Reaction class	Soil-temperature class	Soil-moisture class
JOY	H	VD	SPD	0–5	28	Bm	Mollisols	Udolls	Hapludolls	Aquic Hapludolls	Fine-silty	Mixed	Superactive		Mesic	Udic
JUDA	H	VD	MWD	2–12	9	B	Alfisols	Udalfs	Hapludalfs	Mollic Oxyaquic Hapludalfs	Fine-silty	Mixed	Superactive		Mesic	Udic
JUDSON	H	VD	WD	0–12	84	A, J	Mollisols	Udolls	Hapludolls	Cumulic Hapludolls	Fine-silty	Mixed	Superactive		Mesic	Udic
JUNEAU	H	VD	MWD	0–6	48	B, J	Entisols	Fluvents	Udifluvents	Typic Udifluvents	Coarse-silty	Mixed	Superactive	Nonacid	Mesic	Udic
KALMARVILLE	M	VD	PD	0–1	74	Dr, J	Entisols	Aquents	Fluvaquents	Mollic Fluvaquents	Coarse-loamy	Mixed	Superactive	Nonacid	Mesic	Aquic
KANE	H	VD	SPD	0–3	39	Bm	Mollisols	Udolls	Argiudolls	Aquic Argiudolls	Fine-loamy over sandy or sandy-skeletal	Mixed	Superactive		Mesic	Udic
KARLIN	M	VD	SED	0–75	480	G, H	Spodosols	Orthods	Haplorthods	Entic Haplorthods	Sandy	Mixed			Frigid	Aquic
KARLSBORG	S	VD	MWD	1–30	57	I	Alfisols	Udalfs	Hapludalfs	Arenic Oxyaquic Hapludalfs	Clayey	Smectitic			Frigid	Udic
KASSON	H	VD	MWD	1–9	13	A	Alfisols	Udalfs	Hapludalfs	Mollic Oxyaquic Hapludalfs	Fine-loamy	Mixed	Superactive		Mesic	
KATO	M	VD	VPD	0–2	12	Bm, J	Mollisols	Aquolls	Endoaquolls	Typic Endoaquolls	Fine-silty over sandy or sandy-skeletal	Mixed	Superactive		Mesic	Aquic
KAUKAUNA	H	VD	MWD	0–4	14	E, I	Alfisols	Udalfs	Hapludalfs	Mollic Oxyaquic Hapludalfs	Clayey over loamy	Mixed	Active		Mesic	Udic
KEGONSA	H	VD	WD	0–6	72	B	Alfisols	Udalfs	Hapludalfs	Mollic Hapludalfs	Fine-silty over sandy or sandy-skeletal	Mixed	Superactive		Mesic	Udic
KELLOGG	S	VD	WD	0–12	100	I	Spodosols	Orthods	Haplorthods	Alfic Oxyaquic Haplorthods	Sandy over clayey	Mixed	Active		Frigid	Udic
KELTNER	H	D	MWD	0–30	7	Am	Mollisols	Udolls	Argiudolls	Oxyaquic Argiudolls	Fine-silty	Mixed	Superactive		Mesic	Udic
KENDALL	H	VD	SPD	0–5	5	B, J	Alfisols	Aqualfs	Endoaqualfs	Aeric Endoaqualfs	Fine-silty	Mixed	Superactive		Mesic	Udic
KENNAN	H	D	WD	1–35	1139	F, G	Alfisols	Udalfs	Glossudalfs	Haplic Glossudalfs	Coarse-loamy	Mixed	Superactive		Frigid	Udic
KEOWNS	M	VD	PD	0–3	0	E, J	Inceptisols	Aquepts	Endoaquepts	Mollic Endoaquepts	Coarse-loamy	Mixed	Superactive	Nonacid	Mesic	Aquic
KERT	H	MD	SPD	0–6	308	Dr	Alfisols	Udalfs	Glossudalfs	Aquic Glossudalfs	Fine-loamy	Mixed	Superactive		Frigid	Udic
KESHENA	H	VD	MWD	0–6	17	E, G	Alfisols	Udalfs	Glossudalfs	Oxyaquic Glossudalfs	Fine-loamy	Mixed	Active		Frigid	Udic
KEVILAR	H	VD	MWD	0–12	164	Dr	Alfisols	Udalfs	Hapludalfs	Mollic Hapludalfs	Coarse-loamy	Mixed	Active		Mesic	Udic

(continued)

(continued)

Soil series[a]	Soil YP[b]	Soil Depth[c]	Drainage class[d]	Slope (%)	Soil area (km^2)	Soil region	Order	Suborder	Great group	Subgroup	Particle-size class	Mineral class	CEC Activity class	Reaction class	Soil-temperature class	Soil-moisture class
KEWAUNEE	H	VD	WD	0–45	2002	E, I	Alfisols	Udalfs	Hapludalfs	Typic Hapludalfs	Fine	Mixed	Active		Mesic	Udic
KEWEENAW	S	VD	WD	0–70	1278	H	Spodosols	Orthods	Haplorthods	Alfic Haplorthods	Sandy	Mixed			Frigid	Udic
KEYESVILLE	M	MD	SED	20–65	6	B, Dr	Inceptisols	Udepts	Dystrudepts	Typic Dystrudepts	Loamy-skeletal	Mixed	Active		Mesic	Aquic
KIBBIE	H	VD	SPD	0–6	67	E	Alfisols	Udalfs	Hapludalfs	Aquollic Hapludalfs	Fine-loamy	Mixed	Active		Mesic	Udic
KICKAPOO	H	D	MWD	0–6	71	D, J	Entisols	Fluvents	Udifluvents	Typic Udifluvents	Coarse-loamy	Mixed	Superactive	Nonacid	Mesic	Udic
KIDDER	H	VD	WD	0–35	644	B	Alfisols	Udalfs	Hapludalfs	Typic Hapludalfs	Fine-loamy	Mixed	Active		Mesic	Udic
KINGSVILLE	S	VD	VPD	0–2	70	C, J	Entisols	Aquents	Psammaquents	Mollic Psammaquents		Mixed			Mesic	Aquic
KINROSS	S	VD	PD	0–3	102	G, H, J	Spodosols	Aquods	Endoaquods	Typic Endoaquods	Sandy	Mixed			Frigid	Aquic
KIVA	S	VD	WD	0–45	41	E	Spodosols	Orthods	Haplorthods	Entic Haplorthods	Sandy	Mixed			Frigid	Udic
KNOWLES	H	MD	WD	0–30	76	B	Alfisols	Udalfs	Hapludalfs	Typic Hapludalfs	Fine-silty	Mixed	Superactive		Mesic	Udic
KOLBERG	M	MD	WD	0–12	72	Er, I	Alfisols	Udalfs	Glossudalfs	Haplic Glossudalfs	Fine	Mixed	Active		Frigid	Udic
KOMRO	S	VD	MWD	0–3	14	Dr	Mollisols	Udolls	Hapludolls	Entic Hapludolls	Sandy	Mixed			Mesic	Udic
KOROBAGO	H	VD	SPD	0–3	39	I	Inceptisols	Udepts	Eutrudepts	Aquic Eutrudepts	Coarse-loamy over clayey	Mixed	Active		Mesic	Udic
KOST	S	VD	ED	0–18	2	H	Mollisols	Udolls	Hapludolls	Entic Hapludolls	Sandy	Mixed			Frigid	Udic
KRANSKI	S	VD	SED	2–30	103	C	Alfisols	Udalfs	Hapludalfs	Typic Hapludalfs	Sandy	Mixed			Mesic	Udic
LA FARGE	H	MD	WD	2–35	595	A, Dr	Alfisols	Udalfs	Hapludalfs	Typic Hapludalfs	Fine-silty	Mixed	Superactive		Mesic	Udic
LABLATZ	H	VD	SPD	0–4	1	G, J	Spodosols	Aquods	Epiaquods	Alfic Epiaquods	Coarse-loamy	Mixed	Active		Frigid	Udic
LAMARTINE	H	VD	SPD	0–6	264	B, I	Alfisols	Udalfs	Hapludalfs	Aquollic Hapludalfs	Fine-silty	Mixed	Superactive		Mesic	Udic
LAMBEAU	H	VD	WD	6–20	8	A	Alfisols	Udalfs	Hapludalfs	Typic Hapludalfs	Fine-silty	Mixed	Superactive		Mesic	Udic
LAMOILLE	H	MD	WD	2–50	54	A	Alfisols	Udalfs	Hapludalfs	Typic Hapludalfs	Fine	Mixed	Superactive		Mesic	Udic
LAMONT	H	VD	WD	0–45	3	A	Alfisols	Udalfs	Hapludalfs	Typic Hapludalfs	Coarse-loamy	Mixed	Superactive		Mesic	Udic
LANGLADE	H	VD	WD	0–6	5	G	Alfisols	Udalfs	Glossudalfs	Haplic Glossudalfs	Coarse-loamy	Mixed	Superactive		Frigid	Udic
LAONA	M	MD	WD	5–35	32	G	Spodosols	Orthods	Fragiorthods	Alfic Fragiorthods	Coarse-loamy	Mixed	Superactive		Frigid	Udic
LAPEER	H	VD	WD	0–60	248	B, I	Alfisols	Udalfs	Hapludalfs	Typic Hapludalfs	Coarse-loamy	Mixed	Semiactive		Mesic	Udic
LAPOIN	M	MD	MWD	0–6	7	I	Spodosols	Orthods	Haplorthods	Alfic Oxyaquic Haplorthods	Fine	Mixed	Active		Frigid	Udic
LARA	S	VD	MWD	0–6	2	H	Mollisols	Udolls	Argiudolls	Oxyaquic Argiudolls	Sandy over clayey	Mixed OVER smectitic			Frigid	Udic

(continued)

(continued)

Soil series[a]	Soil YP[b]	Soil Depth[c]	Drainage class[d]	Slope (%)	Soil area (km^2)	Soil region	Order	Suborder	Great group	Subgroup	Particle-size class	Mineral class	CEC Activity class	Reaction class	Soil-temperature class	Soil-moisture class
LAWLER	H	VD	SPD	0–6	20	A, G	Mollisols	Udolls	Hapludolls	Aquic Hapludolls	Fine-loamy over sandy or sandy-skeletal	Mixed	Superactive		Mesic	Udic
LAWSON	H	VD	SPD	0–5	74	Am	Mollisols	Udolls	Hapludolls	Aquic Cumulic Hapludolls	Fine-silty	Mixed	Superactive		Mesic	Udic
LENROOT	S	VD	MWD	0–4	161	H	Entisols	Psamments	Udipsamments	Oxyaquic Udipsamments		Mixed			Frigid	Udic
LEOLA	S	VD	SPD	0–3	47	C, G	Alfisols	Udalfs	Hapludalfs	Aquic Arenic Hapludalfs	Loamy	Mixed	Active		Mesic	Udic
LERCH	M	VD	VPD	0–2	48	I, J	Inceptisols	Aquepts	Epiaquepts	Vertic Epiaquepts	Very fine	Mixed	Active	Nonacid	Frigid	Aquic
LEROY	H	VD	WD	2–30	98	B	Alfisols	Udalfs	Hapludalfs	Typic Hapludalfs	Fine-loamy	Mixed	Active		Mesic	Udic
LILAH	S	VD	ED	0–35	8	Dr	Alfisols	Udalfs	Hapludalfs	Psammentic Hapludalfs	Sandy	Mixed			Mesic	Udic
LINDQUIST	S	VD	SED	0–50	61	G	Spodosols	Orthods	Haplorthods	Lamellic Haplorthods	Sandy	Mixed			Frigid	Udic
LINDSTROM	H	D	WD	1–30	50	Am	Mollisols	Udolls	Hapludolls	Cumulic Hapludolls	Fine-silty	Mixed	Superactive		Mesic	Udic
LINO	S	VD	SPD	0–3	14	H	Entisols	Psamments	Udipsamments	Aquic Udipsamments		Mixed			Frigid	Udic
LOBO	M	VD	VPD	0–1	6	E, I, J	Histosols	Fibrists	Sphagnofibrists	Hemic Sphagnofibrists				Dysic	Frigid	Aquic
LOCKE	H	VD	SPD	0–6	8	A	Alfisols	Udalfs	Hapludalfs	Aquollic Hapludalfs	Fine-loamy	Mixed	Semiactive		Mesic	Udic
LOMIRA	H	VD	WD	0–20	275	B	Alfisols	Udalfs	Hapludalfs	Typic Hapludalfs	Fine-silty	Mixed	Superactive		Mesic	Udic
LONGRIE	M	MD	WD	0–35	150	Er, I	Spodosols	Orthods	Haplorthods	Typic Haplorthods	Coarse-loamy	Mixed	Superactive		Frigid	Udic
LORENZO	M	VD	WD	0–45	60	Bm	Mollisols	Udolls	Argiudolls	Typic Argiudolls	Fine-loamy over sandy or sandy-skeletal	Mixed	Active		Mesic	Udic
LOWS	M	VD	PD	0–2	81	J	Inceptisols	Aquepts	Endoaquepts	Mollic Endoaquepts	Fine-loamy over sandy or sandy-skeletal	Mixed	Superactive	Nonacid	Frigid	Udic
LOXLEY	M	VD	VPD	0–2	1333	J	Histosols	Saprists	Haplosaprists	Typic Haplosaprists				Dysic	Frigid	Aquic
LOYAL	H	VD	MWD	1–12	1157	F	Alfisols	Udalfs	Glossudalfs	Oxyaquic Glossudalfs	Fine-loamy	Mixed	Superactive		Frigid	Udic
LUDINGTON	S	MD	MWD	1–20	170	Dr	Spodosols	Orthods	Haplorthods	Oxyaquic Ultic Haplorthods	Sandy over loamy	Siliceous	Semiactive		Frigid	Udic

(continued)

(continued)

Soil series[a]	Soil YP[b]	Soil Depth[c]	Drainage class[d]	Slope (%)	Soil area (km^2)	Soil region	Order	Suborder	Great group	Subgroup	Particle-size class	Mineral class	CEC Activity class	Reaction class	Soil-temperature class	Soil-moisture class
LUNDEEN	M	MD	WD	2–25	1	F	Inceptisols	Udepts	Dystrudepts	Humic Dystrudepts	Coarse-silty	Isotic			Frigid	Aquic
LUPTON	M	VD	VPD	0–2	1578	J	Histosols	Saprists	Haplosaprists	Typic Haplosaprists				Euic	Frigid	Aquic
LUTZKE	M	VD	SED	2–20	45	B, I	Alfisols	Udalfs	Hapludalfs	Typic Hapludalfs	Loamy-skeletal	Mixed	Active		Mesic	Udic
MAGNOR	H	VD	SPD	0–6	2852	F, G	Alfisols	Udalfs	Glossudalfs	Aquic Glossudalfs	Coarse-loamy	Mixed	Superactive		Frigid	Udic
MAGROC	H	D	SPD	0–4	21	G	Alfisols	Udalfs	Glossudalfs	Aquic Glossudalfs	Coarse-loamy	Mixed	Superactive		Frigid	Udic
MAHALASVILLE	M	VD	PD	0–2	78	Bm, J	Mollisols	Aquolls	Argiaquolls	Typic Argiaquolls	Fine-silty	Mixed	Superactive		Mesic	Aquic
MAHTOMEDI	S	VD	ED	0–45	474	H	Entisols	Psamments	Udipsamments	Typic Udipsamments		Mixed			Frigid	Udic
MAINCREEK	M	VD	SPD	0–3	27	F	Alfisols	Udalfs	Glossudalfs	Aquic Glossudalfs	Coarse-loamy	Mixed	Superactive		Frigid	Udic
MAJIK	S	VD	SPD	0–3	14	Dr	Entisols	Psamments	Quartzipsamments	Aquic Quartzipsamments					Mesic	Udic
MAKWA	M	VD	VPD	0–12	74	J	Inceptisols	Aquepts	Humaquepts	Histic Humaquepts	Loamy-skeletal	Isotic		Nonacid	Frigid	Udic
MANAWA	H	VD	SPD	0–6	956	I	Alfisols	Udalfs	Hapludalfs	Aquollic Hapludalfs	Fine	Mixed	Active		Mesic	Aquic
MANCELONA	S	VD	SED	0–70	146	H	Spodosols	Orthods	Haplorthods	Alfic Haplorthods	Sandy	Mixed			Frigid	Udic
MANISTEE	S	VD	WD	0–50	56	I	Spodosols	Orthods	Haplorthods	Alfic Haplorthods	Sandy over clayey	Mixed	Active		Frigid	Udic
MANITOWISH	S	VD	MWD	0–18	127	G, H	Spodosols	Orthods	Haplorthods	Oxyaquic Haplorthods	Sandy	Mixed			Frigid	Udic
MANN	M	D	VPD	0–2	6	F, J	Mollisols	Aquolls	Epiaquolls	Typic Epiaquolls	Fine-loamy	Mixed	Superactive		Frigid	Aquic
MAPLEHURST	H	VD	SPD	0–3	67	F	Alfisols	Udalfs	Glossudalfs	Aquic Glossudalfs	Fine-silty	Mixed	Superactive		Frigid	Udic
MARATHON	H	D	MWD	2–15	91	Fr	Alfisols	Udalfs	Glossudalfs	Haplic Glossudalfs	Coarse-loamy	Mixed	Superactive		Frigid	Udic
MARCELLON	H	VD	SPD	0–3	19	Bm	Mollisols	Udolls	Argiudolls	Aquic Argiudolls	Fine-loamy	Mixed	Superactive		Mesic	Udic
MARKESAN	H	VD	WD	2–20	34	Bm	Mollisols	Udolls	Argiudolls	Typic Argiudolls	Fine-loamy	Mixed	Active		Mesic	Udic
MARKEY	M	VD	VPD	0–2	1449	J	Histosols	Saprists	Haplosaprists	Terric Haplosaprists	Sandy or sandy-skeletal	Mixed		Euic	Frigid	Aquic
MARKHAM	H	VD	MWD	0–20	132	B	Alfisols	Udalfs	Hapludalfs	Mollic Oxyaquic Hapludalfs	Fine	Illitic			Mesic	Udic
MARSHAN	M	VD	VPD	0–5	124	J	Mollisols	Aquolls	Endoaquolls	Typic Endoaquolls	Fine-loamy over sandy or sandy-skeletal	Mixed	Superactive		Mesic	Aquic

(continued)

(continued)

Soil series[a]	Soil YP[b]	Soil Depth[c]	Drainage class[d]	Slope (%)	Soil area (km^2)	Soil region	Order	Suborder	Great group	Subgroup	Particle-size class	Mineral class	CEC Activity class	Reaction class	Soil-temperature class	Soil-moisture class
MARSHFIELD	M	VD	PD	0–2	418	F, J	Alfisols	Aqualfs	Epiaqualfs	Mollic Epiaqualfs	Fine-loamy	Mixed	Superactive		Frigid	Aquic
MARTINTON	H	VD	SPD	0–6	69	Bm	Mollisols	Udolls	Argiudolls	Aquic Argiudolls	Fine	Illitic			Mesic	Udic
MATHERTON	M	VD	SPD	0–6	15	B, I, J	Alfisols	Aqualfs	Endoaqualfs	Udollic Endoaqualfs	Fine-loamy over sandy or sandy-skeletal	Mixed	Superactive		Mesic	Aquic
MAUMEE	S	VD	PD	0–2	5	Bm, J	Mollisols	Aquolls	Endoaquolls	Typic Endoaquolls	Sandy	Mixed			Mesic	Aquic
MAYVILLE	H	VD	MWD	0–15	163	B, E	Alfisols	Udalfs	Hapludalfs	Oxyaquic Hapludalfs	Fine-silty	Mixed	Superactive		Mesic	Udic
MCHENRY	H	VD	WD	0–30	576	B	Alfisols	Udalfs	Hapludalfs	Typic Hapludalfs	Fine-loamy	Mixed	Superactive		Mesic	Udic
MEADLAND	H	VD	SPD	0–6	255	Fr	Alfisols	Udalfs	Glossudalfs	Aquic Glossudalfs	Fine-loamy	Mixed	Superactive		Frigid	Udic
MECAN	S	VD	WD	2–30	93	C	Alfisols	Udalfs	Hapludalfs	Typic Hapludalfs	Coarse-loamy	Mixed	Active		Mesic	Udic
MECOSTA	S	VD	SED	0–35	2	C	Entisols	Orthents	Udorthents	Typic Udorthents	Sandy-skeletal	Mixed			Mesic	Udic
MEDARY	H	VD	MWD	0–45	10	A	Alfisols	Udalfs	Hapludalfs	Oxyaquic Hapludalfs	Fine	Mixed	Superactive		Mesic	Udic
MEEHAN	S	VD	SPD	0–3	780	C	Entisols	Psamments	Udipsamments	Aquic Udipsamments		Mixed			Frigid	Udic
MEENON	S	VD	SPD	0–3	22	I	Alfisols	Udalfs	Hapludalfs	Aquic Arenic Hapludalfs	Clayey	Smectitic			Frigid	Udic
MENAHGA	S	VD	ED	0–55	2090	H	Entisols	Psamments	Udipsamments	Typic Udipsamments		Mixed			Frigid	Udic
MENASHA	M	VD	PD	0–2	49	J	Mollisols	Aquolls	Epiaquolls	Typic Epiaquolls	Very fine	Mixed	Active		Mesic	Aquic
MENDOTA	H	VD	WD	0–12	131	Bm	Mollisols	Udolls	Argiudolls	Typic Argiudolls	Fine-silty	Mixed	Superactive		Mesic	Udic
MENOMIN	H	VD	MWD	0–3	2	Dr	Alfisols	Udalfs	Hapludalfs	Mollic Hapludalfs	Fine-loamy over sandy or sandy-skeletal	Mixed	Superactive		Mesic	Udic
MENOMINEE	S	VD	WD	6–70	346	E, G, H	Spodosols	Orthods	Haplorthods	Alfic Haplorthods	Sandy over loamy	Mixed	Active		Frigid	Udic
MEQUITHY	M	MD	WD	0–15	33	G	Spodosols	Orthods	Haplorthods	Alfic Haplorthods	Coarse-loamy	Mixed	Superactive		Frigid	Udic
MEQUON	H	VD	SPD	1–3	92	B, J	Alfisols	Aqualfs	Endoaqualfs	Udollic Endoaqualfs	Fine	Mixed	Superactive		Mesic	Udic
MERIDIAN	H	VD	WD	0–20	213	A, C	Alfisols	Udalfs	Hapludalfs	Mollic Hapludalfs	Fine-loamy over sandy or sandy-skeletal	Mixed	Superactive		Mesic	Udic

(continued)

(continued)

Soil series[a]	Soil YP[b]	Soil Depth[c]	Drainage class[d]	Slope (%)	Soil area (km^2)	Soil region	Order	Suborder	Great group	Subgroup	Particle-size class	Mineral class	CEC Activity class	Reaction class	Soil-temperature class	Soil-moisture class
MERIMOD	H	VD	MWD	0–6	38	Dr	Alfisols	Udalfs	Hapludalfs	Mollic Hapludalfs	Fine-loamy over sandy or sandy-skeletal	Mixed	Superactive		Mesic	Udic
MERIT	H	VD	WD	0–6	57	Dr	Alfisols	Udalfs	Hapludalfs	Mollic Hapludalfs	Fine-loamy over sandy or sandy-skeletal	Mixed	Superactive		Mesic	Udic
MERRILLAN	M	MD	SPD	0–6	313	Dr, J	Spodosols	Aquods	Epiaquods	Ultic Epiaquods	Coarse-loamy over clayey	Mixed	Semiactive		Frigid	Udic
METEA	H	VD	WD	0–25	50	B	Alfisols	Udalfs	Hapludalfs	Arenic Hapludalfs	Loamy	Mixed	Active		Mesic	Aquic
METONGA	M	MD	WD	1–60	63	G	Spodosols	Orthods	Haplorthods	Entic Haplorthods	Coarse-loamy	Mixed	Superactive		Frigid	Udic
MIAMI	H	VD	MWD	0–60	480	B	Alfisols	Udalfs	Hapludalfs	Oxyaquic Hapludalfs	Fine-loamy	Mixed	Active		Mesic	Udic
MICHIGAMME	M	MD	WD	1–75	108	G	Spodosols	Orthods	Haplorthods	Fragic Haplorthods	Coarse-loamy	Mixed	Superactive		Frigid	Aquic
MICKLE	H	VD	WD	2–12	23	A	Mollisols	Udolls	Argiudolls	Typic Argiudolls	Fine-silty	Mixed	Superactive		Mesic	Udic
MIFFLIN	H	D	WD	2–30	13	A	Alfisols	Udalfs	Hapludalfs	Typic Hapludalfs	Fine-loamy	Mixed	Superactive		Mesic	Udic
MILACA	H	VD	MWD	2–45	12	D, F, G	Alfisols	Udalfs	Glossudalfs	Oxyaquic Glossudalfs	Coarse-loamy	Mixed	Superactive		Frigid	Aquic
MILFORD	M	VD	VPD	0–2	45	Bm, J	Mollisols	Aquolls	Endoaquolls	Typic Endoaquolls	Fine	Mixed	Superactive		Mesic	Aquic
MILITARY	M	MD	WD	2–30	28	B	Alfisols	Udalfs	Hapludalfs	Typic Hapludalfs	Fine-loamy	Mixed	Active		Mesic	Udic
MILLADORE	H	VD	PD	0–6	137	F	Alfisols	Udalfs	Glossudalfs	Aquic Glossudalfs	Fine-loamy	Mixed	Superactive		Frigid	Udic
MILLINGTON	M	VD	PD	0–2	6	Bm, J	Mollisols	Aquolls	Endoaquolls	Cumulic Endoaquolls	Fine-loamy	Mixed	Superactive	Calcareous	Mesic	Aquic
MILTON	M	MD	WD	0–25	2	B	Alfisols	Udalfs	Hapludalfs	Typic Hapludalfs	Fine	Mixed	Active		Mesic	Udic
MINDORO	S	VD	MWD	0–3	6	Dr	Inceptisols	Udepts	Dystrudepts	Humic Psammentic Dystrudepts		Siliceous			Mesic	Udic
MINOCQUA	M	VD	VPD	0–2	562	J	Inceptisols	Aquepts	Endoaquepts	Typic Endoaquepts	Coarse-loamy over sandy or sandy-skeletal	Mixed	Superactive	Nonacid	Frigid	Aquic
MISKOAKI	H	VD	WD	2–70	15	I	Alfisols	Udalfs	Glossudalfs	Vertic Glossudalfs	Very fine	Mixed	Active		Frigid	Udic
MOBERG	M	D	SED	2–15	26	Fr, H	Inceptisols	Udepts	Dystrudepts	Typic Dystrudepts	Sandy-skeletal	Mixed			Frigid	Udic

(continued)

(continued)

Soil series[a]	Soil YP[b]	Soil Depth[c]	Drainage class[d]	Slope (%)	Soil area (km^2)	Soil region	Order	Suborder	Great group	Subgroup	Particle-size class	Mineral class	CEC Activity class	Reaction class	Soil-temperature class	Soil-moisture class
MONICO	H	D	SPD	0–6	90	G, J	Spodosols	Aquods	Endoaquods	Typic Endoaquods	Coarse-loamy	Mixed	Superactive		Frigid	Udic
MONTELLO	H	VD	MWD	0–6	17	Bm, J	Mollisols	Udolls	Argiudolls	Oxyaquic Argiudolls	Fine	Mixed	Superactive		Mesic	Udic
MONTGOMERY	M	VD	VPD	0–1	60	Bm, J	Mollisols	Aquolls	Endoaquolls	Vertic Endoaquolls	Fine	Mixed	Active		Mesic	Aquic
MOODIG	H	VD	SPD	0–4	40	G, J	Spodosols	Aquods	Epiaquods	Alfic Epiaquods	Coarse-loamy	Mixed	Superactive		Frigid	Aquic
MOPPET	H	VD	MWD	0–3	109	F	Inceptisols	Udepts	Dystrudepts	Oxyaquic Dystrudepts	Coarse-loamy	Mixed	Superactive		Frigid	Aquic
MOQUAH	H	VD	MWD	0–3	91	G, J	Entisols	Fluvents	Udifluvents	Typic Udifluvents	Coarse-loamy	Mixed	Superactive	Nonacid	Frigid	Udic
MORA	H	VD	SPD	0–6	7	G	Alfisols	Udalfs	Glossudalfs	Aquic Glossudalfs	Coarse-loamy	Mixed	Superactive		Frigid	Udic
MORGANLAKE	S	VD	MWD	0–35	46	H	Spodosols	Orthods	Haplorthods	Alfic Oxyaquic Haplorthods	Sandy over loamy	Mixed	Active		Frigid	Udic
MORLEY	H	VD	MWD	1–18	416	B	Alfisols	Udalfs	Hapludalfs	Oxyaquic Hapludalfs	Fine	Illitic			Mesic	Udic
MOROCCO	S	VD	SPD	0–3	67	C, J	Entisols	Psamments	Udipsamments	Aquic Udipsamments		Mixed			Mesic	Udic
MOSEL	H	VD	SPD	0–3	75	B, I	Alfisols	Udalfs	Hapludalfs	Aquollic Hapludalfs	Fine-loamy	Mixed	Superactive		Mesic	Udic
MOSHAWQUIT	S	VD	WD	2–15	3	G	Alfisols	Udalfs	Glossudalfs	Arenic Glossudalfs	Loamy	Mixed	Active		Frigid	Udic
MOSINEE	M	D	WD	2–20	172	Fr	Inceptisols	Udepts	Dystrudepts	Typic Dystrudepts	Loamy-skeletal	Mixed	Superactive		Frigid	Udic
MOUNDVILLE	S	VD	WD	0–6	53	B	Alfisols	Udalfs	Hapludalfs	Oxyaquic Hapludalfs	Sandy	Mixed			Mesic	Udic
MT. CARROLL	H	VD	WD	0–45	49	A, B, Dr	Alfisols	Udalfs	Hapludalfs	Mollic Hapludalfs	Fine-silty	Mixed	Superactive		Mesic	Udic
MUDLAKE	H	VD	SPD	1–6	108	G, J	Spodosols	Aquods	Epiaquods	Alfic Epiaquods	Coarse-loamy	Mixed	Superactive		Frigid	Udic
MUNDELEIN	H	VD	SPD	0–5	79	B, I	Mollisols	Udolls	Argiudolls	Aquic Argiudolls	Fine-silty	Mixed	Superactive		Mesic	Udic
MUNUSCONG	M	VD	PD	0–2	15	G, J	Inceptisols	Aquepts	Epiaquepts	Mollic Epiaquepts	Coarse-loamy over clayey	Mixed	Active	Nonacid	Frigid	Aquic
MUSCATINE	H	VD	SPD	0–5	63	Am	Mollisols	Udolls	Hapludolls	Aquic Hapludolls	Fine-silty	Mixed	Superactive		Mesic	Udic
MUSCODA	S	D	WD	2–20	2	Dr	Alfisols	Udalfs	Hapludalfs	Lamellic Hapludalfs	Coarse-loamy	Mixed	Superactive		Mesic	Udic
MUSKEGO	H	VD	VPD	0–2	7	B, J	Histosols	Saprists	Haplosaprists	Limnic Haplosaprists		Coprogenous		Euic	Mesic	Aquic
MUSSEY	M	VD	VPD	0–2	26	E, I, J	Mollisols	Aquolls	Argiaquolls	Typic Argiaquolls	Fine-loamy over sandy or sandy-skeletal	Mixed	Semiactive		Mesic	Udic

(continued)

(continued)

Soil series[a]	Soil YP[b]	Soil Depth[c]	Drainage class[d]	Slope (%)	Soil area (km^2)	Soil region	Order	Suborder	Great group	Subgroup	Particle-size class	Mineral class	CEC Activity class	Reaction class	Soil-temperature class	Soil-moisture class
MYLREA	H	D	SPD	0–6	98	Fr	Inceptisols	Udepts	Dystrudepts	Aquic Dystrudepts	Coarse-loamy	Mixed	Superactive		Frigid	Udic
MYRTLE	H	VD	WD	2–20	17	A	Alfisols	Udalfs	Hapludalfs	Mollic Hapludalfs	Fine-silty	Mixed	Superactive		Mesic	Udic
NADEAU	M	D	WD	0–45	44	G, H	Alfisols	Udalfs	Hapludalfs	Typic Hapludalfs	Coarse-loamy	Mixed	Active		Frigid	Aquic
NAHMA	M	MD	PD	0–2	3	G, J	Inceptisols	Aquepts	Humaquepts	Histic Humaquepts	Coarse-loamy	Mixed	Active	Nonacid	Frigid	Aquic
NAMUR	M	VS	ED	0–20	97	Er	Mollisols	Udolls	Hapludolls	Lithic Hapludolls	Loamy	Mixed	Active		Frigid	Udic
NAVAN	M	VD	PD	0–3	81	B, I, J	Mollisols	Aquolls	Argiaquolls	Typic Argiaquolls	Fine-loamy	Mixed	Superactive		Mesic	Aquic
NEBAGO	S	VD	SPD	0–4	44	E, G, I	Inceptisols	Udepts	Eutrudepts	Aquic Dystric Eutrudepts	Sandy over clayey	Mixed	Active		Mesic	Udic
NECONISH	S	VD	MWD	0–3	6	H	Spodosols	Orthods	Haplorthods	Oxyaquic Haplorthods	Sandy	Isotic			Frigid	Udic
NEDA	H	VD	MWD	2–30	14	B	Alfisols	Udalfs	Hapludalfs	Mollic Oxyaquic Hapludalfs	Fine-loamy	Mixed	Active		Mesic	Aquic
NEENAH	H	VD	SPD	0–3	43	I	Alfisols	Udalfs	Hapludalfs	Aquollic Hapludalfs	Very fine	Mixed	Active		Mesic	Udic
NENNO	H	MD	SPD	0–6	44	Bm	Mollisols	Udolls	Argiudolls	Aquic Argiudolls	Fine-loamy	Mixed	Active		Mesic	Udic
NEOPIT	H	VD	WD	0–6	26	G	Alfisols	Udalfs	Glossudalfs	Oxyaquic Glossudalfs	Coarse-loamy	Mixed	Superactive		Frigid	Udic
NESTER	H	VD	MWD	0–12	0	E	Alfisols	Udalfs	Glossudalfs	Oxyaquic Glossudalfs	Fine	Mixed	Semiactive		Frigid	Udic
NEWGLARUS	H	S	WD	2–45	430	A	Alfisols	Udalfs	Hapludalfs	Typic Hapludalfs	Fine-silty over clayey	Mixed	Superactive		Mesic	Udic
NEWLANG	S	VD	PD	0–2	25	Dr, J	Entisols	Aquents	Psammaquents	Humaqueptic Psammaquents		Siliceous			Mesic	Udic
NEWOOD	M	VD	MWD	1–30	847	G	Spodosols	Orthods	Haplorthods	Alfic Oxyaquic Haplorthods	Coarse-loamy	Isotic			Frigid	Udic
NEWOT	M	VD	WD	10–45	294	F, G	Spodosols	Orthods	Haplorthods	Alfic Haplorthods	Coarse-loamy	Mixed	Superactive		Frigid	Udic
NEWSON	S	VD	VPD	0–2	840	J	Entisols	Aquents	Psammaquents	Humaqueptic Psammaquents		Mixed			Frigid	Aquic
NICHOLS	H	VD	MWD	0–25	78	E, I	Inceptisols	Udepts	Eutrudepts	Oxyaquic Eutrudepts	Coarse-silty	Mixed	Superactive		Mesic	Udic
NICKIN	H	VD	WD	1–20	60	Am, Dr	Mollisols	Udolls	Argiudolls	Typic Argiudolls	Fine-loamy over sandy or sandy-skeletal	Mixed over siliceous	Active		Frigid	Udic
NOKASIPPI	S	VD	VPD	0–1	5	G, J	Mollisols	Aquolls	Epiaquolls	Typic Epiaquolls	Coarse-loamy	Mixed	Superactive		Frigid	Aquic
NORDEN	H	D	WD	2–65	1077	Dr	Alfisols	Udalfs	Hapludalfs	Typic Hapludalfs	Fine-loamy	Mixed	Superactive		Mesic	Udic

(continued)

(continued)

Soil series[a]	Soil YP[b]	Soil Depth[c]	Drainage class[d]	Slope (%)	Soil area (km^2)	Soil region	Order	Suborder	Great group	Subgroup	Particle-size class	Mineral class	CEC Activity class	Reaction class	Soil-temperature class	Soil-moisture class
NORGO	M	S	WD	2–20	19	Dr	Alfisols	Udalfs	Hapludalfs	Lithic Hapludalfs	Loamy	Mixed	Active		Frigid	Udic
NORTHBEND	H	VD	SPD	0–3	44	Dr	Inceptisols	Udepts	Dystrudepts	Fluvaquentic Dystrudepts	Coarse-loamy over sandy or sandy-skeletal	Mixed	Active		Mesic	Udic
NORTHFIELD	M	S	WD	0–60	149	Dr	Alfisols	Udalfs	Hapludalfs	Lithic Hapludalfs	Loamy	Mixed	Active		Mesic	Udic
NORTHMOUND	M	MD	SED	6–50	5	Dr	Alfisols	Udalfs	Glossudalfs	Haplic Glossudalfs	Loamy-skeletal	Mixed	Superactive		Frigid	Udic
NOSEUM	M	VD	MWD	0–12	7	H	Spodosols	Orthods	Haplorthods	Oxyaquic Haplorthods	Sandy	Isotic			Frigid	Udic
NUXMARUHANIXETE	H	VD	WD	1–12	3	Am	Mollisols	Udolls	Argiudolls	Typic Argiudolls	Fine-loamy	Mixed	Superactive		Mesic	Udic
NYMORE	S	VD	ED	0–25	33	C, H	Entisols	Psamments	Udipsamments	Typic Udipsamments		Mixed			Frigid	Udic
OAKVILLE	S	VD	ED	0–60	80	C, E, H, I	Entisols	Psamments	Udipsamments	Typic Udipsamments		Mixed			Mesic	Aquic
OCKLEY	H	VD	WD	0–30	19	A	Alfisols	Udalfs	Hapludalfs	Typic Hapludalfs	Fine-loamy	Mixed	Active		Mesic	Udic
OCONTO	H	VD	WD	0–20	1	G	Alfisols	Udalfs	Glossudalfs	Haplic Glossudalfs	Coarse-loamy	Mixed	Active		Frigid	Udic
ODANAH	H	VD	WD	0–60	114	I	Alfisols	Udalfs	Glossudalfs	Haplic Glossudalfs	Fine	Mixed	Active		Frigid	Udic
OESTERLE	M	VD	SPD	0–3	397	F, G	Alfisols	Udalfs	Glossudalfs	Aquic Glossudalfs	Coarse-loamy	Mixed	Superactive		Frigid	Udic
OGLE	H	VD	WD	2–18	47	Bm	Mollisols	Udolls	Argiudolls	Typic Argiudolls	Fine-silty	Mixed	Superactive		Mesic	Udic
OKEE	S	VD	SED	1–25	111	C	Alfisols	Udalfs	Hapludalfs	Arenic Hapludalfs	Loamy	Mixed	Active		Mesic	Udic
OMEGA	S	VD	SED	0–25	88	E, G, H	Spodosols	Orthods	Haplorthods	Typic Haplorthods	Sandy	Mixed			Frigid	Udic
OMENA	H	VD	WD	0–50	115	E	Alfisols	Udalfs	Glossudalfs	Haplic Glossudalfs	Coarse-loamy	Mixed	Active		Frigid	Udic
OMRO	H	VD	MWD	1–12	43	E, I	Alfisols	Udalfs	Hapludalfs	Typic Hapludalfs	Clayey over loamy	Mixed	Active		Mesic	Udic
ONAWAY	H	VD	MWD	0–60	739	B, E, I	Alfisols	Udalfs	Hapludalfs	Inceptic Hapludalfs	Fine-loamy	Mixed	Active		Frigid	Udic
ORION	H	VD	SPD	0–5	359	J	Entisols	Fluvents	Udifluvents	Aquic Udifluvents	Coarse-silty	Mixed	Superactive	Nonacid	Mesic	Udic
ORONTO	H	VD	SPD	0–3	24	F, G, J	Alfisols	Aqualfs	Glossaqualfs	Aeric Glossaqualfs	Fine-silty	Mixed	Active		Frigid	Udic
OSHKOSH	H	VD	WD	0–30	174	E, I	Alfisols	Udalfs	Hapludalfs	Typic Hapludalfs	Very fine	Mixed	Active		Mesic	Udic
OSHTEMO	S	VD	WD	0–55	176	C	Alfisols	Udalfs	Hapludalfs	Typic Hapludalfs	Coarse-loamy	Mixed	Active		Mesic	Udic
OSSIAN	M	VD	PD	0–4	146	Bm, I, J	Mollisols	Aquolls	Endoaquolls	Typic Endoaquolls	Fine-silty	Mixed	Superactive		Mesic	Aquic

(continued)

(continued)

Soil series[a]	Soil YP[b]	Soil Depth[c]	Drainage class[d]	Slope (%)	Soil area (km^2)	Soil region	Order	Suborder	Great group	Subgroup	Particle-size class	Mineral class	CEC Activity class	Reaction class	Soil-temperature class	Soil-moisture class
OSSMER	H	VD	SPD	0–3	301	G	Alfisols	Udalfs	Glossudalfs	Aquic Glossudalfs	Coarse-loamy over sandy or sandy-skeletal	Mixed	Superactive		Frigid	Aquic
OTTER	M	VD	PD	0–5	121	J	Mollisols	Aquolls	Endoaquolls	Cumulic Endoaquolls	Fine-silty	Mixed	Superactive		Mesic	Aquic
OTTERHOLT	H	VD	WD	1–20	66	F	Alfisols	Udalfs	Glossudalfs	Haplic Glossudalfs	Fine-silty	Mixed	Superactive		Frigid	Udic
OZAUKEE	H	MD	WD	0–35	255	B	Alfisols	Udalfs	Hapludalfs	Oxyaquic Hapludalfs	Fine	Illitic			Mesic	Udic
PADUS	M	VD	WD	0–45	2373	F, G	Spodosols	Orthods	Haplorthods	Alfic Haplorthods	Coarse-loamy	Mixed	Superactive		Frigid	Udic
PADWET	M	VD	MWD	0–6	39	G	Spodosols	Orthods	Haplorthods	Alfic Haplorthods	Coarse-loamy	Mixed	Superactive		Frigid	Udic
PADWOOD	H	VD	MWD	0–15	55	G	Spodosols	Orthods	Haplorthods	Alfic Oxyaquic Haplorthods	Coarse-loamy	Mixed	Superactive		Frigid	Udic
PALMS	H	VD	VPD	0–6	510	J	Histosols	Saprists	Haplosaprists	Terric Haplosaprists	Loamy	Mixed		Euic	Mesic	Aquic
PALSGROVE	H	D	WD	2–30	659	A	Alfisols	Udalfs	Hapludalfs	Typic Hapludalfs	Fine-silty	Mixed	Superactive		Mesic	Udic
PARDEEVILLE	H	VD	WD	2–30	8	B	Alfisols	Udalfs	Hapludalfs	Mollic Hapludalfs	Coarse-loamy	Mixed	Active		Mesic	Udic
PARKFALLS	M	VD	SPD	0–4	49	G, J	Spodosols	Aquods	Epiaquods	Alfic Epiaquods	Coarse-loamy	Mixed	Superactive		Frigid	Udic
PARTRIDGE	S	MD	SPD	0–3	23	C	Entisols	Psamments	Udipsamments	Aquic Udipsamments		Mixed			Mesic	Udic
PEARL	S	VD	MWD	0–6	49	C, G	Alfisols	Udalfs	Hapludalfs	Arenic Oxyaquic Hapludalfs	Loamy	Mixed	Superactive		Mesic	Udic
PECATONICA	H	VD	WD	0–30	92	B	Alfisols	Udalfs	Hapludalfs	Typic Hapludalfs	Fine-loamy	Mixed	Superactive		Mesic	Udic
PECORE	H	VD	WD	2–35	4	G	Alfisols	Udalfs	Glossudalfs	Haplic Glossudalfs	Fine-loamy	Mixed	Active		Frigid	Udic
PEEBLES	H	VD	MWD	0–6	10	Bm	Mollisols	Udolls	Argiudolls	Oxyaquic Argiudolls	Very fine	Mixed	Active		Mesic	Udic
PELISSIER	M	VD	ED	1–50	41	G	Spodosols	Orthods	Haplorthods	Entic Haplorthods	Sandy-skeletal	Mixed			Frigid	Udic
PELKIE	S	VD	MWD	0–4	37	H, J	Entisols	Psamments	Udipsamments	Oxyaquic Udipsamments		Mixed			Frigid	Udic
PELLA	M	VD	PD	0–3	786	Bm, J	Mollisols	Aquolls	Endoaquolls	Typic Endoaquolls	Fine-silty	Mixed	Superactive		Mesic	Aquic
PENCE	S	VD	SED	0–50	2343	F, G, H	Spodosols	Orthods	Haplorthods	Typic Haplorthods	Sandy	Isotic			Frigid	Aquic
PEPIN	H	VD	WD	2–30	191	A, Dr	Alfisols	Udalfs	Hapludalfs	Typic Hapludalfs	Fine-silty	Mixed	Superactive		Mesic	Udic
PEQUAMING	S	D	SPD	0–4	34	H, J	Spodosols	Aquods	Endoaquods	Argic Endoaquods	Sandy	Mixed			Frigid	Udic
PERCHLAKE	S	VD	SPD	0–3	9	H	Entisols	Psamments	Udipsamments	Aquic Udipsamments		Mixed			Frigid	Udic
PERIDA	S	VD	MWD	0–30	24	H, I	Alfisols	Udalfs	Hapludalfs	Arenic Hapludalfs	Clayey	Smectitic			Frigid	Udic

(continued)

(continued)

Soil series[a]	Soil YP[b]	Soil Depth[c]	Drainage class[d]	Slope (%)	Soil area (km^2)	Soil region	Order	Suborder	Great group	Subgroup	Particle-size class	Mineral class	CEC Activity class	Reaction class	Soil-temperature class	Soil-moisture class
PEROTE	H	VD	WD	2–35	23	G	Alfisols	Udalfs	Glossudalfs	Haplic Glossudalfs	Coarse-loamy	Mixed	Active		Frigid	Udic
PESABIC	H	VD	SPD	0–4	283	G, J	Spodosols	Aquods	Epiaquods	Alfic Epiaquods	Coarse-loamy	Mixed	Superactive		Frigid	Udic
PESHEKEE	M	S	WD	3–75	17	G	Spodosols	Orthods	Haplorthods	Lithic Haplorthods	Loamy	Mixed	Semiactive		Frigid	Aquic
PESHTIGO	H	VD	SPD	0–4	2	G	Alfisols	Udalfs	Glossudalfs	Aquic Glossudalfs	Fine-loamy	Mixed	Active		Frigid	Aquic
PICKFORD	M	VD	PD	0–2	103	I, J	Inceptisols	Aquepts	Epiaquepts	Aeric Epiaquepts	Fine	Mixed	Active	Nonacid	Frigid	Aquic
PILLOT	H	VD	WD	0–18	100	A, F	Mollisols	Udolls	Argiudolls	Typic Argiudolls	Fine-silty over sandy or sandy-skeletal	Mixed	Superactive		Mesic	Udic
PINCONNING	S	VD	PD	0–2	9	E, J	Entisols	Aquents	Epiaquents	Mollic Epiaquents	Sandy over clayey	Mixed	Semiactive	Nonacid	Frigid	Aquic
PLAINBO	S	MD	ED	1–45	188	C, Dr, H	Entisols	Psamments	Udipsamments	Typic Udipsamments		Mixed			Frigid	Udic
PLAINFIELD	S	VD	SED	0–70	2432	C	Entisols	Psamments	Udipsamments	Typic Udipsamments		Mixed			Mesic	Udic
PLANO	H	VD	WD	0–12	1122	Bm	Mollisols	Udolls	Argiudolls	Typic Argiudolls	Fine-silty	Mixed	Superactive		Mesic	Udic
PLEINE	M	VD	PD	0–2	100	G, J	Inceptisols	Aquepts	Humaquepts	Histic Humaquepts	Coarse-loamy	Mixed	Superactive	Nonacid	Frigid	Aquic
PLOVER	H	VD	SPD	0–3	58	F, G	Alfisols	Udalfs	Glossudalfs	Aquic Glossudalfs	Coarse-loamy	Mixed	Superactive		Frigid	Udic
PLUMCREEK	H	VD	WD	12–45	23	Dr	Alfisols	Udalfs	Hapludalfs	Typic Hapludalfs	Fine-loamy	Mixed	Superactive		Mesic	Udic
POINT	S	VD	SPD	1–6	74	Fr	Alfisols	Udalfs	Glossudalfs	Aquic Glossudalfs	Fine-loamy	Mixed	Superactive		Frigid	Udic
POMROY	S	VD	MWD	2–30	27	G	Alfisols	Udalfs	Hapludalfs	Arenic Oxyaquic Hapludalfs	Loamy	Mixed	Superactive		Frigid	Udic
PONYCREEK	S	VD	PD	0–2	136	C, Dr, J	Entisols	Aquents	Psammaquents	Humaqueptic Psammaquents		Siliceous			Frigid	Aquic
PORT BYRON	H	VD	WD	0–30	62	Am	Mollisols	Udolls	Hapludolls	Typic Hapludolls	Fine-silty	Mixed	Superactive		Mesic	Udic
PORTWING	H	VD	MWD	2–6	228	I	Alfisols	Udalfs	Glossudalfs	Oxyaquic Glossudalfs	Fine	Mixed	Active		Frigid	Udic
POSKIN	H	VD	SPD	0–3	108	F	Alfisols	Udalfs	Glossudalfs	Aquic Glossudalfs	Fine-silty over sandy or sandy-skeletal	Mixed	Superactive		Frigid	Udic
POY	M	VD	PD	0–2	123	C, I, J	Mollisols	Aquolls	Endoaquolls	Typic Endoaquolls	Clayey over sandy or sandy-skeletal	Mixed	Active		Mesic	Aquic
POYGAN	M	VD	PD	0–3	456	E, I, J	Mollisols	Aquolls	Epiaquolls	Typic Epiaquolls	Fine	Mixed	Active		Mesic	Aquic

(continued)

(continued)

Soil series[a]	Soil YP[b]	Soil Depth[c]	Drainage class[d]	Slope (%)	Soil area (km^2)	Soil region	Order	Suborder	Great group	Subgroup	Particle-size class	Mineral class	CEC Activity class	Reaction class	Soil-temperature class	Soil-moisture class
PRISSEL	S	VD	MWD	0–60	70	B, Dr	Alfisols	Udalfs	Hapludalfs	Arenic Hapludalfs	Loamy	Mixed	Active		Mesic	Udic
PUCHYAN	S	VD	MWD	2–12	21	B	Alfisols	Udalfs	Hapludalfs	Arenic Oxyaquic Hapludalfs	Loamy	Mixed	Superactive		Mesic	Udic
QUARDERER	H	VD	MWD	0–3	14	Cm	Mollisols	Udolls	Hapludolls	Fluventic Hapludolls	Coarse-silty	Mixed	Superactive		Frigid	Udic
RABE	S	VD	WD	2–35	11	G	Alfisols	Udalfs	Glossudalfs	Arenic Glossudalfs	Loamy	Mixed	Active		Frigid	Udic
RADFORD	H	VD	SPD	0–5	80	Bm	Mollisols	Udolls	Hapludolls	Fluvaquentic Hapludolls	Fine-silty	Mixed	Superactive		Mesic	Udic
RASSET	H	VD	WD	0–40	71	A, Dr	Mollisols	Udolls	Argiudolls	Typic Argiudolls	Coarse-loamy	Mixed	Superactive		Mesic	Udic
REDRIM	M	S	ED	0–6	2	G, H	Spodosols	Orthods	Haplorthods	Entic Lithic Haplorthods	Sandy-skeletal	Mixed			Frigid	Udic
REEDSBURG	H	D	SPD	0–12	34	A, Dr	Alfisols	Udalfs	Paleudalfs	Aquic Paleudalfs	Fine-silty	Mixed	Superactive		Mesic	Udic
RENOVA	H	VD	WD	2–35	31	G	Alfisols	Udalfs	Hapludalfs	Typic Hapludalfs	Fine-loamy	Mixed	Superactive		Mesic	Udic
RIB	M	VD	PD	0–2	110	J	Alfisols	Aqualfs	Endoaqualfs	Mollic Endoaqualfs	Fine-silty over sandy or sandy-skeletal	Mixed	Superactive		Frigid	Aquic
RIBHILL	M	MD	SED	2–40	43	Fr, G	Alfisols	Udalfs	Glossudalfs	Haplic Glossudalfs	Loamy-skeletal	Mixed	Superactive		Frigid	Udic
RIBRIVER	H	VD	MWD	0–3	9	F	Alfisols	Udalfs	Glossudalfs	Oxyaquic Glossudalfs	Fine-silty	Mixed	Superactive		Frigid	Udic
RICHFORD	S	VD	WD	0–20	581	Cm	Alfisols	Udalfs	Hapludalfs	Arenic Hapludalfs	Loamy	Mixed	Superactive		Mesic	Udic
RICHWOOD	H	VD	WD	0–12	53	Am	Mollisols	Udolls	Argiudolls	Typic Argiudolls	Fine-silty	Mixed	Superactive		Mesic	Udic
RIETBROCK	H	D	SPD	1–12	305	Fr	Alfisols	Udalfs	Glossudalfs	Aquic Glossudalfs	Fine-loamy	Mixed	Superactive		Frigid	Udic
RIFLE	M	VD	VPD	0–2	136	J	Histosols	Hemists	Haplohemists	Typic Haplohemists				Euic	Frigid	Aquic
RIMER	H	VD	SPD	0–6	5	A	Alfisols	Udalfs	Hapludalfs	Aquic Arenic Hapludalfs	Loamy	Mixed	Active		Mesic	Udic
RINGWOOD	H	VD	WD	0–12	181	Bm	Mollisols	Udolls	Argiudolls	Typic Argiudolls	Fine-loamy	Mixed	Superactive		Mesic	Udic
RIPON	H	MD	WD	0–12	8	Bm	Mollisols	Udolls	Argiudolls	Typic Argiudolls	Fine-silty	Mixed	Superactive		Mesic	Udic
RITCHEY	M	S	WD	0–60	47	A, 111B, 90B	Alfisols	Udalfs	Hapludalfs	Lithic Hapludalfs	Loamy	Mixed	Superactive		Mesic	Udic
ROBAGO	H	VD	SPD	0–3	30	G, J	Spodosols	Aquods	Endoaquods	Argic Endoaquods	Coarse-loamy	Mixed	Superactive		Frigid	Udic
ROBY	H	VD	SPD	0–7	20	C, Dr	Alfisols	Udalfs	Hapludalfs	Aquic Hapludalfs	Coarse-loamy	Mixed	Superactive		Mesic	Udic
ROCKBLUFF	S	D	ED	30–60	49	Dr	Entisols	Psamments	Quartzipsamments	Typic Quartzipsamments					Mesic	Udic
ROCKBRIDGE	H	VD	WD	6–20	8	A	Alfisols	Udalfs	Hapludalfs	Typic Hapludalfs	Fine-loamy	Mixed	Superactive		Mesic	Udic

(continued)

(continued)

Soil series[a]	Soil YP[b]	Soil Depth[c]	Drainage class[d]	Slope (%)	Soil area (km^2)	Soil region	Order	Suborder	Great group	Subgroup	Particle-size class	Mineral class	CEC Activity class	Reaction class	Soil-temperature class	Soil-moisture class
ROCKDAM	S	VD	MWD	0–3	123	Dr	Spodosols	Orthods	Haplorthods	Entic Haplorthods	Sandy	Siliceous			Frigid	Aquic
ROCKERS	S	VD	SPD	0–6	44	G, J	Spodosols	Aquods	Endoaquods	Argic Endoaquods	Coarse-loamy	Mixed	Active		Frigid	Udic
ROCKLAND	H	VD	WD	18–70	8	G	Inceptisols	Udepts	Eutrudepts	Typic Eutrudepts	Fine-loamy	Mixed	Active		Frigid	Aquic
ROCKMARSH	M	VD	SPD	0–20	4	G	Alfisols	Udalfs	Hapludalfs	Aquollic Hapludalfs	Loamy-skeletal	Mixed	Superactive		Frigid	Udic
ROCKTON	M	MD	WD	0–25	101	Bm	Mollisols	Udolls	Argiudolls	Typic Argiudolls	Fine-loamy	Mixed	Superactive		Mesic	Udic
RODMAN	M	VD	ED	2–70	153	Bm	Mollisols	Udolls	Hapludolls	Typic Hapludolls	Sandy-skeletal	Mixed			Mesic	Udic
RONDEAU	M	VD	VPD	0–1	15	A	Histosols	Saprists	Haplosaprists	Limnic Haplosaprists		Marly		Euic	Frigid	Aquic
ROOT	M	D	VPD	1–4	1	A	Entisols	Aquents	Fluvaquents	Mollic Fluvaquents	Coarse-loamy	Mixed	Active	Nonacid	Mesic	Aquic
ROSCOMMON	S	VD	VPD	0–2	379	J	Entisols	Aquents	Psammaquents	Mollic Psammaquents		Mixed			Frigid	Udic
ROSHOLT	M	VD	WD	0–45	1501	G, F	Alfisols	Udalfs	Glossudalfs	Haplic Glossudalfs	Coarse-loamy	Mixed	Superactive		Frigid	Udic
ROTAMER	H	VD	WD	2–45	129	Bm	Mollisols	Udolls	Argiudolls	Typic Argiudolls	Fine-loamy	Mixed	Active		Mesic	Udic
ROUSSEAU	S	VD	WD	0–70	291	E, H	Spodosols	Orthods	Haplorthods	Entic Haplorthods	Sandy	Mixed			Frigid	Aquic
ROWLEY	H	VD	SPD	0–4	26	Am	Mollisols	Udolls	Argiudolls	Aquic Argiudolls	Fine-silty	Mixed	Superactive		Mesic	Udic
ROZELLVILLE	H	VD	WD	1–12	200	Fr	Alfisols	Udalfs	Glossudalfs	Haplic Glossudalfs	Fine-loamy	Mixed	Superactive		Frigid	Udic
ROZETTA	H	VD	WD	0–25	60	A	Alfisols	Udalfs	Hapludalfs	Typic Hapludalfs	Fine-silty	Mixed	Superactive		Mesic	Udic
RUBICON	S	VD	ED	0–70	742	G, H	Spodosols	Orthods	Haplorthods	Entic Haplorthods	Sandy	Mixed			Frigid	Udic
RUSE	M	S	VPD	0–2	6	E, J	Mollisols	Aquolls	Endoaquolls	Lithic Endoaquolls	Loamy	Mixed	Active		Frigid	Udic
RUSKTOWN	M	VD	MWD	0–3	17	Dr	Alfisols	Udalfs	Hapludalfs	Mollic Hapludalfs	Coarse-loamy	Mixed	Active		Mesic	Udic
SABLE	M	VD	PD	0–2	94	J	Mollisols	Aquolls	Endoaquolls	Typic Endoaquolls	Fine-silty	Mixed	Superactive		Mesic	Aquic
SALTER	H	VD	WD	0–20	32	B	Inceptisols	Udepts	Eutrudepts	Typic Eutrudepts	Coarse-loamy	Mixed	Superactive		Mesic	Udic
SANBORG	H	VD	MWD	0–6	228	I	Alfisols	Udalfs	Glossudalfs	Oxyaquic Glossudalfs	Fine	Mixed	Active		Frigid	Udic
SANTIAGO	H	D	WD	1–45	663	F, G	Alfisols	Udalfs	Glossudalfs	Haplic Glossudalfs	Coarse-loamy	Mixed	Superactive		Frigid	Udic
SARGEANT	H	VD	SPD	0–6	4	F	Alfisols	Udalfs	Glossudalfs	Aquic Glossudalfs	Fine-loamy	Mixed	Superactive		Mesic	Udic
SARONA	H	VD	WD	1–35	3668	G, H	Spodosols	Orthods	Haplorthods	Alfic Haplorthods	Coarse-loamy	Mixed	Superactive		Frigid	Udic
SARWET	H	VD	MWD	0–6	72	G	Spodosols	Orthods	Haplorthods	Alfic Oxyaquic Haplorthods	Coarse-loamy	Mixed	Superactive		Frigid	Udic

(continued)

(continued)

Soil series[a]	Soil YP[b]	Soil Depth[c]	Drainage class[d]	Slope (%)	Soil area (km^2)	Soil region	Order	Suborder	Great group	Subgroup	Particle-size class	Mineral class	CEC Activity class	Reaction class	Soil-temperature class	Soil-moisture class
SATTRE	H	VD	WD	0–12	204	F, G	Alfisols	Udalfs	Hapludalfs	Mollic Hapludalfs	Fine-loamy over sandy or sandy-skeletal	Mixed	Superactive		Mesic	Udic
SAYBROOK	H	VD	MWD	0–20	33	Bm	Mollisols	Udolls	Argiudolls	Oxyaquic Argiudolls	Fine-silty	Mixed	Superactive		Mesic	Udic
SAYLESVILLE	H	VD	WD	0–40	49	B	Alfisols	Udalfs	Hapludalfs	Typic Hapludalfs	Fine	Illitic			Mesic	Udic
SAYNER	S	VD	ED	0–60	1069	G, H	Spodosols	Orthods	Haplorthods	Entic Haplorthods	Sandy	Mixed			Frigid	Aquic
SCHAAT CREEK	M	VD	PD	0–2	1	I, J	Alfisols	Aqualfs	Endoaqualfs	Aeric Endoaqualfs	Fine	Mixed	Active		Frigid	Aquic
SCHAPVILLE	M	MD	MWD	2–35	1	Am	Mollisols	Udolls	Argiudolls	Oxyaquic Argiudolls	Fine	Mixed	Active		Mesic	Udic
SCHWEITZER	M	MD	WD	6–70	9	G	Spodosols	Orthods	Fragiorthods	Alfic Fragiorthods	Coarse-loamy	Mixed	Superactive		Frigid	Udic
SCOBA	M	VD	MWD	0–6	19	G	Alfisols	Udalfs	Glossudalfs	Haplic Glossudalfs	Coarse-loamy	Mixed	Superactive		Frigid	Udic
SCONSIN	H	MD	MWD	0–6	240	G	Alfisols	Udalfs	Glossudalfs	Oxyaquic Glossudalfs	Coarse-loamy	Mixed	Superactive		Frigid	Udic
SCOTAH	S	VD	MWD	0–3	47	Dr	Entisols	Psamments	Udipsamments	Typic Udipsamments		Mixed			Mesic	Udic
SCOTT LAKE	M	VD	MWD	0–6	201	G	Alfisols	Udalfs	Glossudalfs	Oxyaquic Glossudalfs	Coarse-loamy	Mixed	Superactive		Frigid	Udic
SEATON	H	VD	WD	0–60	1042	A, Dr	Alfisols	Udalfs	Hapludalfs	Typic Hapludalfs	Fine-silty	Mixed	Superactive		Mesic	Udic
SEBBO	H	VD	MWD	1–6	19	Dr	Alfisols	Udalfs	Hapludalfs	Mollic Hapludalfs	Fine-loamy	Mixed	Superactive		Mesic	Udic
SEBEWA	M	VD	VPD	0–3	286	Bm, J	Mollisols	Aquolls	Argiaquolls	Typic Argiaquolls	Fine-loamy over sandy or sandy-skeletal	Mixed	Superactive		Mesic	Aquic
SECHLER	M	VD	SPD	0–3	4	Cm	Inceptisols	Udepts	Dystrudepts	Aquic Humic Dystrudepts	Coarse-loamy	Siliceous	Superactive		Mesic	Udic
SEDGWICK	S	VD	SPD	0–15	97	G, I, J	Spodosols	Aquods	Epiaquods	Alfic Epiaquods	Coarse-loamy over clayey	Mixed	Active		Frigid	Udic
SEELYEVILLE	M	VD	VPD	0–15	1595		Histosols	Saprists	Haplosaprists	Typic Haplosaprists				Euic	Frigid	Aquic
SELKIRK	H	VD	SPD	0–6	0	E	Alfisols	Udalfs	Glossudalfs	Aquic Glossudalfs	Fine	Mixed	Semiactive		Frigid	Udic
SEWARD	S	VD	MWD	0–6	7	B	Alfisols	Udalfs	Hapludalfs	Arenic Oxyaquic Hapludalfs	Loamy	Mixed	Active		Mesic	Udic
SHAG	M	VD	PD	0–2	3	F, G, J	Mollisols	Aquolls	Endoaquolls	Typic Endoaquolls	Coarse-silty	Mixed	Active		Frigid	Aquic
SHAWANO	S	VD	ED	0–35	337	E, J	Entisols	Psamments	Udipsamments	Typic Udipsamments		Mixed			Frigid	Aquic
SHERRY	M	VD	VPD	0–3	243	Fr, J	Alfisols	Aqualfs	Endoaqualfs	Udollic Endoaqualfs	Fine-loamy	Mixed	Superactive		Frigid	Aquic

(continued)

(continued)

Soil series[a]	Soil YP[b]	Soil Depth[c]	Drainage class[d]	Slope (%)	Soil area (km^2)	Soil region	Order	Suborder	Great group	Subgroup	Particle-size class	Mineral class	CEC Activity class	Reaction class	Soil-temperature class	Soil-moisture class
SHIFFER	H	D	SPD	0–3	62	Dr	Alfisols	Udalfs	Hapludalfs	Aquollic Hapludalfs	Fine-loamy over sandy or sandy-skeletal	Mixed	Superactive		Mesic	Udic
SHULLSBURG	H	MD	SPD	1–25	2		Mollisols	Udolls	Argiudolls	Aquic Argiudolls	Fine	Mixed	Superactive		Mesic	Udic
SHIOCTON	H	VD	SPD	0–3	231	E, I	Mollisols	Udolls	Hapludolls	Aquic Hapludolls	Coarse-silty	Mixed	Superactive		Frigid	Udic
SILVERHILL	H	D	WD	1–6	11	Dr	Alfisols	Udalfs	Hapludalfs	Ultic Hapludalfs	Coarse-loamy	Siliceous	Active		Mesic	Udic
SIMESCREEK	S	VD	ED	0–3	9	Dr	Entisols	Psamments	Quartzipsamments	Typic Quartzipsamments					Frigid	Udic
SIOUXCREEK	M	MD	SPD	0–3	2	Dr	Ultisols	Udults	Hapludults	Aquic Hapludults	Coarse-loamy over sandy or sandy-skeletal	Mixed	Active		Frigid	Udic
SIREN	H	VD	SPD	0–3	9	G	Alfisols	Udalfs	Glossudalfs	Aquic Glossudalfs	Fine-loamy over clayey	Mixed OVER smectitic	Superactive		Frigid	Udic
SISSABAGAMA	S	VD	MWD	0–6	6	H	Entisols	Psamments	Udipsamments	Oxyaquic Udipsamments		Mixed			Frigid	Udic
SISSON	H	VD	WD	0–50	166	B, E, I	Alfisols	Udalfs	Hapludalfs	Typic Hapludalfs	Fine-loamy	Mixed	Semiactive		Mesic	Udic
SKOG	M	VD	MWD	0–3	2	H	Entisols	Orthents	Udorthents	Oxyaquic Udorthents	Sandy-skeletal	Mixed			Frigid	Udic
SKYBERG	H	VD	PD	0–2	6	G, J	Alfisols	Aqualfs	Epiaqualfs	Mollic Epiaqualfs	Fine-loamy	Mixed	Superactive		Mesic	Aquic
SLIMLAKE	S	VD	MWD	0–3	11	H	Inceptisols	Udepts	Dystrudepts	Oxyaquic Dystrudepts	Sandy	Mixed			Frigid	Udic
SMESTAD	S	VD	SPD	0–3	16	G	Mollisols	Udolls	Argiudolls	Aquic Argiudolls	Coarse-loamy over clayey	Mixed over smectitic	Superactive		Frigid	Udic
SODERBECK	M	D	SPD	0–2	0	G	Mollisols	Udolls	Hapludolls	Aquic Hapludolls	Loamy-skeletal	Mixed	Active		Frigid	Udic
SOGN	M	VS	SED	0–20	284	Am	Mollisols	Ustolls	Haplustolls	Lithic	Loamy	Mixed	Superactive		Mesic	Ustic
SOLONA	H	VD	SPD	0–3	562	E	Mollisols	Udolls	Argiudolls	Aquic Argiudolls	Coarse-loamy	Mixed	Superactive		Frigid	Udic
SOONER	H	VD	SPD	0–3	7	Dr	Alfisols	Udalfs	Hapludalfs	Aquollic Hapludalfs	Fine-loamy over sandy or sandy-skeletal	Mixed	Superactive		Mesic	Udic

(continued)

(continued)

Soil series[a]	Soil YP[b]	Soil Depth[c]	Drainage class[d]	Slope (%)	Soil area (km^2)	Soil region	Order	Suborder	Great group	Subgroup	Particle-size class	Mineral class	CEC Activity class	Reaction class	Soil-temperature class	Soil-moisture class
SOPERTON	H	MD	WD	15–35	44	G	Spodosols	Orthods	Fragiorthods	Alfic Fragiorthods	Coarse-loamy	Mixed	Active		Frigid	Udic
SPARTA	S	VD	ED	0–40	222	Cm	Mollisols	Udolls	Hapludolls	Entic Hapludolls	Sandy	Mixed			Mesic	Udic
SPEAR	H	VD	SPD	0–3	14	G	Alfisols	Udalfs	Glossudalfs	Aquic Glossudalfs	Coarse-silty	Mixed	Superactive		Frigid	Udic
SPENCER	H	VD	MWD	0–12	364	F	Alfisols	Udalfs	Glossudalfs	Oxyaquic Glossudalfs	Fine-silty	Mixed	Superactive		Frigid	Udic
SPIDERLAKE	H	VD	MWD	0–3	45	G	Spodosols	Orthods	Haplorthods	Alfic Oxyaquic Haplorthods	Coarse-loamy	Mixed	Superactive		Frigid	Udic
SPINKS	S	VD	WD	0–70	8	B	Alfisols	Udalfs	Hapludalfs	Lamellic Hapludalfs	Sandy	Mixed			Mesic	Udic
SPOONERHILL	S	VD	MWD	2–6	46	H	Inceptisols	Udepts	Dystrudepts	Oxyaquic Dystrudepts	Sandy	Mixed			Frigid	Udic
SPRINGSTEAD	M	VD	MWD	1–6	142	G, H	Spodosols	Orthods	Haplorthods	Oxyaquic Haplorthods	Sandy	Isotic			Frigid	Udic
ST. CHARLES	H	VD	WD	0–30	909	A, B	Alfisols	Udalfs	Hapludalfs	Typic Hapludalfs	Fine-silty	Mixed	Superactive		Mesic	Udic
STAMBAUGH	H	VD	WD	0–25	315	G	Spodosols	Orthods	Haplorthods	Alfic Haplorthods	Coarse-silty over sandy or sandy-skeletal	Mixed	Superactive		Frigid	Udic
STANBERRY	M	VD	MWD	1–30	171	G	Spodosols	Orthods	Haplorthods	Alfic Oxyaquic Haplorthods	Coarse-loamy	Mixed	Superactive		Frigid	Udic
STENGEL	S	VD	SPD	0–3	6	I	Alfisols	Udalfs	Hapludalfs	Aquic Arenic Hapludalfs	Clayey	Smectitic			Frigid	Udic
STINNETT	H	VD	SPD	0–4	52	G	Alfisols	Udalfs	Glossudalfs	Aquic Glossudalfs	Coarse-loamy	Mixed	Superactive		Frigid	Udic
STRONGHURST	H	VD	SPD	0–6	4	A, J	Alfisols	Aqualfs	Endoaqualfs	Aeric Endoaqualfs	Fine-silty	Mixed	Superactive		Mesic	Udic
STURGEON	H	VD	SPD	0–1	37	J	Entisols	Fluvents	Udifluvents	Aquic Udifluvents	Coarse-silty over sandy or sandy-skeletal	Mixed	Superactive	Nonacid	Frigid	Udic
SULTZ	S	VD	WD	0–65	227	H	Spodosols	Orthods	Haplorthods	Entic Haplorthods	Sandy	Mixed			Frigid	Udic
SUMMERVILLE	M	S	WD	0–45	192	Er, I	Inceptisols	Udepts	Eutrudepts	Lithic Eutrudepts	Loamy	Mixed	Active		Frigid	Udic
SUNIA	S	VD	MWD	0–3	3	H	Entisols	Psamments	Udipsamments	Oxyaquic Udipsamments		Mixed			Frigid	Udic
SUPERIOR	H	VD	MWD	2–50	71	I	Spodosols	Orthods	Haplorthods	Alfic Oxyaquic Haplorthods	Coarse-loamy over clayey	Mixed	Active		Frigid	Udic
SYLVESTER	H	MD	WD	2–12	12	Am	Mollisols	Udolls	Argiudolls	Typic Argiudolls	Fine-silty over sandy or sandy-skeletal	Mixed	Superactive		Mesic	Udic

(continued)

(continued)

Soil series[a]	Soil YP[b]	Soil Depth[c]	Drainage class[d]	Slope (%)	Soil area (km^2)	Soil region	Order	Suborder	Great group	Subgroup	Particle-size class	Mineral class	CEC Activity class	Reaction class	Soil-temperature class	Soil-moisture class
SYMCO	H	VD	SPD	0–3	345	E, G, I	Alfisols	Udalfs	Hapludalfs	Aquollic Hapludalfs	Fine-loamy	Mixed	Active		Mesic	Udic
SYMERTON	H	VD	MWD	0–10	7	Bm	Mollisols	Udolls	Argiudolls	Oxyaquic Argiudolls	Fine-loamy	Mixed	Superactive		Mesic	Udic
TACOOSH	M	D	VPD	0–2	2	J	Histosols	Hemists	Haplohemists	Terric Haplohemists	Loamy	Mixed		Euic	Frigid	Aquic
TAMA	H	VD	WD	0–20	621	Am	Mollisols	Udolls	Argiudolls	Typic Argiudolls	Fine-silty	Mixed	Superactive		Mesic	Udic
TARR	S	VD	SED	0–60	727	C, Dr	Entisols	Psamments	Quartzipsamments	Typic Quartzipsamments					Mesic	Udic
TAWAS	M	VD	VPD	0–2	109	E, J	Histosols	Saprists	Haplosaprists	Terric Haplosaprists	Sandy or sandy-skeletal	Mixed		Euic	Frigid	Aquic
TAYLOR	H	VD	MWD	0–45	1	I	Alfisols	Udalfs	Glossudalfs	Aquertic Glossudalfs	Very fine	Smectitic			Frigid	Udic
TEDROW	S	VD	SPD	0–6	87	B	Entisols	Psamments	Udipsamments	Aquic Udipsamments		Mixed			Mesic	Aquic
TELL	H	VD	WD	0–40	45	A	Alfisols	Udalfs	Hapludalfs	Typic Hapludalfs	Fine-silty over sandy or sandy-skeletal	Mixed	Superactive		Mesic	Udic
THACKERY	H	VD	MWD	0–6	5	A	Alfisols	Udalfs	Hapludalfs	Aquic Hapludalfs	Fine-loamy	Mixed	Active		Mesic	Udic
THERESA	H	MD	WD	0–30	589	B, I	Alfisols	Udalfs	Hapludalfs	Typic Hapludalfs	Fine-loamy	Mixed	Superactive		Mesic	Udic
TILLEDA	H	VD	WD	1–35	235	E	Alfisols	Udalfs	Glossudalfs	Haplic Glossudalfs	Fine-loamy	Mixed	Active		Frigid	Udic
TIMULA	H	VD	WD	2–60	9	A	Inceptisols	Udepts	Eutrudepts	Typic Eutrudepts	Coarse-silty	Mixed	Superactive		Mesic	Udic
TINT	S	VD	MWD	0–6	28	Dr	Entisols	Psamments	Quartzipsamments	Typic Quartzipsamments					Mesic	Udic
TINTSON	S	VD	MWD	0–6	9	Dr	Entisols	Psamments	Quartzipsamments	Oxyaquic Quartzipsamments					Mesic	Udic
TIPLER	M	VD	WD	0–3	175	G	Spodosols	Orthods	Haplorthods	Alfic Oxyaquic Haplorthods	Coarse-loamy	Mixed	Superactive		Frigid	Udic
TODDVILLE	H	VD	MWD	0–6	94	Am	Mollisols	Udolls	Argiudolls	Typic Argiudolls	Fine-silty	Mixed	Superactive		Mesic	Udic
TONKEY	M	VD	VPD	0–2	9	G, J	Inceptisols	Aquepts	Endoaquepts	Mollic Endoaquepts	Coarse-loamy	Mixed	Semiactive	Nonacid	Frigid	Udic
TOTAGATIC	S	VD	VPD	0–2	88	H, J	Entisols	Aquents	Fluvaquents	Typic Fluvaquents	Sandy	Mixed			Frigid	Aquic
TOURTILLOTTE	S	VD	MWD	0–15	3	H	Entisols	Psamments	Udipsamments	Oxyaquic Udipsamments		Mixed			Frigid	Aquic
TRADELAKE	H	VD	MWD	1–35	4	G, H	Alfisols	Udalfs	Glossudalfs	Oxyaquic Glossudalfs	Coarse-loamy over clayey	Mixed over smectitic	Superactive		Frigid	Udic
TREMPE	S	D	ED	0–12	17	Cm	Mollisols	Udolls	Hapludolls	Entic Hapludolls	Sandy	Mixed			Mesic	Udic

(continued)

(continued)

Soil series[a]	Soil YP[b]	Soil Depth[c]	Drainage class[d]	Slope (%)	Soil area (km^2)	Soil region	Order	Suborder	Great group	Subgroup	Particle-size class	Mineral class	CEC Activity class	Reaction class	Soil-temperature class	Soil-moisture class
TREMPEALEAU	H	MD	WD	0–6	3	Cm	Mollisols	Udolls	Argiudolls	Typic Argiudolls	Fine-loamy over sandy or sandy-skeletal	Mixed	Superactive		Mesic	Udic
TROXEL	H	VD	WD	0–4	105	Bm	Mollisols	Udolls	Argiudolls	Pachic Argiudolls	Fine-silty	Mixed	Superactive		Mesic	Udic
TULA	M	VD	SPD	0–4	179	G, J	Spodosols	Aquods	Fragiaquods	Argic Fragiaquods	Coarse-loamy	Mixed	Superactive		Frigid	Aquic
TUSCOLA	H	VD	MWD	0–12	11	B	Alfisols	Udalfs	Hapludalfs	Aquic Hapludalfs	Fine-loamy	Mixed	Active		Mesic	Udic
TUSTIN	S	VD	WD	0–35	37	E, I, J	Alfisols	Udalfs	Hapludalfs	Arenic Hapludalfs	Clayey	Mixed	Active		Mesic	Udic
TWINMOUND	S	MD	ED	6–50	77	Dr	Entisols	Psamments	Quartzipsamments	Typic Quartzipsamments					Frigid	Udic
URNE	M	MD	WD	2–65	712	Dr	Inceptisols	Udepts	Eutrudepts	Dystric Eutrudepts	Coarse-loamy	Mixed	Active		Mesic	Udic
VALTON	H	D	WD	1–45	606	A	Alfisols	Udalfs	Paleudalfs	Mollic Paleudalfs	Fine-silty	Mixed	Superactive		Mesic	Udic
VANCECREEK	M	VD	PD	0–2	12	Dr, J	Mollisols	Aquolls	Endoaquolls	Fluvaquentic Endoaquolls	Fine-silty	Mixed	Superactive		Frigid	Aquic
VANZILE	H	VD	MWD	0–6	192	G	Spodosols	Orthods	Haplorthods	Alfic Haplorthods	Coarse-silty over sandy or sandy-skeletal	Mixed	Superactive		Frigid	Udic
VARNA	H	VD	MWD	1–18	95	Bm	Mollisols	Udolls	Argiudolls	Oxyaquic Argiudolls	Fine	Illitic			Mesic	Udic
VASA	H	VD	MWD	0–6	13	A	Alfisols	Udalfs	Hapludalfs	Mollic Oxyaquic Hapludalfs	Fine-silty	Mixed	Superactive		Mesic	
VEEDUM	M	MD	PD	0–2	254	Dr, J	Inceptisols	Aquepts	Epiaquepts	Humic Epiaquepts	Fine-loamy	Mixed	Superactive	Acid	Frigid	Udic
VESPER	M	D	PD	0–2	265	Dr, J	Inceptisols	Aquepts	Epiaquepts	Humic Epiaquepts	Fine-loamy over sandy or sandy-skeletal	Mixed	Superactive	Acid	Frigid	Aquic
VILAS	S	VD	ED	0–55	1162	G, H	Spodosols	Orthods	Haplorthods	Entic Haplorthods	Sandy	Isotic			Frigid	Udic
VIRGIL	H	VD	SPD	0–6	19	B, J	Alfisols	Aqualfs	Endoaqualfs	Udollic Endoaqualfs	Fine-silty	Mixed	Superactive		Mesic	Udic
VLASATY	H	VD	MWD	1–12	214	A	Alfisols	Udalfs	Glossudalfs	Aquic Glossudalfs	Fine-loamy	Mixed	Superactive		Mesic	Udic
WABENO	H	MD	MWD	1–18	265	G	Spodosols	Orthods	Fragiorthods	Alfic Oxyaquic Fragiorthods	Coarse-loamy	Mixed	Superactive		Frigid	Udic
WACOUSTA	M	VD	VPD	0–2	154	Bm, J	Mollisols	Aquolls	Endoaquolls	Typic Endoaquolls	Fine-silty	Mixed	Superactive		Mesic	Aquic
WAINOLA	S	VD	SPD	0–4	279	E, H, J	Spodosols	Aquods	Endoaquods	Typic Endoaquods	Sandy	Mixed			Frigid	Aquic
WAKEFIELD	M	S	MWD	1–18	54	G	Spodosols	Orthods	Fragiorthods	Alfic Oxyaquic Fragiorthods	Coarse-loamy	Isotic			Frigid	Udic

(continued)

(continued)

Soil series[a]	Soil YP[b]	Soil Depth[c]	Drainage class[d]	Slope (%)	Soil area (km^2)	Soil region	Order	Suborder	Great group	Subgroup	Particle-size class	Mineral class	CEC Activity class	Reaction class	Soil-temperature class	Soil-moisture class
WAKELEY	S	VD	VPD	0–2	30	H, I, J	Entisols	Aquents	Epiaquents	Aeric Epiaquents	Sandy over clayey	Mixed	Semiactive	Nonacid	Frigid	Aquic
WALLKILL	M	VD	VPD	0–3	30	J	Inceptisols	Aquepts	Humaquepts	Fluvaquentic Humaquepts	Fine-loamy	Mixed	Superactive	Nonacid	Mesic	Aquic
WARMAN	S	VD	VPD	0–2	2	G, H, J	Mollisols	Aquolls	Endoaquolls	Typic Endoaquolls	Coarse-loamy over sandy or sandy-skeletal	Mixed	Superactive		Frigid	Aquic
WARSAW	H	VD	WD	0–15	226	Bm	Mollisols	Udolls	Argiudolls	Typic Argiudolls	Fine-loamy over sandy or sandy-skeletal	Mixed	Superactive		Mesic	Udic
WASEPI	M	VD	SPD	0–6	61	B, I	Alfisols	Udalfs	Hapludalfs	Aquollic Hapludalfs	Coarse-loamy	Mixed	Semiactive		Mesic	Udic
WASHTENAW	M	D	PD	0–2	8	J	Entisols	Aquents	Fluvaquents	Aeric Fluvaquents	Fine-loamy	Mixed	Active	Nonacid	Mesic	Aquic
WATSEKA	S	VD	SPD	0–4	13	Bm	Mollisols	Udolls	Hapludolls	Aquic Hapludolls	Sandy	Mixed			Mesic	Udic
WATTON	H	D	WD	0–40	0	G	Alfisols	Udalfs	Glossudalfs	Haplic Glossudalfs	Fine-loamy	Mixed	Semiactive		Frigid	Udic
WAUCONDA	H	VD	SPD	0–5	3	B, J	Alfisols	Aqualfs	Endoaqualfs	Udollic Endoaqualfs	Fine-silty	Mixed	Superactive		Mesic	Aquic
WAUKEGAN	H	VD	WD	0–12	4	A, F	Mollisols	Udolls	Hapludolls	Typic Hapludolls	Fine-silty over sandy or sandy-skeletal	Mixed	Superactive		Mesic	Udic
WAUPACA	M	D	PD	0–2	38	E, J	Entisols	Aquents	Fluvaquents	Mollic Fluvaquents	Coarse-silty	Mixed	Superactive	Nonacid	Frigid	Aquic
WAUSEON	M	VD	VPD	0–2	31	E, J	Mollisols	Aquolls	Epiaquolls	Typic Epiaquolls	Coarse-loamy over clayey	Mixed OVER illitic	Superactive		Mesic	Aquic
WAUTOMA	S	VD	PD	0–2	50	C, J	Entisols	Aquents	Epiaquents	Mollic Epiaquents	Sandy over clayey	Mixed	Active	Nonacid	Mesic	Aquic
WAYKA	M	MD	SPD	0–4	3	G, J	Spodosols	Aquods	Epiaquods	Typic Epiaquods	Coarse-loamy	Isotic			Frigid	Udic
WAYMOR	H	VD	WD	0–35	233	E, I	Alfisols	Udalfs	Glossudalfs	Haplic Glossudalfs	Fine-loamy	Mixed	Active		Mesic	Aquic
WEGA	H	D	SPD	0–3	15	E, J	Entisols	Fluvents	Udifluvents	Aquic Udifluvents	Coarse-silty	Mixed	Superactive	Nonacid	Frigid	Udic
WESTVILLE	H	VD	MWD	2–30	82	B	Alfisols	Udalfs	Hapludalfs	Typic Hapludalfs	Fine-loamy	Mixed	Superactive		Mesic	Udic

(continued)

(continued)

Soil series[a]	Soil YP[b]	Soil Depth[c]	Drainage class[d]	Slope (%)	Soil area (km^2)	Soil region	Order	Suborder	Great group	Subgroup	Particle-size class	Mineral class	CEC Activity class	Reaction class	Soil-temperature class	Soil-moisture class
WHALAN	M	MD	WD	0–25	290	B, E, G	Alfisols	Udalfs	Hapludalfs	Typic Hapludalfs	Fine-loamy	Mixed	Superactive		Mesic	Udic
WHEATLEY	S	VD	VPD	0–3	7	J	Entisols	Aquents	Psammaquents	Mollic Psammaquents		Mixed			Frigid	Aquic
WHISKLAKE	M	VD	SPD	0–3	54	G, J	Spodosols	Aquods	Endoaquods	Argic Endoaquods	Coarse-loamy over sandy or sandy-skeletal	Mixed	Superactive		Frigid	Aquic
WHITEHALL	H	VD	MWD	0–3	14	Dr	Mollisols	Udolls	Argiudolls	Typic Argiudolls	Fine-silty over sandy or sandy-skeletal	Mixed	Superactive		Mesic	Udic
WICKWARE	H	VD	WD	6–30	37	Dr	Alfisols	Udalfs	Glossudalfs	Haplic Glossudalfs	Fine-silty	Mixed	Superactive		Frigid	Udic
WILDALE	H	VD	WD	2–45	104	Dr	Alfisols	Udalfs	Paleudalfs	Mollic Paleudalfs	Fine	Mixed	Active		Mesic	Udic
WILDWOOD	M	VD	VPD	0–1	10	I, J	Inceptisols	Aquepts	Humaquepts	Histic Humaquepts	Very fine	Smectitic		Nonacid	Frigid	Aquic
WILL	M	MD	PD	0–3	3	Bm, J	Mollisols	Aquolls	Endoaquolls	Typic Endoaquolls	Fine-loamy over sandy or sandy-skeletal	Mixed	Superactive		Mesic	Aquic
WILLETTE	H	VD	VPD	0–2	128	J	Histosols	Saprists	Haplosaprists	Terric Haplosaprists	Clayey	Illitic		Euic	Mesic	Aquic
WINDWARD	S	VD	SED	2–12	9	Dr	Entisols	Psamments	Quartzipsamments	Lamellic Quartzipsamments					Mesic	Udic
WINNEBAGO	H	VD	MWD	2–30	25	Bm	Mollisols	Udolls	Argiudolls	Typic Argiudolls	Fine-loamy	Mixed	Superactive		Mesic	Udic
WINNECONNE	H	VD	MWD	0–12	115	E, I	Alfisols	Udalfs	Hapludalfs	Mollic Oxyaquic Hapludalfs	Very fine	Mixed	Active		Mesic	Udic
WINNESHIEK	M	MD	WD	0–35	9	B	Alfisols	Udalfs	Hapludalfs	Mollic Hapludalfs	Fine-loamy	Mixed	Superactive		Mesic	Udic
WINTERFIELD	S	VD	SPD	0–4	65	Dr, H	Entisols	Psamments	Udipsamments	Aquic Udipsamments		Mixed			Frigid	Udic
WITHEE	H	VD	SPD	0–6	1186	F	Alfisols	Udalfs	Glossudalfs	Aquic Glossudalfs	Fine-loamy	Mixed	Superactive		Frigid	Udic
WORCESTER	M	VD	SPD	0–3	192	G, H, J	Spodosols	Aquods	Endoaquods	Argic Endoaquods	Coarse-loamy	Mixed	Superactive		Frigid	Aquic
WORMET	M	VD	SPD	0–3	65	G, H, J	Spodosols	Aquods	Endoaquods	Typic Endoaquods	Sandy	Mixed			Frigid	Aquic
WORTHEN	H	VD	WD	0–12	83	J	Mollisols	Udolls	Hapludolls	Cumulic Hapludolls	Fine-silty	Mixed	Superactive		Mesic	Udic
WORWOOD	H	VD	SPD	0–3	14	G, J	Spodosols	Aquods	Epiaquods	Alfic Epiaquods	Coarse-loamy	Mixed	Superactive		Frigid	Aquic
WOZNY	M	VD	VPD	0–2	43	G, J	Alfisols	Aqualfs	Epiaqualfs	Typic Epiaqualfs	Coarse-loamy	Mixed	Superactive		Frigid	Aquic

(continued)

(continued)

Soil series[a]	Soil YP[b]	Soil Depth[c]	Drainage class[d]	Slope (%)	Soil area (km^2)	Soil region	Order	Suborder	Great group	Subgroup	Particle-size class	Mineral class	CEC Activity class	Reaction class	Soil-temperature class	Soil-moisture class
WURTSMITH	S	VD	MWD	0–12	102	H	Entisols	Psamments	Udipsamments	Oxyaquic Udipsamments		Mixed			Frigid	Udic
WYEVILLE	S	VD	SPD	0–3	76	C, E	Alfisols	Udalfs	Hapludalfs	Aquic Arenic Hapludalfs	Clayey	Mixed	Active		Mesic	Aquic
WYKOFF	M	D	WD	2–20	34	G	Alfisols	Udalfs	Hapludalfs	Typic Hapludalfs	Fine-loamy	Mixed	Superactive		Mesic	Udic
WYOCENA	M	VD	WD	2–45	363	C, G	Alfisols	Udalfs	Hapludalfs	Typic Hapludalfs	Coarse-loamy	Mixed	Semiactive		Mesic	Udic
YAHARA	M	VD	SPD	0–4	94	Bm, I	Mollisols	Udolls	Hapludolls	Aquic Hapludolls	Coarse-loamy	Mixed	Superactive		Mesic	Aquic
ZEBA	M	MD	SPD	0–3	6	G, J	Spodosols	Aquods	Endoaquods	Argic Endoaquods	Coarse-loamy	Mixed	Active		Frigid	Udic
ZITTAU	H	VD	SPD	0–3	71	E, G, I	Alfisols	Udalfs	Hapludalfs	Aquollic Hapludalfs	Clayey over sandy or sandy-skeletal	Mixed	Active		Mesic	Udic
ZURICH	H	VD	MWD	0–35	85	B, E, I	Alfisols	Udalfs	Hapludalfs	Oxyaquic Hapludalfs	Fine-silty	Mixed	Superactive		Mesic	Udic

[a]Soil series in bold face are prime agricultural soils if 0–3 or <6% and drained, moderately eroded, and infrequently flooded
[b]Soil yield potential for corn (Laboski and Peters 2006)
[c]Soil depth class (depth to a root-restricting layer): *VD* Very deep (>1.5 m); *D* Deep (1.0–1.5 m); *MD* Moderately deep (0.50–1.0 m); *S* Shallow (0.25–0.50 m); *VS* Very shallow (<0.25 m)
[d]Drainage class: *ED* Excessively drained; *SED* Somewhat excessively drained; *WD* Well drained; *MWD* Moderately well drained; *SPD* Somewhat poorly drained; *PD* Poorly drained; *VPD* Very poorly drained

Appendix J
Published Soil Surveys for Wisconsin

County	Date	Archived PDF online	Web Soil Survey (generated from official soil data)
Adams County	1920	Yes	No
Adams County	1984	Yes	No
Adams County	Current	No	Yes
Apostle Is. Nat. Seashore	2014	Yes	No
Ashland County	2006	Yes	No
Ashland County	Current	No	Yes
Barron County	1958	No	No
Barron County	2001	Yes	No
Barron County	Current	No	Yes
Bayfield Area	1912	Yes	No
Bayfield County	1961	Yes	No
Bayfield County	2006	Yes	No
Bayfield County	Current	No	Yes
Brown County	1929	No	No
Brown County	1974	No	No
Brown County	Current	No	Yes
Buffalo County	1915	Yes	No
Buffalo County	1962	Yes	No
Buffalo County	Current	No	Yes
Burnett County	2006	Yes	No
Burnett County	Current	No	Yes
Calumet County	1925	No	No
Calumet and Manitowoc Counties	1980	Yes	No
Calumet and Manitowoc Counties	Current	No	Yes
Carlton Area, MN-WI	1906	Yes	No
Chippewa County	1989	No	No
Chippewa County	Current	No	Yes
Clark County	2000	Yes	No
Clark County	Current	No	Yes
Columbia County	1913	Yes	No
Columbia County	1978	Yes	No
Columbia County	Current	No	Yes
Crawford County	1930	No	No
Crawford County	1961	Yes	No
Crawford County	Current	No	Yes
Dane County	1915	Yes	No

(continued)

J.G. Bockheim and A.E. Hartemink, *The Soils of Wisconsin*,
World Soils Book Series, DOI 10.1007/978-3-319-52144-2

(continued)

County	Date	Archived PDF online	Web Soil Survey (generated from official soil data)
Dane County	1978	Yes	No
Dane County	Current	No	Yes
Dodge County	1980	Yes	No
Dodge County	Current	No	Yes
Door County	1918	Yes	No
Door County	1978	Yes	No
Door County	Current	No	Yes
Douglas County	2006	Yes	No
Douglas County	Current	No	Yes
Dunn County	1975	Yes	No
Dunn County	2004	Yes	No
Dunn County	Current	No	Yes
Eau Claire County	1977	Yes	No
Eau Claire County	Current	No	Yes
Florence County	2004	Yes	No
Florence County	Current	No	Yes
Fond du Lac County	1913	Yes	No
Fond du Lac County	1973	Yes	No
Fond du Lac County	Current	No	Yes
Forest County	2005	Yes	No
Forest County	Current	No	Yes
Grant County	1961	Yes	No
Grant County	Current	No	Yes
Green County	1922	Yes	No
Green County	1974	Yes	No
Green County	Current	No	Yes
Green Lake County	1922	No	No
Green Lake County	1977	Yes	No
Green Lake County	Current	No	Yes
Iowa County	1912	Yes	No
Iowa County	1962	Yes	No
Iowa County	Current	No	Yes
Iron County	2006	Yes	No
Iron County	Current	No	Yes
Jackson County	1922	Yes	No
Jackson County	2001	Yes	No
Jackson County	Current	No	Yes
Janesville Area	1902	Yes	No
Jefferson County	1914	Yes	No
Jefferson County	1979	Yes	No
Jefferson County	Current	No	Yes
Juneau County	1913	Yes	No
Juneau County	1991	Yes	No
Juneau County	Current	No	Yes
Kenosha and Racine Counties	1919	Yes	No
Kenosha and Racine Counties	1970	Yes	No
Kenosha and Racine Counties	Current	No	Yes
Kewaunee County	1913	Yes	No
Kewaunee County	1980	Yes	No
Kewaunee County	Current	No	Yes

(continued)

(continued)

County	Date	Archived PDF online	Web Soil Survey (generated from official soil data)
La Crosse County	1913	Yes	No
La Crosse County	1960	Yes	No
La Crosse County	2006	Yes	No
La Crosse County	Current	No	Yes
Lafayette County	1966	Yes	No
Lafayette County	Current	No	Yes
Langlade County (state)	1947	No	No
Langlade County	1986	Yes	Yes
Langlade County	Current	No	No
Lincoln County	1996	Yes	No
Lincoln County	Current	No	No
Manitowoc County	1926	No	Yes
Marathon County	1989	Yes	No
Marathon County	2003	Yes	No
Marathon County	Current	No	Yes
Marinette County (Reconnaissance)	1911	Yes	No
Marinette County	1991	Yes	No
Marinette County	Current	No	Yes
Marquette County	1975	Yes	No
Marquette County	Current	No	Yes
Menominee County	1967	No	No
Menominee County	2004	Yes	No
Menominee County	Current	No	Yes
Milwaukee County	1918	Yes	No
Milwaukee and Waukesha Counties	1971	Yes	No
Milwaukee and Waukesha Counties	Current	No	Yes
Monroe County	1923	No	No
Monroe County	1984	Yes	No
Monroe County	Current	No	Yes
North Pt. of North-Central WI (Recon.)	1914	Yes	No
North Pt. of Northwestern WI (Recon.)	1914	No	No
Northeastern Wisconsin (Recon.)	1913	Yes	No
Oconto County	1988	Yes	No
Oconto County	Current	No	Yes
Oneida County (state)	1959	No	No
Oneida County	1993	Yes	No
Oneida County	Current	No	Yes
Outagamie County	1921	Yes	No
Outagamie County	1978	Yes	No
Outagamie County	Current	No	Yes
Ozaukee County	1970	Yes	No
Ozaukee County	Current	No	Yes
Pepin County	1964	Yes	No
Pepin County	2002	No	No
Pepin County	Current	No	Yes
Pierce County	1923	No	No
Pierce County	1968	No	No
Pierce County	2006	Yes	No
Pierce County	Current	No	Yes
Polk County	1979	Yes	No

(continued)

(continued)

County	Date	Archived PDF online	Web Soil Survey (generated from official soil data)
Polk County	Current	No	Yes
Portage County	1906	Yes	No
Portage County	1917	Yes	No
Portage County	1978	Yes	No
Portage County	Current	No	Yes
Price County	2006	Yes	No
Price County	Current	No	Yes
Racine County	1907	Yes	No
Richland County	1959	Yes	No
Richland County	2006	Yes	No
Richland County	Current	No	Yes
Rock County	1920	Yes	No
Rock County	1974	Yes	No
Rock County	Current	No	Yes
Rusk County	2006	Yes	No
Rusk County	Current	No	Yes
Sauk County	1925	No	No
Sauk County	1980	Yes	No
Sauk County	Current	No	Yes
Sawyer County	2006	Yes	No
Sawyer County	Current	No	Yes
Shawano County	1982	Yes	No
Shawano County	Current	No	Yes
Sheboygan County	1924	No	No
Sheboygan County	1978	Yes	No
Sheboygan County	Current	No	Yes
South Pt. of North Central WI (Recon.)	1917	Yes	No
St. Croix County	1978	Yes	No
St. Croix County	Current	No	Yes
Superior Area, WI-MN	1905	Yes	No
Taylor County	2005	Yes	No
Taylor County	Current	No	Yes
Trempealeau County	1927	No	No
Trempealeau County	1977	Yes	No
Trempealeau County	Current	No	Yes
Vernon County	1928	No	No
Vernon County	1969	Yes	No
Vernon County	Current	No	Yes
Vilas County	1988	Yes	No
Vilas County	Current	No	Yes
Viroqua Area	1904	Yes	No
Walworth County	1920	No	No
Walworth County	1971	Yes	No
Walworth County	Current	No	Yes
Washburn County	2006	Yes	No
Washburn County	Current	No	Yes
Washington County	1971	Yes	No
Washington County	Current	No	Yes
Washington and Ozaukee Counties	1926	No	No
Waukesha County	1912	Yes	No

(continued)

(continued)

County	Date	Archived PDF online	Web Soil Survey (generated from official soil data)
Waupaca County	1920	Yes	No
Waupaca County	1984	Yes	No
Waupaca County	Current	No	Yes
Waushara County	1909	Yes	No
Waushara County	1989	Yes	No
Waushara County	Current	No	Yes
Winnebago County	1927	No	No
Winnebago County	1980	Yes	No
Winnebago County	Current	No	Yes
Wood County	1917	Yes	No
Wood County	1977	Yes	No
Wood County	Current	No	Yes

Source Natural Resources Conservation Service [http://www.nrcs.usda.gov/wps/portal/nrcs/surveylist/soils/survey/state/?statId=WI (accessed 01-07-16)]

References

Adhikari K, Hartemink AE (2015) Digital Mapping of topsoil carbon content and changes in the Driftless Area of Wisconsin, USA. Soil Sci Soc Am J 79:155–164

Adhikari K, Hartemink AE, Lal R (2017) Soil carbon stock dynamics under landuse and climate change in Wisconsin, USA Geoderma

Allan RJ, Hole FD (1968) Clay accumulation in some Hapludalfs as related to calcareous till and incorporated loess on drumlins in Wisconsin. Soil Sci Soc Am Proc 32:403–408

Allgood FP, Gray F (1978) Utilization of soil characteristics in computing productivity ratings of Oklahoma soils. Soil Sci 125:359–366

Appraisal Institute (2000) The appraisal of rural property, 2nd edn. Appraisal Institute and American Society of Farm Managers and Rural Appraisers, Chicago, IL, 525 pp

Arriaga FJ, Lowery B (2005) Spatial distribution of carbon over an eroded landscape in southwest Wisconsin. Soil Tillage Res 81: 155–162

Attig JW, Bricknell M, Carson EC, Clayton L, Johnson MD, Mickelson DM, Syverson KM (2011) Glaciation of Wisconsin. Wisconsin Geological and Natural History Survey Educational Series 36

Baker RG, Maher LJ, Chumbley CA, Van Zant KL (1992) Patterns of Holocene environmental change in the Midwestern United States. Quat Res 37:379–389

Baldwin M, Kellogg CE, Thorp J (1938) Soil classification. Soils and Men. U.S. Department of Agriculture Yearbook. U.S. Government Printing Office, Washington, DC, pp 979–1001.

Ballagh TM, Runge ECA (1970) Clay-rich horizons over limestone—illuvial or residual? Soil Sci Soc Am J 34:534–536

Bartlein PJ, Webb T III, Fleri E (1984) Holocene climatic change in the northern Midwest: pollen-derived estimates. Quat Res 22:361–374

Baxter FP, Hole FD (1967) Ant (*Formica cinerea*) pedoturbation in a prairie soil. Soil Sci Soc Am Proc 31:425–428

Beaudette DE, O'Geen AT (2010) An iPhone application for on-demand access to digital soil survey information. Soil Sci Soc Am J 74:1682–1684

Beaudette DE, Roudier P, O'Geen AT (2013) Algorithms for quantitative pedology: a toolkit for soil scientists. Comput Geosci 52:258–268

Beatty MT (1991) Soil science at the University of Wisconsin-Madison. A history of the department 1889–1989. University of Wisconsin-Madison, Madison.

Beaver AJ (1966) Characteristics and genesis of some bisequal soils in eastern Wisconsin. Ph.D. Dissertation, Department of Soil Science, University of Wisconsin, Madison

Beek KJ (1978) Land evaluation for agricultural development. International Institute for Land Reclamation and Improvement, Wageningen, publication 23. From soil survey interpretation to land evaluation, pp 29–42

Bernabo JC, Webb T III (1994) Changing patterns in the Holocene pollen record of Northeastern North America: a mapped summary. Quat Res 8:64–96

Beyer L, Pinkpank K, Wriedt G, Bölter M (2000) Soil formation in coastal continental Antarctica (Wilkes Land). Geoderma 95:283–304

Black RF (1964) Periglacial phenomena of Wisconsin, north-central United States. In: VI international Congress on Quaternary, vol IV, pp 21–38

Black RF (1965) Ice-wedge casts of Wisconsin. Wis Acad Sci Arts Lett 56:117–148

Black R (1970) Residuum and ancient soils of the driftless area of south-western Wisconsin, I, Pleistocene geology of southern Wisconsin. Geological Society of America Annual Meeting Field Trip Guide, pp I1–I12

Bockheim JG (1987) Soils in a hemlock-hardwood ecosystem mosaic in the Southern Lake Superior Uplands. Can J For Res 27:1147–1153

Bockheim JG (1997) Soils in a hemlock-hardwood ecosystem mosaic in the Southern Lake Superior Uplands. Can J For Res 27:1147–1153

Bockheim JG (2001) Genesis of bisequal soils on acidic drift in the upper Great Lakes region, USA. Soil Sci Soc Am J 67:612–619

Bockheim JG (2003) Genesis of bisequal soils on acidic drift in the upper Great Lakes region, USA. Soil Sci Soc Am J 67:612–619

Bockheim JG, Marshall JG, Kelsey HM (1996) Soil-forming processes and rates on uplifted marine terraces in southwestern Oregon, USA. Geoderma 73:39–62

Bockheim JG (2005) Soil endemism and its relation to soil formation theory. Geoderma 129:109–124

Bockheim JG (2012) Origin of glossic horizons in Cryalfs of the eastern Rocky Mountains, USA. Geoderma 187–188:1–7.

Bockheim JG (2014) Soil Geography of the USA: a Diagnostic Horizon Approach. Chapter 27. The future of Soil Taxonomy. Springer, NY.

Bockheim JG (2015) Distribution and origin of glossic horizons in soils of the western Great Lakes Region, USA. Geoderma Reg 5:226–236

Bockheim JG, Gennadiyev AG (2000) The role of soil-forming processes in the definition of taxa in Soil Taxonomy and the World Soil Reference Base. Geoderma 95:53–72

Bockheim JG, Gennadiyev AN (2015) General state soil maps in the USA. Geoderma 253:78–89.

Bockheim JG, Haus N (2013) Soil endemism and its importance to taxonomic pedodiversity. In: Ibanez JJ, Bockheim JG (eds) Pedodiversity. CRC Press, Boca Raton, pp 373–380

Bockheim JG, Schliemann SA (2013) Soil richness and endemism across an environmental transition zone in Wisconsin, USA. Catena 113:89–94

J.G. Bockheim and A.E. Hartemink, *The Soils of Wisconsin*, World Soils Book Series, DOI 10.1007/978-3-319-52144-2

Borchardt GA, Hole FD, Jackson ML (1968) Genesis of layer silicates in representative soils in a glacial landscape of southeastern Wisconsin. Soil Sci Soc Am Proc 32:399–403

Bouma J (1973) Use of physical methods to expand soil survey interpretations of soil drainage conditions. Soil Sci Soc Am Proc 37:413–421

Brevik EC, Hartemink AE (2010) Early knowledge and the birth and development of soil science. Catena 83:23–33

Brugham RB, Owen B, Kolesa L (2004) Continental-scale climate forcing factors and environmental change at Glimmerglass Lake in the Upper Peninsula of Michigan. The Holocene 14:807–817

Bryson RA (1966) Air masses, stream lines, and the boreal forest. Geogr Bull 8:228–269

Buol SW, Hole FD (1959) Some characteristics of clay skins on peds in the B horizon of a Gray-Brown Podzolic soil. Soil Sci Soc Am Proc 23:239–241

Buol SW, Hole FD (1961) Clay skin genesis in Wisconsin soils. Soil Sci Soc Am Proc 25:377–379

Cain JM (1990) Soil survey interpretations for forestry—a Wisconsin experiment. Soil Surv Horiz 31:13–16

Campbell HC (1906) Wisconsin in three centuries, 1684–1905. Century History Company, New York

Chadwick OA, Chorover J (2001) The chemistry of pedogenic thresholds. Geoderma 100:321–353

Chamberlin TC (1877) Geology of Wisconsin. Survey of 1873–1877, vol II. Commissioners of Public Printing, Beloit

Chamberlin TC (1882) General map of the soils of Wisconsin. Wisconsin Geological Survey, Beloit

Chamberlin TC, Salisbury RD (1886) Preliminary paper on the Driftless Area of the upper Mississippi Valley. Government Printing Office

Chendev YG (2001) Natural evolution of forest-steppe soils in the southwest of the central Russian upland during the Holocene. Eurasian Soil Sci 32:495–506

Clark N (1940) Soil erosion. Farmers and government together can whip it. Extension Service of College of Agriculture, The University of Wisconsin, Madison

Clayton L (1984) Pleistocene geology of the Superior region, Wisconsin. Wisconsin Geological and Natural History Survey Information Circular No. 46. Madison, WI.

Clayton L, Attig JW (1989) Glacial Lake Wisconsin. Geol Soc Am Mem 173:80

Clayton L, Attig JW, Mickelson DM (2001) Effects of Pleistocene permafrost on the landscape of Wisconsin, USA. Boreas 30:173–188

Conrey GW (1935) Soil type as a basis for land appraisal in the Muskingum Watershed Conservancy District. Soil Sci Soc Am Proc B16:42–43

Cottam G, Loucks OL (1965) Early Vegetation of Wisconsin. University of Wisconsin-Extension, Geological and Natural History Survey.

Curtis JT (1959) Vegetation of Wisconsin: an ordination of plant communities. University of Wisconsin Press, Madison

Darmody RG, Thorn CE, Dixon JC, Schlyter P (2000) Soils and landscapes of Karkevagge, Swedish Lapland. Soil Sci Soc Am Proc 64:1455–1466

Davis AM (1977) The prairie-deciduous forest ecotone in the Midwest. Ann Assoc Am Geogr 67:204–213

Davis MB, Woods KD, Webb SI, Futyma RP (1986) Dispersal versus climate: expansion of Fagus and Tsuga into the upper Great Lakes region. Vegetatio 67:93–103

Devitt T (1988) Down to earth with Francis D. Hole. Orion 7:56–59

Ditzler C (2003) Endangered soils. National Cooperative Soil Survey Newsletter No. 25, Nov 2003, pp 1–2

Drohan PJ, Farnham TJ (2006) Protecting life's foundation: a proposal for recognizing rare and threatened soils. Soil Sci Soc Am J 70:2086–2096

Dury GH, Knox JC (1971) Duricrusts and deep-weathering profiles in southwestern Wisconsin. Science 174(4006):291–292

Evans DM, Hartemink AE (2014) *Terra Rossa* catenas in Wisconsin, USA. Catena 123:148–152

Evans DM, Hartemink AE (2014) Digital soil mapping of a red clay subsoil covered by loess. Geoderma 230–231:296–304

Follmer LR (1983a) Sangamonian and Wisconsinan pedogenesis in the Midwestern United States. In: Wright HE (ed) Late Quaternary environments of the Unites States. University of Minnesota Press, Minneapolis, pp 138–144

Follmer L (1983b) The geomorphology of the Sangamon surface: its spatial and temporal attributes. In: Thorn C (ed) Space and time in geomorphology. Allen & Unwin, Boston, pp 117–146

Foss JE, Rust RH (1968) Soil genesis study of a lithologic discontinuity in glacial drift in western Wisconsin. Soil Sci Soc Am Proc 32:393–398

Frazier BE, Lee GB (1971) Characteristics and classification of three Wisconsin Histosols. Soil Sci Soc Am 35:776–780

French HM (1996) The periglacial environment, 2nd edn. Longman, Essex, England

Frolking TA (1978) The upland red clays of the Driftless Area of the Southwestern Wisconsin: their genesis, distribution, and geomorphic significance. Department of Soil Science, UW Madison, Madison

Frolking TA (1989) Forest soil uniformity along toposequences in the loess-mantled Driftless Area of Wisconsin. Soil Sci Soc Am J 53:1168–1172

Frolking TA, Jackson ML, Knox JC (1983) Origin of red clay over dolomite in the loess-covered Wisconsin driftless uplands. Soil Sci Soc Am J 47:817–820

Frolking S, Roulet NT, Moore TR, Richard PJH, Laovie M, Muller SD (2001) Modeling northern peatland decomposition and peat accumulation. Ecosystems 4:479–498

Fujinuma R, Bockheim J, Balster N (2005) Base-cation cycling by individual tree species in old-growth forests of Upper Michigan, USA. Biogeochemistry 74:357–376

Gaikawad ST, Hole FD (1961) Characteristics and genesis of a Podzol soil in Florence County, Wisconsin. Wis Acad Sci Arts Lett Trans 50:183–190

Gersmehl PJ (1980) Productivity ratings based on soil series: a methodological critique. Prof Geogr 32:158–163

Gessel SP, Harrison RB (1999) A short history of forest soils research and development in North America. In: Steen HK (ed) Forest and wildlife science in America. A history Forest History Society, Durham

Gibbons FR, Haans JCFM (1976) Dutch and Victorian approaches to land appraisal. Soil Survey Paper Number 11, Netherlands Soil Survey Institute, Wageningen.

Graumlich LJ, Davis MB (1993) Holocene variation in spatial scales of vegetation pattern in the upper Great Lakes. Ecology 74:826–839

Grossman EB, Mladenoff DJ (2008) Farms, fires, and forestry: disturbance legacies in the soils of the northwest Wisconsin (USA) Sand Plain. For Ecol Manag 256:827–836

Guo Y, Gong P, Amundson R (2003) Pedodiversity in the United States of America. Geoderma 117:99–115

Gwartney T (2015) Estimating land values. Assessor, Greenwich. http://www.henrygeorge.org/ted.htm. Accessed 6 Jan 2016

Habecker MA, McSweeney K, Madison FW (1990) Identification and genesis of fragipans in Ochrepts of north central Wisconsin. Soil Sci Soc Am J 54:139–146

Hartemink AE (2012) Some noteworthy soil science in Wisconsin. Soil Horiz 53:20–26

Hartemink AE (2015) The use of soil classification in journal papers between 1975 and 2014. Geoderma Reg 5:127–139

Hartemink AE, Bockheim JG (2012) An inverted horizon soilscape in Wisconsin. Soil Horiz. doi:10.2136/sh12-07-0021

Hartemink AE, Minasny B (2014) Towards digital soil morphometrics. Geoderma 230–231:305–317

Hartemink AE, Lowery B, Wacker C (2012) Soil maps of Wisconsin. Geoderma 189–190:451–461

Hartemink AE, Krasilnikov P, Bockheim JG (2013) Soil maps of the world. Geoderma 207:256–267

Heide K (1984) Holocene pollen stratigraphy from a lake and small hollow in north central Wisconsin, USA. Palynology 8:3–19

Helms D (2010) Hugh Hammond Bennett and the creation of the soil conservation service. J Soil Water Conserv 65:37A–47A

Hempel J, Micheli E, Owens P, McBratney A (2013) Universal soil classification system report from the International Union of Soil Sciences Working Group. Soil Horiz 54. doi:10.2136/sh12-12-0035

Hogan JD, Beatty MT (1963) Age and properties of Peorian loess and buried paleosols in southwestern Wisconsin. Soil Sci Soc Am Proc 27:345–350

Hole FD (1956) Soil survey of Waukesha County, Wisconsin: an introductory report. Wisconsin Geological and Natural History Survey Bulletin No. 81, Madison, WI

Hole FD (1961) A classification of pedoturbations and some other processes and factors of soil formation in relation to isotropism and anisotropism. Soil Sci 91:375–377

Hole FD (1968a) Aeolian silt and sand deposits of Wisconsin. Soil Survey Division, Wisconsin Geological and Natural History Survey, University of Wisconsin-Extension

Hole FD (1968b) Soils of Wisconsin. University of Wisconsin Extension, Madison

Hole FD (1968c) Soils of Wisconsin. University of Wisconsin Extension. Madison, WI

Hole FD (1974a) Wild soils of the Pine-Popple Rivers basin. Wis Acad Sci Arts Lett Trans 62:37–50

Hole FD (1974b) Some relationships between forest vegetation and podzol B horizons in soils of Menominee Tribal Lands, Wisconsin, U.S.A. Sov Soil Sci 7(6):714–723

Hole FD (1975) Some relationships between forest vegetation and Podzol B horizons in soils of Menominee tribal lands, Wisconsin, U.S.A. Sov Soil Sci 7(12):714–728

Hole FD (1976) Soils of Wisconsin. The University of Wisconsin Press, Madison

Hole FD (1980) Soil guide for Wisconsin land lookers. Geological and Natural History Survey. University of Wisconsin-Extension, Madison

Hole FD, Hironaka M (1960) An experiment in ordination of some soil profiles. Soil Sci Soc Am Proc 24:309–312

Hole FD, Lee GB (1955) Introduction to the Soils of Wisconsin: a general scheme of classification of the principal soils of Wisconsin. Wisconsin Geological and Natural History Survey Bulletin No. 79, 48 pp

Huddleston JH (1984) Development and use of soil productivity ratings in the Unites States. Geoderma 32:297–317

Hupy CM, Yansa CH (2009a) The last 17,000 years of vegetation history. In: Schaetzl RJ, Darden JT, Brandt D (eds) Michigan geography and geology. Pearson Publishing; Boston, pp 91–105

Hupy CM, Yansa CH (2009b) Late Holocene vegetation history of the forest tension zone in central Lower Michigan, USA. Phys Geogr 30:205–235

Irvin BJ, Ventura SJ, Slater BK (1997) Fuzzy and isodata classification of landform elements from digital terrain data in Pleasant Valley, Wisconsin. Geoderma 77:137–154

Isard SA, Schaetzl RJ, Andresen JA (2007) Soils cools as climate warms in the Great Lakes region: 1951–2000. Ann Assoc Am Geogr 97:467–476

Jacobs PM, Knox JC, Mason JA (1997) Preservation and recognition of middle and early Pleistocene loess in the Driftless Area, Wisconsin. Quat Res 47:147–154

Jacobs PM, Konen ME, Curry BB (2009) Pedogenesis of a catena of the Farmdale-Sangamon geosol complex in the north central United States. Paleaeogeog Palaeoclim Palaeoecol 282:119–132

Jacobs PM, Mason JA, Hanson PR (2012) Loess mantle spatial variability and soil horizonation, southern Wisconsin, USA. Quat Int 265:43–53

Jenny H (1941) Factors of soil formation. McGraw-Hill Book Co., New York

Johnson JE, Bockheim JG, Cain JM (1993) Forest soils of Wisconsin: an overview. University of Wisconsin Cooperative Extension Publication G3452

Kaczorek D, Sommer M, Andruschkewitsch I et al (2004) A comparative micromorphological and chemical study of "Raseneisenstein" (bog iron ore) and "Ortstein". Geoderma 121:83–94

Kellogg CE (1930) Preliminary study of the profiles of the principal soil types of Wisconsin. Wisconsin Geological and Natural History Survey Bulletin No. 77A, 113 pp

King FH (1895) The soil—its nature, relations, and fundamental principles of management. Macmillan and Co., New York

Klingelhoets AJ, Westin FC (1954) Soil survey and land valuation for tax purposes. South Dakota Agricultural Experiment Station. Paper 106. http://openprairie.sdstate.edu/agexperimentsta_circ/1060. Accessed 6 Jan 2016

Klingelhoets AJ (1972) Use of soils data in local land use control ordinances increases. Soil Surv Horiz 13:16–19

Klingelhoets AJ (1978) Soil surveys in Wisconsin help implement farmland preservation legislation. Soil Surv Horiz 19:3–4

Knox JC (2006) Floodplain sedimentation in the upper Mississippi Valley: natural versus human accelerated. Geomorphology 79: 286–310

Laboski CAM, Peters JB (2009) Nutrient application guidelines for field, vegetation, and fruit crops in Wisconsin. University of Wisconsin Extension, A2809.

Langohr R (2001) Facing problems in the discipline of soil classification. Conclusions based on 35 years practice and teaching. In: Micheli E, Nachtergaele FO, Jones RA, Montanarell L (eds) Soil classification. Soil Bureau Research Report No. 7. Office for Official Publications of the European Communities, Luxembourg, pp 15–26

Langton JE, Lee GB (1964) Characteristics and genesis of some organic soil horizons as determined by morphological studies and chemical analysis. Trans Wis Acad Sci Arts Lett 53:149–158

Lambert JM, Loebel TJ (2015) Paleoindian colonization of the recently deglaciated Great Lakes: mobility and technological organization in eastern Wisconsin. PaleoAmerica 1:284–288

Lee GB, Horn ME (1972) Pedology of the Two Creeks section, Manitowoc County, Wiscosin. Trans Wisc Acad Sci Arts Lett 60:183–199

Lindeburg KS, Almond P, Roering JR, Chadwick OA (2013) Pathways of soil genesis in the Coast Range of Oregon, USA. Plant and Soil. doi:10.1007/s111040121566-z

Long CJ, Power MJ, McDonald B (2011) Millennial-scale fire and vegetation history from a mesic hardwood forest of southeastern Wisconsin, USA. J Quat Sci 26:318–325

Madison FW, Lee GB (1965) Some mineralogical characteristics of sandy soils in Wisconsin. Wisc Acad Sci Arts Lett Trans 54:223–230

Madison FW, Gundlach HF (1993) Soil regions of Wisconsin (map). University of Wisconsin-Extension/Wisconsin Geological and Natural History Survey, Madison

Mahaney WC (2015) Clay mineralogical evidence of a bioclimatically-affected soil, Rouge River basin, South-Central Ontario, Canada. Geomorphology 228:189–199

Maher LJ Jr (1982) The palynology of Devils Lake, Sauk County, Wisconsin. In: Knox JC, Clayton L, Mickelson DM (eds)

Quaternary history of the Driftless Area, Wisconsin. Geological and Natural History Survey Field Trip Guidebook, vol 5, pp 119–125
Maher LJ Jr, Mickelson DM (1996) Palynological and radiocarbon evidence for deglaciation events in the Green Bay Lobe, Wisconsin. Quat Res 46:251–259
Marbut CF (1928) A scheme for Soil Classification. In: 1st International Congress Soil Science Com. 5, Proceedings and Papers, vol 4, pp 1–31
Margurran AE (1988) Ecological diversity and its measurement. Princeton University Press, Princeton, NJ, p 179
Mason JA, Milfred CJ, Nater EA (1994) Distinguishing soil age and parent material effects on an Ultisol of north-central Wisconsin, USA. Geoderma 61:165–189
Michéli E, Láng V, Owens PR, McBratney A, Hempel J (2016) Testing pedometric evaluation of taxonomic units on soil taxonomy—a step in advancing towards a universal soil classification systems. Geoderma 264:340–349
Morgan CLS, Norman JM, Wolkowski RP, Lowery B, Morgan GD, Schuler R (2000) Two approaches to mapping plant available water: EM-38 measurements and inverse yield modeling. In: Robert PC, Rust RH, Larson WE (eds) Proceedings of the 5th international conference on precision agriculture, Bloomington, Minnesota, USA, 16–19 July, 2000. American Society of Agronomy, Madison, pp 1–13
Motew MM, Kucharik CJ (2013) Climate-induced changes in biome distribution, NPP, and hydrology in the Upper Midwest U.S.: a case study for potential vegetation. J Geophys Res Biogeosci 118:248–264
Muckenhirn RJ, Dahlstrand NP (1947) Soils of Wisconsin. Soil Survey Division, Wisconsin Geological and Natural History Survey, and the Agricultural Experiment Station, University of Wisconsin, Madison
Muñoz SE, Gajewski K, Peros MC (2010) Synchronous environmental and cultural change in the prehistory of the northeastern United States. Proc Natl Acad Sci 107:22008–22013
Natural Resources Conservation Service (2007) History of Wisconsin Soil Survey. USDA, NRCS
Natural Resources Conservation Service (2016a) Web soil survey. http://websoilsurvey.sc.egov.usda.gov/App/HomePage.htm. Accessed 14 Mar 2016
Natural Resources Conservation Service (2016b) Official soil series descriptions. https://soilseries.sc.egov.usda.gov/osdname.asp. Accessed 14 Mar 2016
Nielsen GA, Hole FD (1963) A study of the natural processes of incorporation of organic matter into soil in the University of Wisconsin Arboretum. Wis Acad Sci Arts Lett Trans 52:213–227
Nielsen GA, Hole FD (1964) Earthworms and the development of coprogenous A1 horizons in forest soils of Wisconsin. Soil Sci Soc Am J 28:426–430
NRCS (2006) Major Resource Regions and Major Land Resource areas of the United States, the Caribbean, and the Pacific Basin. U. S. Dep. Agric. Handbook 296
Nygard IJ, McMiller PR, Hole FD (1952) Characteristics of some Podzolic, Brown Forest, and Chernozem soils of the northern portion of the Lake States. Soil Sci Soc Am Proc 16:123–129
Odgers NP, Sun W, McBratney AB, Minasny B, Clifford D (2014) Disaggregating and harmonising soil map units through resampled classification trees. Geoderma 214:91–100
Olson GW, Hole FD (1967) The fragipan in soils of northeastern Wisconsin. Trans Wis Acad Sci Arts Lett 56:173–184
Park SJ, McSweeney K, Lowery B (2001) Identification of the spatial distribution of soils using a process-based terrain characterization. Geoderma 103:249–272
Park SJ, Almond P, McSweeney K, Lowery B (2006) Fragipan formation within closed depressions in southern Wisconsin, United States. J Korean Geogr Soc 41:150–167
Pastor J, Aber JD, McClaugherty CA (1982) Geology, soils and vegetation of Blackhawk Island, Wisconsin. Am Midl Nat 108:266–277
Pauli JN, Zuckerberg B, Whiteman JP, Porter W (2013) The subnivium: a deteriorating seasonal refugium. Front Ecol Environ 11:260–267. doi:10.1890/120222
Phillips JA (1968) Use of soil surveys in land valuation for tax assessment. PhD diss., Iowa State Univ., Ames, IA. http://lib.dr.iastate.edu/rtd. Accessed 6 Jan 2016
Quandt L, Watts FC (1995) State soils of the United States. J Soil Water Conserv 50:340–342
Quideau SA, Bockheim JG (1996) Vegetation and cropping effects on pedogenic processes in a sandy prairie soil. Soil Sci Soc Am J 60:536–545
Ranney RW, Beatty MT (1969) Clay translocation and albic tongue formation in two Glossoboralfs of westcentral Wisconsin. Soil Sci Soc Am Proc 33:768–775
Richardson JL, Hole FD (1979) Mottling and iron distribution in a Glossoboralf-Haqlaquoll hydrosequence on a glacial moraine in northwestern Wisconsin. Soil Sci Soc Am J 43:552–558
Ruhe RV (1956) Geomorphic surfaces and the nature of soils. Soil Sci 82:441–455
Ruhe RV (1965) Quaternary paleopedology. In: Wright HE, Frey DG (eds) Quaternary of the United States. Princeton University Press, Princeton, pp 755–764
Schulte LA, Mladenoff DJ, Burrows SN, Sickley TA, Nordheim EV (2005) Spatial controls of pre-Euro-American wind and fire disturbance in northern Wisconsin (USA) forest landscapes. Ecosystems 8:73–94
Shi X, Zhu A-X, Burt JE, Qi F, Simonson D (2004) A case-based reasoning approach to fuzzy soil mapping. Soil Sci Soc Am J 68:885–894
Simonson RW (1989) Historical highlights of soil survey and soil classification with emphasis on the United States, 1899–1970. In: International Soil Reference and Information Centre (ISRIC), Wageningen, Netherlands. Technical paper 18
Smith HTU (1949) Periglacial features in the Driftless Area of Wisconsin. J Geol 57:196–215
Soil Survey Staff (1960) Soil Classification, a Comprehensive System, 7th Approximation. U.S. Government Printing Office, Washington, DC
Soil Survey Staff (1999) Soil Taxonomy: a basic system of soil classification for making and interpreting soil surveys, 2nd edn. Agriculture Handbook No. 436, USDA Natural Resources Conservation Service, U.S. Government Printing Office, Washington, D.C.
Soil Survey Staff (2014) Keys to soil taxonomy, 12th edn. U.S. Department Agriculture, Natural Resources Conservation Service, Washington, D.C.
Stolt MH, Needelman BA (2015) Fundamental changes in soil taxonomy. Soil Sci Soc Am J 79:1001–1007
Sudduth KA, Kitchen NR, Wiebold WJ, Batchelor WD, Bollero GA, Bullock DG, Clay DE, Palm HL, Pierce FJ, Schuler RT, Thelen KD (2005) Relating apparent electrical conductivity to soil properties across the north-central USA. Comput Electron Agric 46:263–283
Sun W, McBratney AB, Hempel JW, Minasny B, Malone BP, D'Avello T, Burras L, Thompson JA (2010) Digital harmonisation of adjacent analogue soil survey areas—4 Iowa counties. In: Proceedings of the 4th global workshop on Digital Soil Mapping IUSS, Rome
Syverson KM, Colgan PM (2011) The Quaternary in Wisconsin: a review of stratigraphy and glaciation history. In: Ehlers J, Gisbard PL, Hughes PD (eds) Quaternary glaciations—extent and chronology. Developments in Quaternary Science, vol 15. Elsevier, Amsterdam, pp 537–552
Tiurin IV (1930) Genesis and classification of forest-stepe and forest soils. Second international congress of soil science. ISSS, Leningrad

Trimble SW, Lund SW (1982) Soil conservation in the Coon Creek Basin, Wisconsin. J Soil Water Conserv 37:355–356

Trotta LC, Cotter RD (1973) Depth to bedrock in Wisconsin. Wisconsin Geology & Natural History Surv, Madison

Tsai H, Hseu Z-Y, Kuo HY et al (2016) Soilscape of west central Taiwan: its pedogenesis and geomorphic implications. Geomorphology 255:81–94

Van Lanen HAJ, Hack-ten Broeke MJD, Bouma J, De Groot WJM (1992) A mixed qualitative/quantitative physical land evaluation methodology. Geoderma 55:37–54

Watson H, Hartemink AE (2015) Rapid changes in sandy soils under intensive agriculture in Wisconsin. Soil Horiz 1–6 doi:10.2136/sh14-11-0014

Wayne WJ, Guthrie RS (1993) Permafrost and periglacial winds around the Wisconsinan ice margin in the northern plains of the USA. In: Permafrost, Sixth International Conference on Permafrost Proceedings vol 1, Beijing, pp 694–699

Weaver AJ, Eby M, Fanning AF, Wiebe EC (1998) Simulated influence of carbon dioxide, orbital forcing and ice sheets on the climate of the Last Glacial Maximum. Nature 394:847–853

Webb SL (1987) Beech range extension and vegetation history: pollen stratigraphy of two Wisconsin lakes. Ecology 68:1993–2005

Webb T III (1974) A vegetational history from northern Wisconsin: evidence from modern and fossil pollen. Am Midl Nat 92:12–34

Webb T III, Bryson RA (1972) Late- and postglacial climatic change in the northern Midwest, USA: quantitative estimates derived from fossil pollen spectra by multivariate statistical analysis. Quat Res 2:70–115

Whitson AR (1927) Soils of Wisconsin. Bulletin no. 68, Soil Series no. 49. Wisconsin Geological and Natural History Survey, Madison.

Whitson AR, Walster HL (1918) Soils and soil fertility. Web Publishing Co., St. Paul Minnesota

Wilde SA (1946) Forest soils and forest growth. Chronica Botanica Company, Mass

Wilde SA, Wilson FG, White DP (1949) Soils of Wisconsin in relation to silviculture. Wisconsin Conservation Department, Madison

Wisconsin Department of Natural Resources (2016) Forest land owners. http://dnr.wi.gov/topic/ForestLandOwners/openToPublicApp.html. Accessed 14 Mar 2016

Wisconsin Legislative Reference Bureau (2015) State of Wisconsin 2015–2016 Blue Book. Madison, Wisconsin

Wisconsin Department of Natural Resources (2012) Silica sand mining in Wisconsin. Online report: http://wisconsingeologicalsurvey.org/silica-sand.htm. Accessed 26 Sept 2014

Wurman E (1952) The geological material: its impact on soil profile characteristics in west central Wisconsin. Wis Acad Sci Arts Lett Trans 50:191–202

WICCI (2011) Wisconsin's changing climate: impacts and adaptation, 2011. Wisconsin Initiative on Climate Change Impacts. Nelson Institute for Environmental Studies. University of Wisconsin-Madison and the Wisconsin Department of Natural Resources, Madison, Wisconsin.

Wisconsin Geological and Natural History Survey (1976) Glacial deposits of Wisconsin: sand and gravel resource potential. University of Wisconsin-Extension, Madison

Wisconsin Geological and Natural History Survey (2005) Bedrock geology of Wisconsin. University of Wisconsin-Extension, Madison

Wisconsin Initiative on Climate Change Impacts (WICCI) (2011) Wisconsin's changing climate: impacts and adaptation. Nelson Institute for Environmental Studies. University of Wisconsin-Madison and Wisconsin Department of Natural Resources, Madison, Wisconsin

Wright HE, Winter TC, Patten HL (1963) Two pollen diagrams from southeastern Minnesota: problems in the regional late-glacial and postglacial vegetation history. Geol Soc Am Bull 74:1371–1396

Yaalon DH (1971) Soil-forming processes in space and time. In: Yaalon DH (ed) Paleopedology—origin, nature and dating of Paleosols. Israel University Press, Jerusalem, pp 29–39

Zhu J, Morgan CLS, Norman JM, Yue W, Lowery B (2004) Combined mapping of soil properties using a multi-scale tree-structured spatial model. Geoderma 118:321–334

Index

J.G. Bockheim and A.E. Hartemink, *The Soils of Wisconsin*,
World Soils Book Series, DOI 10.1007/978-3-319-52144-2

V

W

Y

MIX
Papier aus verantwortungsvollen Quellen
Paper from responsible sources
FSC® C105338

If you have any concerns about our products,
you can contact us on
ProductSafety@springernature.com

In case Publisher is established outside the EU,
the EU authorized representative is:
Springer Nature Customer Service Center GmbH
Europaplatz 3, 69115 Heidelberg, Germany

Printed by Libri Plureos GmbH
in Hamburg, Germany